AN ATLAS OF ORCHID POLLINATION

AMERICA, AFRICA, ASIA AND AUSTRALIA

AN ATLAS OF ORCHID POLLINATION

America, Africa, Asia and Australia

N.A. VAN DER CINGEL

CRC Press
Taylor & Francis Group
Boca Raton London New York

CRC Press is an imprint of the
Taylor & Francis Group, an **informa** business
A BALKEMA BOOK

CRC Press
Taylor & Francis Group
6000 Broken Sound Parkway NW, Suite 300
Boca Raton, FL 33487-2742

ISBN-13: 978-9-054-10486-5 (hbk)

Publisher's Note

The publisher has gone to great lengths to ensure the quality of this reprint but points out that some imperfections in the original copies may be apparent.

Visit the Taylor & Francis Web site at
http://www.taylorandfrancis.com

and the CRC Press Web site at
http://www.crcpress.com

Contents

Acknowledgements

We all stand on each other's shoulders
Anonymous

This book would probably never have been written if the book on European orchids was not received so positively, and if several people had not stimulated me to 'finish the job' by extending it with a second book on non-European orchids or better, orchids from the A-continents: The Americas, Africa, Asia and Australia (a lot of orchids from Europe are shared by Asia or other regions).

It obviously was an enormous task that I had ahead when I decided to write this book. Even for a retired biologist who has seemingly plenty of time to do so. Seemingly, for being retired often means, as in my case, to enter a busy life of pursuing a lot of activities that had to wait for the period of post-professional life.

Among those who encouraged me I especially thank Kees Boele, who was the first to tell me that the title of the first book implied a supplementary volume.

This book could never have been written without the professional help and opportunities offered by librarians and libraries. The library of the Department of Biology of Groningen University, and of the Herbaria at Leiden and Kew, were indispensible in this respect and I am grateful to the staffs of these institutions for what they did or let me do. The librarian of the Herbarium of Sydney should be added to this short list.

Many persons helped me by giving or sending publications on request: I mention, in gratitude, the following people: Jim Ackerman, Tad Bartareau, Bob Bates, Mark Clements, James Indsto, Ken Inoue, Steve Johnson, Donald Jones. Gerald McCraith, L. Anders Nilsson, Joseph Patt, Linda and Timothy Reeves, Rodrigo Singer, D.J. Smedley, Ian St George, Warren Stoutamire, Naoto Sugiura, Wal Upton.

And I undoubtedly forgot to mention some people by name who sent me their papers in the period before I intended to write the book and, as the editor of the Dutch Orchid Association (NOV), asked for their publications. They contributed as well.

Others (or sometimes the same people) sent me pictures or slides: Bob Bates, Paul Boeren, Andrea Cocucci, Jim Comber, Ken Inoue, James Indsto, Steve Johnson, Jac. Mulder, Rodrigo Singer, Warren Stoutamire, Naoto Sugiura, Wal Upton, Ed van Wijngaarden. Thanks to their contributions the book is better illustrated than I had hoped for.

Some (other) people introduced me to the orchids of their country Australia. This not only gave me the right perspective on the ecology and habitats of the orchids found, but unforgettable experiences as well. I especially thank Tad Bartareau, Colin Bower, Bill Lavarack, Len Lawler, Grahame Muller, John Roberts and above all, Wal and Jill Upton.

James Indsto got the help of members of the Orchid Species Society of NSW Ltd and the Australasian Native Orchid Society, Sydney Group, who have made their often rare and valuable plants available for photography. I am indebted to them as well. These people include: David Banks, Sid Batchelor, Marie Bradd, Bev De Brincat, Jim Brydie, David Butler, Ian Chalmers, Bernard Cooley, Jeane Deane, Bill Ditchfield, Madge Errington, Michael Ferenczi, J. and P. Fetherston, Jeanette Fraser, C. and J. Fulcher, G. Hansen, Mike Harrison, Michael Hynes, D. and J. Johnson, Bruce Lonnon, Daryl Smedley, Peter Taylor, Gerry Walsh and B.V. Williams.

A lot of my travelling was done in the last decade and that contributed greatly to a good understanding of the phenomena described in my book as well as giving me a lot of nice experiences in travelling with many people interested in studying nature. This was made possible by the members of the Royal Dutch Society for Natural History and their Travel Committee (KNNV), whose chairman I have been for eight years and who let me be a guide on so many journeys around the world. All this in fact originated from my experiences in my former professional life at Groningen University.

A word of thank also to the publisher and his staff, who once again did a lot of work to guarantee the quality of this publication.

Last but not least I thank my wife Jo for all the conditions she shaped for the work on this book and all the liberties she gave me for travelling around the world.

Nelis A. van der Cingel

Introduction

I was never more interested in any subject in my life than this of orchids
Charles Darwin in a letter to Hooker, 13 Oct. 1861

Those who are interested in orchids and enjoy cultivating or photographing them in the wild or even sample them there for cultivation in garden or greenhouse, rarely know the facts of life by which the orchids survive through time, the facts of pollination. Those who do know, share this interest with the greatest of all biologists, Charles Darwin (Fig. 1), who wrote a nice book on this subject called: *Fertilisation of Orchids*: *The various contrivances by which orchids are fertilised by insects* (1862). It is a classic and today still relevant for every student of orchid pollination (Plates 1 and 2, p. 12).

In 1966 L. van der Pijl and Calaway H. Dodson wrote a second book on the subject in commemoration of the Centenary of Darwin's publication: *Orchid flowers, their pollination and evolution*. It was a real update. But since 1966 extensive professional research and many discoveries by amateurs have revealed an enormous amount of interesting and important new facts.

In 1995 the author wrote *An Atlas of Orchid Pollination*: *European Orchids*. It was received with very positive comments. This Atlas tried to relate in a coherent manner the theory and practice of orchid pollination. The first part dealt with the history, evolution, morphology, typology, mechanics, chemistry, geometry, ethology, economy, genetics, language and research methodology of orchid pollination. In the second part, a systematic account was given of all the details known of the pollination of European orchids. As the first two books mentioned covered also the orchids of non-European origin, a logical step was to compose a second book to complete the updated picture of knowledge of orchid pollination. Writing such a second book was not an easy task however, because, even in the computerised era in which we live it is not easy to get hold of all the relevant information stored in different parts of the world and scattered in tens of magazines in many languages.

It was even more difficult to find adequate illustrations of pollination phenomena. There are many pictures, scattered around the world, but it is not an easy task to lay your hands on them.

Nevertheless this book, although maybe not entirely complete, is the result of a thorough tracing of the existing literature, both historical and recent and of sufficient help from photographers around the world.

I refer those interested in a more complete story of the theory of orchid pollination to the first part of the *Atlas of Orchid Pollination*: *European Orchids* and to the *Complete references*. In the next chapter I will present merely a summary of the different aspects.

In this Atlas I have paid more attention to the pollinators than in my first book and this forms the content of the second chapter. As a trained ethologist, I can see the weaknesses in some research work when it comes to the interpretation of insect behaviour. Students of pollination should at least be aware of the pitfalls of it.

It is evident that much more is known about pollination of orchids in some continents than in other parts of the world. In general more is known about terrestrials of temperate regions with the exception of South America (although there are some promising developments in the study of orchid pollination in this large continent), compared to epiphytes from tropical and subtropical regions. It may be obvious that research in pollination of epiphytes is much harder to carry out. Nevertheless a lot of investigations have been performed in Middle America and (subtropical) Australia. Especially Central Asia, Japan and South America seem underdeveloped as to this kind of knowledge on orchid pollination. This may in part be due not to lack of observations, but to inaccessibility of the literature.

But certainly a lot more remains to be revealed by future investigations. A lot of the orchid research has been done by enthusiastic 'amateurs', mainly in the field of taxonomy. Pollination studies have been scarce, however. This might be due partly to a lack of experience in the methodology of pollination research, partly to a lack of opportunities to witness pollination. Systematic studies need great patience, careful attention and some ingenuity in asking the right questions and developing methods to get answers. In my opinion here is a field full of opportunities for both professional biologists and lay people. The latter can be especially helpful in simple observations.

The significance of research in this area is obvious to those interested in evolution and ecology. In both fields the orchids play a major role because of the rapid evolutionary developments taking place in this family and because of the vulnerability of orchid populations in the wild. I hope that many readers and users of this book will be stimulated to make their own observations, to write them down and to report them in adequate and accessible journals and magazines. The more we know the better are the chances for conservation of orchid habitats of sufficient size. The more we know the better we will be able to recognise patterns and to generalise our knowledge into satisfying theories concerning the evolution and coevolution of species.

Moreover, orchids are fascinating. Orchids are beautiful creatures. And there is a lot of mystery in them as well as in our fascination for them. In my introduction to the first book I wrote extensively about this attraction, caused by the beauty and mysteriousness of orchids (the mystery of orchidiocy as it is called sometimes) and I refer the reader to what I wrote there about the relationship to eroticism and art and about the metaphoric qualities of orchids.

In fact this book was born out of the author's editorship of 'Orchideeën', the magazine of the Dutch Orchid Society. After a special issue on pollination of orchids in 1980 the author regularly published short notes on the subject and gained an overall picture of the field. His background as a broadly trained biologist and as a professional educationalist both motivated and guided him. It is his hope, therefore, that the reader may recognise both the love and the skill with which this book was written.

Figure 1. Charles Darwin (1809-1882) (crayondrawing by Samuel Lawrence in 1853).

CHAPTER 1

A summary of the theory of orchid pollination

There is grandeur in this view of life, with its several powers, having originally breathed into a few forms or into one; and that, whilst this planet has gone cycling on according to the fixed laws of gravity, from so simple a beginning endless forms most beautiful and most wonderful have been and are being evolved

Charles Darwin

1.1 THE HISTORY OF ORCHID POLLINATION

Fertilisation of plants has been understood only for about three centuries. Before there was much superstition and misunderstanding. Rudolf Jakob Camerarius (Fig. 2) of Tübingen in Germany was the first to analyse (around 1690) the role of the sex-organs in flowers. He detected the male role of *stamens* and the female function of the *pistil.* Carl Linnaeus (1707-1778; Fig. 3) based his taxonomy of plants on the sexual organs. J.G. Koelreuter (1733-1806; Fig. 4) published about insect pollination and Christian Konrad Sprengel (1750-1816) discovered the roles of the different characters of flowers in the pollination process like nectar, flower colour and *nectar guides*, as well as the *adaptations* to insects that prevent self-fertilisation. Real understanding of the process of *cross-fertilisation* had to wait until the rediscovery of the work on the genetics of plants of Gregor Johann Mendel, around 1900 (Fig. 5).

The exact roles of *instinctive* and/or *learning behaviour* of pollinators in the processes were only analysed effectively in our age. It is only recently that we understand the typical character of orchid *anthecology*, a continuous story of defrauding pollinators and of survival in difficult circumstances.

The most famous story of all is possibly the discovery of the phenomenon of *pseudocopulation* by Pouyanne who was a judge and president of the court of Sidi-Bel Abbas in Algeria. He studied the pollination of *Ophrys speculum* (*=O.ciliata*) and *Ophrys lutea* as a leisure occupation during 20 years and published the results of his studies from 1916-1923 in scientific magazines. This was met with disbelief by the laboratory biologists, who could not imagine insects being deceived by such a superficial resemblance. Today, the study of the pollination of the genus *Ophrys* forms the basis of its *taxonomy*.

1.2 THE EVOLUTION OF ORCHID POLLINATION

Most *fossil* findings (Fig. 6a) ascribed to the orchid family are disputed. However, from the area of the Bodensee (Germany) there is a find of a fossil plant from the Upper *Miocene*, about 15 million years old, that clearly shows a typical orchidaceous flower and fruit (Fig. 6b). It is named *Eoorchis miocaenica.* It shows that the orchids are a recent family, which today is still very much under rapid development. Modern concepts of evolution theory stress the occurrence of rapid developments and *speciation* in the evolution of recent families. Many problems of orchid taxonomy and the many disagreements about *classification* in the orchid family reflect its youth, boundaries between taxa being vague and allowing easy *hybridisation.* We know next to nothing about the evolution of pollinator-behaviour in relation to orchids. The present theory of evolution, called the *modern synthesis*, uses facts from the study of *population dynamics*, *taxonomy*, *morphology*, *genetics*, *embryology*, *plant-geography* and of many more branches of biology to build up a picture of the evolution of the family.

The orchid flower has evolved from a lily-like ancestor by changing several important characteristics.

The evolution of a species is determined by the *genetic variance* present in populations. To keep up variance species need time to adapt to changing circumstances and room. Speciation can be *sympatric*

Figure 2. Rudolf Jakob Camerarius (1665-1721).

Figure 3. Carl Linnaeus (1707-1778) (pastel by G. Lundberg).

Figure 4. J.G. Koelreuter (1733-1806).

Figure 5. Gregor Johann Mendel (1822-1884).

Figure 6. Some fossil finds of orchids or supposed orchids. A) *Protorchis monorchis*: leaves, tuber and roots. Eocene, Monte Bolca, Italy (after Massalonga 1859). B) *Eoorchis miocaenica*. Upper Miocene, Bodensee, Germany (after Buttler 1986).

or *allopatric*. In sympatric evolution species split up into different taxa while living in the same area: different pollinators could be responsible for this and many ecological factors could influence such speciation. In allopatric evolution populations are split up in geographically separated taxa. *Genetic drift*, in which small parts of a population become isolated, could be one of the causes of such creation of new species, through divergence of the genetic composition of small samples of a population by chance alone.

The interdependence between orchids and pollinators is such that the evolution of one may influence that of the other. This is what we call *co-evolution*. However, with orchids the picture is very often complicated by the one-sidedness of the relationship, with the orchid being dependent on the pollinator and not vice versa. This is the direct consequence of an evolution from reward to deceit. The orchids, as inhabitants of environments poor in nutrition have evolved pollination systems for deceit of the pollinator. This saves the investment of substances in the production of rewards.

It is supposed that the first orchids secreted nectar from a *spur* and that they were scented, just like many orchids today. The original pollinators, probably bees, were thus rewarded.

Sexual attraction by fragrance in combination with a reward for pollination services led probably to *rendezvous* flowers: both sexes visited, the males trying to mate on the flowers. Some stimuli from the flowers could have led to copulatory behaviour *with* these flowers in the absence of females. From here on the development of *mimicry* of females by the flower could have started by coevolutionary processes, the resemblance of the flower to an insect keeping pace with the steadily improving capacity of the insect to detect differences between the real insect and the mimic.

A special case is that of *euglossinophily* (Plate 4, p. 12) or euglossine-pollinated orchids. More than 600 orchid species belong to the male euglossine-pollination system. More than 200 bee species may act as visitors (Gerlach & Schill 1991). Orchids and euglossine bees seem to have a relatively ancient relationship. The orchids possibly evolved from nectar-producing flowers that produced strong floral fragrances. Male euglossine bees are attracted to the floral fragrances, collect these with their front tarsi and store them in the hind tibia. In the process they often pollinate the flowers they visit. It is suggested that the male bees modify the floral fragrances, translocate them to the mandibular glands where they may be further modified, and secrete the products as female sex attractants (Williams 1980).

Some orchid species attract both male and female euglossine bees, which feed on nectar. There are orchid species which have *nectaries* but fail to produce nectar. Such species may attract more males than (misled) females or even only male euglossine bees, depending on the amount of fragrance produced.

Other lines of evolution have exploited the food drive instead of the *sexual motivation* to 'lure' insects. Insects can for example be deceived in the phase of life in which they learn about food sources. Inexperienced bees, making a reconnaissance in the environment, pop into the flowers without being rewarded by nectar. In this period, the insects are learning fast where to patrol in order to find food. By spreading out in the habitat and mimicking in colours and sometimes in form rewarding flowers the orchids are able to mislead the bees. Other examples of systems of *deception* will be discussed later. Many hypotheses have been developed to explain the problems of keeping up *fitness* in spite of reduced visitation due to lack of reward. Insights from pollination and breeding system data have played an important role in the recent improvements in classification. By their nature orchids are ideal subjects for testing hypotheses relating to pollination ecology and evolution (Catling & Catling 1991).

There are several non-flower characters that have influenced and determined survival and thus orchid evolution, like extremely small and numerous seeds, combined with *mycotrophy*, the seeds being able to develop only with the help of symbiotic fungi, and *asexual* propagation by tubers.

Although there is some discussion whether the ancestral orchids were *terrestrial* or *epiphytic*, most evidence suggests a terrestrial origin (Ackerman 1983).

1.3 THE MORPHOLOGY OF ORCHID POLLINATION

In general plants produce superfluous amounts of reproduction materials and are relatively careless with the handling of it. In fact, this is an enormous waste. This is even the case with most insect-pollinated plants. Orchids, however, prevent visiting insects from using the costly *pollen* as food. The pollen is stored in small packets, the *pollinia*, which are transported by the insect as a whole in such a way that they cannot normally be taken for food. Aggregation of pollen into pollinia seems to be an adaptation to the increasing potential seed productivity of orchid flowers. The energetic provision for this intensification was obtained by a strong economization of the whole reproductive process under the conditions of a decrease of the frequency of visitation by pollinators (Nazarov & Gerlach 1997).

Orchids have evolved from lilylike ancestors (Fig. 7) by at least four important changes, adaptations that made their flowers precision-instruments specifically for insect pollination.

If we compare the flower diagram (in which the different parts are projected in whorls, Fig. 8) of an orchid with that of a lily we see these differences, which are all related to the manner of pollination:

1. The development of a *labellum.* In lilies we see two whorls of three flower-leaves that are alike. The outer, normally called *sepals* and the inner, the *petals*, are because of this likeness called *tepals.* In orchids one of the petals has a divergent form. This is the labellum, which serves as a landing-platform for insects and/or forms a kind of flag that attracts the insects and leads it via nectar-guides to the reward (if there is any), which in itself is a complex adaptation. The development of the labellum makes the flower *zygomorphic*: just like the insects themselves it can only be divided in one way into mirroring halves.

2. *Resupination.* This is a rotation of the flower

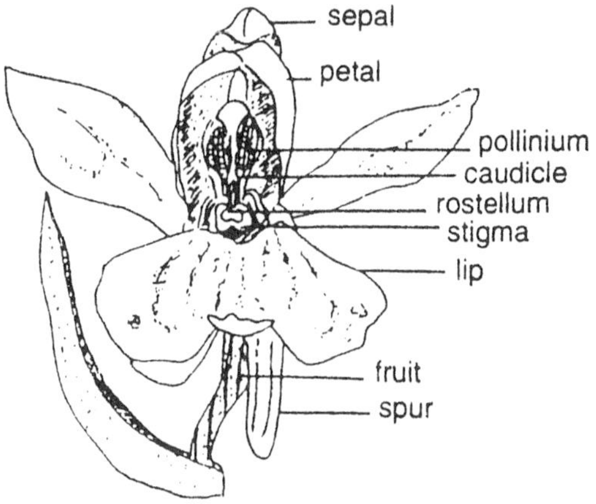

Figure 7. Typical orchid flower (after Van der Cingel 1986b).

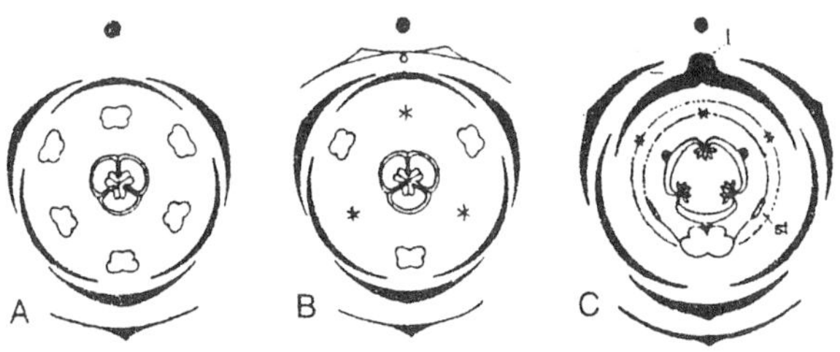

Figure 8. Flower diagrams. A) Lily, *Ornithogalum* (Liliaceae, Liliiflorae). B) *Iris* (Iridaceae, Liliiflorae). Note reductions in stamens. C) Orchid before resupination (non-twisted, non-resupination), l = labellum, st = staminodium (after Eichler 1875).

by 180° in relation to the *bract*, so that the lip becomes the lowest part. The labellum has to turn 180° to be able to function as a landing place and the torsion can be seen from the ridges and furrows of the fruit: the flower is resupinate.

3. The reduction of the stamens. We can see in the diagram that from both whorls of stamens in the lilies only one stamen (in some subfamilies there are two or three) is left in orchids. This stamen is fused with the pistil. The *anther* may be split up into two pollinia, such that there seem to be two anthers. The pollinia are solid masses, or packages of pollen, which become glued as a whole to the insect because of its own stickiness or a special sticky pad, the *viscidium*. Sometimes there are remnants of other stamens, called *staminodia*.

4. Pollinia and *stigma* are combined in a special structure, the *column* or *gynostemium*, that prevents the visiting insect from pollinating the flower with its own pollen and thus promoting cross-fertilisation. The fusion of stamen or anther and pistil is a typical characteristic of orchids. Sometimes, as in *Cypripedium* and in *Spiranthes*, the stamen and pistil are only partly united. The column has a special organ at the top, considered to be a sterile part of the pistil by some investigators or a separate new organ by others. It is called the *rostellum* and the pollinia lie behind it or perhaps fixed to it. In some species no rostellum is present and the pollinia simply stick to a stigmatic fluid, which is smeared at the back of the pollinator. In some species the visiting insect touches the rostellum while backing out of the flower and the viscid rostellum picks up the pollinia from the anther. In more advanced orchids the pollinia have threadlike appendices, the *caudicles*, that are attached to the rostellum by means of the *viscidium*. Still more advanced orchids have a *stipe*, a non-sticky part of the column which connects pollinia and viscidia. Orchids with stipes have also caudicles which connect the pollinia to the stipe.

The above description of the orchid flower is a very generalised one. Many variations of this pattern exist and form the basis of orchid classification.

Anthers, columns and pistils and their parts may differ considerably: the variations in form, size, number, position, texture, composition (stickiness) are adaptations to different pollinator species. They all have an evolutionary background and an ecological significance in the speciation process.

Different pollinators have different characteristics, different behaviour and different powers of *perception*. All these differences are reflected in the adaptations of the flowers; for example, colours of the labellum may differ in relation to the wavelengths of the light that a particular insect can perceive. Some pollinators receive the pollinia on their proboscis, others on their eyes, their heads, backs or legs which relates to the position they adopt on a flower or the route they take.

1.4 THE TYPOLOGY OF ORCHID POLLINATION

1.4.1 *Syndromes*

Vogel (1954), Faegri & Van der Pijl (1966) and Proctor & Yeo (1973) have used the concept of *pollination syndrome*, sets of characters that represent adaptations to particular classes of pollinators. Although the significance of this concept has been overestimated it is a useful way of looking at flowers, because it organises bodies of facts into recognisable 'pictures'. It makes us aware of the similarities between flowers visited by the same type of pollinator and it can supply us with hypotheses in cases of unknown modes of pollination of orchid species (see Plates 3-14, pp 12 and 13).

Bee-pollination syndrome

'Bee' is a collective name for a group of Hymenoptera which may differ widely in their behaviour towards flowers. It comprises social as well as solitary bees, nectar feeders and oil-, perfume- and pollen-gatherers: even wasps have to be included in the syndrome. Moreover, bees do not only visit orchids with their special morphology, but many other kinds of flowers as well. Nor are they *oligotrop(h)ic* and *constant* since they may visit several different flowers in one round. *Sympatrical* orchids have overcome the dangers of hybridisation by different deposition of the pollinia on the pollinator.

Plants may use different strategies to attract bees (as well as other insects), involving deception, general rewards and exclusive or unique rewards (Ackerman 1986b).

Despite all the things mentioned, the syndrome description shows its usability in detecting potential bee-pollinated or *melittophilous* orchids (Plates 3 and 4, p. 12).

Moth-pollination syndrome

Flowers that are pollinated by butterflies and moths differ from each other, due to the diurnal or nocturnal lifestyles of the pollinators. Sphinxes (Hawkmoths), differ from smaller moths in feeding behaviour, so that we may discriminate flowers to a certain extent on this basis. Flowers adapted to moths (e.g. those belonging to the Noctuidae) are called *phalaenophilous* and those adapted to Sphingids *sphingophilous*. The Sphingidae are good flyers and need no landing platform. In orchids however there is no easily observable difference between these two classes, because a lip is always present (Plate 5, p. 12).

Butterfly-pollination syndrome

Butterfly flowers are called *psychophilous* (note the connection with psyche, the soul, which was thought to fly like a butterfly).

Butterflies fly by daytime and are generally less able flyers than larger moths, needing a landing platform since they lack the ability to hover because their body structure dictates a much lower wing beat speed. They are able to see colours, though their visible spectrum may differ from ours. They seem to prefer vivid flowers and among orchids several flowers adapted to butterflies show contrasting and striking colours, many of them containing an element of red (Plate 6, p. 12).

Fly-pollination syndrome(s)

Most families of flies (Diptera) are not primarily adapted as flower visitors. Some, however, like the Syrphidae, Bombylidae and some Tachinidae are and flowers adapted to these flies are called *myophilous*.

However most flowers attracting these flies draw bees, butterflies and beetles as well. They are usually shallow, produce abundant nectar in superficial nectaries and have a sweet scent. There is, however, another story to tell about Diptera pollinating flowers, and this story is highly relevant to orchids. Orchids tend to reduce investment in rewards and try to attract pollinators by deceiving them: among those deceived are flies, mosquitoes etc. The plants involved exploit different drives of these insects, such as oviposition in or food uptake from decaying substances, dung or carrion and are called *sapromyophilous* (see Section 1.4.2).

Some people propose an even more differentiated system of fly-pollination syndromes. Sapromyophily should be divided in simple and complex sapromyophily (the first being open flowers, the latter being fly traps, Plate 7, p. 13) and *mycetomyophily* (mimetes of fungi).

Some flowers attract relatively long-tongued syrphids, for whom pollen is an important food source and these could be better described as *myiomelittophilous*, but this category is not so relevant for orchids. Others combine the characteristics of psychophily or sphingophily with myophily and the term *rhinomiophily* has been suggested. They are pollinated by nemestrinid and tabanid flies.

Flies are particularly abundant in higher altitudes where bees are less common. Since most of the smaller flies do not move over long distances, the chances for local speciation may be relatively higher in fly-pollinated orchids than in bee- or lepidopteran-pollinated ones (Christensen 1994).

Beetle-pollination syndrome

It seems that recently more and more indications have been found for regular pollination of orchid-species by beetles. These concern in part pseudocopulation (see below) and partly exploitation of the food-drive, either by rewarding or by deceiving visiting beetles. According to Van der Pijl & Dodson (1966) the few reports on beetle pollination apparently do not show a trend towards adaptation of orchids to beetles as pollinators. Reinhard et al. (1991) state that cases of pseudocopulation involve mistakes or erratic behaviour by the beetles, caused by the lack of specificity of the scents emitted by the orchid.

Both statements need more convincing proof. It may well be that e.g. the dactylorchids from Europe that depend on beetles (*cantharophily*), have evolved food-bodies (hairs) as adaptations to *anthophilous* beetles (Plate 8, p. 13). It might as well be possible that *Ophrys* species like *O.holoserica* attract beetles because they emit fragrances that include substances which stimulate beetles or flies as well.

Bird-pollination syndrome

Ornithophilous flowers have vivid colours and no

odour, obviously because birds depend primarily on vision and mostly have almost no sense of smell (Plate 9, p. 13). *Anthesis* of these more or less tubular flowers is diurnal. Since birds hover in front of the flowers that are visited for their nectar, a lip to land on is not needed. The nectar is usually abundant and produced in medium-length, broad tubes and spurs, with capillary systems in the container. To prevent damage by the strong beaks of birds they have a hard wall and protective ridges etc. In ornithophilous orchids there is usually a barrier to entrance of the beak. Most bird-pollinated orchids have a heavy *callus* or strong fold in the lip which closes the floral tube, so that the bird has to push its beak against the column to enter it and take up pollinia. Apart from hummingbirds there are several other families of birds that visit flowers, like sunbirds, sugarbirds and honeyeaters, flower-peckers and lorikeets, to mention only the most important. In orchids, the hummingbirds are most prominent, although there are reports of pollination by sunbirds in the Old World and South Africa, as well as by honeyeaters in Australia.

An interesting feature of hummingbird-pollinated flowers is that nearly half of these have pollinia of a blue, grey or brown colour instead of yellow ones. Another third part has cream or greyish white pollinia. This seems an adaptation to make the pollinia cryptic in order to avoid stimulating the birds to clean their beaks before effecting pollination (Dressler 1971).

The pollination syndromes are summarised in Table 1.

1.4.2 *Deceit-flowers and unusual or unique reward flowers*

Orchids have evolved different types of floral deception.

Although deceptive flowers are not unusual in plants, according to Vogel (1978b) this is as a rule only partial, because a small quantity of pollen as true reward will usually be gathered, either from the reproductive anthers or from special food pollen. The exception occurs apparently in the orchids, where a total gamble is made in these modes of flower *parasitism*.

In the first place there are flowers which deceive insects by exploiting their sexual drive. The syndrome is called pseudocopulation and it is questionable whether it is really deceit, for one could argue that there may be 'reward' in the stimulation

Table 1. Pollination syndromes (after Wyatt 1983).

Syndrome	Pollinators	Anthesis	Colours	Odours	Flower depth	Nectar guides	Rewards
Cantharophily	Beetles	Day and night	Variable, usually dull	Strong, fruity or aminoid	Actinomorph, flat to bowl-shaped	None	Pollen or food bodies
Sapromyophily	Carrion and dung flies	Day and night	Purple-brown or greenish	Strong, often of decaying protein	None, or deep if traps involved	None	None
Myophily	Syrphids and bee flies	Day and night	Variable	Variable	None to moderate	None	None or pollen or nectar
Melittophily	Bees	Day and night or diurnal	Variable but no pure red	Present, usually sweet	None to moderate	Present	Nectar (41.6%) and pollen; open or concealed
Spingophily	Hawkmoths	Nocturnal or crepuscular	White or pale to green	Strong, usually sweet	Deep, narrow tube or spur	None	Ample nectar (22.1%); concealed
Phalaenophily	Small moths	Nocturnal or crepuscular	White or pale to green	Moderately strong, sweet	Moderately deep tube	None	Nectar; concealed
Pyschophily	Butterflies	Day and night or diurnal	Bright red, yellow or blue	Moderately strong, sweet	Deep narrow tube or spur	Present	Nectar (22.8%); concealed
Ornithophily	Birds	Diurnal	Bright red	None	Medium-length, wide tubes or spurs with barred entrance	Absent	Nectar abundant (25.4%); concealed

of the sex drive as such. We have to be careful however, since we can be easily trapped into what is called *anthropomorphism*, by postulating a reward. It is not known what is rewarding to insects, since real sexual satisfaction by sperm ejaculation is never seen in pseudocopulation in *Ophrys* flowers. It is however known from the ichneumoid wasps (*Lissopimpla excelsa*) which copulate with the Australian orchid *Cryptostylis leptochila*. Anyway, in terms of survival value the behaviour is not rewarding at all. On the contrary!

One could argue that pseudocopulation in itself is an anthropomorphic concept. Kullenberg (pers. comm.) opposes the use of the word and prefers more neutral terms like 'stimuli' and 'answers'. Also in the Australian species pollinated by thynnid wasps, for example, the male insects do not always attempt to copulate with the flowers, but to take the 'pseudofemale' to a place they use to copulate with the females.

There are orchids which exploit the sexual drive in a different, less direct way. The most indirect way is to lure male bees during mate-seeking flights, which in fact is exploitation of the hunger drive during periods of sexual patrolling. Several of these orchids imitate rewarding plants without offering nectar. It is possible to call this type of flower a 'rendezvous' flower.

Still another form of '*sexploitation*' is seen in orchids that attract male euglossine bees which gather fragrances (perfumes), consisting of terpenoids and aromatic compounds from the flowers to use these as *sex-attractants* (*pheromones*). We cannot call this kind of exploitation deceit, because the bees are rewarded with something that is important to them and affects their fitness in a positive sense.

The fragrance is collected with the tarsal brushes of the front legs, transported with help of the middle legs to the swollen tibia of the hind legs. It is now believed that the bees store the collected fragrances in their hind legs, modify them by glands in their tibial store-organs (Vogel 1966), translocate some of the modified compounds to their mandibular glands, where they may be modified again into larger compounds (Dressler 1982), and use mandibular gland secretions to mark mating sites to which females are attracted (*lek behaviour*). The blends of compounds could consist of generalised attractants, modifiers and repellents. Some of the components may be long-distance attractants, whereas others probably are short-range releasers of mating behaviour in the females (Williams 1982).

According to Williams (1978) the euglossine bees pollinate at least 3000 species of orchids. Such orchids, to be found in tropical America, show different contrivances to bring insects to effective pollination. I will not deal with this pluriformity here, but in a more appropriate place in the relevant systematic chapter. Euglossine bees of both sexes may also visit flowers for nectar. We therefore should discriminate between *andro-euglossophilous* and *gynandro-euglossophilous* (typical bee-pollinated) flowers and syndromes. The andro-euglossine flowers, which only attract males and are exclusively pollinated by them, are very fragrant, lack nectar, and attract no other groups of bees or insects. Williams (1982) has given an excellent overview of the phenomena of euglossine pollination and deals also with the morphology and classification of these neotropical bees.

Many orchids deceive inexperienced bees and bumblebees in early spring, before the animals have developed *flower fidelity* or a degree of constancy. If the insects perform orientation flights in the environment they try out flowers as potential food sources. They will soon learn that some orchids do not offer rewards, but will on average need more than one unsuccessful experience to know this. An example is *Calypso bulbosa*. The principle of exploiting the foraging behaviour of naive bees is fairly common and is often probably supported by certain aspects of mimicry, especially *guild mimesis* or *nonmodel mimicry*. In case of deceptive exploitation it is also called *nonmimic deception.*

Mimicry is a common and universal trait among orchids (Plates 10-12, p. 13). It depends on the absence or presence of a reward whether we speak of mimicry as recognised by Müller or Bates. Sometimes the pollinator receives a substitute for pollen in the form of hairs or papillae. It is a matter of definition and degree to call this deceit since this *pseudopollen* is a kind of facsimile food. We should be aware of the existence of gradual transitions between no reward at all and rewards which lead to preferential choice on the part of the visiting insects. The possibilities range from 0-100%, so there is a gradual transition between deceit and reward.

In fact there is hardly a limit to the exploitation of insect behaviour. All behaviour that lends itself to it seems to be useful for the development of adaptations by orchids:

In *pseudoantagonism* as shown by *Oncidium* (*Tolumnia*) species, territorial bees defend areas in the neighbourhood of these orchid flowers against other insects. Sitting on high observation-posts they fly towards intruders to chase these. Flowers dancing in a breeze are attacked as if they were intruders. The bees then touch the viscidia and pick up pollinia on their foreheads.

In *prey-imitation* as is known from the European orchid *Epipactis veratrifolia* (Plate 14, p. 13) the flowers have structures on the labellum that imitate aphids. Hoverflies, defending territories around the orchid, chase other males but try to copulate with females. These females try to lay eggs in the pseudo-aphids. The females are probably stimulated by odours from the flower.

Related to prey-imitation is the exploitation of egg-laying behaviour or feeding behaviour by mushroom- or toadstool-imitating orchids (*mycetophily*). It ranges from the imitation of the odours of the substrate on which eggs are laid to the optical (and tactile) imitation of mushrooms. In cases where substrates for *oviposition* are simulated for attraction of insects that lay eggs in dung or carrion this is called sapromyophily. The smell is usually fetid. This chemical mimicry is effected by volatiles like ammonia and other amines (Dafni 1984).

In *Cypripedium calceolus* the instinct for nesting behaviour is exploited. The flower emits fragrances which are used by male bees to odour-mark objects in order to attract females and probably attract the males themselves as well, causing them to aggregate. One substance acts as a signal on the nest site, releasing landing responses.

There are also orchid flowers (*Serapias* species e.g.) which offer shelter to solitary bees by imitating nest holes (Plate 13, p. 13). The animals spend the night in the flowers where the average temperature inside is slightly higher than the temperature of the surrounding air. In the morning solar radiation may heat the flower to 3°C above the ambient temperature. Sometimes a nesthole is mimicked and no real shelter offered (*Ophrys helenae* (Paulus 1990, but see Molnàr 1998).

In many cases it seems that a combination of two or more of the phenomena mentioned is the order of the day (see Table 2).

In looking at syndromes we should be aware of the level of pollinator specifity: low pollinator specifity occurs in flowers with readily-accessible nectar, whereas high specifity is associated either with nectar in a deep tube or with deceptive flowers (Catling & Catling 1991).

1.5 THE MECHANICS OF ORCHID POLLINATION

Plants show all kind of movements. There are those movements during growth that we seldom observe, but which become particularly obvious when photo-

Table 2. Orchid flower syndromes in relation to exploited insect drives.

Syndrome	Exploited insect drive	Relation to: a. Insects b. Other plants	Attraction	Reward
Floral mimicry (Müller)	Food foraging	a. *Mutualism* b. Mutualism or *commensalism*	Visual (olfactory)	Nectar (pseudopollen)
Floral mimicry (Bates)	Food foraging	a. *Parasitism* b. Resource parasitism	Visual (olfactory)	None
Rendezvous flowers	Sexual	a. *Facilitation*	(Visual) olfactory	None
Pseudocopulation	Sexual	a. Parasitism	Visual, olfactory, tactile	None
Prey or substrate imitation	Egg-laying	a. Parasitism	Visual, olfactory, tactile	(Nectar)
Food imitation	Food foraging	a. Parasitism	Visual, olfactory	None (pseudofood)
Shelter imitation	Sleep/warmth (sexual?)	a. Parasitism or mutualism	Visual, olfactory (tactile)	Hiding
Nest/nesting imitation	Territorial behaviour	a. Parasitism	Visual, olfactory	None
Pseudoantagonism	Territorial behaviour	a. Parasitism	Visual	None
Perfume sampling	Sexual	a. Mutualism	Olfactory	Pheromone precursors

graphs are taken intermittently (time-lapse photography) and than projected as a continuous film. These growth movements are regulated by plant hormones and directed by gravity or light. There are also movements of a more sudden nature, either passively (from the point of the plant), by means of hinges or joints and manipulated by pollinating insects or active movements, caused by turgor changes or *hygroscopy*. I shall not deal here with the dynamics of growth, nor with *flower induction* or *photoperiodism* and comparable phenomena and will restrict myself to the subjects of resupination, deposition of pollinia and movements of the labellum.

1.5.1 *Resupination*

The great majority of orchids have the labellum turned downward by 180°, to make the labellum a landing platform for pollinators or to fulfil other functions in the pollination process, such as positioning the insect for pseudocopulation and consequent deposition of pollinia. Initially the labellum is in an *adaxial* position, forming the inner, median sepal. It originally occupies the upper side of the flower, while in the bud. After opening the position on the lower side of the flower is a derived position seen in most cases from the twisting of the ridges on the *ovary*. The turning of the lip to its final position may be genetically fixed, which is the normal case, and it is brought about by torsion of the pedicel or the *gynaecium*. Resupination may also be brought about by the weight of the flowers on a slender inflorescence, causing it to nod or making it pendant, e.g. flowers of *Cypripedium*. In that case the resupination often disappears with respect to the total view of the inflorescence. Another possibility is that the flower is directed by gravity: this can be shown by experiments in which the gravitational forces are excluded.

Some orchids show a non-resupinate position with the labellum above. This again can be caused by different mechanisms, e.g. torsion of 360°. This phenomenon is clearly a secondary derivation from the resupinate position and probably an adaptation to hovering visitors and/or advertisement with a type of flag.

1.5.2 *The deposition of the pollinaria*

The orchid flower has developed by means of centralisation as a precision instrument for deposition of the pollen. In most orchids only one anther was left as a result of this evolution, one anther in one place in the median plane opposite the labellum, such that it could deposit its pollen on the most advantageous side of the pollinator. In this way efficiency in deposition and in reception by the pollinator was attained. The pollen moreover became packed in one mass, the *pollinarium*, which led to an ‘all or none’-strategy. If the pollinarium reaches a stigma it will fertilise numerous ovules, without wastage of pollen on the way, by being eaten, for example. The same centralisation process incorporated the stigma into the gynostemium or column, greatly increasing the efficiency of the reception by the flower. Thus during one visit an insect can take away new pollen and deposit pollen from a flower visited before. In the column the anther overhangs the top of the column and the stigma is lying underneath. A special organ, the rostellum, lying on top of the column (sometimes roofing the stigma) has been developed to separate both functions and prevents self-pollination. In several orchids (in Orchioideae for example) a viscid disc, called the viscidium (in pairs viscidia), is formed by the rostellum and is sometimes covered by a membrane, the *bursicula*. The viscidium may either be freely exposed or, if it is covered, may be freed more or less explosively by the touch of the visiting insects. In primitive Neottioidae and in the Cypripedioidae the

Plates

1. Hothouse at Down House, Kent, after a recent renovation. Charles Darwin used to nurse his orchids here for pollination studies. Photo N.A.van der Cingel, 1998 (p. 1).
2. *Angraecum sesquipedale*, the orchid of which Darwin hypothesised (the length of the proboscis of) its pollinator (see Chapter 6). Photo N.A.van der Cingel, 1967, Kew Gardens (pp 1 and 155).
3. Typical bee-pollinated flower: *Eulophia speciosa*. Photo N.A.van der Cingel, 1989, Inhaca Island, Mozambique (p. 8).
4. Euglossinophilous flower: *Catasetum viridiflavum*. Photo N.A.van der Cingel (pp 5, 8 and 87).
5. Typical moth-pollinated flower: *Brassavola martiana*. Photo N.A.van der Cingel, 1984, Suriname (p. 8).
6. Typical butterfly-pollinated flower: *Anacamptis pyramidalis* (with *Aporia crataegi*). Photo N.A.van der Cingel, 1990, Aveyron, France (p. 8).

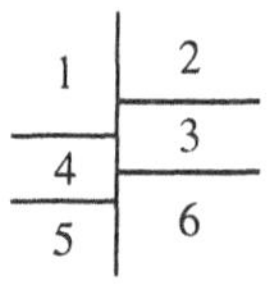

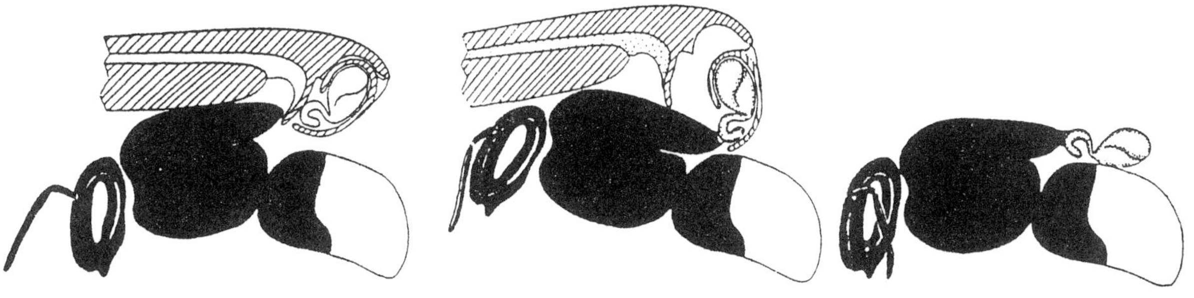

A B C

Figure 9. Mechanism of attachment of a pollinarium in a simple bee-pollinated flower (e.g. *Cattleya* spec.). A) Position of bee in relation to the column after the bee has completely entered the flower. B) As the bee backs out, the dorsal thorax pushes against the rostellar flap and sticky liquid from the stigma cements the base of the pollinarium to the bee. C) Position of the pollinarium as the bee flies to another flower (drawing by G.P.Frymire).

pollen is still powdery, but in higher forms the pollen grains are cemented together by viscin threads and divided into a number of pollinia. The pollinia may bear threads that are concentrated into a *caudicle*, which may greatly help in transportation. Sometimes there is a straplike pedicel (stipes) between pollinium and viscidium, but they may also be connected directly. Pollinia, stipes and viscidium together comprise the pollinarium.

Pollinaria can be attached to the visitor when it leaves the flower, as is the case in lower orchids, (Fig. 9) but in some higher orchids the pollinarium is attached upon entry.

The viscid matter with which the pollinia stick to the insect has the quality of setting, like a cement, hard and dry in a few minutes. During transportation the stipes can bend, leading to the pollinia being directed straight into the stigma of the next flower visited. This bending can be a matter of seconds or may last up to several hours, preventing *geitonogamy* and enhancing the likelihood of cross-fertilisation.

Cross-fertilisation is also promoted by anther cap retention (Catling & Catling 1991). In e.g. *Tipularia discolor* the cap firmly surrounds the pollinarium for 8-30 minutes after removal. A pollinarium with a cap attached is unlikely to adhere to a stigma because the elastic strength of the pollen stalk is sufficient to pull it away.

The receiving stigma is viscid and will break the elastic threads by which the packets of pollen grains are tied together and thus leave some of them on the stigma. This means that parts of a pollinium as a rule can be applied to several stigmas. There are many variations on this theme of the mechanics of pollination movements.

In self-pollinating (*autogamous*) flowers we can find many adaptations to this kind of reproduction (Fig. 10), in which mechanical means (e.g. raindrops) may play a prominent role (Catling 1991).

For more detailed descriptions of the processes that take place when an insect enters a flower and picks up the pollinia or deposits the pollen I refer to Darwin (1888).

1.5.3 *Movements of the labellum*

In many tropical epiphytic and terrestrial Australian

<table>
<tr><td>7</td><td rowspan="2">8</td><td rowspan="2">9</td></tr>
<tr><td>10</td></tr>
<tr><td colspan="3">11</td></tr>
<tr><td>12</td><td>13</td><td>14</td></tr>
</table>

Plates

7. Typical fly-pollinated flower: *Bulbophyllum orectopetalum*. Photo N.A.van der Cingel, January 1997, Khao Sok, Thailand (pp 8 and 14).
8. Beetle-pollinated flower: *Dactylorhiza fuchsii*. Photo N.A.van der Cingel, 1985, Austria (p. 8).
9. Bird-pollinated flower: *Rodriguezia secunda*. Photo N.A.van der Cingel (pp 9 and 14).
10. Floral mimicry complex: *Epidendrum radicans*. Photo N.A.van der Cingel (p. 10).
11. Floral mimicry complex: *Asclepias curassavica*. Photo N.A.van der Cingel (p. 10).
12. Floral mimicry complex: *Lantana camara*. Photo N.A.van der Cingel (p. 10).
13. Shelter imitation: *Serapias orientalis*. Photo N.A.van der Cingel, 1999, Cyprus (p. 11).
14. Prey imitation: *Epipactis veratrifolia*. The epichile has been removed. The hypochile shows dark callus-like swellings imitating aphids. Photo N.A.van der Cingel, 1999, Cyprus (pp 11 and 164).

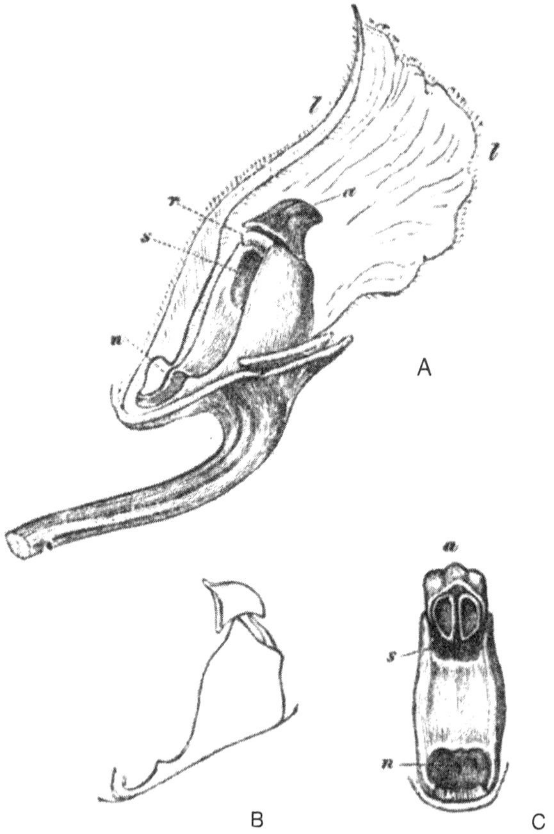

Figure 10. *Dendrobium chrysanthum.* The elastic labellum and ditto anther filament function as a mechanism for self-pollination in case a visiting insect fails to remove the pollinium. A) Lateral view of flower, with the anther in its proper position, before the ejection of the pollinia. All the sepals and petals are removed except the labellum, which is longitudinally bisected. B) Outline of column, viewed laterally, after the anther has ejected the pollinia. C) Front view of column, showing the empty cells of the anther, after it has ejected its pollinia. The anther is shown hanging too low down, and covering more of the stigma than it really does. a = anther, l = labellum, n = nectary, r = postellum, s = stigma (after Darwin 1888).

orchids we find movable labellae, which in one way or another play a role in the pollination process by influencing the behaviour of the pollinator (Plate 7, p. 13 and Plate 15, p. 28). In Eurasian orchids one example of such a mechanism is presented by *Epipactis palustris*. This mechanism has often been disputed by different investigators of this species. The bipartite labellum of this orchid is so shaped that the distal (*epichile*) half is joined to the basal half (*hypochile*) by a narrow hinge.

Ilona Müller analysed movie-pictures of pollinating bees and saw that the honeybee holds on to the hypochile with its forelegs while standing at the epichile with its hind legs. The labellum moves up and down like a seesaw caused by the movements of the bee. If the bee presses itself deeper into the flower the labellum presses back when the bee seizes the upper flowerparts with its four front legs, and catapults itself upwards like a trampoline-athlete, visiting flower after flower with the same experience.

Darwin described several examples of irritable labellae, e.g. of the Australian *Pterostylis longifolia* (Fig. 11).

1.6 THE CHEMISTRY OF ORCHID POLLINATION

Orchids produce chemical substances in their flowers especially for pollination purposes. Firstly there are rewards, of which the best known is nectar: secondly, various fragrant substances to attract pollinators.

1.6.1 *Nectar*

Nectar is easy to produce and also relatively cheap since any green plant has at its disposal the raw material glucose, produced in leaves. It can transport it to the nectaries, where it can be used as such

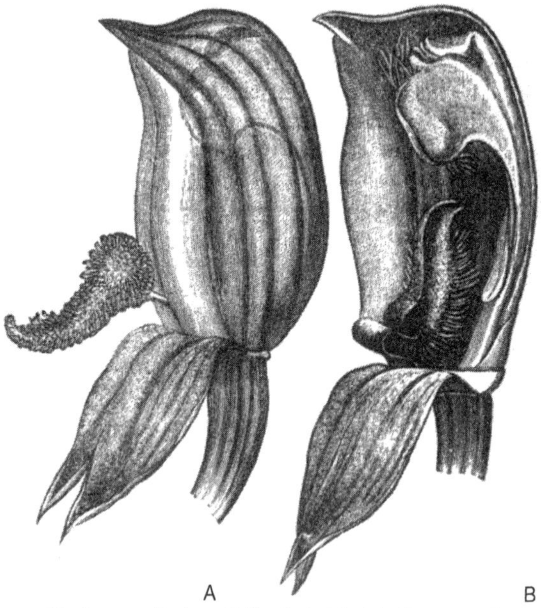

Figure 11. *Pterostylis longifolia.* Irritable labellum. A) Flower in its natural state: the outline of the column is dimly seen within. B) Flower with the near lateral petal removed, showing the column with its two shields, and the labellum in the position which it occupies after having been touched (after Darwin 1888).

or can be simply converted into other sugars. The translocation of sugars and other products from the leaf to other organs of the plant takes place via the sieve tubes of the phloem. From the content of the solution the amino acids and other non-sugar components are removed, although slight amounts of lipids and amino acids as well as other substances can be found in low concentrations in nectar. Nectar is as a rule a mixture of sucrose, glucose and fructose. The mixture tends to be constant in a given species but differs in different species according to the type of pollinator. Flowers that are visited by lepidopterans (hawkmoths and butterflies) are generally rich in sucrose (50-99.9%) or even *sucrose-dominant* (less than 0.1% hexoses). Nectars of flowers visited by long-tongued bees and wasps are usually sucrose-rich, whereas those pollinated by short-tongued bees and flies are generally rich in hexoses (10-49.9% sucrose) or even *hexose-dominant* (less than 10% sucrose). This figure is also typical for short-tubed bee- and butterfly-flowers. The above is a generalised picture. It seems logical to suppose that these classes of nectar reflect pollinator preferences.

Apart from diversity in composition there may also be differences in concentration, volume and time of production. The concentration can be influenced by the relative humidity and vapor pressure. Many insects have adapted in morphology to flowers in relation to nectar feeding, e.g. by elongation of mouthparts.

Stigmatic exudates may be important as rewards. They contain lipids, amino acids, phenolics, alkaloids and anti-oxidants. Sometimes also free sugars are present. The normal function of the exudate is to capture and hold pollen and to foster its germination. It seems that they can also protect the stigma and offer resistance against desiccation. The exudates have hitherto received little attention as rewards.

The labellae of orchids may bear sugar-containing parenchymatic tissues which are supposed to act in some cases as a reward. Sometimes these tissues have the shape of hairs that mimic pollen.

Extrafloral exudates are also well known in orchids. Jeffrey et al. (1970) analysed sugar content in both floral and extrafloral exudates. Both may act as attractants for ants which feed on them and repel grazers.

1.6.2 *Fragrance*

Plants invest in fragrances to attract pollinators, which saves energy (production of sex cells) in comparison with wind pollination. These substances need to be attractive and have to fulfil some function in insect life. This can be the advertising of food or even the apparent advertising. It can also be the arousal of sexual, social or nesting behaviour or any other kind of conduct that is influenced or led by scents. These fragrances have to be specific, at least in orchids, in order to attract specific pollinators. The plant communicates with the pollinator by means of chemicals which must be perceived by the insect's sense organs. The fragrances are produced in special places called *osmophores* (Figs 12 and 13; Plate 16, p. 28). In places where scents are produced it is demonstrated that starch, the substance that functions in the plant as storage for glucose, and as such the basic form of energy, disappears. Even complicated fragrances are built up from simple substances like sugars and amino acids.

There are many chemically different types of fragrant substances like mono-terpenes, aromatic substances derived from benzene (e.g. vanillin), nitrogen-compounds, i.e. aminoids and amines, carbohydrates, methyl esters and derivatives of fatty acids.

From the osmophores of euglossine-pollinated orchids more than 50 compounds have been isolated and identified, consisting of monoterpenes and simple aromatic compounds. In other species many compounds have also been analysed. Some

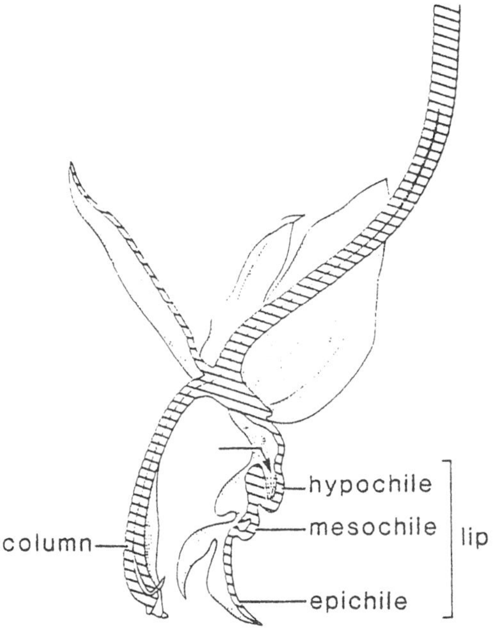

Figure 12. Morphology of a typical *Stanhopea* flower (bisected) with the osmophore (arrow) located in the hypochile of the lip (from Stern et al. 1987).

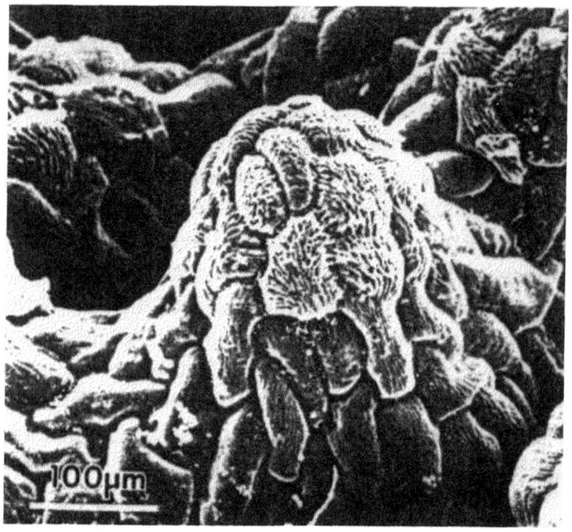

Figure 13. Scanning electron micrograph of osmophore of *Stanhopea wardii* (from Stern et al. 1987).

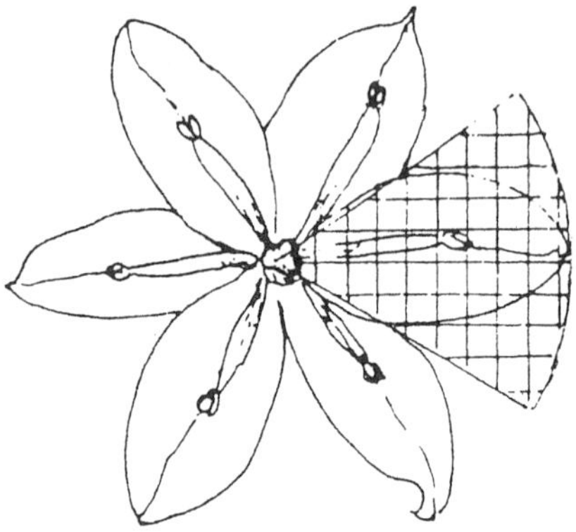

A

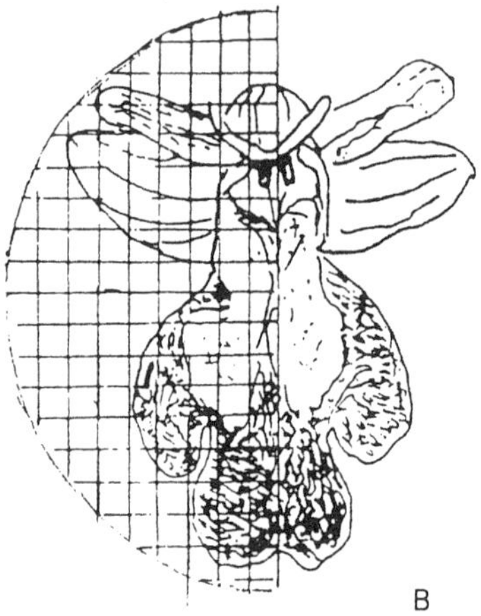

B

Figure 14. Zygomorphy: the chequered parts contain in principle all information with which the flowers can be described. A) Bilateral symmetric flower. B) Radial symmetric flower (after Van der Cingel 1983).

orchid species alone contain over 50 of these substances in their fragrance. (Kaiser 1993).

It is to be expected that with modern instruments to analyse fragrances additional compounds will be identified in the future. The human nose is, according to Kaiser, still the best instrument to detect fragrances.

1.7 THE GEOMETRY OF ORCHID POLLINATION

1.7.1 *Zygomorphy*

Orchid flowers are zygomorphic as an adaptation to insects. Bilateral symmetric flowers contain more information per cm^2 than radial ones. (Davenport & Kohanzadeh 1982). To describe a zygomorphic flower we need a more detailed description than with a multilateral flower; zygomorphic flowers are by any interpretation more complex (Fig. 14).

The significance of this is that by developing zygomorphic flowers the plant can attain a higher degree of differentiation and specificity. Zygomorphic flowers are, as a result of this, easier to discriminate from one another.

Mathematically it can be argued and calculated that movement reinforces this effect and, because flower-visiting insects should recognise flowers when flying this really is important. All kind of information can play a role in recognition. The act of recognition can be phased. At a distance an insect may react to colour and/or scent, while close by other elements may come into play, for example patterns of spots on the lip. Pollinators may develop search images of the flowers they have learned to prefer. A search image is an awareness directed towards a certain form or structure. The observed structure is what psychologists call a 'Gestalt'. The '*Gestalt*' principle is often stated as: 'the whole is more than the sum of the parts'. In terms of information theory: the whole contains more bits of information than the sum of bits of information from the separate parts. The relations between the parts, for example the patterns of coherence between the different parts themselves, contain information. However there is no doubt that while such a phenomenon as a search image does exist, in relation to pollination it is merely a hypothesis.

1.7.2 *The geometry of co-adaptation*

There is still another aspect of zygomorphy that

needs attention if we look at orchids, because pollinators are zygomorphic as well. In fact the zygomorphy of flowers is derived from that of animals by adaptation. Flowers which depend on pseudocopulation by mimicking females of the pollinating species, need to be zygomorphic. Pseudocopulation as in *Ophrys* is unthinkable with radial flowers because the pollinating animals have to position themselves very precisely in the right place and with the right orientation.

All those characteristics of the flower like size, colour, scent, form and texture (hairiness) cooperate in cajoling the insect in the right place in order to remove pollinia.

This matching of flowers and pollinators is not limited to Ophrys-flowers or, generally speaking, to flowers which depend on pseudocopulation. As remarked, most orchids are precision-instruments, developed in the processes of coevolution of pollinator and flower, in which the orchid especially had to adapt in size for all floral elements to match the dimensions of the pollinator.

1.7.3 *Biometry*

To conclude this topic I should remind the reader of the facts of numerical taxonomy or *biometry*. The taxonomy of orchids nowadays is still to a large extent determined by measurements, at least in Europe. Different populations are compared by carefully measuring all the flower parts, up to 30 or more characters. Decisions on the status of species and *subspecies* are in principle based on these measurements.

1.8 THE ECONOMY OF ORCHID POLLINATION

1.8.1 *Economic foraging behaviour of pollinators*

Survival is principally a matter of efficiency, taking the right decisions at the right time. On the basis of what is learned during orientation flights, the insects can afterwards make decisions about foraging that are economically sensible, that is to say those that yield the best results in relation to the time spent.

Orientation, change of direction and behaviour on the inflorescences are all determined by economical decisions (Waddington 1983).

Where the insects have a choice of food sources they tend to develop *fidelity*, which is beneficial to both plant and insect. Decisions seem to be made on the basis of cost-benefit calculations, in which the amount, constancy and 'caloric worth' of the reward are negotiated. Besides this there are also some inherited preferences.

Floral density and floral morphology may influence choices.

In social insects, individual decisions can be based on information received from other animals.

1.8.2 *Economy of plants – The adaptive nature of floral traits*

Plants have to economise their investments in flowers: energy and food used in the production of petals and sepals, colours and nectar rewards cannot be used for seed production. This is especially true for plants that are adapted to poor environments, like most orchids. It seems that in many orchids, investment in seeds is spent at the cost of vegetative propagation, which may result in fewer and smaller plants on the same clone in the next year.

For plants to be effective as attractors and exploiters of insects, their floral traits ought to influence the visiting behaviour of pollinators in many instances and in many ways. These floral traits range from factors like flowering time through to colour and rewards (Waser 1983).

Plants compete for pollinators. Also the morphologies of flowers and their visitors have to match for effective exploitation.

Constancy, the tendency of animals to restrict their visits to flowers of a single species, even when these occur intermixed with other equally accessible and rewarding flowers, is beneficial to the plant because it limits the chance of receiving strange pollen. It can be beneficial to the insect as well and therefore it may be expected that flowers evolve characteristics that promote flower constancy by becoming more accessible to certain species than to others.

As to rewards, in orchid pollination it is nectar especially that should concern us, because pollen has evolved in the pollinaria to prevent pollinators from using it as food.

Flowers of insect-pollinated plant species vary greatly in the amounts and concentrations of nectar.

Because so many orchids deceive pollinators by not rewarding, the characteristics of the orchid flowers in attracting pollinators need special attention.

We need also population studies over long periods to gain more insight into all phenomena that influence plant fitness.

Computer models and population studies done so far indicate that maximum fitness is attained at intermediate pollination levels (a.o. Calvo 1990c).

1.8.3 *Competition and facilitation for pollinators among plants*

We can examine how in a population the behaviour of insects is regulated and what the consequences will be for the plants either from the viewpoint of insects (to find out about the optimal foraging behaviour) or from the viewpoint of the plant (to find out about the optimal adaptation to insects), but we may also look at the interactions between plants to find regularities in patterns of cause and effect, or irregularities.

Interaction for pollinator visitation may occur as *competition* for the same pollinators, as *facilitation* (as for example in *Müllerian mimicry*) in which the presence of one species increases visitation to another species to its profit or at no cost to the second species (called *mutualism* and *commensalism* respectively), or as *resource parasitism* (one species draws pollinators away from another, resulting in enhanced visitation for the first plant species in the presence of the second plant species, as is the case in *Batesian mimicry*).

Theoretical reflections suggest that interaction can change from facilitation to competition with increasing floral density, as a consequence of saturation in pollinator numbers. Aspects of the life history of plants will influence the interactions (Rathcke 1983).

In any habitat it will be important that pollinator numbers are maintained over long periods. Early flowering plants help maintain starting populations of pollinators and through that influence the survival of plants which flower later in the season (Real 1983b).

The study of all these phenomena is still in its infancy. Nevertheless its value is tremendous and we may expect a lot of insight from careful observations and experiments.

1.9 THE GENETICS OF ORCHID POLLINATION

The laws of Mendelian inheritance are well known and will be assumed in what follows. However, the genetics of pollination biology are of interest to any student of the title subject of this book. The genetics of populations raise important notions and, of late, some new methods have been developed to study variance in populations. The results of genetic investigations of orchid populations can be valuable in view of the current need for conservation and management of populations in nature reserves and on wasteland.

1.9.1 *Gene pool*

The whole set of genes in a breeding population is called a *gene pool* and in each population there is a certain amount of genetic variation or variability which is often denoted by one word: *variance*. Individuals may contain many divergent characters and this variance is the basis for adaptation in critical or changing circumstances. This is the very nucleus of the evolution theory. *Natural selection* has to be considered as a mechanism and is far from being a freely acceptable or rejectable theory and certainly not a belief or tautology, as some people claim. In any system of replication with mutation, natural selection must occur.

The larger a population is, the larger in general its variance and the greater also its flexibility and adaptability. Now that many populations become smaller and smaller under the influence of ever changing land-use and exploitation of natural reserves, and large areas are often divided into smaller ones, there is a serious risk of severe loss of variance, even to a dangerous extent. Populations which are too small run the risk of dying out in changing circumstances.

Two important concepts of population genetics need some initial explanation. Firstly *genetic erosion* which is the disappearance of genetic variance by loss of parts of the population. Secondly, as certain genes become rarer in a population chance may come into play. In this case genes can disappear accidentally through death or removal of individuals. This phenomenon is called *genetic drift* and is defined as change in gene frequencies over generations, resulting from chance rather than from selection.

1.9.2 *Variance – Sexual reproduction and vegetative propagation*

Cross-fertilisation promotes the distribution of variance in a population. Self-fertilisation tends to diminish variance and, especially if it is continuous, leads to inbreeding. Therefore we often find some cross-fertilisation even in populations that depend normally on self-fertilisation.

Asexual propagation, via tubers or by axillary bulbils, also limits variance, because it leads to the formation of clones of which the individuals are genetically alike. Because so many terrestrial orchids reproduce vegetatively, there is a question of whether populations can maintain variance and how small they can be to do so.

Peakall & James (1989) studied variance in an Australian orchid population of the species *Leporella fimbriata*. They found 67-80% for genetic individuality and for the whole population they estimated even higher figures for *outcrossing*: 89-97%. It is believed that three factors promote *panmixis*, that is the thorough mixing of *genotypes* in populations of this orchid. First, ants function as pollen vectors, and in spite of their short mean flight distances they cross the borders of clones sufficiently to maintain some outcrossing. Second, seeds are dispersed by wind, creating a seed flow. This flow seems to be extensive because of the small and numerous seeds so characteristic for orchids. Seedling establishment therefore seems to be random, leading to random interclonal matings. Thirdly, the products of selfing are able to maintain themselves to a lesser extent than products of cross-fertilisation (e.g. through reduced *self-compatibility*). The elimination of self progeny may lead to the appearance of panmictic adult populations. It seems that *Leporella fimbriata* is able to maintain a considerable amount of variance in populations. If this research is applicable to terrestrial orchids the consequences are great.

1.9.3 *Pollen flow*

Peakall (1989) also studied the transport of pollen with a new technique in one of the Australian orchid species, *Prasophyllum fimbria*, by marking individual pollen grains within the pollinia. The pollinia were then followed in several populations. It was also possible to detect in this way the origin of pollen deposited on the stigmas of labelled flowers. Peakall used six different colours.

The technique seems to approach actual pollen flow in a natural population and is fast as well as cheap (compared to radioactive labelling) and therefore is promising for analyses of pollen flow, probably including phenomena like genetic erosion and genetic drift.

1.9.4 *Fathers and mothers*

Analyses like the above have mainly been oriented toward the question of where a certain flower succeeds in depositing pollen (*male function*).

To a much lesser extent the question has been answered concerning the distribution of origins (genetically and in distance) of pollen received by stigmas (*female function*).

Piper & Waite (1988) investigated pollen export and pollen import in *Epipactis helleborine* in a population in Sussex in relation to number and density of open flowers borne by an inflorescence and concluded that in this *hermaphrodite* flower, secondary floral characters are predominantly male in function.

Catling & Catling (1991) mention the interesting case of *Platanthera* species with a flower morphology that leads to exploitation of more than one pollinator by forcing them to approach the flower from one side. This can be explained in terms of *sexual selection*. Bateman (1948) has suggested that maternal fitness is resource-limited, whereas paternal fitness is limited by ability to achieve fertilisations (pollination-limitation). Consequently sexual selection in plants is expected to involve an adaptive extension of the male role that increases fitness. In hermaphroditic plants this may lead to the evolution of features that increase the number of occasions during which pollinating agents can pick up pollen. Sexual selection therefore can operate by restricting the amount of pollen that is removed during a single pollination and by extending the period of pollen presentation.

The examples above show how complex the study of *population dynamics* is and that we are just beginning to ask some intriguing questions about the genetic consequences of population structure in relation to pollinator behaviour.

1.9.5 *Hybridisation*

The influence of hybridisation is the opposite of inbreeding. Successful hybridisation may lead to enrichment of nature, providing it with new material to act upon. Although it may eventually lead to new species, hybridisation also has its limitations: the pollinator-plant geometry may fail, leading to nonreproduction.

Hybridisation in orchids is of a special nature. Because of their dependence on *mycorrhiza* orchids have developed very small seeds which lack *endosperm*. This phenomenon has made it possible for orchids that are not closely related and differ considerably in genetic composition to hybridise, and thus a lot of hybrids occur in nature. Complete hybrid swarms have been regularly found.

Nilsson (1983) extensively studied populations of *Platanthera bifolia* and *P.chlorantha*, including hybrids, to find out why these sympatric species keep isolated as distinct species. Hybrids are in principle fertile and not rare, so that the isolation of the two gene pools of the species is at first glance amazing and thus one would expect substantive *introgression*. Nilsson found that hybridisation is in most areas relatively rare because of seasonal, ethological and/or morphological isolation mechanisms. The effects of the hybridisation slowly fade away. Interbreeding between bastards is very rare, because the male gene material of the intermediate hybrids gets lost and does not reach the stigmas of other *Platanthera* individuals. The pollinaria occasionally adhere to moths but as a rule they do not adhere at the right place, either the proboscis (as in *P.bifolia*) or the eyes (as in *P.chlorantha*). The two gene pools of these species interplay always at a low level.

This example should suffice to show that different isolation mechanisms tend to reduce hybridisation and that where it takes place, the hybrids remain isolated from the populations of the parent-species by the same or different isolation mechanisms.

1.9.6 *Pollination dynamics*

The study of what happens to genes in plant populations is full of pitfalls, although many methods have been developed. The complex nature of gene-flow has led many investigators to develop process-oriented and pattern-oriented computer models (Handel 1983, Lertzmann & Gass 1985, Peakall 1989, Peakall & James 1989). With these models many factors can be varied and the consequences studied. This is of great help in the formulation of hypotheses that can be tested in field experiments.

The actual methods used to study gene flow are diverse, as seen from the experiments mentioned above. They vary from simple observations of pollinator behaviour or measures of plant morphologies to sophisticated methods with the use of chemical and radioactive labelling. These are applied as well to mark genes that can be studied in progeny. However the correlation between the method and the actual gene flow is not always clear and is the subject of much debate.

The movement patterns of pollinators can be quite different with enormous implications for gene flow, especially where different types of pollinators visit the same species.

In orchids *pollinator limitation* on seed set seems to be a normal phenomenon. The activities of pollinators may influence gene flow patterns in relation to plant density. Extraneous pollen is more likely to increase in a population of small size, increasing variance.

Selection alone does not determine what happens to genes from introduced pollen. The joint effects of pollinator behaviour, deposition rates of compatible pollen, clonal growth rate and clone size influence the internal distribution of the various genotypes present in the population in relation to populations with clonal growth.

The picture of the dynamics of progeny genotypes within populations is complicated by differences in seed set and seed abortion rates, variation in seed and seedling mortality rates and variation in fitness of offspring in relation to *xenogamy*, geitonogamy and self-fertilisation, as well as lowered pollen-pistil compatibility among near neighbours.

Subtle changes in local habitats affecting microclimates may also affect population dynamics.

In conclusion: modern studies of pollination biology have made it evident that the distribution of genes, particularly as expressed in the amount of cross-pollination, is very much influenced by the geometry of the population. Different sizes, densities and shapes of populations of a species may elicit different responses from pollinators and hence differences in the amount of transferred pollen. Populations may be unique in their pollen milieu and consequently in their development and evolution. It probably will take a long time before we are able to generalise our knowledge on *pollination biology* in such a way that we can take intelligent decisions on the conservation of our orchid species. As long as we lack such knowledge we should be very careful with what we have.

1.10 THE RESEARCH METHODOLOGY OF ORCHID POLLINATION

Dafni (1992), noting that only a few publications have been devoted to the methodology of research in the field of Pollination Ecology, cites Macior who stated:

'Our present knowledge of visual, olfactory and taste perception of pollinators must be related experimentally to the structure and function of pollination mechanisms. Circumstantial evidence between spectral reflectance from corollas and insect vision, between chemical composition of nectar and pollinators' taste perception, and other similar correlations, is valuable as a starting point for testing the significance of such putative adaptations. More precise observational techniques of pollinator behaviour are also required to document the operation of contemporary pollination mechanisms and the need to avoid the observational errors of the past.'

Cox and Knox (mentioned in Dafni's *Pollination ecology: A practical approach* 1992) have suggested the application of Koch's postulates concerning the methodology of research into infectious diseases and their spread to research in pollination ecology in the following manner:

1. Pollen transfer from anther to vector must be demonstrated.

2. Pollen transport by the vector must be demonstrated.

3. Pollen transfer from vector to stigma must be demonstrated.

4. Pollen deposited by vector must be shown to affect fertilisation.

Dafni adds some postulates to include other aspects of pollination ecology to attain a consistent framework of methodology of the field:

5. Flower advertisement (visual as well as olfactory) must be demonstrated to be perceived and used by the pollinating vector.

6. Flower reward must be demonstrated to be consumed (or used) by the vector as an integral part of the pollination process.

7. The relative contribution of pollen and ovules to the next generation has to be demonstrated as a result of the pollination process.

8. Interrelationships between different vectors involved in pollination have to be demonstrated at a community-based level.

I would, in turn, add several postulates which refer to hypotheses in the pollination ecology of certain orchids, such as:

9. Sexual stimulation by flower scents has to be demonstrated (Plates 17-20, p. 28).

10. Sleeping behaviour in flowers has to be demonstrated.

In general, if a particular behaviour pattern is thought to contribute towards pollination, every step of this behaviour has to be analysed and has to be demonstrated to be relevant or irrelevant for the attainment of the results claimed.

Moreover, every morphological structure or character of a flower, suspected of contributing to pollination, has to be shown to influence the pollinating insect.

It is obvious that sometimes claimed results omit this type of evidence, which demonstrates that there is still a long way to go before we understand all the phenomena observed in orchid pollination.

CHAPTER 2

Pollinators

Wiser far than human seer
Yellow-breeched philosopher!
Seeing only what is fair
Sipping only what is sweet
Thou dost mock at fate and care
Leave the chaff, and take the wheat
Emerson, *The Humble-Bee*

This book is on pollination of orchids and therefore something more should be said about pollinators. About their taxonomy and about the characters that are important in relation to pollination.

The pollinators of orchids belong to the Insecta and to the Aves: insects and birds. Pollination by bats or other mammals is unknown in orchids. Pollinations by spiders are at best accidental.

The insects form by far the largest class of all animals with more than a million species, subdivided into 20-30 orders, the number depending on the chosen classification.

Relevant for this review are only a few of the orders: Hymenoptera (bees, many kinds of wasps, ants), Diptera (flies and mosquitoes), Lepidoptera (moths, hawkmoths, butterflies), Coleoptera (beetles). Beside these a few insects from the Thysanoptera (thrips), Heteroptera (bugs) and Homoptera (aphids) have been recorded. Isoptera (termites) probably also play a role in the pollination of an Australian orchid.

2.1 INSECTS

2.1.1 *Hymenoptera*

There are over 110,000 species known in this third largest order, usually grouped in two suborders: Symphyta and Apocrita. Most Hymenopteran pollinators of orchids belong to the Apocrita, especially to the section Aculeata.

From the Symphyta ('waistless' wasplike hymenopterans), sawflies belonging to the Tenthredinidae (one of the six superfamilies), have been observed as visitors and pollinators of *Listera ovata* and other orchids. The Apocrita ('waisted' hymenopterans) are subdivided in Parasitica and Aculeata. Of the Parasitica (mainly parasitic wasps without a sting) several species from the Ichneumonidae are also found on *Listera ovata*, as well as on other orchids. They visit much the same flowers as the short-tongued Diptera. An Australian species of Ichneumon *Lissopimpla excelsa* (*L.semipunctata*) is involved in pseudocopulation in *Cryptostylis*.

Chalcid wasps and gall-wasps have also been observed as orchid pollinators.

Especially relevant here are several superfamilies from the Aculeata, like the Scolioidea, Vespoidea, Sphecoidea and above all the Apoidea. Less important are the Formicoidea (ants).

Scoliid wasps and sphecoid wasps are among the pseudocopulating pollinators of the genus *Ophrys* from Europe and several Australian orchids. Vespoids are among the pollinators of *Epipactis* species.

To the Apoidea belong the Apidae, social bees (Bombinae = bumble-bees, Apinae = honey-bees, Meliponinae = the stingless bees of the tropics) and all kinds of solitary bees (Euglossidae = euglossine bees, Andrenidae, Ceratidinae, Xylocopidae, Megachilidae, Halictidae, Colletidae, Anthophoridae, Melittidae). These groups contain the really important pollinators of orchids, both in numbers (about 50%) and in diversity of pollination systems.

Ants (Formicoidea) do like nectar and collect it from many flowers, including orchids. Their winglessness makes them crawl up stems and this method makes cross-pollination unlikely. They are not good pollinators and are often harmful visitors.

Many epiphytic orchids in the tropics have adapted to this by excreting extrafloral nectar and profit from the relationship with ants by gaining protection against herbivores. Some cases of orchid pollination by pseudocopulating ants are known from Australia.

2.1.2 *Diptera*

With over 100,000 known species and more than 100 families this is the fourth order in size. Three (some classifications have two) suborders subdivide this large group: Nematocera, to which all kinds of mosquitoes belong (with, among others, Mycetophilidae as one out of 11 flower-visiting families that contains pollinators of orchids), Brachycera and Cyclorrhapha (both together called flies), the second suborder comprising Bombyliidae and Empididae as well as other families, the third, for example, Syrphidae, Drosophilidae, Tachinidae and Muscidae all of which may pollinate orchids.

2.1.3 *Lepidoptera*

With over 100,000 species the second largest order. Many superfamilies and families of moth, butterflies and hawkmoth have species that visit orchids, such as Papilionoideae (Papilionidae, Lycaenidae, Satyridae and Pieridae), Pyraloidea (Pyralidae), Noctu(o)idea (Notodontidae, Noctuidae), Gemetroidea (Geometridae), Sphingoidea (Sphingidae), Zygenoidea (Zygenidae).

2.1.4 *Coleoptera*

The largest order, comprising 250,000-300,000 species. Only a few families visit flowers for food and among these we find the pollinators of orchids. Most important families in this respect are: Cerambycidae, Scarabeidae, Cantharidae, Oedemeridae, Elateridae and Chrysomelidae. In general beetles may be very destructive to the reproductive organs of flowers.

2.2 HOW DO INSECTS EXPLOIT FLOWERS?

Insects that visit flowers normally do so to acquire food; pollen and nectar, oil and pseudopollen. The plant offers food to gain reproductive benefits. The relation is mutualistic, but in terms of energy the balance of profit is on the insect's side. In orchids, (although not exclusively so, there are some other plant families that do the same) there are also species that produce special fragrances, precursors of sex-pheromones.

Orchids are exceptional in the sense that most of the species have flowers which attract pollinators without rewarding them with food. Exploitation is reversed, the plant one-sidedly benefiting from the activities of the insect. This is a kind of parasitic relationship. Anthropomorphically this is called fraud or deception. The flower exploits the instinctive or learning behaviour relating to sex, food, comfort and reproduction (in a wide sense) in diverse ways, 'seducing' the insect to copulate, to search for food, hiding or mates, to lay eggs, to defend territories by attacking 'intruders' etc. All this poses the question of how insects are equipped to behave towards flowers as they do.

2.2.1 *Nectar collecting*

How do insects collect nectar from flowers? They do so with their mouthparts that sometimes are highly modified and adapted for that purpose. The simplest way is to lick nectar with the tongue, the more 'sophisticated' way is to suck nectar by a special apparatus, like the long proboscis of a butterfly.

Mouthparts of insects are constructed of a very adaptable and variable material, the exoskeletal chitinous cuticle. The early insects had a kind of chewing tool, such as we still find in cockroaches (Fig. 15).

In bees the proboscis consists of five elements (labiomaxillary complex). Central in this is a hairy licking instrument, the glossa (tongue), which is surrounded by the other four elements to form a tube, with which the bee sucks up nectar that passively rises by capillary action. After licking up the bee pulls its tongue back into the suction tube, which is connected to a pump in the head that actively pumps the fluid higher up with the help of muscles, which are attached to the cuticle in the pharynx. The structure works like a pipette in the sucking phase. The increase in volume of the pharynx reduces the pressure and consequently the fluid is taken in. Bees can, in case the nectar forms a large drop, dip the suction tube into the liquid and pump it up without licking. In the inactive state the proboscis is folded back like a clasp knife. The mandibles can be used as a gnawing apparatus as well for eating pollen and even for purposes like fighting and building the comb.

Lepidopterans do not eat pollen. Some species

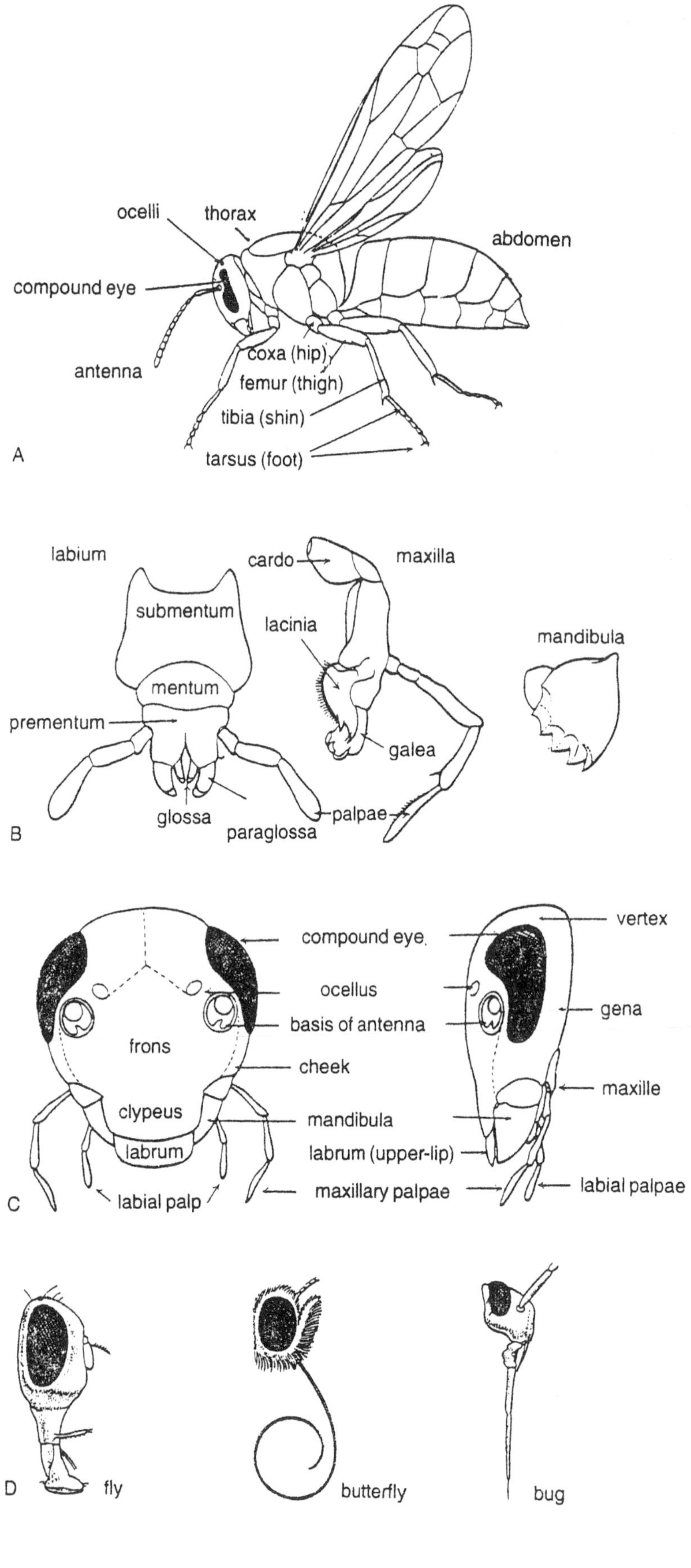

Figure 15. Insect morphology. A) Typical insect body. B) Morphology of separate mouth parts of an insect (cockroach). C) Front- and side-view of an insect-head. D) Three modifications of mouthparts for uptake of fluid food (after Chinery 1983).

even don't eat at all and merely live a short period as imagines to reproduce only. But those that feed only use liquids. The mandibles and labium are reduced strongly. The middle appendages (galea) of the maxillae form the only functional mouthparts. They consist of two elongated tubes surrounding a central hollow space. The tubes contain muscles, nerves and cavities filled with blood. The two halves are neatly linked by hooks and small teeth to form a flexible watertight tube that can be rolled out and up. The rolling up of the proboscis occurs passively like a steel spring. A rod in the outer wall of each proboscis half has elastic properties, composed as it is of resilin, a rubberlike protein substance. This material saves energy and therefore the tongue can do without many space-consuming muscles. To extend the tongue, small muscles are needed. They are helped by the hydraulic force of the blood pumped into the cavities. A muscular valve at the beginning of the two halves can close off the tube, causing the pressure needed to extend the proboscis when the blood is pumped into it. The suction pump operates in the same way as in bees by expanding a part of the pharynx which leads to lower pressure (vacuum).

The tip of the proboscis is armed with numerous fine spines which, it is thought, are used for cutting open the tissues of nectarless flowers to release the fluid, as butterflies are occasionally seen sucking at such flowers. This is observed in *Epidendrum anceps*, from the flowers of which a fluid is released that is held in copious supply by thin-walled cells lining the nectar cavity.

In flies the proboscis is a multifunctional tool, constructed of three tubes, one within the next. The inner one, a modification of the labrum, is the food-sucking tube, the one surrounding this an elongation of the pharynx. This one produces saliva to dissolve food (or as in mosquitoes to prevent coagulation of blood of the victim). The third encloses the other two and is formed by the labium. Mandibles and maxillae are strongly reduced and form e.g. the piercing spines (mosquitoes). The end of the labial sucking tube in flies is equipped with fine canals which open into larger collecting channels, from which the uptaken food is transported into the digestive tract by a pharyngeal pump. The labellum is covered with sense organs which are capable of determining food consistency and chemical composition. The feet of flies have also sense organs which send information about the chemical composition of the fluids, e.g. the nectar concentration, on the landing surface. There is a hierarchy between the sensory cells of the antennae and feet, the fly not beginning to suck until the substance, qualified as acceptable by the feet has passed a second test by the sensory hairs at the proboscis.

Very small teeth around the mouth opening of flies can be used to fragment large food particles or scrape pollen out of flowers.

In all orders of nectar feeders we find species with long, middle or short proboscis lengths (Fig. 15) It is the length of the sucking proboscis that determines which flowers can be exploited effectively.

2.2.2 *Visual capacities of insects*

The facet eyes of insects produce a mosaic image. Honeybees have about 5500 facets, house-flies 3300.

The spatial resolution is comparable to the graininess of a film, the size of the grains depending on the density of the *ommatidia*. The part of the visual field in front of and below the animal, which is very important during flower visits, has a very fine 'grain' representation, since in this part the so-called visual axes of neighbouring ommatidia are very small (Barth 1991).

Insects recognise patterns: geometric correlations. Pattern recognition involves processes of generalisation and classification. The brain selects the biologically important features (contour length, inclination, size and density) as a result of the complex interconnections between the nerve cells. Honeybees and wasps prefer highly subdivided structures when foraging, but low-contour figures which resemble the entrance hole of the beehive or nest when flying home. Flowers represent important pattern complexes. The configuration of the nectar guides of many flowers come close to the ones found to be most attractive to bees in laboratory tests. Flying bees are able to compare patterns during their approach to flowers with stored information. Pollen and anthers often have conspicuous colours and function as signals to the insect. It is nowadays believed that guide marks may be pollen imitations or pollen models, especially in cases in which the pollen is hidden. Even when nectar is the chief (or as in many orchids the only) attraction for the insects, the signalling action of the structures imitating anthers and pollen may be retained.

Lepidopterans, like Hymenopterans, see colours. They strongly depend on vision for guidance during flights. Many have a preference for either the blue or the yellow colours.

Some butterflies seem to prefer flower colours which are similar to their own colours and the preference is thought to have become transferred from choosing a mate to choosing flowers. They have a wide spectral range, from ultraviolet to red, but a low resolving power. The temporal resolution is high, the flicker-fusion frequency (above which images appear continuous, as in movies) is 150 images per second (40 images per second in man, 10 times that value in honeybees). Hawkmoths have a well-developed colour sense even in extremely dim light. Skipper butterflies (Hesperidae) have *bichromatic vision* instead of *trichromatic*. They prefer flowers in the range blue-magenta and yellow. The hummingbird hawkmoth (*Macroglossum stellatatum*) is red-blind and this red-blindness is also found in some bees, including bumblebees and honeybee. They perceive blue-green and ultraviolet as distinct (Fig. 16) (Proctor et al. 1996).

Diptera can distinguish colours when feeding and in general tend to visit white, pink, yellow and green flowers most readily (Proctor et al. 1996). Visits to purple and blue flowers are commonly made by the longer-tongued genera of the Brachycera and Cyclorrhapha.

In general, most insects are sensitive to ultraviolet but have no sensitivity to red, the visible spectrum ranging from 300-650 nm (whereas in man this is from 400-750 nm).

Colour vision has been demonstrated in some beetles.

2.2.3 *Assimilation of scent information*

Scent is a characteristic of flowers as important as or even more so than vision. Scent marks are more common than visual guide marks. The approach of flowers to the food sources is usually marked with odour, including the visible nectar marks (including UV patterns). Flowers produce a mosaic of smells to bees and the insect can discriminate different quantities and qualities at close range. Scent marks stand out from the rest of the flower. The olfactory acuity of the honeybee, however, is not much different from that of humans. They perform slightly better in tests in discerning flower scents, but worse in discerning fatty acids. Moths perform much better, especially if they are equipped with large and complex antennae, like the male silkmoths, which smell even concentrations of sex pheromone (bombykol) more than one million times more diluted than the average threshold for humans or bees permits.

The bee antennae, which are essentially a kind of pore plates, contain about 6000 olfactory sensilla, sensory cells of which the dendrites run to the pores. An odour substance may increase or decrease the electrical potential of the membrane of the nerve cell or leave it unchanged, which leads to either a change or not of the frequency with which the impulses are sent to the brains. Odour substances are excitatory, inhibitory or ineffective. Each scent corresponds to a specific pattern of unexcited, excited or inhibited sense cells and therefore a limitless number of distinct patterns can be produced. Since the two antennae do perceive and locate odours separately a bee can orient itself towards an odour source by perception of quantitatively or qualitatively different stimuli of the two antennae at each moment: three-dimensional smell-

Figure 16. A) Ultraviolet photograph of *Dendrobium chrysanthum* flowers. B) *Dendrobium chrysanthum*, the same flowers in visible light (black and white photography) (photos J.Indsto, Sydney).

ing or *tropotaxis*. Humans are only capable of *klinotactic* orientation by sniffing at various parts of an object one after another. In the case of long range orientation with differences in concentration too small for tropotactic or klinotactic orientation, bees seem to rely on their memory of earlier perceived concentrations. On flowers, after landing, they are capable of discriminating sequential perceived scents to orient themselves and find the food source.

Flies combine visual and olfactory stimuli in orienting themselves towards food sources. In the antennae of blowflies we find hairs, either within a furry mass of other sensilla on the surface of the antennae or recessed in small pits. They have three different types of flower-odour receptors, responding to different types of alcohol or to terpenes. Aphids also have pore plates. Lamellicorn flower-visiting beetles have either one of two different types of pore plates or hairlike sensilla. Butterflies have olfactory hairs on their antennae with different overlapping reaction spectra as in the pore plates of bees. Male silkmoths have 2600 pores per olfactory hair through which the molecules are sieved. They diffuse to the pores and through the 13,000 fine pore tubules which each hair contains to the sensory cells, but these react very specifically to the sexual attractant produced by female moths.

In beetles the sense of smell plays a large part in feeding, and some spicy scents are associated specifically with beetles.

2.3 BIRDS

Bird-pollination is a widespread phenomenon, extending in latitude from the tip of South America up to Alaska in North America. We find it also in Africa, Australia and Asia. Only Europe and Asia north of the Himalayas are excluded. It occurs up to an altitude of 4000 m. in East Africa and in the Andes mountains (Proctor et al. 1996).

Flower visiting has been recorded for as much as about 50 bird families. Flowers have adapted by coevolution in respect to food supply (nectar composition and content), guide marks, colour, shape, size and positioning.

Birds using nectar combine this diet with insects for their protein. It is even hypothesised that nectar-feeding has evolved from looking into flowers for insects or spiders. Birds can be very efficient and fast pollinators, showing considerable constancy to certain species of flowers. But birds can also be nectar thieves which rob nectar by piercing the flowers. Proctor, Yeo and Lack publish an impressive list of the main flower-visiting families, representing about 900 bird species. But only a few of these are known to pollinate orchids. Best known and most important as orchid pollinators are hummingbirds (Trochilidae). In 1963 a beautiful picture-book was published, based on drawings dating from about 1878 by Montes de Oca entitled 'Hummingbirds and orchids of Mexico'. Every picture shows a combination of a hummingbird species and an orchid species, suggesting a relationship between the two, especially in those pictures in which a hummingbird approaches orchid flowers as if in search for nectar. But it is unclear which drawings are based on real observations. Certainly most if not all combinations sprang from the artist's fantasy.

Hummingbirds have straight or slightly curved (up- or downward) needle-shaped bills. During nectar feeding the tip of the bill is opened to allow the bifurcated tongue to move rapidly in and out. The lobes form a thin lamina, rolled up lengthwise to build a slender tube. Nectar is believed to be held in these tubes by capillary force until the tongue is withdrawn into the beak and then is sucked up and swallowed. The end of this tube is frayed and this probably increases capillarity (cf. the bee-tongue).

The strong learning capacities of hummingbirds allow them to develop and change colour preferences in conformity with the temporary availability of nectar sources. Preferred colours are red, orange and orange-yellow, hues that cannot be seen as such by bees, which is an advantage for the development of mutual relations between birds and flowers. The bill-length of most North American hummingbirds are much alike, so there is no specific relationship, flowers being visited and pollinated by several species of birds. This, however, is different in tropical America where bill-lengths and flower shapes, sizes and colours differ more, which leads to specific adaptations towards certain species. Hummingbirds have small feeding territories and defend these against other individuals and species.

Hovering in front of the flowers, which is the most efficient way for acquiring nectar, is very energy consuming though. This limits the maximum size of the birds. They can reduce energy consumption under cold conditions at rest by becoming torpid with a lowered, but still controlled temperature.

Pollination of orchids (*Satyrium*) by sunbirds

(Nectariniidae, Plate 21, p. 28) has been observed in South Africa (Johnson 1996) and in south-east Asia (*Dendrobium* species). Honeyeaters (Meliphagidae) have been observed on orchids in Australia.

Sunbirds and honeyeaters, although not related to hummingbirds, nor mutually, show *convergent* characters.

2.4 THE ETHOLOGY OF ORCHID POLLINATORS

Ethology is the study of animal (and human) behaviour. It is relevant to the study of orchid pollination because insect and bird pollination can be best understood by insight into the laws that govern the behaviour of the pollinators and the specific facts of insect or bird life. To know for example that bees do not discriminate certain colours or that they see 'colours' that we do not perceive (except when we use instruments sensitive to radiation outside the visible spectrum) makes us aware of phenomena that are otherwise overlooked or misinterpreted. The fact that, for example, animals when foraging have a kind of general model in their heads of what they look for is relevant for understanding the working of flowers that are mimicking other flowers, and so on.

2.4.1 *Organisation of behaviour*

Behaviour is the result of the animal's reacting to stimuli. These stimuli can be external and internal. External stimuli may:

a) Arouse internal drives;

b) Elicit responses;

c) Serve to orientate the animal as it responds.

Behaviour is the product of activity in the nervous system. External stimuli are perceived with the sense-organs which are adapted to receive certain specific stimuli: sense-organs can be seen as sensitive 'filters' of great complexity. Stimuli are transmitted to the central parts of the neural system where they are coordinated and integrated and eventually transformed into commands for action that are executed by the glands and muscles. Each level can be very complex.

The central nervous system is hierarchically ordered, being able to sum up and value stimuli in different ways. Summation is possible for stimuli coming at different times (temporal summation) and for stimuli from different places (spatial summation) (Manning 1973, Tinbergen 1958, Gould & Gould 1992).

2.4.2 *Motivation*

The same stimulus does not always evoke an identical response when there are intervening parameters in the animals' existence, such as fatigue, maturation, learning and motivation. Sometimes very slight stimuli may invoke powerful response, while at other times stronger stimuli are needed to evoke the same response. Motivational factors tend to organise the behaviour into tendencies, so that it leads to the achievement of particular goals.

Different motivational systems interact with one another: specific motivations are called drives. Hunger and sexuality are important drives in relation to pollination.

Three different stages may be observed in the behaviour connected with motivation.

1. Searching for the goal, *appetitive* behaviour.

2. Behaviour oriented around the goal once it has been found. For example the landing on the flower and the brushing of fragrances in euglossine-pollinated orchids: these actions are called *consummatory* acts.

Plates

15. *Bulbophyllum orectopetalum*. Mobile labellum. Note the difference in position with Plate 7. Photo N.A.van der Cingel, January 1997, Khao Sok, Thailand (p. 14).
16. *Stanhopea wardii*. An orchid with osmophores. Compare Figure 12 for the place of the osmophores. Photo N.A.van der Cingel (p. 15).
17. Researcher Colin Bower preparing bait experiments with *Chiloglottis formicifera* and related species. Photo by N.A.van der Cingel, August 1998, Appin, NSW, Australia (p. 21).
18. *Neozeleboria* no 41, just caught. Photo N.A.van der Cingel (p. 21).
19. *Chiloglottis formicefera* flowers used in baiting. Photo N.A.van der Cingel (p. 21).
20. Several thynnine wasps (*Neozeleboria*) attracted by *Chiloglottis formicifera* flowers. Photo N.A.van der Cingel (p. 21).
21. *Nectarinia amethystina*. Pollinator of some *Satyrium* species. Photo N.A.van der Cingel, December 1995, Knysna, South Africa (p. 28).

15	16
17	18
17	20
19	21

Plates 15-21

3. A phase of 'rest' (or change in motivation) following the achievement of the goal.

All three phases are probably important in relation to pollination phenomena and should be distinguished carefully.

In searching flowers insects may use different stimuli than when orientating, e.g. odours and colours respectively. Phases of rest may be important for cross-fertilisation e.g. if they lead to time lapses between two visits. This may be especially relevant in cases where sexual motivation is at issue. Here the performance of the behaviour may reduce the drive.

2.4.3 *Discrimination and perception*

What animals can perceive depends on the sensitivity of their sense organs – a matter of quality and of quantity. Butterflies are known to have an amazing sensitivity to odour perception, allowing differences of single molecules seemingly to be detected. Thresholds can be very low indeed: in this sense quality is measured quantitatively.

Quality in quite a different sense is the extent or range of perception. Insects are insensitive to some colours we perceive, e.g. bees and most other insects are insensitive to red, but they see ultra-violet, for which we are blind. Some flowers show ultraviolet nectar guides and this image of a flower for bees can be quite different from our own image of the same object. Colour vision in insects is in general trichromatic, different receptor cells being sensitive to green, blue and ultra-violet light.

Moths are sensitive to infrared signals. This fact has hardly been recognised hitherto. This may even influence the perception of scent, because scent molecules emit infrared frequencies when they resonate, so this may influence the location of food as well as of mates. Sphingid and noctuid moths are able to orientate towards flowers by odours only, as was experimentally proved by Brantjes (1978), but mostly they use visual cues in combination with odour.

Bees are known to possess sensitivity to the plane of polarisation of light. They can make use of this in locating the sun's position, even when the sun is behind the clouds because the scattering process which creates the blue sky also produces polarisation which varies with the position of the sun.

Other sense organs of insects also have different properties compared with those of humans. Insects live in a world of quite different perception from ours.

Apart from different sensory capacities, animals have differently organised neural systems. Already mentioned is the fact that they may respond only to parts of the stimuli present and that their reaction is under the influence of specific drives.

Selective responsiveness to sign-stimuli is of great adaptive significance. This orders the behaviour in the sense that the animal reacts to the adequate stimulus at the right moment and in the right way. The other side of the picture is the inflexibility in the reactions which can sometimes lead to false responses. It means that insects sometimes can be misled easily and it is this rigidness of reaction that is exploited by orchids (see also comments on learning).

An interesting statement concerning the adaptive significance of flowers is made by Wyatt (1983). He mentions the suggestion of Raven that the red colour of hummingbird-pollinated plants is a mechanism to reduce visits by nectar-robbing bees, whose vision is insensitive to red. It should make us aware of the possible occurrence of selective factors which are wrongly ascribed to the actual pollinators.

2.4.4 *Instinct*

The background of this rigidness of behaviour is its inherited character. It usually is called instinctive. The terms 'inherited' and 'instinct' do not, however, very much clarify the picture and should be avoided. Behaviour is determined by genes as well as by environmental factors so therefore we can better speak of *fixed action patterns*. The measure to which such stereotyped actions are susceptible to

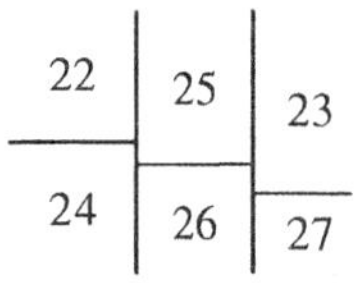

Plates
22. *Cypripedium montanum*. Photo J.Mulder, June 1995, Husk Lake, Quesnel (p. 38).
23. *C.reginae*. Photo J.Mulder, June 1995, Dorcas Bay (p. 38).
24. *C.calceolus* var. *pubescens*. Photo J.Mulder, June 1995, Dorcas Bay (p. 38).
25. *Epipactis helleborine* with wasp (*Dolichovespula*). Photo W.P.Stoutamire (p. 43).
26. *Pogonia ophioglossoides*. Photo J.Mulder, June 1995, Dorcas Bay (p. 49).
27. *Platanthera ciliaris* with *Papilio palamedes*. Photo W.P.Stoutamire (p. 58).

learning processes may differ but is generally very low.

We must also be aware that this kind of behaviour can be very complex. In the first place there are many reactions that in fact are long chains of several actions, each of which is elicited by a different stimulus. For example, the copulatory behaviour of a bee on an orchid flower consist of several steps, such as landing followed by orienting and then making copulatory movements, guided subsequently by scents, colours and patterns of hairs on the labellum. So different senses, vision, smell and touch, are involved in a strict order and combination. Another example of behaviour with a chain character is the collection of fragrances in euglossine bees. The bees, after being attracted by the scent, land on the flower, move to the area of maximum odour production, brush the surface with the tarsal brushes of the front legs, launch into the air, transfer the collected fragrance to the tibia of the hind legs and than return to the flower to repeat the sequence.

In the second place recognition of relevant situations can be based on a complex interplay of features. This complex stimulus situation is perceived as a 'Gestalt', a whole being the summation of several aspects that are not perceived successively, but simultaneously.

One of the examples is the 'search image' that animals tend to have of their food and sexual mates. This non-analytical perception of complex things is highly relevant to orchid pollination when 'misleading' is under discussion.

There is still another important complicating factor in animal recognition relevant to pollination. It is the sensitivity to what is called a *supernormal* stimulus. If one (or some) of the stimuli are exaggerated, the animal may respond to the stimulus situation still better than to the normal stimulus – even if some other details of the complex situation are less stimulating than normal. If offered a choice between two comparable situations the animal will react with preference to the one that combines the necessary stimuli in the 'strongest' way. So in the case of pseudocopulation some stimuli of the flower have in certain cases been shown to be a stronger stimulus than the female can offer. It may be clear that this sensitivity to supernormal stimuli has played an important role in the coevolution of orchids and pollinators.

A last factor to mention is the importance of motivation in reacting to stimuli: the higher the motivation, the lower the stimulus can be and still elicit a response. Therefore, even imperfect mimicry may work in cases where the drive is very highly aroused. So, even in *Ophrys* flowers, which resemble the female imperfectly, it is possible to lure the male when the females have not yet emerged, as is often the case. Normally the males appear two weeks in advance of the females. Nevertheless in this period the motivation for mating may well be high enough to elicit responses to *Ophrys* flowers that normally cannot be a match, compared to the female as a stimulus.

2.4.5 *Learning behaviour*

Many types of learning are known, like habituation, classical conditioning, instrumental or operant conditioning (trial and error), latent learning, insight learning and imprinting.

Habituation is involved in cases of a repeated stimulus that is not associated with a reward. If an orchid is offering no reward and can be discriminated from the models it mimics, the potential pollinator will soon be habituated to this stimulus and no longer visit it. In a population of that orchid many flowers might depend on comparatively few pollinators, so this kind of learning is one of the reasons why sometimes few flowers bear fruits. Especially early in the season, when insects are learning their environment by trial and error, this habituation will adjust the animal rapidly to its environment, not wasting its time on unrewarding plants. A *conditioned reflex* is created by the association of a new stimulus with a reward (or a punishment). Insects learn to associate the colour of a flower with a nectar reward and so try every flower with the same colour. They need 5-6 experiences to become 'programmed' for such an association of colour and a reward. Colours that are the most likely food signals give the best responses but the reward has to follow within 2-3 seconds of the colour signal. Reinforcement is needed because bees have a *short-term memory* and a *long-term memory*, just like humans. It is, however, clear that we rarely see conditioned reflexes in pure form in nature. Bees do not just learn to associate the colour and the reward, for at the same time they are learning the position of the flowers with respect to their nest or hive and establishing at what time of the day the nectar secretion is highest. If a visitor leaves his or her scent in the flower the next visitor learns to avoid landing by habituation, etc. So learning in nature can be very complex, a mixture of simultaneous and successive occurrences of different learning processes. Processes of generalisation lead the animal to include similar stimuli, like flowers of slightly different colour and form in cases where it is rewarded for doing so: the opposite will happen, as we saw, in the case of no reward. Karl von Frisch has shown by discrimination experiments that bees are very sensitive to small differences in colour and form (Gould & Gould 1992).

Latent learning is the association of indifferent stimuli or situations without patent reward. This kind of learning comes into play when animals explore new surroundings. Many insects make special orientation flights during which they make a memory-picture of the area in relation to landmarks or to the sun. Information gained during these flights can be employed later when searching for food or a mate. Many Hymenopterans, the group to which so many pollinators of orchids belong, have such orientation abilities.

It is very difficult to prove *insight learning* in insects, because we can be misled by certain capacities to perceive stimuli that we ourselves miss. So detour experiments which show learning by insight in rats, can possibly be explained by orientation to the sun in cases of ability to detour in insects.

Imprinting is, in fact, not a separate category of learning as was accepted in the past. It is characterised as fast learning in young animals at a highly sensitive period in their development. The classical example is that of the geese of Lorenz that became attached to him as their 'mother', in the absence of the true mother. It was considered to be an irreversible form of learning, but this has been proven to be incorrect. It is now recognised that imprinting-like processes are also important in insect learning.

The conclusion is that insects are highly developed animals and not just simple reflex-automata since their behaviour can be highly complex. Students of pollination have to be constantly aware of this; if not, they run a serious risk of misinterpretation of pollination phenomena.

It should be mentioned here as a warning that insect species can differ enormously in behaviour. Social insects have distinct habits and capacities from solitary species.

CHAPTER 3

Classification of orchids

The Orchidaceae have all the earmarks of a group in active evolution; species, genera, tribes, and subtribes are all difficult to delimit. As such, I believe it is an excellent group for studying evolution

Robert L. Dressler, *Phylogeny and Classification of the Orchid Family*

3.1 TAXONOMY – THE SCIENCE OF CLASSIFICATION

Classification is the science of naming organisms and ordering them in a system. It is based on comparing characters and weighing these. Similarities and differences are used to determine relationships between organisms. This science is called *systematics* or also, but usually with a less wider scope, *taxonomy*.

Systematics is the study of organisation, history and evolution of living organisms.

The ultimate objective of taxonomy is creating a natural system which reflects the relationships of different kind of organisms in an evolutionary sense. The concrete objects of classification are not individual organisms, but groups of organisms called taxons.

A *taxon* is a *group of any rank* that is sufficiently distinct to be worthy of being assigned to a definite category (species, genus, family, order etc.). A taxon always refers to specified organisms and must be described under a designated name (Mayr 1970). The objectives of the study of systematics are:

- The inventory and description of biodiversity;
- To understand and explain biodiversity;
- To develop a stable and accessible classification.

Systematics in the widest sense is a complex science with at least five subsciences, each with its own methodology (Roos 1998):

- Description, classification, naming and identification of organisms;
- Phylogeny (study of evolutionary history and phylogenetic relationship);
- Historical biogeography (study of distribution patterns in relation to phylogeny, biogeography, geology and physical geography);
- Historical ecology (study of the origin of interactions between taxa;
- Biosystematics (study of species formation and processes of evolutionary changes).

3.1.1 *Taxonomical methods*

The oldest classical method of taxonomy is the careful qualitative comparison of differences in morphological characters and weighing their importance for species definition. It is still in use on a large scale for the analysis, description, and revision of genera. Its essential method is observation.

Numerical taxonomy, in which measures of morphological differences (*biometry*) are used as the primary basis of discrimination of species, is also still widely in use, especially by many amateur orchidologists.

Cladistic analysis or phylogenetic systematics is the modern approach to classification. Its objective is to develop a system which reflects the phylogeny of species.

Classical, numerical and cladistic taxonomy are not independent methods. Observation and measuring are the basic tools of any taxonomical science.

3.1.2 *Species concepts*

Species are real entities. They exist, but change with time, they evolve. Evolution as a process involves modification in characters or traits.

Whereas early generations of biologists believing in the creation of species have thought them invariable, we now know that variability is an essential characteristic of species. Through this variability species may even gradually develop into a different, a new species or split up into more than one new species. They evolve on a geological scale.

The inherent problem therefore is the delimitation of a species.

Different species are related, because by evolution each species has developed out of other species and if the birth of life was a unique phenomenon all species are related (as is most clear from the *cytotaxonomical* or *chemotaxonomical* point of view). Species are closely related to other species if their shared evolution by common ancestry has lasted until recently. The longer ago the separation, the more distant the relationship is. The more recent the moment of separation into different species has been, the more difficult it is to discriminate between species. What we may expect therefore is confusion and complications. Different species may come into existence by allopatric or sympatric processes of speciation. Separated populations on the edge of being split up into different species may even come into close contact again and merge to become one species anew.

So it is really difficult to define a species. Most biologists agree that a species is defined by a gene pool.

As long as populations exchange genes the individuals of those populations belong to the same species. But even this definition has its own problems. For different species may hybridise, without, however, the species involved losing their identity. Nevertheless new species may develop from hybridisation and become different from both parent species. Orchids, by the way, are peculiar as to hybridisation in that they easily hybridise on levels of unusual large distance of relatedness. Even species of different genera may do so.

Nowadays we are faced with different species concepts. These may be morphological, biological (based on reproduction), ecological and phylogenetic. The background of this is in part differing methodology in data collection. Some taxonomists, at one end of the spectrum, rely primarily on *karyological* and chemical methods, and base their taxonomy on DNA- or enzyme-analysis, whereas others, at the opposite end, stick to herbarium examination as the basis of their taxonomy.

Good taxonomists use both methods and also the ones in between, including geographical and ecological data, for example.

Besides those differences in collecting data there are also differences in dealing with them.

Differences of opinion on the demarcation of species are in essence reducible to differences in objectives. Those who think in terms of evolution, tend to look for similarities, whereas those who neglect the evolutionary side tend to look for differences (Van der Meijden 1996). This is of course a bit of a caricature, but is not without any foundation. Usually this difference is expressed as one between *lumpers* and *splitters*.

An example to illustrate the differences in outcome of a classification between splitters and lumpers is that in Europe Delforge (1994) recognises over 375 species, whereas Buttler (1996) has 216 and Sundermann (1975) only 101.

An observation the author has made in looking at the different approaches of splitters and lumpers is probably caused by a differing emphasis on coevolutionary processes. Several modern orchidologists look at the evolution of orchid species as guided and directed by pollination through insects. Several investigators are entomologists of origin and are trained in taxonomical methods that are in use in zoology. They recognise new species of orchids (and of pollinating insects!) on the basis of differences in scents and in pollinators. Other taxonomists deny the usefulness of such characteristics and as botanists occupy themselves with herbarium practise and therefore cannot accept such differences as a basis for a taxonomy.

The concept of so-called *biospecies* as used by Paulus, for example, is derived from the zoologist Ernst Mayr's (Mayr 1970) definition of the *biological species concept* as compared to the *typological* species concept which goes back to Plato and Aristotle (and was used by Linnaeus) and the *nominalistic* species concept of Occam and his followers (among whom was Charles Darwin). In the typological concept the observed diversity of the universe reflects the existence of a limited number of 'universals' or types (eidos). Variation is the result of imperfect manifestations of the idea implicit in each species.

The nominalists deny the existence of 'real' universals. Only individuals exist and species are manmade abstractions, invented in order that we may refer to great numbers of individuals collectively.

Ernst Mayr's biological species concept stresses the fact that species consist of populations and that they have reality and an internal genetic cohesion owing to the historically evolved genetic program that is shared by all members of the species. According to this concept the members of a species constitute a reproductive community. A multitude of devices ensures intraspecific reproduction in all organisms. The species is also an ecological unit that, regardless of the individuals composing it, interacts as a unit with other species with which it shares the environment. Species are groups of interbreeding natural populations that are reproductively isolated from other such groups.

So we are faced with definitions that use or stress different criteria. These are based on morphological differences, reproduction, ecology and history (evolution), usually in some combination. But neither criterion is unambiguous. Morphologically indistinguishable organisms may be infertile in crosses (sibling species) and morphologically clearly differing organisms may be fertile when crossed (polytypical species), etc.

3.1.3 *Cladistic analysis*

Cladistics is the dominant approach nowadays to the analysis of the patterns that the evolutionary process has taken (Arditti 1992). This methodology is used to construct so-called *cladograms.*

Cladograms can be seen as hypotheses of evolutionary history. Their use is intended to avoid and eliminate intuition and arbitrary views. A cladogram is a graphic reflection of the relations between taxa. It shows in principle that a species *a* is more related to a species *b* than to a species *c*. That is to say: *a* and *b* have a more recent ancestor in common than each of them with *c*. From this common ancestor they inherited unique characters, called synapomorphic characters. *Apomorphic* means derived and denotes an advanced state or type. The prefix 'syn' refers to the sharing.

A *synapomorphy* is an evolutionary new character developed in the most recent common ancestor. For example, if we compare orchids which are supposed to have lily-like ancestors with lilies, the labellum is a synapomorphy of the orchids not present in lilies. A more primitive character developed in an earlier ancestor and therefore shared by other taxa than the most recent ancestor is called a plesiomorphic character or *plesiomorphy*. For example in the above case of lilies and orchids the existence of two whorls, of three petals and three sepals is a plesiomorphic character. But if we compare only orchids mutually, the labellum, a modified petal, is a plesiomorphic trait. Petal and labellum represent alternate states of the same trait. The relation between two taxa is always observed in relation to a third taxon. A cladogram with more than three taxa is always reducible to a series of three-taxon statements.

The most reliable method of determining which state of a character is plesiomorphic is by outgroup comparison. This is done by identifying the closest relative of the group under study (called the ingroup) that is not a member of it, and the character state it exhibits is then considered the primitive one of the ingroup. Errors in outgroup determination can cause problems with this procedure.

A group with one or more synapomorphies and consisting of an ancestor and all its descendants is called *monophyletic*. Monophyletic groups are the building elements of a cladogram, because they are a direct reflection of evolution. By examining a whole series of traits in a set of organisms and making estimates regarding which traits are *homologous*, a cladogram can be constructed.

In phylogenetic systematics the principle of *parsimony* is used as a methodological criterion. Choose the most simple solution if there is more than one. In other words: choose the hypothesis that best suits all the observations, i.e. the cladogram that needs the smallest amount of steps. By using the parsimony criterion it is possible to make an objective and testable choice between the different three-taxon statements. Cladistics, however, is not the panacea for all problems inherent in earlier approaches. For the many pitfalls and drawbacks of this methodology, especially in relation to orchid taxonomy, see Arditti (1992).

3.1.4 *An example to illustrate cladistics*

An example of a cladistic analysis in relation to pollination of orchids is the study of the phylogeny and radiation of pollination systems in the African orchid genus *Disa* by Johnson et al. (1998).

They constructed a cladogram for 27 species, using 44 morphological traits (Fig. 17). Pollination systems were then mapped onto the phylogeny in order to analyse pathways of floral evolution. Shifts from one pollination system to another are believed to have been a major feature of the evolutionary diversification of *Disa*. The genus is characterised by radiation in pollinating systems through adaptation to nearly all major groups of pollinating insects, like butterflies, moths, sphingids; carpenter, leafcutter, mason, anthophorid and halictid bees; sphecid and pompilid wasps; muscid, nematoceran and tanglewing flies, as well as horseflies (Fig. 18).

Some species, moreover, are autogamous. In total 19 different specialised systems of pollination were found in these 27 *Disa* species. Several similar pollination systems evolved more than once in unrelated clades. It is suggested that a few dominant pollinator species in a region may be sufficient to generate diversification in plants through repeated

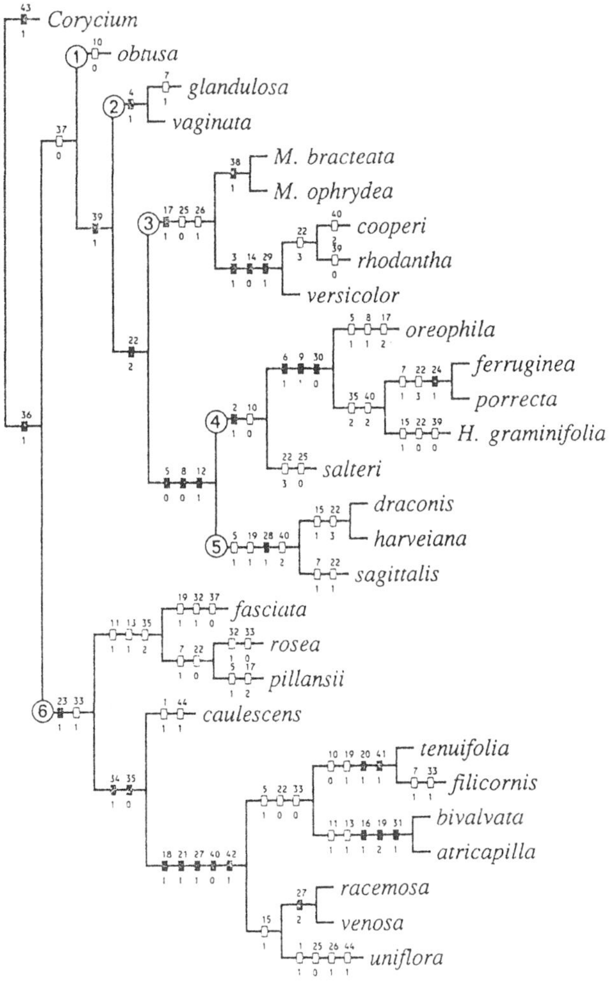

Figure 17. Cladogram of *Disa* species (Johnson et al. 1998). Solid bars represent apomorphies that evolved only once, whereas hollow bars represent characters that are *homoplasious* (i.e. appear to have evolved more than once). The number above each bar refers to the morphological characters used in the cladistic analysis. They bear on floral as well as non-floral traits. The number below the bar refers to the character state [e.g. present (0) or absent (1), small (0) or large (1) etc.]. The species are indicated by the specific epithet only. *Corycium* was used as an outgroup.

floral shifts that never retrace the same pathways. This example illustrates how a cladistic analysis of morphological traits may produce explanations and understanding of pollination relations and generate hypotheses concerning its evolution.

3.2 SOME CLASSIFICATIONS OF THE ORCHIDACEAE

There are several well-known classifications of Orchidaceae, which in some respects differ considerably. They reflect in their differences the methodology used. There is not yet a stable situation in orchid taxonomy and in the near future many changes are to be expected. Recently, for example, changes in names and numbers of recognised orchid genera and species in Australia have 'stirred up the country'. There are special reasons for the fact that in orchidology such a situation exists (see Arditti 1992). Although this situation is a nuisance to many people interested in orchids, there is nothing to worry about. Science is an ongoing human activity. To prevent misunderstandings about which taxon is meant, most investigators refer to the authors who played Adam's role in naming the organisms studied.

A special situation in orchid taxonomy, compared to other fields of plant taxonomy, is probably the fact that besides professional taxonomists many less trained amateurs participate and have their influence. This enthusiasm is to be valued positively, for it contributes considerably to our knowledge of existing biodiversity. But in the end professionals should decide on more definite taxonomies, especially in regard to higher levels of the hierarchy of classifications. The contributions of amateurs are most worthwile in the field of mapping and describing biodiversity. And, not to forget, in observing and describing pollinations observed!

Widely used in orchidology are the classifications of Schlechter and Dressler. Both have been revised recently.

Schlechter published his standard work in its first version in 1915. The second edition appeared in 1927 and was a bestseller with great influence. It was recently revised by a team of German orchidologists. This revision is based on the methodology used by Schlechter for his first draft. It is a dichotomic system and presents all taxa in keys for determination purposes at the cost of the natural character of the system. The conservative character of it is recognised by the revisors. The system is primarily based on morphological characters valued as phylogenetic adaptations, but still is an arrangement of *polyphyletic* and *paraphyletic* grades, or a 'horizontal' classification, according to Dressler (1993).

Dressler's classification, published in its original form in 1981 and revised in 1993, is presented as a transitional one that incorporates as much of the new information based on new methods of classification as possible. It incorporates new knowledge on chromosome numbers and often refers to polli-

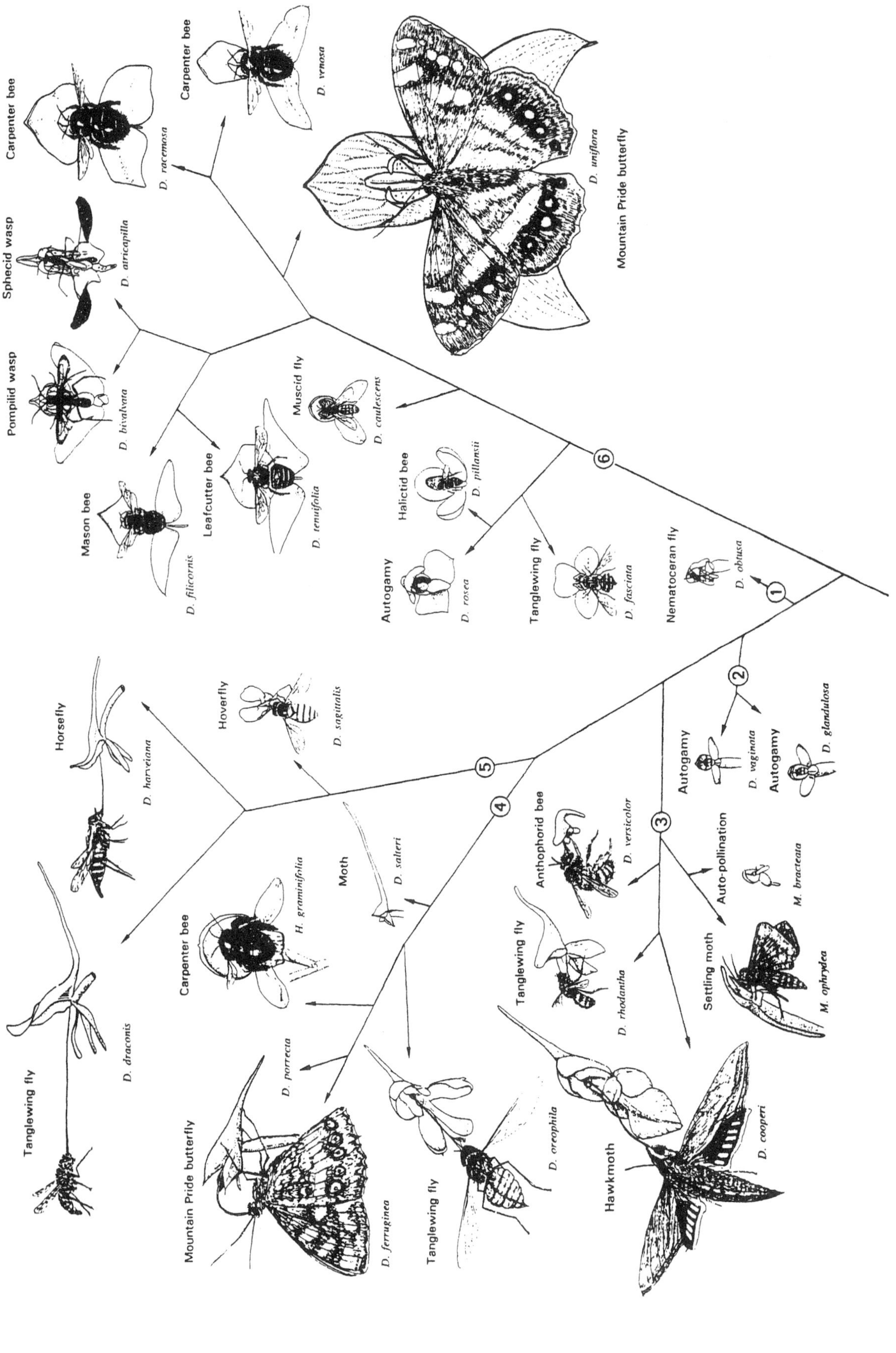

Figure 18. Phylogeny of pollination systems in *Disa* species. The phylogeny is based on the cladogram in Figure 17. The numbers for the clades are the same as in Figure 17.

nation phenomena, without however yet being able to draw inferences, which indeed may be difficult, if not impossible, except on the level of species. In Dressler's own words it is a beginning of a phylogenetic classification (at least as far as the advanced Epidendroideae is concerned). It stresses the gaps in our knowledge. Dressler's classification is probably the most widely used and this book will not be an exception, as the reader will note in the following chapters.

Burns-Balogh & Funk (1986) made a cladistic analysis in which 68 apomorphies were grouped into 42 transformation series. The resulting cladogram was used to develop a classification and a natural key to the tribes of the family. Seven subfamilies (Neuwiedioideae, Apostasioideae, Cypripedioidea, Spiranthoideae, Neottioideae, Orchidoideae and Epidendroideae) were divided into 20 tribes, the majority of which are defined by synapomorphies.

These three examples may suffice to illustrate that a lot of work is being done to improve orchid classifications, as well as the fact that much more work is still ahead.

CHAPTER 4

Orchids of North America

For even if knowledge of the true and original Tree of Life ... is lost, the plants nevertheless remain and renew their flowers, and with gratitude enduring through the years they shall always exhale the sweet memories of your names, and make them more lasting than marble, so that they will outlive those of kings and heroes. For riches vanish, the most stately mansions fall into decay, the most prolific families die out sooner or later; the mightiest states and the most flourishing kingdoms may be overthrown; but the whole of nature must be obliterated before the genera of plants disappear and he be forgotten who held the torch aloft in botany

C. Linnaeus

In the area north of the Mexican border, consisting of the United States, Canada and Greenland, according to Williams & Williams (1983) 177 species of orchids are found and some of these species show geographical races or subspecies, and varieties. Most of these are terrestrial, as is to be expected in view of the temperate or even cold climate. In Florida, with a subtropical climate, several epiphytes are to be found. For practical reasons I will follow the order and nomenclature of Willams and Williams *Field Guide to Orchids of North America*. Superimposed on this, however, I will use Dressler's classification of orchids (Dressler 1993). I leave out species that have been recorded for Florida only once and that have not been rediscovered since. However, Catling & Catling (1991) give more species than Williams and Williams (and sometimes the same species with different names) and where they or other authors on pollination present relevant information concerning species I have included these data.

Orchids with two or three fertile anthers

CYPRIPEDIOIDEAE

Of the four genera of lady-slippers, with together over 120 species, only the genus *Cypripedium* is represented in North America. Atwood places *C.arietinum* in a separate genus *Criosanthes*, but I leave it where it has traditionally been placed.

Genus **Cypripedium**

The genus *Cypripedium* is represented in North America by 11 (or 12) species, almost half of the existing species in the world, all growing in temperate and cold regions. The flowers are typical insect traps (Plates 22-24, p. 29), attracting bees with their odours, which may signify different complex signals. Only the Japanese *Cypripedium debile* is exceptional in possessing a strong mushroom odour and therefore probably attracting fungivore flies. The myophilous syndrome is the usual one in the related tropical genus *Paphiopedilum.* But see also *C.reginae.*

The morphology of the flowers and the mode of pollination were clearly described by Charles Darwin: 'there is no rostellum, the three stigmas are fully developed and confluent, two anthers of the inner whorl are fertile and the single anther normally present in other orchids is only rudimentary in this genus, which relates to the presence of two exits to the labellum pouch. The pollen grains are not tied together by 'elastic' bands, nor furnished with a caudicle or cemented into a waxy mass. The pollen is itself viscid and the stigma dry and slightly convex, so as to rub off more effectively the glutinous pollen adhering to the insect's body. The stigmatic surface is scattered with minute, rigid, sharp pointed papillae, directed forwards, which act as a brush. The inner surface of the labellum is clothed with hairs, the tips of which secrete little drops of slightly viscid fluid. Insects enter through the large opening of the 'shoe' and crawl out through one of the two small orifices. The insect first brushes the stigma and afterwards one of the anthers, so that it cannot leave pollen on the stigma until it has already been smeared with pollen from another flower'.

Case (1994) investigated genetic variation in five related species of *Cypripedium* (among taxa and among populations) by *electrophoretic analysis* of *allozyme* variation, in which *alleles* are inferred from enzyme phenotypes (*Cypripedium calceolus, C.candidum, C.arietinum, C.acaule, C.reginae*). Most populations contain very low levels of *heterozygosity*, except *C.calceolus.* The data reveal an extremely wide range in the degree of genetic relatedness among these taxa. The species are clearly demarcated and genetically distinct. They obviously have experienced a long period of *phylogenetic* separation. The low genetic diversity within these species and among populations of the species probably points in the direction of derivation from genetically depauperate progenitor populations and later glacial disturbances, that may have reduced their geographical range, restricting them to small, isolated populations. It makes the species vulnerable.

According to Klier et al. (1991) who analysed allozymes of 17 populations of *Cypripedium calceolus* var. *pubescens* and nine of *C.candidum* at five sites with hybrid swarms of these species, there is a bidirectional gene flow wherever the two species are sympatric. The data and morphological facts and ecological considerations suggest that the prairie ecotype of *C.pubescens* may have arisen as a direct consequence of its acquisition of genetic information from *C.candidum.*

The glandular hairs of Cypripedium plants (to which some people show allergic reactions) contain volatile lipidlike fragrant substances that probably attract some insects and repel others and larger less volatile lipid molecules which protect against predators by their toxic properties (Holman et al. 1981).

The *trichomes* on the ovary are a possible site of synthesis and secretion of the floral scent. (Swanson et al. 1980)

Van der Pijl & Dodson (1966) mention bees of the genera *Andrena* or *Megachile* as pollinators of *Cypripedium*, attracted by 'deceit'.

Not every known species has information published about its pollination.

Stoutamire (1967) produced a table containing the names of all insect visitors observed until 1967 on *Cypripedium* species.

Cypripedium acaule

Demographic data (Gill 1989) suggest that population establishment in this species may be viewed as a series of improbable events beginning with a *founder* event and leading to a relatively low degree of heterozygosity.

According to Davis (1986) bumblebees are the normal pollinators of *Cypripedium acaule*, the pink Lady's slipper orchid.

Fruit formation appears to be pollinator-limited. Pollination intensity (on average less than 10%) seems independent of floral density. One of the bumblebee queens observed by Davis was *Psythirus ashtoni* with pollen mass on her thorax.

The pollen of *C.acaule* is more granular and less viscid than in the other American species and the stigma slightly concave (Darwin 1888). Stoutamire (1967) captured a specimen of *Bombus vagans* with pollen on its thorax. He mentions *B.borealis* as a pollinator observed by R. and K. Dreisbach in Michigan.

Plowright et al. (1980) found several bumblebee queens of *Bombus vagans* carrying pollen masses on the dorsal surfaces of their thoraces and one *B.terricola* queen visiting a flower. In comparing pollen removal in different habitats they found that removal was significantly lower in areas sprayed with fenitrothion in New Brunswick compared to unsprayed control areas, probably due to direct mortality of the bees as a result of the spraying. The control areas had, respectively, 5.5% and 9.9% removal.

A 13-year field study of a population of *C.acaule* by Gill & Stoutamire (1990) has revealed a consistent pattern of fruiting failure because of ineffective pollination. Only 2.33% of 902 flowers had naturally set fruit in that period. The pattern is reported to be the same everywhere. The species is self-compatible as is proved by *hand pollination*, but not autogamous. There is obviously *depression of inbreeding* in this species.

Primack & Hall (1990, 1994) studied the *costs of reproduction* of this species under field conditions in two populations in eastern Massachusetts with high density of plants and low density of bumblebees, resulting in about 2% of the flowers developing fruits. Manipulation of fruit set by hand pollination leads (on average) to 90% fruiting. Repeated treatment during four successive years resulted in lower growth (10-13% decrease in leaf area) and a lower flowering rate in the third and fourth year (5-16% decrease). Effects of defoliation, successive fruiting and fire are all additive in a simple way. Plants apparently have to be put under severe physiological stress in order for a cost of reproduction to become evident in the following year.

However, Calvo & Horvitz (1990) state that cost of reproduction and pollinator limitation have a complex influence on plant fitness. If pollination is uncertain, then taking advantage of a good year might be the best way a plant can deal with that uncertainty, even if a cost is involved.

Cypripedium (=Criosanthes) arietinum

Van der Pijl & Dodson (1966) present a photograph of a bee crawling from under the anther of *C.arietinum.*

Stoutamire (1967) observed two bees of the genus *Lasioglossum* (*Dialictus*), popping into a flower, staying inside for 1-2 minutes and forcing their way out of an anther opening, carrying pollen masses. One of these was identified as *Lasioglossum coeruleum.*

Cypripedium calceolus

The nominate form of this species grows in Europe and Asia. In North America several subspecies are recognised: var. *pubescens*, the most widespread subspecies with a variety of habitats, var. *parviflorum*, a form with small flowers from swamps and bogs and in north-eastern US and adjacent Canada, and var. *planipetalum* from limestone barrens in northern Newfoundland, eastern Quebec and Ontario. The divergence of different taxa is supposed to be correlated with *geographic isolation* caused by *Pleistocene* glaciations and this in turn may have led to different pollinators.

The pollination of the nominate species was extensively investigated in Europe. Although there may be differences between the pollination of the North American and European taxa the account below contains information relevant for the understanding of the pollination of *C.calceolus*.

H. Müller (in Darwin 1888) was the first to observe that *C.calceolus* is fertilised in nature in the manner described by Darwin, by bees belonging to five species of *Andrena*. Robertson reported visits by bees of the genus *Ceratina* in North America (Stoutamire 1967). Stoutamire collected, besides bees of the genus *Ceratina*, bees of the genera *Evylaeus*, *Agapostemon*, *Apis*, *Osmia*, *Halictus* and *Lasioglossum* (*Dialictus*) with or without pollen smears on their thoraxes from within the flowers. He found also Diptera inside. Yellow crab spiders exploit this situation of abundant visitors. A complete pollination was observed by him in var. *pubescens* in Michigan by *Ceratina calcarata*.

Attraction

There has been much debate in the past about the reward for pollinators, the hairs in the labellum being thought to function as food. These hairs secrete oil but there are, however, no reliable observations of feeding on hairs or other secretions in the flowers. Nilsson (1978) suggests that these hairs may retain pheromones from the visiting bees or by their strong light refractive ability may stimulate photoactive behaviour of pollinating bees under the stigma in addition to the light windows in the labellum. They may also lower friction against the bottom for creeping visitors. The *window-theory* mentioned has been much disputed.

All recent opinions stress 'deceit'. The patterns of crimson spots on the staminode and on the veins in the bottom are believed to act as false nectar guides.

The close-range attraction by scents (see below) is an interesting lure phenomenon. Nilsson found more than nine floral fragrance substances – among them some monoterpenes, sesquiterpenes, several acetates and an ether. Acetates, as present in the flower, are also found in the cephalic pheromone secretions of the visiting *Andrena* bees. These are used by males to odourmark objects in order to attract females and probably also attract the males themselves, causing them to aggregate. Alphafarnasene is also present in the abdominal *Dufour glands* of the bees and serves as a lining material in nests. Its volatile part is a signal on the nest site, releasing landing responses, that are probably inherited (and therefore irresistible).

Linalool may be associated by the bees with food.

Visitors

Over 60 visiting insect species have been observed by Nilsson on the island of Öland, alighting on floral parts. Other investigators also mention many different insects, among them beetles, butterflies, flies and hoverflies, bumblebees and many kinds of solitary bees. Flowers are eaten by slugs and snails: we should be aware that visitors form a different class from pollinators.

Pollinators

Of the many visitors, few are real pollinators: these are mainly medium-sized bees and especially female *Andrena* bees.

Bees of the genera *Lasioglossum*, *Halictus* and *Andrena* were captured by Nilsson, with pollen smears.

Approach
The majority of bees fly directly and aim straight towards the labella. *Long-range attraction* does not seem to be influenced by odours. During the final stage of the approach bees may perform undulating flights, indicating nearby stimulation by scents. Daumann (1968) did experiments from which he concluded that long-range attraction is entirely regulated by eye-sight and the yellow colour, especially, has been proved to be attractive to bees.

Alighting
Bumblebees have, according to Nilsson, the highest rate of alighting. They land on the lateral margins of the front opening and search for food, but normally leave without entering. Medium-sized Andrenas, among them females of *A.haemorrhoa*, are the most abundant. They land either on the margins or fly directly into the central opening near the staminode, which they try to reach with out-stretched legs, as if trying to get a foothold. They then fall to the bottom of the inflated pouch.

Larger *Lasioglossum* bees as well as small bees may also try to alight in the flower and the staminode seems to be especially attractive to them.

Entering
The smaller the size of the bees alighting, the higher the proportion of labellum entries. Sometimes bumblebees inspect the inner, without losing control over their movements, whereas large *Andrena* bees often crawl into the flower and tumble down from the margins because they lose grip: medium-sized *Andrena* bees also may lose their grip if they do not enter directly.

Captivity and escape
Small flies creep around undisturbed inside the labellum and may fly out again, but bees immediately return to the entrance. Bumblebees and large Andrenas crawl to the margins or staminode, and lift themselves out easily, but the smallest bees entering leave without sufficient contact with the stigma and therefore appear to fail as effective pollinators.

Medium sized bees fail to climb out and seem initially to panic, then after some minutes they calm down and put their wings to rest. Then, more methodically, they try to pass the stigma and force themselves more or less upward under the curving of the stigma. They pass the narrow passage with the anther in the same way, picking up pollen on the anterior part of their *mesonotum*, finally drawing themselves out with their claws behind the margins. The time spent in the flower may vary from a few seconds up to some days, but normally they stay less than 10 minutes.

Bees leaving after creeping out, normally sit down on the petals or fly to leaves of the surrounding vegetation for grooming and then leave the place. About 10% of the bees were seen to directly enter a second labellum. Although self-pollination by insects backing out cannot be excluded, this phenomenon is considered to be rare. Moreover, according to Daumann, the plants are, to a great extent, *self-sterile*.

It seems that the medium sized Andrenas, although being the best pollinators of the species, alight and enter less often than expected for effective pollination, whereas there are more frequent visits by bumblebees and solitary bees too large and too small respectively for the pollination role.

Fruit set
Nilsson found fruit set to vary from 0-25%. Detto found in a German locality 65%, but in general seed pods set seems to be very low. The same is observed for *C.acaule*. On average 2%, the low percentage raises questions about the lack of development of a strategy for self-pollination. The costs of reproduction however seem to be high, as the plants in the years after fruit is set pay their due with smaller leaves and fewer flowers. This might prove to be a general phenomenon in orchids. Many orchids seem to be pollination-limited.

Moreover, as a result of one successful pollination, orchids can produce numerous dust-like, wind-dispersed seeds. The potential for long-distance seed dispersal probably facilitates geographic isolation among populations as well as producing founder events. Isolated populations founded by one or a few individuals may also contain reduced levels of genetic variation relative to the source population, especially if population growth afterwards remains slow (Case 1994).

Tremblay (1994), in comparing results of hand pollinated flowers and naturally pollinated flowers in *C.calceolus* var. *pubescens*, found that multi-paternally sired capsules had significantly smaller seeds with a higher germination rate than uni-paternal and self-sired ones. Although multi-paternal pollinations are rare in nature the higher germination rate could affect fitness as well as the direction of selection.

Catling & Catling (1991) state that in much of North America fruit set is less than 10%, but that in the Rocky Mountains of Alberta and British Columbia it is more than 90% in both large and small populations.

Variation

Local *anthecotypes* may differ in the narrowness of the passage for the pollinating bees and therefore have been changed slightly as an adaptation to pollinators of differing size. There are striking chemical differences between the fragrances of European *C.calceolus* and the North-American var. *parviflorum*. This variety should probably be recognised as a different biospecies and is adapted to bees other than Andrenas and, there are also differences in floral morphology. Robertson (1928, in Catling & Catling 1991) reported two bees, a *Ceratina* spec. and *Zaodontomerus* spec. as visitors. In 1981 *Cypripedium kentuckiense* (Atwood 1985) was separated from the *C.calceolus* complex as a different species.

Cypripedium candidum

Stoutamire observed pollination by *Andrena placida* only once in spite of numerous visits to flowering colonies. The species hybridises occasionally with the syntopic *C.candidum*. Catling and Knerer mention *Andrena* and Halictid bees (Vogt 1990, Catling & Catling 1991). The sticky pollen attaches to the dorsal surface of the thorax.

Bender (1985) observed pollination by at least seven genera of solitary bees during a three-year study.

Cypripedium fasciculatum

Stoutamire mentions the similarity between this Rocky Mountain species and *C.debile* of Japan, which species imitates mushrooms by a fungose structure of the labellum, and with its mushroom odour probably attracts fungivore flies (Vogel 1978). There are, however, no reports on the pollination of *C.fasciculatum*.

Cypripedium kentuckiense

Covell & Medley (1986) describe the trapping of a moth *Polychrysia morigera* in the lip of *C.kentuckiense* in Tennessee. The moth is extending its range from the west and was new to the state.

Cypripedium passerinum

Catling (1990) mentions self-pollination for this species of coniferous forest, wooded ravines and littorals of streams and lake-shores, that is distributed from Alaska to the Hudson Bay and southwards to north-western Montana. In these areas there is a paucity of insects. The anthers develop in contact with the stigma.

Cypripedium reginae

Stoutamire (1967) observed a medium-sized black bee which flew directly to the flower, alighted on the edge of the opening and dropped to the base of the labellum. It crawled up the base after exploring the long hairs, and forced its way out under the anther. He mentions observations of Guignard (1886, 1887) of *Megachile centuncularis* and *M.melanophaea* as pollinators with similar behaviour (the latter removed pollen but chewed its way out and is possibly not a legitimate pollinator). Van der Pijl & Dodson mention Robertson (1887) for what seems the same observation. See also Dodson (1966) who recalls Guignard (1886). Even Lepidoptera and Coleoptera have been reported as visitors entering the labellum or crawling across anther and stigma. They may however not be considered as regular pollinators. The European skipper is frequently trapped, but cannot escape, so does not act as pollinator (Catling & Catling 1991).

Vogt (1990) found 12 syrphid flies (*Syrphus torvus*) as the main pollinators, and two flower beetles (*Trichiotinus assimilis*) as occasional ones. Beetle pollination was also reported by Smith as early as 1863 (Catling & Catling 1991).

In Vogt's study in Vermont over 90% of the pollination was apparently performed by syrphid flies. He detected several trapped flies by audible cues: high pitched whines were noticeable. Usually it took the insects about 5-10 minutes of struggle to escape. Some syrphids with large pollen smears were collected. Vogt also collected 35 dead skippers in 28 flowers. These observations contradict previous ones and it may be that the syrphid flies have been overlooked before.

The sticky pollen in all cases observed was smeared on the insect's thorax.

Primitive orchids with a single fertile anther

NEOTTIEAE

Note: Dressler (1981) first placed the Neottieae within the Orchidoideae, but in his Classification of 1993 he separates them as one of the tribes of a group of primitive orchids with one fertile anther. Alternate classifications have included the Neottieae in a subfamily Neottioideae or placed within the Spiranthoideae.

NEOTTIEAE: LIMODORINAE

There are 51 species in four genera world-wide. Especially the splitters among European orchidologists recognise many more species in the genus *Epipactis*.

Genus Cephalanthera

Cepahalanthera austinae

Only one species of this mainly European and Asian genus with about 15 species is found in North America, *Cephalanthera austinae* (syn. *Eburophyton austinae*). It is the only member of the genus that is saprophytic, lacking chlorophyll in stem and leafs. Darwin described the morphology of *Cephalanthera*, calling it a 'degraded *Epipactis*'. There is no rostellum and the terminal portion of the labellum, the epichile, is usually frosted with globular papillae of a yellow or orange colour, and within the cup there are several transversely wrinkled, longitudinal ridges of the same colour. These ridges are often gnawed by visiting insects and are considered to be pseudopollen. The pollen is extremely friable and readily adheres to any object. The species are therefore (facultative) autogamous. According to Burns-Balogh et al. (1987) there is a large discrepancy between column structure and the type of pollination in *Cephalanthera*. This could possibly be explained by the evolutionary history of the genus. The loss of the rostellum by reduction could be due to the advantages for a pioneering autogamous race. During expansion of the race increasing heterozygosity by *entomogamy* could again become a positive feature. When the rostellum was lost, however, the evolution back towards entomogamy could have taken a different route via evolution of the perianth. The route chosen in this case may have been that of mimicking co-blooming entomogamous taxa. The European species are known to mimic foodplants of wasps, solitary bees and bumblebees. It is interesting to note that with the mimicking of bell-shaped or rose-shaped flowers in *Cephalanthera* a reduction of the zygomorphic character of orchid flowers has occurred. There is an evolution back to *actinomorphy*. It should be added that the development of deceit is a widespread feature in orchids and several characteristics of the family have made this possible via different routes. It is not known, however, whether *C.austinae* is also a mimic of a foodplant of pollinating insects or is entirely autogamous (Catling 1990). The orchid is found in the shade of coniferous forests in the mountains of Washington, Oregon, Idaho, and California. Self-pollination is observed by means of falling of a friable pollen mass, but also pollination by Halictid bees, the pollinaria attaching to the dorsal thorax (Kipping 1971).

Genus Epipactis

This genus is widespread in Europe, Asia and Africa with many species. These species are either *allogamous*, mostly pollinated by wasps and to a lesser extent by bees or syrphids, or autogamous and even cleistogamous. The autogamous species are characterised by their reduced flower characters and are usually seen as species adapted to areas poor in pollinators. The allogamous species offer nectar in open nectaries. The labellum is divided in a basal part, the epichile and in a distal part, the hypochile. As wasp-pollinated flowers they often have reddish-brown colours. In North America two species are found, one of which is also widespread in Europe. *Epipactis* species have a rather short, massive column structure, there is no labellum spur, but the hypochile is concave and nectariferous.

Epipactis gigantea

Wiefelspütz (1970) mentions pollination by syrphid flies as observed by Dr Ross and documented in pictures in the National Geographic, vol. 130, nr 6, 1966. The pollinia were attached to the back of the syrphids. The species is facultative autogamous. The species ranges from British Columbia to California, Arizona and New Mexico and grows along gravel shores of lakes and rivers, on sandbars in streams and in seepage areas.

Epipactis helleborine

It is generally believed to be an introduced species from Europe, which has spread in Canada since 1879 where it is pollinated by wasps, just like in its area of supposed origin (Plate 25, p. 29). In Europe its pollination was extensively investigated by different people.

Visitors and pollinators

Most observers in Europe agree on the facts of pollination with wasps with long heads, *Dolichoves-*

pula species being the legitimate pollinators. *Paravespula* species with short heads are of much lesser importance. According to Wiefelspütz (1970), wasps primarily depend on eyesight and are attracted by brown, dull red and yellow-green colours. With their short mouthparts they need easy accessible superficial nectar.

Halictus bees and honeybees, (hover)flies and beetles are among observed visitors that only accidentally pollinate. *Halictus* bees, for example, are too small to find a good place on the short epichile and thus keep their heads far from the column. After carrying off pollinia some of these visiting animals free themselves from this load, sitting on the perianth or on surrounding vegetation, where hoverflies may be found feeding on them.

Bumblebees have been observed drinking from the nectar, but with their long mouthparts they take it up without touching the rostellum.

According to Vöth (1982) the pollinating wasps are males and old females, the latter no longer searching for animal prey. The plants flower when the populations of wasps are rich in individuals and just before the queens emerge. The worker-females are then feeding larvae with animal food and are not interested in nectar. They fly around without looking at the orchids and if placed on them they flee. Old worker-females have given up feeding in this way and live on nectar and plant juices in their dotage.

In Ontario four species of wasps were observed as potential pollinators, feeding on the flowers and bearing pollinia (Judd 1970). These were *Vespula arenaria*, *V.consobrina*, *V.vidua* and *Polistes fuscatus*. Most wasps were female workers, but some males were observed and/or collected as well. Some wasps were found dead in spider webs (*Enogplognatha ovata*) on the flowers. Pollinia were stuck to the hairs of the head on the lower frons and the clypeus.

No autogamy

Ilona Müller (1988) did experiments with nets to find out about facultative autogamy and concluded that under natural circumstances no autogamy took place. Although the species is self-compatible, autogamy must be very rare if, indeed, it occurs at all, and then probably only when for example small insects aid it. The species therefore has to be considered as strictly allogamous. Statements about autogamy in literature may be false because of confusion due to the taxonomic problems with this genus, where the observers have not discriminated between *E.helleborine* s.s. and closely related autogamous subspecies or species.

Wasp behaviour

Wasps may behave in very peculiar fashion. Szlachetko observed wasps in Poland landing on the labellum, where they proceed to lap nectar stored in the hypochile. Because they are much larger than the flowers, they clasp neighbouring flowers with their limbs and, in this way cause more pollinia to be deposited on their limbs and on the ventral side of their abdomens. After emptying the hypochile they move on to other flowers and take more nectar. After this, they become very sleepy, so that they can be held in the hand. Some wasps had their abdomen chewed, leading Szlachetko to conclude that some toxic or drug substance must be present in the nectar. Other investigators have found similar behaviour, wasps even falling down and sluggishly or clumsily trying to climb into the plants again.

Analysis of the nectar by Müller showed this to be sucrose-dominant. Some alcohol is present in the nectar as ethanol, but could hardly be demonstrated. Yeast-cells and bacteria are abundant. The exact nature of the substances that cause this peculiar behaviour is not known.

NEOTTIEAE: LISTERINAE

About 29 species are known in two genera, all from the northern hemisphere. They are closely related to the Limodorinae.

Genus **Listera**

The twayblades are represented in North America with eight species. They are characterised by a wingless column and an anther that is borne on the back of the column near the apex with two powdery pollinia. There is no spur and the nectar is presented on the surface of the labellum. The flowers are generally green or greenish. At anthesis, the access to the stigma is blocked, making the flowers *protandrous*. The rostellum is sensitive and when touched, releases the pollinia explosively. These are cemented to the insect's body by the fluid that is first expelled, separately from the pollinia, and hardens as soon as it is exposed to the air. One to two days after the pollinia are removed, the rostellum lifts by hygroscopic movement and exposes the receptive stigma. The temporary inaccessibility of the stigma promotes cross-fertilisation.

Listera cordata

Ackerman & Mesler (1979) studied the pollination biology of *Listera cordata* in 1976-1970 in Northern California. The morphology of the pollination-mechanism strongly resembles that of *L.ovata*, as described by Darwin.

L.cordata has two or three small hairy points projecting from the middle of the crest of the labellum: Darwin confessed not to know their function. These form, however, the touch-sensitive trigger of the explosion mechanism in which the ejection of glue and the release of pollinia are independent but synchronous processes. Ackerman and Mesler succeeded in extracting the glue with a needle without catching the falling pollinia. The rostellum fluid does not contact the tips of the pollinia as it is thrust from the labellum, unlike the mechanism Darwin states for *L.ovata*.

The flap-like rostellum blocks the access to the stigma and lifts one or two days after removal of the pollinia to expose the receptive stigma (*protandry*).

Attraction

Pollinators are attracted by both odour and nectar, the latter running in a narrow median nectary along almost the entire length of the labellum. The scent is clearly repellent to the human nose.

Visitors and pollinators

Several Hymenoptera and Diptera as well as Coleoptera have been reported as visitors, of which representatives of the first two groups act as pollinators. The spectrum of visitor-groups resembles that of *L.ovata*, but the species are much smaller, in conformity with the smaller size of *L.cordata*. Small fungus gnats especially are witnessed as pollen vectors and their behaviour strongly resembles that which is described for *L.ovata*. Ackerman and Mesler do not want to designate the flowers of *Listera cordata* as 'fungus-gnat-flowers', but prefer to call them 'small insect-flowers'. The scent, however, points in the direction of adaptation to the attraction of *sapromyophilous* or mycetophilous insects, although no oviposition takes place and the main reason for visiting is to feed on nectar and both male and female gnats where captured on the flowers. An analysis of the chemical components of the scent could be worthwhile.

Argue (1996) mentions as hymenopteran pollinators nectar-seeking wasps.

Fruit set, autogamy and clone-formation

Artificial self-pollination produces capsules with well-developed embryos but the species is, according to Ackerman and Mesler, not autogamous. Löjtnant & Jacobsen (1976) however state that the fructification of *L.cordata* in Greenland is so abundant, and the capsules swell so quickly, that it is common to find swollen capsules in the lower part of the spike and newly opened flowers in the upper part. This effective and fast pollination and fruiting indicate, in their view, autogamous/*apogamous* flowers. Fruit set is reported as varying from 61-78% of flowers: flowers persist for some time after the ovary swells.

Listera ovata

The European Twayblade has in recent years been recorded from south-eastern Ontario on an island in Lake Huron. Its pollination biology has been studied by Darwin already and more extensively by Nilsson (1981). Hover-flies, bees, ants, beetles and many other insects have been observed as visitors.

Pollinators

The most effective pollinators are relatively strong insects of about 8-12 mm length, with short mouthparts and hairless medium-sized heads. The pollinators need to be strong because of the problems of countering the adhesion of the viscid fluid and withdrawal of the pollinia, as well as handling the weight of the pollinia and carrying these to other racemes.

The movement of the rostellum to its final position as described above, in general takes longer than the time the pollen vector remains on the same raceme. Pollination occurs when the insects press their heads against the column in trying to remove the innermost nectar at the place where the furrow is broadest. The animals are induced to remain in the same position and make small sideward movements with the head, thus rubbing the pollen over the stigma.

Fruit set

Fruit set varied in Nilsson's investigations from 13-70%, on average 37%, which is low in view of the number of visitors.

There is considerable variation in fruit set between individual plants.

Autogamy and clone-formation

The pollen in plants that are not visited becomes friable but stays in its place. Artificial self-pollination leads to seed formation, but the embryos are slightly reduced in size.

The occurrence of spontaneous autogamy is doubtful, but opinions about this differ. Probably there is insect-aided auto-pollination by thrips, as Darwin already suggested and Nilsson suspected from his experiments.

Thrips may infrequently transfer pollen from anthers to stigma and, since *L.ovata* is self-compatible this may contribute to fruit set.

L.ovata forms clones easily which means that a certain amount of inbreeding will occur.

Other **Listera** species

No data are found about the pollination of the other species.

TRIPHOREAE

There are 28 species known in this tribe, primarily from tropical America, but some species extend into North America. Some forms are autogamous, but in other species pollination by small bees is supposed or observed.

Genus Triphora

Five species occur in the US of which four are confined to Florida and one has a wider distribution in the East. All species are small, succulent, fragile plants with alternating leaves and flowers with three-lobed lips having three crests in the centre. Some species are autogamous; otherwise not known, except for one observation of bumblebee pollination in *Triphora trianthophora* (see below), but the flower structure suggests pollination by small bees (Dressler 1981). According to Williams (1994), all species of *Triphora* which are not self-pollinating exhibit gregarious flowering of short-lived flowers, which she believes to enhance their chances of cross-pollination. She observed over a six year period on a daily basis at 25 sites (1988) and 72 sites (1993) only two visitors on two different occasions. A small bumblebee (*Bombus* spec.) entered for a few seconds and backed out with pollinia attached. A much smaller and slender bee (*Hylaeus* spec.) did not remove pollinia. The 1-4 flowers per plant are available to pollinators for only one day. Many buds get lost by dropping off due to underdevelopment or by consumption of plants or buds by chipmunks etc., and from those that flower very few ripen. Plants can exist for several years at a time in a subterranean, tuberous condition. They reproduce vegetatively by producing up to 17 secondary tubers, which remain in leaf litter (in the climax hardwood forests which they inhabit), without reaching the soil substrate. It is supposed that the species lives primarily saprophytically in the subterranean phase by extracting food from the beech trees with its mycorrhizae.

The combination of asexual propagation as the main means of reproduction with rare seed production ensures conservation of a certain genetic variability, but makes the species vulnerable.

Lownes (1920) observed pollination by halictid bees and Medley (1979) confirmed this observation. He found *Augochlora pura* to be the principal pollinator in Indiana (Dressler 1993).

According to Medley the mass-flowering is synchronized for subsequent developing flowers. Some populations separated by substantial distance are also synchronous. It appears therefore that synchrony is between and within populations. In this way floral display is maximised. This system is of course dependent on learning behaviour by reward and it appears that ridges on the column of *T.trianthophora* secrete nectar, and sometimes fluid can be seen at the base of the lip (Catling & Catling 1991).

VANILLEAE: VANILLINAE

A pantropical subtribe of about 120 species in five genera of which the genus *Vanilla* occurs in North America.

Genus Vanilla

Five species occur in the US in the southern part of Florida. They are climbing plants, that initially start as terrestrial, but as mature plants lose contact with the soil.

According to Dressler (1981) most articles about *Vanilla* tell of pollination by *Melipona beechei* in Mexico, but authors have copied each other until the source seems hopelessly lost. In the American tropics, *Vanilla* is frequently pollinated by much larger bees of the genus *Eulaema*, this fact arousing doubt concerning effective pollination by a small bee such as *Melipona beechei*. *Eulaema* specimens occasionally are observed with a flat, triangular pad of pollen on the scutellum, which Dressler believes to be pollen of *Vanilla*.

Vanilla planifolia is the producer of the vanilla bean of commerce. It is produced by artificial pollination of the flowers.

According to Van der Pijl & Dodson (1966) the stigmatic fluid may be so abundant in *Vanilla* that it reaches the pollinia, inducing germination. This fact is overlooked by Catling in his survey on autogamy in orchids (but mentioned in Catling & Catling 1991). It has been assumed that the labellum hairs and scales are 'grazed' on by bees. They are thin-walled and filled with plasma, oil and sugar (food hairs).

Primitive orchids of uncertain classification

POGONIINAE

A subtribe of 64 terrestrial or saprophytic species in five genera, of which three occur in North America. The status of the Pogoniinae within the Vanilleae is uncertain.

Genus **Cleistes**

Cleistes divaricata
The species, formerly classified as *Pogonia*, is distributed from southern New Jersey to northern Florida and to south-eastern Texas, including the mountain areas of eastern Tennessee and Kentucky.

Gregg (1989) studied var. *bifaria* (syn. *C.bifaria*) of this species in a West Virginia meadow, where it quadrupled in numbers between 1980 and 1986 in an acidic, nutrient-deficient habitat.

The aerial, upright stems may be vegetative with one to three leaves; when flowering they typically bear one flower, one cauline leaf, and a floral bract. The long-lived flowers contain no reward and operate as bee-foodflower mimics, exploiting naive bees. The protruding labellar apex bears a central crest and may serve as pseudopollen. A hinged anther drops clumps of pollen tetrads onto the dorsal thorax of *Bombus* and *Megachile* workers. The anther, after being opened, will snap back into place, giving the flower an appearance of not having been disturbed. Capsules with 120-11,000 seeds result from xenogamous and geitonogamous pollinations. Pollinators are scarce and occasionally severe herbivory of flowers and capsules may lower chances of reproduction. Fruit set ranged from 21-87%, but only 50% of seeds reached maturity. The species is not autogamous, although self-compatible. Vegetative multiplication is a normal feature.

An interesting trait which this species shares with deceptive orchids with the same syndrome is colour *polymorphism*, with flowers ranging from pale pink to deep rose and with different contrasting mottled or striped labellum patterns. This could prevent bees from learning to avoid the flowers before they have tried out several. *C.divaricata* has, moreover, variable sepal configurations that might function in the same way.

Gregg (1990) captured 12 bees in West Virginia and 21 in North Carolina with pollen tetrads of *C.divaricata* on the body and examined another 116 bees under a dissection microscope. She found pollen tetrads of *Cleistes* in the hind tibian pollen baskets of 25 bumblebees. In nine of these bees the total of pollen masses packed were one-third to almost all *Cleistes* tetrads.

Thus, she concludes, some individual bees have learned to exploit an orchid that usually deceives pollinators. This is the first time that collection of orchid pollen by a pollinator is documented.

A higher percentage of bees in North Carolina carried large amounts of *Cleistes* pollen in corbiculae than in West Virginia (Gregg 1991b). This fact, in combination with floral fragrance and other data, supports a conclusion that in North Carolina natural selection has favoured a reward strategy for pollinator attraction, whereas in West Virginia the predominant strategy appears to be deceit.

The pollen tetrad masses that fall on the dorsal thorax allow more than one bee to receive pollen from a single flower. Hand-pollination experiments (Gregg 1991a) demonstrated that capsules produced from a flower's first three pollen dumps were significantly larger and heavier and contained more seeds than produced by later pollen deposits with fewer tetrads. Decreasing pollen dosage did, however, not affect fruit set, capsule development, individual seed weight, or percentage of healthy looking embryos per fruit. As the set of naturally pollinated plants was only 38% (compared to ca 95% in artificially pollinated plants) production of seed is pollinator-limited. The efficacy of infrequent visits is, however, maximised, because a flower's first pollen releases contain the most pollen. Low-dosage capsules contained about five times more *pericarp* per seed and flower fading was as slow as in unpollinated controls (flowers remain fresh for 10-12 days), thus increasing opportunity for additional pollination. In conclusion, *Cleistes divaricata* employs two additional strategies when low dosage pollination occurs: delayed flower fading and maturation of capsules with small numbers of seeds. Putting resources into pericarp, a photosynthetic organ capable of replenishing at least some of its own energy requirements, is apparently a workable alternative for this orchid, which expends no energy on floral fragrance or nectar. Timing of anthesis may be an additional adaptation to pollinator scarcity. *C.divaricata* flowers when other flower species are scarce and young bees begin to develop foraging preferences.

We may also look to the above phenomena from the viewpoint of sexual selection. Gregg (1991a) suggests that paternal success may be enhanced, since:

- A pollen donor may fertilise a number of different individuals,
- Genetically superior individuals have more than one chance to act as pollen donors,
- Attractive plants can repeatedly donate pollen, and
- The opportunities for cross-pollination are increased.

Genus Isotria

This is a genus of two species in the eastern US. The species have a different morphology of flowers and column.

Isotria verticillata

This is the more common of the two. It has multicoloured, fragrant flowers with apparent nectar guides, but no nectar.

It flowered short, from 4-7 days and had a total flowering period of 13 days, measured in a population investigated by Mehrhoff (1983). The species is self-compatible in experimental pollinations, but there were no signs of autogamy or *agamospermy*. Only 21% of control shoots formed capsules, reflecting the percentage in natural populations. But very few capsules reached maturity, the majority being aborted or destroyed by predators.

Pollinators are solitary bees (seven species were observed belonging to the Andrenidae, Anthophoridae and Halictidae) which pollinate in the usual way. The dorsal surface of the bee's thorax is wetted by stigmatic fluids. As the bee backs out past the rostellar flap, the posterior portion of the upper thorax catches the anther cap and swings it downward, depositing a pollen mass on the wetted thorax. Several bees can pick up pollen from the same flower, since this is not transported in intact pollinia.

Besides the pollinators several visitors were observed that were either too small or too large to pollinate. Some small insects were found in spider webs on the orchid racemes.

Isotria medeoloides

This is considered to be the rarest orchid in temperate North America. Only about 20 populations are known, with altogether approximately 500 plants.

The species self-pollinates. After initial flower opening, the anther drops downward, out of its socket. The pollen begins to extrude out of the sacs, forming a mass projecting from the anther. Both processes lead to eventual contact between pollen and stigma. The percentage of naturally pollinated flowers is 83%, almost four times higher than in *I.verticillata*. The flower has a reduced attractant system: smaller flowers, less ornamented, greenish-white, scentless, and lacking nectar guides. Pollinator scarcity and flower scarcity (one or two widely separated flowers) may have resulted in a low probability of effective pollen transfer. Mehrhoff supposes that since floral visitation is not excluded by the mechanical selfing system, the reduced attractiveness probably results from positive selection, because in a small population with one or two flowers pollinators would completely eliminate reproductive output if pollen was picked up, but not transported to another flower.

The high pollination rate suggests that other factors are responsible for the rarity of the species.

Genus Pogonia

Pogonia ophioglossoides

The genus consists of only three species, one of which only occurs in North America with a distribution in the eastern part, from Ontario and Newfoundland up north and down south to Florida and south-eastern Texas. It grows in marshes, sphagnum bogs, wet meadows. Related species are known from Japan and China.

Thien & Marcks (1972) studied its floral biology in North Wisconsin, together with that of *Arethusa bulbosa* and *Calopogon tuberosus*. These three largely sympatric species with overlapping flowering periods, possess a number of floral characters that suggest convergence. They remain distinct species as a result of isolating mechanisms. The convergence is suggested by: size of the flowers, the pink-magenta perianth, lips with similar yellow-white brushes, and strong absorption of ultraviolet light by the lip brushes. No hybrids have been detected in nature. All three species are visited and pollinated by bees.

Thien and Marcks observed *Bombus ternarius*, *B.terricola*, *B.vagans* and *Megachile melanophaea* on these orchid species, *Pogonia ophioglossoides* being pollinated by the three *Bombus* species. Occasionally bees would visit two or three of the species in sequence. The pollen adheres to different parts of their bodies, thus largely preventing hybridisation. The pollinia of *Arethusa bulbosa* are placed on the back part of the thorax. The pollen of *P.ophioglossoides* is placed upon the head and in *Calopogon tuberosus* on the top part of the abdomen.

In *P.ophioglossoides* (Plate 26, p. 29) the bees land on the fleshy-bearded hairs on the lip. The flowers are pollinated by queens and workers of *Bombus* and other genera of bees. The bees after landing on the lip crawl towards its base where the column and lip form a small tube containing a small quantity of nectar.

The bee upon withdrawal brushes against the fringed anther exposing the pollen. Once some pollen is released the anther snaps back into its original position. Insects visiting a second flower place the pollen on the stigma located in front of the anther cap. Interspecific crosses involving *P.ophioglossoides* as the female parent are dubious, since it appears that some populations of the species may be apomictic. Crosses involving *P.ophioglossoides* as the male parent with *A.bulbosa* and *C.tuberosus*, however, indicate 0-1% of the seed with embryos, probably due to number and size differences in chromosomes.

It is indicated that *P.ophioglossoides* is self-compatible.

The ultra-violet pattern mentioned may signify the location of the nectar. A large amount of pink flowering plants and shrubs in the bogs where the three orchid species were investigated are pollinated by bumblebees, and these seem to be preconditioned to respond to pink flowers.

Capsule set in *P.ophioglossoides* varied from 10-100%. It is supposed that the floral convergence between the three species may act to increase the efficiency of pollination, particularly since capsule set is low for many northern species of orchids.

More advanched orchids with a single fertile anther

SPIRANTHOIDEAE

TROPIDIEAE

A pantropical tribe of about 43 species in two genera.

Genus **Tropidia**

A small genus mainly from East Asia, with only one species in the American tropics and West Indies, ranging into southern Florida: *Tropidia polystachya*. Pollination unknown.

CRANICHIDEAE: GOODYERINAE

A widespread, mainly tropical subtribe from especially tropical Asia with 476 species in 35 genera in two tentative alliances.

Genus **Goodyera**

This genus has about 25 species world-wide, of which four species occur in North America. These are nectar-producing orchids, that in some respects resemble *Spiranthes* species with their spiral, mostly white flowers. Bumblebee pollination has been reported (Ackerman 1975, Kallunki 1976). Fruit set is usually high.

Barclay-Estrup et al. (1991) studied flowering sequences of mixed populations of three species in Ontario and found differences between north and south. However, consequences for reproductive biology are unknown as yet. Kallunki (1981) studied isolation mechanisms operating in mixed populations and concluded that hybridisation is kept to levels low enough to allow the species to maintain integrity (or identity).

Goodyera pubescens
Pollinated by long-tongued bees (Fig. 19), to which the pollinaria adhere on the mouthparts (Kallunki 1981, in Catling & Catling 1991).

Goodyera repens
This species is also known from most European countries.

Figure 19. Goodyera pubescens. Plant about natural size. 1 = enlarged flower, 2 = sectioned flower, 3 = column, 4 = pollinia, 5 = pollen tetrad, 6 = mature seed (drawing by B.Ames).

Morphology

Goodyera repens is closely related to *Spiranthes*. Some taxonomists put it in a separate subtribe, but it could also be incorporated into the Spiranthinae. Darwin related the species to *Epipactis*, because most of the characters that concern pollination resemble *Epipactis*. According to him *Goodyera* is interesting as a connecting link between several very distinct forms: 'In no other member of the Neotteae (note: Darwin used another classification) is there so near an approach to the formation of a true caudicle; and it is curious that in this genus alone the pollen-grains cohere in large packets, as in the Ophreae (=Ophrydae). If the nascent caudicles had been attached to the lower ends of the pollinia, and they are attached a little beneath their summits, the pollinia would have been almost identical with those of a true *Orchis*. In the rostellum being supported by sloping sides, which wither when the viscid disc is removed, in the existence of a membranous cup or *clinandrium* between the stigma and anther, and in some other respects, we have a clear affinity with *Spiranthes*. In the anther having a broad *filament* we see a relation to *Cephalanthera*. In the structure of the rostellum, with the exception of the sloping sides, and in the shape of the labellum, *Goodyera* resembles *Epipactis*. *Goodyera* probably shows us the state of the organs in a group of Orchids, now mostly extinct, but the parents of many living descendants.'

Pollinators
Darwin mentions that a Mr Thomson observed many bumblebees of the species *Bombus pratorum* in the north of Scotland, visiting the flowers with pollen-masses attached to their probosces. Summerhayes records: 'Nectar is secreted into the cup-like lower part of the lip, while the flowers are sweetly scented, no doubt to attract the few insects likely to be flying in a pine-wood. Bumble-bees are the visiting insects, several kinds of these having been seen carrying pollinia from the flowers. The mechanism is practically the same as in *Spiranthes*, the flower tube being at first only wide enough to allow the pollinia to be removed on the insect's proboscis. Later the aperture is widened by the downward movement of the lip, and pollinia may then be introduced so as to make contact with the receptive stigma. This is so very sticky that some of the small packets of pollen are broken off and left behind, this process being repeated in many flowers before all the pollen has been removed.' Van der Pijl and Dodson mention Müller's observation of *Bombus* on *Goodyera*, but state that he expressed doubt as to whether it was a legitimate pollinator. This doubt seems confirmed by observations of Vöth. Recently Vöth found no indications for the claimed autogamy (Hagerup 1952). In Austria he studied some populations and found *Bombus pascuorum* to be interested. By experiment he showed these bumblebees to be able to accidentally take away pollinia. *Lasioglossum morio*, however, is considered to be the legitimate pollinator, at least in Europe. This bee is of the right size and shows adequate pollinating behaviour.

Kallunki, however, found *Bombus perplexus* with pollinia on their probosces visiting *G.repens* var. *ophioides*. This variety does normally not self-fertilise and is not agamospermous.

Seed production
Summerhayes states: 'There seems no doubt that pollination is quite effective, as many ripe seed pods may be seen on wild plants. It is doubtful, however, if seed production is very important for the maintenance of the species in any given locality, this being probably more efficiently done by the extensive method of vegetative multiplication. Seeds probably serve mainly to enable additional suitable localities, such as plantations or young pinewoods, to be occupied by the species.'

Goodyera tesselata and **G.oblongifolia**
According to Kallunki (1976) the first species is of hybrid origin and is intermediate between *G.oblongifolia* and *G.repens* var. *ophioides*. It is an allotetraploid species, resulting from hybridisation, probably during early post-Pleistocene.

Kallunki observed bumblebees visiting the flowers. The species is normally not self-fertilised and not agamospermous.

G.oblongifolia is either *parthenocarpic* or self-compatible and infrequently self-pollinated. Ackerman (1975) found parthenocarpic fruits in one inflorescence of this species in an experiment with insect-exclusion and found also some poly-embryonic seeds, which phenomenon is also known from *G.tesselata* and *G.pubescens*. Ackerman observed pollination by *Bombus occidentalis*. He mentions observation by Kipping of pollination by *Bombus vosnesenskii* and visitation of *B.mixtus*.

Genus Erythrodes

A genus with about a hundred terrestrial and semi-epiphytic species. Only one species, *Erythrodes querceticola*, occurs in the south-eastern US, in Florida. The short rounded spur and the arrangement and colour of the flowers point to long-tongued bees or small Lepidoptera as pollinators. No data about pollination.

Genus Zeuxine

This genus is mainly found in tropical Asia. One species found its way to Florida, where it was introduced and became naturalised. Pollination data unknown. It shows agamospermy by *adventitious embryony* (Maheshwari 1952, in Catling & Catling (1991).

CRANICHIDEAE: PRESCOTTIINAE

A tropical American subtribe of about 100 species in seven genera of which one species is known from North America.

Genus Prescottia

Only one species of this tropical genus occurs in the US, in Florida: *Prescottia obligantha.* The mode of pollination is unknown. Catling mentions autogamy for *P.stachyodes* from Mexico.

CRANICHIDEAE: SPIRANTHINAE

A primarily American subtribe of over 400 species in 30-40 genera. The number of genera is disputed, Burns-Balogh recognises only 16. Revision seems needed.

Genus Cyclopogon

Cyclopogon cranichoides

This species (previously classified as *Spiranthes*) is confined to Florida. Calvo (1990) studied the reproduction of the species in a population in South Florida and marked flowering individuals. He measured leaf area, number of flowers, and number of fruits during four years. The number of plants with aboveground growth steadily decreased by about 20% each year. In some cases there was evidence that disappearance, especially of small plants, was due to transition in a subterranean state. Fruiting success was pollinator-limited with 26-62% in different years, as was evident from hand pollinations that produced 96%. Fruit set had no effect on growth or reproduction the following year. Pollinators were not observed, they are unknown. Thus, long-term fruiting success may also be pollinator-limited. The number of flowers is strongly size-dependent.

Genus Centrogenium (*syn.* Electroplectis)

This is a genus founded in 1920 by Schlechter. Originally there were 24 genera in Schlechter's classification of the Spiranthinae. In the latest edition of Schlechter's *Die Orchideen* (Part 1A, 1992) this is one of 31 genera within the Spiranthinae. Dressler has 28 genera. *Centrogenium* is a genus of about ten tropical American species. Only one occurs in North America, in southern Florida: *Centrogenium setaceum*. Pollinators unknown.

Genus Dichromanthes

Dichromanthes cinnabarinus, a species with vermilion flowers, is probably pollinated by hummingbirds (Catling & Catling 1991).

Genus Sacoila

The odourless(!) *Sacoila lanceolata* var. *lanceolata* has brick-red flowers and is probably pollinated by hummingbirds in parts of its range. The long slender viscidium is well adapted to the long, slender bill of hummingbirds, but unlike some other bird-pollinated orchids, the pollen masses have not adapted by becoming grey in colour to match the bird's bill colour. Agamospermy by adventitious embryony with poly-embryonic seeds is observed in Florida (Catling & Catling 1991a).

Genus Spiranthes

With about 27 species this genus of mainly terrestrial orchids of temperate regions of over 300 species throughout the world is abundantly represented in North America. The inflorescence has several to many spiral flowers ('ladies tresses').

The flowers, often white, greenish white, or creamy, are more or less tubular in appearance and commonly have a deep nectary united with the ovary, sometimes with a dominant chin or spur. Catling (1982, 1983) studied breeding systems and pollination of north-eastern species and Catling & Brown (1983) investigated morphometrics and ecological isolation in sympatric species in south-western Ontario. Most species are adapted to pollination by long-tongued bees, e.g. *Bombus* spp. and Megachilidae. The adaptation is attained by:

- A long, flat viscidium which attaches the pollinia readily to the flat galea of the pollinator's proboscis;
- Nectar secreted into the base of the floral tube;
- Protandrous flowers and sequential flowering beginning at the base.

Bees moving up the spike from the older female to the younger male flowers (Fig. 20) act initially as pollen donors and later as pollen receivers; in this way cross-fertilisation is promoted. Halictine bees occasionally act as pollinators by visiting the flowers upside down so that the pollinia are inconspicuously attached to the lower side of the *prementum.* Data available from Catling's study on 15 taxa fail to establish pollinator specificity as significant in speciation, except perhaps with respect to *S.lucida,* which is largely pollinated by halictine bees. The above diagnosis creates the problem of how sympatric species maintain species integrity and how they are are isolated.

To determine the extent to which three sympatric species (*S.cernua*, *S.ochroleuca*, *S.magnicamporum*) are distinct and isolated, Catling and Brown collected morphological and ecological data. The

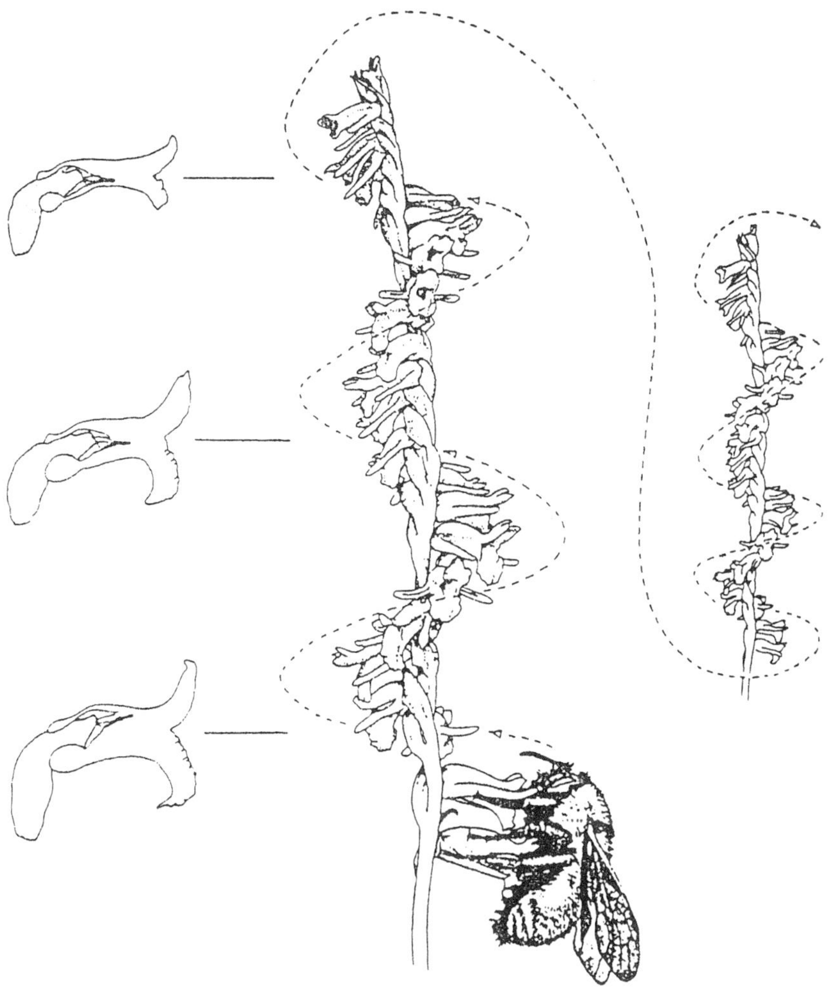

Figure 20. Diagram of pollination of *Spiranthes*. The direction of flying is indicated. The oldest flowers below are protandrous and functionally male (after Catling 1983).

morphological data consisted of 36 measured characters per plant (*morphometry*). Soil *pH*, flowering time and preference for moist sites (of one of the species) proved to be isolating factors. Other isolating mechanisms may also contribute.

Catling discusses previously recorded pollinators and visitors and presents a table. He also produces tables for pollinators and visitors observed in his investigation. In summary the data are shown in Table 3.

Protandry

Depending upon the developmental stage an inflorescence may be functionally male (early flower), female (late flower), or bisexual (peak flower, full anthesis), but when bisexual its older and lower, functionally female flowers are visited first since the bee pollinators move from the base upward. Self-pollination and geitonogamous pollination are consequently decreased and cross-pollination is promoted.

Various explanations for the *acropetal* movement (= upward from the base towards the apex) of bees in inflorescences have been put forward. Catling's observations indicate that there is often more nectar in older flowers at the base than in younger flowers at the top. If reward decreases upwardly in a spike, there is less waste of energy by moving upward. According to Corbet et al. (1981) bees simply move upward when the flowers are visited head up and downward when visited head down.

Darwin was the first to describe the mechanism of protandry in *Spiranthes*. The change in the position of the column is attained by movement of the column itself. In north-eastern species protandry involves movement of both lip and column, the lip turning more downward and the column moving back from the lip.

Pollination involving halictine bees was almost exclusively associated with *S.lucida*. Halictines occasionally pollinate other *Spiranthes* species, espe-

Table 3. Summary of data on pollination of north-eastern North American taxa of *Spiranthes* (after Catling 1983).

	Bombus pollination*	Halictine pollination
S.casei var. *casei*[1]	–	?
S.casei var. *novaescotiae*[1]	–	
S.cernua var. *cernua*[1]	**	
S.cernua var. *odorata*[1]	+[4]	
S.lacera var. *gracilis*	–	?
S.lacera var. *lacera*	**	?[5]
S.laciniata	–	
S.lucida		**
S.magnicamporum[1]	–	
S.ochroleuca[1]	–[6]	
S.ovalis[2]	–	?
S.parksii		
S.praecox	–	
S.romanzoffiana[3]	**	
S.tuberosa	–	?
S.vernalis	**[7]	

*Also some other Apidae and Megachilidae. ** = good data base; + = limited data base; – = extrapolation based upon comparative morphology; ? = possibility based on unknown floral morphology or relatively few records of unknown significance. [1]Taxa demonstrating agamospermy. The agamospermy is by adventitious embryony, the embryos (one or more) being formed by proliferation of the *micropylar* inner *integument* beginning at the time of the first division of the *megasporocyte*. Adventitious embryony in north-eastern *Spiranthes* taxa is autonomous but not necessarily obligate and can be readily detected. [2]Largely autogamous. [3]In the British Isles pollinated by bees. In British Columbia by long-tongued bees (*Bombus*, *Psithyrus*, *Apis*). [4]Pollinated by *Bombus americanorum* in Florida (Catling & Catling 1991). Synonym *S.odorata*. [5]Catling & Catling (1991a) mention smaller bees such as *Hoplitus* and *Megachile* as *co-pollinators* in southern Ontario, the pollinarium adhering to the mouthparts. [6]Catling & Catling (1991a) mention sexual reproduction in New York and Pennsylvania and pollination by *Bombus pennsylvanicus* in north-eastern US. [7]Pollinated by honey-bees in Georgia, the viscidium attaches the pollinarium to the mouthparts.

cially those with relatively small and widely open flowers.

Spiranthes casei
Catling observed pollination by a halictid bee *Dialictus versans* in Wisconsin. The halictine bee *D.versans* was also collected with the pollinia of *S.casei* var. *casei* in Southern Ontario, the viscidium attaching between the compound eyes.

Spiranthes cernua
Pollinated by bumblebees (Luer 1975, Catling 1983).

Antlfinger (1991) studied the population biology of this species near Lincoln, Nebraska during a five year period. 174 individuals were marked and censused. Depletion curves show a constant rate of loss in this long-lived perennial. The species may be dormant for one year or more. In five years about 20% died. Reproduction is size-dependent. On average, less than 30% flowers in a given year and individuals which flower are the most likely to flower also the next year. Reproductive effort was not decreased by previous production. As a facultative apomictic species, the cost of reproduction may be worth paying, because asexual (agamospermic) seed set is high even if pollinator visitation (see Table 3) is low. In sexual reproduction (19.2-33.6%) the seeds are mono-embryonic, but in asexual reproduction the seeds are poly-embryonic if not extruded (Schmidt & Antlfinger 1989). Gene flow from related allopatric species adapted to differing habitats and climates has given rise to great morphological variability in this species (Sheviak 1990).

Spiranthes lucida
Pollinated by halictid bees (Catling 1983). It is unique in having nectar on the roof of the floral tube, on the undersurface of the column, and has a pollinarium with an oval viscidium that attaches to the clypeus below the antennae, i.e. between the eyes of Halictid and Andrenid bee pollinators that have short mouthparts. The flowers, moreover, are not protandrous. Pollinia attached to the chin. The flowers are approached upside down. The halictine bee *Augochlorella striata* behaved differently from the *Bombus* pollinators in other *Spiranthes* species. They crawl much less on the spike, are not systematic, and do not visit all of the open flowers in a spike. Visits to a second flower on the same spike

always involved flight. The column and short claw of the lip of *S.lucida* seem to be adapted to the mouthparts of halictines.

Spiranthes magnicamporum
Hapeman (Internet 1998) observed a *Bombus nevadensis* ssp. *americanorum* pollinating this species in a prairie in south-eastern Wisconsin. After visiting a number of inflorescences, the bee vigorously tried to scratch at the pollinia on its proboscis. In doing so it fell to the ground and was caught.

Spiranthes ovalis var. **erostellata**
This autogamous variety from the eastern US was described by Catling in 1983. The absence of a rostellum results in self-pollination.

Spiranthes romanzoffiana
Larson & Larson (1987, 1990) observed five species of pollinating bees in this species at three sites in Canada. These were *Bombus terricola occidentalis*, *B.bifarius*, *B.flavifrons*, *Psithyrus insularis* and *Apis mellifera*. Hitherto only the first bumblebee was a known pollinator (Catling 1983). Pollinators varied from site to site. This phenomenon might be more usual in other orchids as well (e.g. this is also known from the European *Epipactis palustris*, see Van der Cingel 1995) and since most investigations have been limited to one population this can easily give a distorted picture. *Bombus bifarius* was the most abundant pollinator, visiting an average of 68 plants per hour at an isolated patch of 12 orchids. This high revisitation rate is unusual for orchids, suggesting that nectar production is continuous and/or copious. Continual rewards may ensure pollination of all flowers, increase outcrossing, maintain bee 'loyalty', and ensure pollination under non-optimum conditions.

In a 1990 study on the basis of the same observations of 1983 Larson and Larson conclude that the bees primarily visited the most conspicuous plants with the tallest spikes and longest inflorescences. Bee visits were less influenced by the number of open flowers and the distances between neighbouring plants. The bees performed non-repeated foraging sequences 80% of the time, yet intensely revisited four of the most 'conspicuous' plants.

The viscidium is attached to the tongue.

Spiranthes diluvialis
This rare orchid from Utah, Ute lady's tresses, not mentioned by Williams and Williams, and first described by Sheviak in 1984, grows in moist, alluvial soils, usually on primary and secondary flood plains of perennial rivers and streams (Sipes & Tepedino 1995). It is thought to have originated as the *allopolyploid* offspring of a cross between *S.romanzoffiana* and *S.magnicamporum*. It exhibits a mixed-mating system in which the degree of selfing depends in part upon the abundance of pollinators visiting the flowers. Outcrossing is promoted by protandry and by the acropetal way of movement of long-tongued bees on the inflorescences. The self-compatible flowers depend on a pollen vector for reproduction, since no autogamy or agamospermy is observed. Bumblebees are the most important pollinators, especially the large long-tongued *Bombus morrisoni* and *B.fervidus*. Long-tongued *Anthophora* bees also carry pollinia, but their importance is not clear. It is the first time that anthophorid bees have been reported as pollinators of *Spiranthes* species. Furthermore a tiphiid wasp *Myzinum* was seen to carry pollinia.

Within flowers, male and female functions overlap after about two days when the stigma is exposed if pollinia have not already been removed. It is supposed that when visitor rates are low, more flowers reach the female phase with the pollinarium still in place. When flowers reach the female phase with pollinia intact, the outcrossing mechanism of protandry and acropetallic flowering will break down, and pollinators will more frequently carry out self-pollination or geitonogamous pollination.

CRANICHIDEAE: CRANICHIDINAE

A tropical American subtribe of 152 species in nine genera, of which a few species extend into temperate America.

Genus Cranichis

Another tropical genus with one representative in Florida *C.muscosa*, which is reported to be autogamous in Puerto Rico by Ackerman (Catling 1990).

Genus Ponthieva

A group of both terrestrial and epiphytic orchids from Central and South America, represented in Florida by *Ponthieva racemosa*, a species growing in coniferous woodland on limestone.

Pollination: The species is reportedly visited by halictid bees (Luer), but Dressler found that the lip produces oil, rather than nectar, so that the pollinators are thought to be oil-gathering anthophorid bees.

ORCHIDOIDEAE

ORCHIDEAE: ORCHIDINAE

A primarily northern hemisphere subtribe of about 370 species (some splitters recognise considerably more species) in 34 genera in four tentative alliances. Most of the species occur in Europe. The genus *Platanthera* is widespread in North America.

Orchis and related genera
No data were found about the pollination of the single species of the genus *Dactylorhiza* (*D.aristata*).

Amerorchis rotundifolia (syn. *Orchis rotundifolia*), a species from open tundra, limestone barrens and (in the southern part of its area) coniferous forests is a high northern species, distributed from Greenland and Alaska to Montana and New York. The flowers are nectar-free and scentless. There are no observations on the pollination of the species and in most localities investigated there are no signs of fructification, the species mostly reproducing vegetatively.

Genus Coeloglossum

As in Europe, this genus is represented with one (and the same) widespread species: *C.viride*. From European observations it is known that this species excretes nectar in a short spur from which it flows over the lip (Van der Pijl & Dodson 1966). According to Füller (1980) the column is small and the spur entrance so narrow that Lepidoptera are likely candidates as pollinators. There are however no records of this and Hagerup (1952) mentions that the entrance to the spur is closed by a thin membrane which can easily be penetrated by insects with strong, biting mandibles but scarcely by butterflies and rather small Diptera, and in fact ichneumoid wasps have mainly been observed. Peitz mentioned (in Füller) Ichneumoids, Apidae and *Cantharis* (Coleoptera). Godfery (in Van der Pijl & Dodson) observed *Tenthredopsis* species – (wasp), Silén saw in Finland *Cryptus* (wasp), *Tipula* (mosquito) and *Cantharis* ssp. (Coleoptera). Hagerup supposes autogamy for var. *viride* of the Faroes.

Nilsson (1981) remarks that, in comparison with *Listera ovata*, in *C.viride* no nectar trail has evolved on the labellum, although it is pollinated by similar insects. As a reason for this he mentioned the determining presence of viscidia instead of an explosive rostellum.

Genus Galearis

Galearis spectabilis (syn. *Orchis spectabilis*), the only representative of the genus may, according to Dieringer (1982) be pollinated by long-tongued bees. *Bombus* queens of several species (*B.vagans*, *B.fervidus*, *B.nevadus auricomus*, *B.pennsylvanicus*) are reported as pollinators for Ohio, the pollinaria attaching to the tongue or frons below the eyes. Robertson (1924) already reported long-tongued bees (*Bombus griseocollis* and *B.pennsylvanicus*) as visitors. Dieringer observed pollinia removal by *B.vagans* only. Auclair (1972) mentions *Bombus separatus* and *B.americanorum*, following Van der Pijl & Dodson, who refer to Robertson (1928). Auclair supposed the pollinators to be largely non-specific, but the facts found hitherto point in the direction of long-tongued *Bombus* species only.

According to Dieringer self-pollination can take place when a bumblebee enters, removes pollinia and re-enters the same flower. But self-compatibility is low (4%). Fruit production varied from 9-11% in Dieringer's investigations, whereas caged flowers did not fruit.

Genus Leuchorchis

Leucorchis albida var. **straminea**
The nominate form *Leuchorcis albida* (synonyms a.o.: *Habenaria albida*, *Platanthera albida*, *Pseudorchis albida*) is a European species from the northern mountainous parts of the mainland. The species has white flowers, is nectariferous and possesses a sweet scent. In Europe it is probably insect pollinated, a small tineid moth being observed by Fritsch (Van der Pijl & Dodson 1966), whereas Godfery (same source) saw visiting solitary bees (who did not remove pollinia) and Darwin noticed high quotes of pollinium removal. Fruit set may approach 92.5%, which is obviously due to spontaneous self-pollination (Füller 1978). The narrow spur entrance points in the direction of diurnal butterflies, according to Füller.

In North America the variety *straminea* is known

from southern Greenland and Newfoundland. It has yellowish green to yellowish white flowers and is said to have a still stronger smell than ssp. *albida*, although in Greenland not as persistent as in *Platanthera hyperborea* (Löjtnant & Jacobsen 1976). Here it lives under circumstances where self-pollination is to be expected as the rule, because insects are scarce. In Greenland visiting insects have not been observed but the fructification is abundant and is supposed to be *cleistogamous*, although Harmsen did not find any sign of self-pollination.

Genus **Platanthera**

Williams and Williams mention 22 species. However, since they published their guide at least one new species has been described. Patt et al. (1989) speak of about 30 species. This difference in numbers may be due to different opinions about the status of taxa.

The pollinators are Lepidopterans and the flowers are phalaenophilous, sphingophilous or psychophilous. White or greenish flowers are pollinated by nocturnal or crepiscular moths, whereas purple or yellow-orange flowers are pollinated by diurnal Lepidopterans. Several of these also have fringed labellae, which seems to be an adaptation to visually oriented visitors. Hymenoptera like bumblebees and Diptera like Empididae may also pollinate some species, mostly as co-pollinators.

Some species have a distribution in Boreal areas and are dependent on mosquitoes for pollination and/or are self-pollinating. Since the morphology of the heads of mosquitoes, as well as their habits, are similar to those of small moths, a shift has occurred towards mosquito pollination in the evolution of these species of *Platanthera*. They all have small greenish flowers. Lepidoptera and mosquitoes are scaly or hairy insects and expose small surfaces for effective attachment of viscidia. Moths and butterflies probe the floral spur for nectar while resting on the labellum. In this position the sticky viscidia may contact the eyes of the insect or the proboscis. This has led to two types of column, corresponding to the two available surfaces. One column is rather large with widely separated viscidia and long caudicled pollinia, the other is narrower and small with close viscidia and short caudicled pollinia. These types may coexist in sympatric species that share pollinators, as in the European species *P.chlorantha* and *P.bifolia*.

For atttachment to compound eyes the two pairs of pollen masses are separate (*hemipollinaria*) and widely divergent with two separate viscidia whereas for attachment to a proboscis they are either separate or a single unit. When hemipollinaria are deposited on the probosces, as in *P.leucophaea* and *P.psycodes*, a wide range of proboscis lengths can be exploited because pollinaria are deposited on the proboscis in the correct position for subsequent stigmatic contact. When pollinaria are deposited on the eyes, fewer species can act as pollinators because a specific proboscis length and distance across the eyes is needed (Catling & Catling 1991a).

The pollinaria are removed from the anther as the insect withdraws. Once attached, pollinia rotate downward and inward, generally within 60 seconds as one side of each caudicle dries, until they are positioned directly in front of the head. The *massulae* are brushed over the stigmatic surface at a next visit, thus effecting pollination. There are some indications that species pollinated by larger Lepidoptera have more restricted distributions than similar species pollinated by smaller Lepidoptera (Catling & Catling 1991a). Pollination by larger butterflies and moths is most prevalent in the more southern regions, whereas pollination by smaller insects is most prevalent in arctic and alpine regions.

Catling & Catling (1991a) also found indications for sexual selection in *Platanthera* species (see below, *P.dilatata*).

Platanthera blephariglottis

Cole & Firmage (1984) studied the floral ecology of this species in central Maine during 1979-1981. This orchid is found throughout the eastern US and grows in damp, acidic substrates or in sphagnum bogs. The pollination mechanism and some pollinators have been reported by Smith & Snow (1976).

The flowers are white to cream-coloured and have nectar-containing spurs. Nectar guides and UV-reflectance patterns are absent. They are weakly fragrant during day and night. Nectar concentration is extremely low: 13.8%.

The flowering period averages 38 days at the end of the flowering season. Inflorescences bloom sequentially and each flower remains open 14 days on average. They are receptive to pollen for up to ten days and the pollen remains viable for at least five days.

Pollinators and visitors

Cole and Firmage observed Lepidoptera, Diptera, Coleoptera and Hymenoptera as visitors. Insects with pollinaria attached to their bodies: see Table 4.

Table 4. Insects observed with pollinaria of Platanthera blephariglottis (Cole & Firmage 1984).

Lepidoptera:	Sphingidae	*Hemaris* sp.
	Hesperiidae	*Epargyreus clarus, Polites coras, P.mystic, Euphyes ruricola*
	Lycaenidae	*Strymon melinus*
	Pieridae	*Colias philodice, Artogeia rapae*
Hymenoptera:	Apidae	*Bombus vagans, B.fervidus*

Smith and Snow observed, apart from moths, which they say are the main pollinators, and some bumblebees, the butterflies *Papilio troilus* and *Danaus plexippus* with pollinaria on their heads, which species were classified by Cole and Firmage as visitors only. Smith and Snow also found the honeybee to be a probable pollinator. Catling & Catling (1991a) observed pollination by *Danaus plexippus* in Ontario, and note that the pollinaria attach to the compound eyes.

Butterflies are more successful at extracting nectar from *P.blephariglottis* than bumblebees. Some individuals of *Epargyreus clarus* appeared to establish 'territories' near large groupings of flowering orchids. These individuals would dart at and chase other butterflies entering the area.

Pollination

Generally pollinators remained on individual inflorescences longer than needed for pollinaria to complete movement and be thus involved in geitonogamy. This phenomenon however occurred in two ways: a vector carrying pollinia returned to the same inflorescence after a previous visit or remained long enough to have the pollinia rotated before leaving the inflorescence. Xenogamy is promoted by the pollination mechanism, but geitonogamy and autogamy no doubt occur. Experiments by Smith and Snow proved that the species is self-compatible.

Pollinator limitation and fruit set

Fruit set and fruit production did not seem to be limited by availability of resources. The reproductive success was constrained during the study of Cole and Firmage by scarcity of pollinators. Average capsule set varied from 32.2% in 1979 to 80.2% in 1980. Artificial pollination led to increased fruit set. There was no difference in capsule set between orchids growing in open and semi-open habitats. Olfactory as well as visual cues may play a role.

Flower colour, long receptivity, an extended flowering period and survival over several growing seasons help ensure that individual plants of *P.blephariglottis* reproduce successfully even with extended periods of unfavourable weather or low pollinator activity.

An 11 year study of a Maine population (Firmage 1990) has shown that the average plant flowers approximately 40% of the year it lives. Each plant produces a new ramet each summer. Plants flowering the next year show a positive relationship between ramet weight and the number of flowers produced. A negative relationship exists between the weight of fruits produced and the size of the new ramet. Percent fruit-set is highest for plants found near the shrub fringe but not hidden by shrubs and trees, a result of both pollinator availability and inflorescence predation. Flowering frequency is affected by different factors.

Platanthera chapmanii

Pollinated by long-tongued butterflies, especially swallowtails, the pollinaria attaching to the proboscis (Folsom 1984). The species is intermediate to *P.cristata* and *P.ciliaris* and has for years been considered the naturally occurring hybrid of these two species. It must however be regarded as a separate species, differing in morphology from *P.* x *channellii*, the putative hybrid between the two species mentioned (Folsom 1985b).

Platanthera ciliaris

This is a species of wet open meadows, pine savannahs, moist clearings, seepage slopes or forest and road edges.

Robertson & Wyatt (1985, 1990a, 1990b) studied different populations during 1983 and 1984 in respectively a mountainous and a coastal-plain area of this widespread eastern species. The species is not autogamous, and is pollinated by large butterflies. They found evidence for different pollination ecotypes. In the mountains, *Papilio troilus* was the most frequent and effective visitor and pollinator, whereas in the coastal plain, *P.palamedes* was the predominant pollinator (Plate 27, p. 29). In the mountains *P.glaucus*, *Battus philenor* and rarely, *Phoebis sennae* also pollinated. Coastal plain pollinators included also *Ph.sennae* and, rarely, *P.troilus*.

Smith & Snow (1976) also found *P.troilus* as by

far the most frequent visitor in Booth Lake Bog, Berrien County, Michigan. They found that wherever *P.blephariglottis* and *P.ciliaris* grow together, hybrids may occur and estimated the amount of hybrids in the study area as about 3%. Case saw in southwestern Michigan the hybrids outnumber the parents.

The lengths of the probosces of these butterflies differ, *P.troilus* having a proboscis that is on average more than 5 mm shorter. Correspondingly, the spur lengths of the mountainous and coastal-plain yellow-fringed orchids differ by about 2 mm. The rates of removal and insertion of pollinia were higher in the mountains (over 80%) than in the coastal plain (ca 65%). Fruit set was significantly greater in mountain (ca 85%) than in coastal-plain (ca 65%) populations. The authors suppose that *selection pressure* exerted through pollinator proboscis length has resulted in pollination ecotypes of the species. The shorter spurred mountain plants appear to be in equilibrium with their short-tongued butterfly pollinators, receiving high levels of effective pollination and achieving high levels of fruit set. The plants from the coastal plain, however, have flowers with longer spurs that are sub-optimal in their adaptation to their visitors and pollinators: very long-tongued butterflies do not always make contact with the pollinia. Mean spur length in the coastal plain was approximately 3 mm shorter than the length of the average pollinator proboscis. Therefore effective pollinator service and fruit-set are reduced. Long-tongued butterflies may 'thieve' nectar without rendering the service of pollination. Reciprocal-transplant studies, however, did not yield conclusive evidence for a genetic basis for the difference in spur lengths, although the phenotypic plasticity as measured could not account for the whole difference.

The viscidium of the orchids adheres to the compound eye of the pollinators.

The nectar concentrations of *P.ciliaris* are relatively low, compared to other butterfly-pollinated plants, ca 23% in the mountains and ca 19% for the coastal plain, compared to ca 29% on average for 50 investigated butterfly-pollinated plants.

Papilio troilus is a mimic of *Battus philenor*, which is also observed as a scarce pollinator in the mountain area. *Phoebis sennae* was seen in both mountains (but here only once) and coastal plain, carrying a single pollinium or (rarely) few pollinia respectively. *Xylophanes tersa*, a sphinx moth, is a visitor after dusk, probing nectar, but was never observed to carry pollinia. Its proboscis measures 32 mm and is obviously to long for pollination purposes in the case of *P.ciliaris*.

P.ciliaris is self-compatible, but several mechanisms promote cross-pollination. Fruit-set appears to be pollinator-limited, because hand pollination results in higher levels of fruit-set than the natural levels measured.

Platanthera cristata

Pollinated primarily by *Bombus pennsylvanicus*, the pollinaria attaching to the head (Folsom 1984).

Platanthera flava

P.flava var. *herbiola*, formerly *Habenaria flava*, is pollinated by mosquitoes and butterflies, the pollinaria attaching to their probosces (Stoutamire 1971, Catling & Catling 1991a). This species of periodically flooded woodlands in the eastern US and Canada grows in muddy depressions in deciduous forest areas with large mosquito populations. The flowers smell sweet and have spurs of 5-7 mm. long. The labellum bears a prominent callus in front of the nectary opening which partially obstructs the centre of the flower. There is also an obstructing callus projecting downward from the roof of the nectary. This creates a double nectary entrance with clearly defined canals under each viscidium. The *Aedes* mosquitoes are forced by the obstructions to insert the proboscis into one of the openings. The pollinarium attaches to the shaft rather than to the centre of the compound eye. Mosquitoes carrying pollinaria visit other flowers leaving massulae on the stigma of each flower visited.

Marilyn Light (1998) mentions the pyralid moth *Anageshna primordialis* as pollinator, receiving pollinia at the base of the proboscis. She collected an inflorescence of *P.flava* with 15 dead moths, males and females, entangled in a spider's web.

Platanthera grandiflora and **P.psycodes**

These two species of purple fringed orchids have been the subjects of discussions about their status since 1913 and until 1974. They are eastern North American species, ranging from Newfoundland to the Appalachians. *P.psycodes* has a much wider westward distribution (Iowa). Wet meadows, swamps and seepage areas are their habitats.

They differ in morphology (Plates 28 and 29, p. 68) and are seasonally, mechanically and spatially isolated. *P.psycodes* is smaller-flowered and has a sweeter scent than *P.grandiflora*, and it flowers later in a given area, although blooming periods overlap. The column with two pollinia of ca 2 mm, their viscidia separated by a space of 1-1.5 mm, partially enclosed by projecting column wings. The nectary opening is oblong and a projection from the roof produces two lateral openings, one beneath each viscidium. *P.grandiflora* has pollinia of 3 mm long, laterally unconcealed viscidia and a funnel-shaped nectary opening. It is reported to be visited by hawkmoths, which remove pollinaria on the compound eyes, whereas *P.psycodes* is visited by smaller butterflies (Moldenke, in Stoutamire 1974). Guignard (Stoutamire) mentions observations of Smith in which the diurnal hawkmoths *Haemmorhagis thysbe*, *H.difformis* and the swallowtail *Papilio asterias polyxenes* removed *P.psycodes* pollinaria, which were attached near the base of the proboscis. Stoutamire himself observed pollination of *P.psycodes* by the skipper *Polites mystic* (Plate 30, p. 68) and the hawkmoth *Haemmorhagis thysbe*. One specimen of the latter carried pollinaria of both *P.psycodes* and *P.lacera*. Sky Feller (Hapeman, Internet 1998) has recently observed the same group of pollinators.

Here again we meet a closely related, partly sympatric species pair of which one has a pollination mechanism adapted to medium- to large-sized Lepidoptera that attach pollinaria to the compound eyes and the other to the same class of diurnal lepidopteran pollinators that attach pollinaria to the proboscis. Nilsson describes extensively how hybridisation is prevented by the differing mechanisms for the European sympatric species pair *P.bifolia* and *P.chlorantha* that share pollinators. Although hybrids and hybrid swarms are not rare in these species, they are isolated from the parent species by mechanical factors. In hybrids the viscidia will often neither attach to the proboscis of noctuid moths, nor to the eyes of sphingids. Moreover hybrids tend to disappear by introgression. In this way both species of a pair maintain species integrity.

Platanthera integrilabia

This species is adapted to sphingid moth pollination and, according to Zettler & Fairley (1990), is pollinator-limited with a very low fruit set of 7% in Georgia (Pleasants & Moe 1993) to 20% in South Carolina. Zettler et al. (1996) however recently observed three day-flying Lepidoptera (*Epargyreus clarus*, *Papilio glaucus* and *P.troilus*), carrying pollinia of this orchid on their compound eyes and these obviously acted as pollinators. Fruit set was over 50% in the population investigated (McMinn County, Tennessee). They did not document nocturnal pollinations during several years of investigations, but believe at least 10 sphingid taxa to be better suited for the removal of both pollinia based on range of eye width, geographical abundance, seasonal flight time and available larval food sources. The fragrance contains over 50% of linalool.

The species is now seriously endangered.

Platanthera lacera and P.peramoena

These species are closely related to the two species pairs mentioned above. The six species form a natural group with distinct labellum morphology from other *Platanthera* species. Stoutamire has proposed two evolutionary lines, one marked by eye-deposited pollinaria and the other by deposition on the proboscis. Sheviak and Bowles, however, suggest that the similar differences between the species of the pairs are the products of selection in response to similar pressures. This parallel development seems more logical, since the same splitting up of one species into two on the basis of different mechanisms for pollinarium attachment seems to have taken place independently in Europe and Japan. It could have arisen from isolation by ecological factors like diurnal and nocturnal pollinators.

P.peramoena, a purple, but fringe-less orchid, and a mid-eastern species, appears, according to Stoutamire (1974) (but see Hapeman 1997), to be adapted to diurnal Lepidoptera, with pollinium attachment to the eyes. *P.lacera*, a north-eastern species with creamy green fringed flowers has proboscis-deposition of pollinaria and seems adapted to crepuscular-nocturnal Lepidoptera. *P.lacera* shows the closest morphological relationship with *P.psycodes* in terms of column structure.

P.peramoena and *P.lacera* should, for the rest, not be considered a species pair. *P.peramoena* is reportedly pollinated by *Papilio glaucus* (but see Hapeman, below), the hemipollinaria attaching to the compound eye, *P.lacera* by a diurnal sphinx moth (see above), the pollinarium becoming attached to the proboscis.

Hapeman (1997) studied the pollination of *P.peramoena* in Pennsylvania. The primary pollinator (with 15 visits and carrying several pollinia) observed was *Hemaris* (=*Haemmorhagis*) *thysbe* (Sphingidae). It seems that the orchid has adapted to this sphingid moth because it shows a morphological correspondence to the flowers. Mean width of the compound eyes is identical to the distance between the viscidia and the mean length of the spur was slightly shorter than the length of the spur. Other Lepidoptera observed as (potential) pollinators were: *Danaus plexippus* (Nymphalidae), *Papilio troilus* (Papilionidae), *Hemaris diffinis* (Sphingidae). One individual of each of these species was observed bearing pollinia. As visitors some other Lepidoptera were seen. *Hyles lineata* served as a nectar thief with a longer proboscis than needed to remove pollinia.

Rotation of the pollinia after removal takes 30 seconds.

Homoya mentions *Papilio troilus* as pollinator in an Illinois population.

Platanthera leucophaea and **P.praeclara**

The western and eastern prairie fringed orchids form a species pair, only recognised as such since 1986. Sheviak and Bowles had studied populations of the species *P.leucophaea* and concluded that this species ought to be split into two species with different pollination mechanisms, morphology, and geographic distribution.

Distribution is roughly in and around the lake states in habitats like damp meadows, lake margins, sphagnum bogs.

P.praeclara is very similar to *P.leucophaea* in gross morphology, but differs in a number of floral characters. Most of these reflect the larger size of the flowers of the former species. Other characters include flower colour (in *P.praeclara* lip and petals are creamy white with faint greenish sepals, in *P.leucophaea* the flowers are markedly whiter), floral fragrance (in *P.leucophea* more spicy), petal shape and sepal size. Flowers are generally fewer and the inflorescence shorter in *P.praeclara.*

The plants are sphingophilous and fragrant at night, deeply fringed and with extremely long nectariferous spurs. The column structures are different, dictating different pollination mechanisms. Pollinaria placement is limited to either the proboscis or the eyes of the pollinators. Robertson identified the Sphingidae *Xylophanes tersa* and *Eumorpha achemon* as pollinators of *P.leucophaea.* The pollinaria are deposited on the proboscis and usually were removed one at a time (hemipollinaria), as the viscidia are only 2 mm apart and the moths direct the proboscis to one side or the other. Pollination in *P.praeclara* has not been reported, but it is probably pollinated by long-tongued sphinx moths.

Sheviak and Bowles collected *Sphinx eremites* at *P.leucophaea* with one pollinarium at the base of the proboscis. Also *Manduca sexta* may act as pollinator. Pollinaria of *P.leucophaea* bend forward after being picked up, assuming a position nearly parallel to the proboscis, and with the pollinia only slightly elevated. This position allows insertion between the closely-spaced rostellum lobes as is necessary to contact the stigma.

Those of *P.praeclara* rotate laterally, either to the left or to the right, assuming a central position above and slightly forward of the moth's head, probably on the moth's eye. This position permits contact with the stigma without interference from the wide-spread rostellum lobes. On the basis of an analysis of the length of mouth parts and distance of the eyes of sphinx species of the prairie region, Sheviak and Bowles selected four *potential pollinators* and one potential nectar 'thief'.

Platanthera media

Pollinated by butterflies (*Erebia epipsodea, Vanessa virginiensis*), moths (*Trichodestra dodii, Aletia oxygala, Cucullia intermedia* a.o) and bees (*Bombus appositus, B.flavifrons, B.occidentalis, Psithyrus suckleyi, P.insularis*) in Colorado, the viscidium attaching to the tongue (Catling & Catling 1989, sub *P.huronensis*). See also below about the upheld lip and hemipollinaria removal. Syrphidae are observed as visitors.

Platanthera obtusata

P.obtusata is pollinated by several species of small- and medium-sized moths and mosquitoes (Plate 31, p. 68) (Stoutamire 1968, 1974, Voss & Riefner 1983, and others). Thien (1969) observed that especially the species *Aedes communis* and *A.canadensis* carried pollinia of *P.obtusata* in one Wisconsin population, whereas other mosquitoes did not and were only visitors. In another population in Manitoba, Twinn et al. (1948) and others observed mosquitoes visiting at sundown and here the percentage of females of several *Aedes* species (*A.nigripes, A.punctor, A.excrucians, A.cinereus*) increased progressively with the number of days from the first flowering. The pollinia were found attached to the eyes. Stoutamire (1968) states that due to the lateral approach to the nectary only one pollinium is removed per visit (see also above, *P.dilatata*). Dexter found mosquitoes carrying three or four pollinia. Gorham (1976) observed not less than 16 species of *Aedes* and two species of moths with pollinia of *P.obtusata.*

The mechanism attracting insects is not known. According to Raup (1930) the flowers scent, but Stoutamire (1986) could not detect odour. Stoutamire notes that mosquitoes orient themselves in relation to CO_2-sources. Carbon dioxide is one of the respiration products of plants. Not all species of mosquitoes seem capable of acquiring nectar and pollinating, and of all the species involved in the pollination only a few species do most of it in natural populations.

Platanthera orbiculata

Reported to be pollinated by a large moth, *Sphinx drupiferarum* (Van der Pijl & Dodson 1966).

Platanthera sparsiflora

The var. *sparsiflora* of this species is pollinated by a moth with the pollinia attaching to the proboscis (Kipping 1971).

Platanthera stricta and related species

Patt et al. (1985, 1989) studied the pollination biology of *P.stricta* (Plate 32, p. 68). The slender bog orchid is distributed from the Aleutian Islands and southern Alaska to northern California and eastward to Colorado. It is locally common in mountains in wet meadows, seepages and margins of streams. The flowers are fragrant and attract a wide array of insects. It is pollinated by a diverse assemblage of short-tongued insects. Main pollinators are *Eustroma fasciata* (Lepidoptera: Geometridae) (pollinaria attach to the compound eye), *Bombus flavifrons* and *B. melanopygus* (Hymenoptera: Apidae) (pollinaria attach near the base of the tongue), an undescribed species of *Greya* (Lepidoptera: Prodixidae), and several species of *Empis*, *Rhamphomyia*, and *Anthepiscopus longipalpus* (Diptera: Empididae) (in both classes of pollinators the pollinaria attach to compound eyes or probosces). Patt et al. (1985, 1989) present a long list of insect visitors and casual pollinators and some infrequent co-pollinators. These co-pollinators vary greatly in body size and foraging behaviour, but share the possession of relatively short mouthparts: viz. 0.5-5.0 mm long. The morphology and arrangement of the flowers especially suit bumblebees and short-tongued settling moths. The floral spur length is similar to the length of the insect's mouthparts. However, the relatively distant spatial arrangement of the flowers, together with the concealment of the nectar within a spur seems incongruous with that of flowers visited by empidids and microlepidoptera. Both these classes of small pollinators seem to prefer plants with closely-spaced flowers that have exposed nectaries. The extrafloral glucose present on the raceme may be an adaptation to these pollinators, retaining them on the inflorescence until they locate the floral spur entrance.

P.stricta possesses characteristics that facilitate pollination by insects with disparate morphologies and floral foraging behaviours, e.g. fragrance attractive to empidids, a sparkling and shimmering appearance of the inflorescence attracting Diptera and moths, a mixture of similar proportions of sucrose, fructose and glucose, rewarding to a broad range of pollinators, and an architecture of the pollination chamber that permits efficient pollinaria deposition and pollen removal to occur with the various pollinators.

The floral fragrance consists largely of monoterpene aldehydes and alcohols, monoterpene hydrocarbons and aromatic aldehydes and alcohols (Patt et al. 1988), as such a common blend in this genus and in most orchids (Kaiser 1993), specificity usually being attained by qualitative and quantitative differences.

It is reasonable to suppose that the cool weather conditions, which make pollinators inactive, have led to long-lived flowers. The blooming period of a raceme may last ca 35 days and single flowers are receptive during 14 days. The ability to produce seed via facultative autogamy and geitonogamy enables this species to reproduce after a single pollinator visit.

Capsule set, being 52% in the population studied, is low compared to hand-pollinated orchids (90%) and therefore capsule set in *P.stricta* seems to be pollinator-limited. Since a large proportion of the seeds appears to be sterile, seed-set seems pollen-limited as well, the amount of the pollen massulae deposited on the stigma being too low for full seed-set.

P.stricta seems to have evolved from a normally Lepidoptera-pollinated *Platanthera* species towards a species that is pollinated by a diverse group of short-tongued anthophilous insects as a consequence of adaptation under conditions of unpredictable weather and pollinator availability.

P.dilatata (Plate 33, p. 69), *P.limosa*, and *P.sparsiflora* are members of a complex (section *Limnorchis*), which is one of the taxonomically most difficult ones within *Platanthera* and which is closely related to *P.stricta* and *P.hyperborea*. They hybridise, and intermediate forms are found regularly.

P.dilatata in particular readily hybridises with *P.hyperborea*.

Patt & Merchant (J.M. Patt, pers. comm. 1989) observed butterflies visiting the inflorescences of *P.dilatata* in the sub-alpine zone of the Baily Range in the central portion of the Olympic National Park. The orchid produces white flowers with long floral spurs and emits a spicy-sweet odour. Vogt (1990) observed visitation by *Papilio glaucus* on *P.dilatata* in Woodstock, Vermont.

P.dilatata var. *dilatata* is pollinated by noctuid moths at St Anthony in Newfoundland, Bruce Co, Ontario and in Oregon. In the Canadian Rockies the most important pollinators appear to be large butterflies, including *Papilio glaucus canadensis*, *P.gothica* and *Vanessa cardui* (Catling & Catling 1991).

In Oregon Larson (1992) observed the noctuid moth *Discestra oregonica* as pollinator. The moths selected racemes at random, using visual cues. They probed few-to-many flowers before moving. The large size of the moths caused them to grasp several adjacent *perianths*. In probing, the moths used a variety of orientations. Pollinia were attached to the dorsal side of the proboscis. From the description of probing behaviour it follows that probably *D.oregonica* are not very efficient as pollinators. Depth of insertion varied by millimeters. Most of the flowers were unpollinated, as judged by the non-recurved position of the lip.

P.dilatata var. *leucostachys* is pollinated by noctuid moths, the pollinaria attaching to the proboscis (Kipping 1971, in Catling & Catling 1991a).

Catling & Catling (1991a) collected data on removal of hemipollinaria in *P.dilatata*. The upcurved lip in this and closely related species forces the pollinator to approach and probe the flowers from a side in order to obtain the maximum of nectar. Especially in young flowers with the lip held up the percentage of removed single hemipollinaria is significantly larger than that of both hemipollinaria removed. Only older flowers have proportionally more double removals. This means that older flowers probably have loaded two different pollinators. This may improve the chances of cross-pollination, but may also be explained in terms of sexual selection, paternal selection increasing male fitness.

The same upheld lip is found in *P.hyperborea* (but see below about autogamy) and *P.media*.

Comparable features that restrict pollen acquisition by pollinators possibly could be the tubercle on the lip of *P.flava* and the projection of the column in *P.hookeri*, together with the upcurved lip. Also in *P.leucophaea* hemipollinaria are often removed one at a time. Sexual selection can operate by restricting the amount of pollen that is removed during a single pollinator visit and by extending the period of pollen presentation in a flower, thus increasing the number of potential matings.

Platanthera tipul(ari)oides var. **behringiana**

In Japan varieties and subspecies of this orchid species are pollinated by Lepidoptera, the pollinaria attaching to either the upper or lower surface of the proboscis (Inoue 1983).

Self-pollination and pollination by mosquitoes in **Platanthera**

Catling mentions, in an overview on auto-pollination with a classification of the autogamy phenomena, self-pollination in:

- *P.clavellata* by means of falling down of friable pollen on the stigma and structural modification of the rostellum;
- *P.chorisiana*, falling in the class where rostellum and/or clinandrium are absent or dissolving. Pollinated by an Oedemerid beetle in Japan (Inoue 1983);
- *P.hyperborea* (Plate 34, p. 69), characterised by falling out of pollen or entire pollinia and bending of the caudicle.

P.chorisiana and *P.hyperborea* are northern species living in areas with a cool climate and hence fewer and less active insects (mainly mosquitoes) than in more temperate areas.

P.clavellata is widely distributed, lives more southernly and grows in swampy forests, along wooded streams and creeks, in bogs and on margins of lakes, all places where pollinators may be scarce.

According to Löjtnant & Jacobsen (1976) *P.hyperborea* is regularly autogamous in Greenland. The small pollinia easily crumble and are inclined to fall out of the thecae and down on the stigma, even with the slightest vibration. They do not confirm Hagerup's claim of cleistogamy and believe that methodical errors have led Hagerup to a wrong conclusion. However, they state that this species is potentially *entomophilous* in Greenland. There is no nectar, but the internal hairs of the short spur resemble glandular hairs. The scent is *Dianthus*-like and is especially persistent in the evening, according to Thien, who states that moths (Geometridae, genus *Xanthorhoe*) as well as female mosquitoes (genus *Aedes*, particularly *A.communis*) pollinate the orchid in northern Wisconsin (*Aedes pullatus* is mentioned by Catling & Catling (1991a) for Great Whale River, Northwest Territories). It is remarkable that pollination occurred during the last few days of the long flowering period in spite of the presence of the pollinators all the time. The lag is probably due to the pollinators using other food sources and switching to the orchid after these cease to flower. This behaviour of the pollinators reminds one of the development of searching images as described in ethological literature (Manning 1972). On average 18% of the pollinia were removed with 14.2% capsule set.

The reduced phalaenophilous syndrome of *P.hyperborea* as demonstrated by the short spur, the small flowers, the absent or poor nectar production, the greenish colour of the flowers and the small fragile pollinia make the species an example of a development towards obligatory autogamy, at least in Greenland. It is possible that there are differences between the populations in the different parts of the distribution area regarding the degree of entomophily and autogamy in this polymorphic species.

P.chorisiana is reportedly pollinated by an oedomerid beetle.

Genus **Piperia**

A group of 3-7 species [Williams & Williams (1983) have three, Ackerman (1971) has four, and Catling & Catling (1991a) have seven species], which were also previously treated under *Habenaria* and which are characterised by a basal group of leaves that often disintegrate before the flower spike matures and by the presence of two tubers in the root system. They are confined mainly to western US. The pollination syndrome is that of moth pollination.

Protandry has been reported in this genus. It is caused by age-dependent lip movement. When flowers first open, the lip is horizontal and close to the column. The access to the spur is narrow then, causing the proboscis of a pollinator, during probing, to contact a viscidium. Upon withdrawal this adheres to the proboscis and pollinium removal is effected. The narrow orifice seems to prevent pollen deposition. As it ages, the lip lowers and the orifice becomes broader. A pollinating insect probably will not remove pollinia because the viscidia are positioned well above the lip (cf. *Goodyera* and *Spiranthes*). Pollinators work upward on the inflorescence, first pollinating the lower flowers and then removing pollinia from the upper ones (protandrous inflorescence).

Piperia elegans

Kipping observed pollination by noctuid moths (*Autographa*) in California, the pollinaria attaching to the proboscis. Ackerman captured another pollinating noctuid, *Chrysaspedia nichollae* (Ackerman 1977).

Piperia transversa

Ackerman (1977) observed pollination by a geometrid moth (*Thallophaga* cf. *taylorata*), the pollinaria attaching to the proboscis.

Piperia unalascensis

Kipping observed pollination by three pyralid and two plume moths (*Oidaematophorus* sp. and *Platyptilia* sp.) in the Sierra Nevada, and Stoutamire mentioned pollination by a geometrid moth species (*Eupithecia*) in upper Michigan, the pollinaria attaching to the proboscis (Ackerman 1977). Spur length differences among pollinators may be sufficient to isolate this species from the other longer-spurred taxa. Phenological and geographical factors as well as odour differences may act together as isolating mechanisms, although flowering periods overlap considerably in some species.

ORCHIDEAE: HABENARIINAE

World-wide about 930 species are known in 23 genera in two tentative alliances. The genus *Habenaria* is very widespread, occurring in all parts of the world, but is probably of African origin.

Genus **Habenaria**

Several species of the genus *Platanthera* have formerly been placed within the genus *Habenaria.* This pantropical genus, with about 600 species, is characterised by projecting anther caudicles and paired stigmatic processes lateral to or beneath the nectary opening. However, the temperate species of northern America, included in the genus, have a single median stigma above the nectary orifice (Stoutamire 1974) as is typical for the approximately 200 species of *Platanthera*. It seems that confusion about the assignment of species to these genera is still going.

In tropical *Habenaria* species the projecting viscidia require either a compensating hygroscopic movement or inferior *stigmatophores* in position to receive the often pendent pollen masses. The projecting *antherophores* with terminal viscidia probably serve to attach pollinaria to the compound eyes of visitors, or the projecting viscidia are removed by the forelegs of hawkmoths, as in the African *Habenaria polyphylla. Habenaria* species have typical phalaenophilic flowers with long spurs, often white or greenish in colour. The four species, *H.distans*, *H.quinqueseta*, *H.repens* and *H.odonto-*

petala all grow in Florida. The second and third species have an even wider distribution and are found from Carolina to Texas. No details are known to me about their pollination.

EPIDENDROIDEAE: CYMBIDIOID PHYLAD

MALAXIDEAE

This cosmopolitan tribe contains 960 species in six genera. *Malaxis* has about 300 terrestrial species, *Liparis* 350, mainly epiphytic.

Genus Malaxis

The bog orchids are represented in North America by eight species. Observations make clear that flies are the most frequent visitors and pollinators

Malaxis monophyllos
This is a temperate and boreal species of eastern central Europe and Scandinavia. In North America two varieties are known: *brachypoda* and *diphylla*, the latter variety being non-resupinate and found on the Aleutian islands. The pollination of the variety *brachypoda* was investigated by Reeves & Reeves (1985a). Pollinator insects were identified and their behaviour described but the short note gives no details. No hybrids are known. In Europe, specimens with one or two leaves have been recorded from the same localities. Füller (1966) mentions mosquitoes (after Ziegenspeck). Ziegenspeck himself observed up to 65% seed set, whereas in var. *diphylla* he found 3-30%.

Malaxis paludosa
This species (syn. *Hammarbya paludosa*) is also found in Europe, where its pollination was investigated by Darwin (1988). Reeves & Reeves (1984) did the same in Minnesota.

'The flowers of the *Malaxis*, though so small and inconspicuous, are highly attractive to insects. This was shown by the pollinia having been removed from all the flowers on the spike which I examined, excepting one or two close under the buds. In some older flower-spikes every single pollinium had been carried away' (Darwin 1888).

Darwin observed seed-production in 13 out of 21 flowers on one spike. Reeves and Reeves collected insects in a population in Minnesota, USA with aspirator and net and, for night time collection, a lantern. The plants are inconspicuous which makes them extremely difficult to find and the flowers are small. The blue-green striped lip is thought to probably serve as a 'nectar guide' for insects. They noticed that, after opening of the buds, the lip folds down to a position perpendicular to the plane of the sepals, parallel to the column. A second notable change is the upward bending of the lowermost sepal. This movement is interpreted as providing a landing-platform for insects carrying pollinia. Removal of pollinia sometimes precedes this movement.

The green flowers emit a sweet 'cucumber-like' scent, strongest during the first period of flowering. There is some nectar at the base of the lip adjacent to the base of the column. The green perianth is believed to contribute significantly to the photosynthetic production of this tiny plant.

They found one fungus-gnat (Diptera-Mycetophilidae) with pollinaria and they collected mosquitoes (*Aedes* spec., Culicidae), which were considered to be too large to pollinate. Other Diptera and some Hymenoptera were also caught without pollinia. Nearly all pollinia became removed. Fruit set was considerable: 20.5% of all flowers, 64% of all plants in 1983, lower however in 1984. For fruit set external pollen vectors are needed. The species reproduces also by bulbils from the leaves.

Malaxis unifolia
This species grows sympatrically with the two species mentioned above. Reeves & Reeves (1985b) studied flower morphology, flowering phenology and fruit set in a northern conifer swamp population. Pollinating insects were identified and their behaviour described, but the short note gives no details. There are no reports about pollination for the other species, but the flower syndrome of all the species is similar.

In southern Arizona *M.soulei*, *M.ehrenbergii*, *M.corymbosa* and *M.tenuis* are expected to be pollinated by fungus or gall gnats because of the size of the flowers and timing of pollination. A different type of fly may be involved in the pollination of *M.maclaudi*, which has a strong smell of fish (Christensen 1994).

Genus Liparis

The three North American species (*L.elata*, *L.loeselii*, *L.lili(i)folia*) belong to a large genus that is distributed throughout the world in both temperate and tropical areas. One of the three species is also found in Europe as the only representative in that area: *Liparis loeselii*. Distributed in north-eastern US and adjacent Canada in bogs and damp sandy meadows or sand-dunes, it shows little variation and hybrids are unknown.

According to Catling & Catling (1980, 1991a) in Ontario, New York and Wisconsin sexual reproduction is by rain-assisted auto-pollination by means of rotation of pollinaria. Experiments showed that autogamy increased by a factor of four in flowers watered from above with a rain-like spray.

Füller (1966) refers to Ziegenspeck and Kirchner, mentioning regular self-fertilisation, although they do not exclude pollination by mosquitoes.

Reinhard et al. (1991) suppose obligatory autogamy.

The species reproduces strongly by vegetative means.

L.lilifolia (Plate 35, p. 69) is reportedly pollinated by *Hylemyia juvinalis* (Anthomyiidae) (Christensen 1994). Catling & Catling (1991a) mention visitation by a fly and a bug.

Calypsoeae

About 35 species in nine genera, north temperate and subtropical.

Genus **Aplectrum**

Aplectrum hyemale

The species flowers early in spring and its phenology is closely synchronized with the seasonal periodicity of the deciduous forest. Auclair (1972) suggested that a variety of insects is likely to be involved. Hogan (1983) observed in Illinois only visiting bees of the species *Dialictus oblongus* as potential pollinators. Five bees were seen entering, but none were observed with pollinia. Nevertheless 82.1% of the flowers set fruit. In unmanipulated, bagged flowers, the percentage was 70.6%, indicating that autogamy occurs. In emasculated, bagged flowers 47.1% set fruit, which suggests agamospermy. Also low levels of variability point in the same direction.

Genus **Calypso**

Calypso bulbosa

This is a monotypic genus, circumboreal in distribution. In Europe var. *bulbosa* is rarely found in Scandinavia and North Russia. In Japan a var. *japonica* is found. In North America two varieties are found, *occidentalis* and *americana*. The variety *americana* resembles the European variety *bulbosa* closely. This similarity is important regarding pollination. There are no hybrids known with other species.

Morphology

The flowers are slipper-shaped (Plate 36, p. 69) and have a strong, perfume-like fragrance, especially when fresh. Later the scent may disappear. The front part of the labellum is flat and is white or pinkish with yellow hairs and stripes. The yellow hairs resemble pollen. The saccate part of the lip has purple stripes and dots on a whitish background. At the base of the labellum are deceptive nectar-spurs, the flower being nectarless. The anther is located at the underside of the column, just in front of a small ridge, and is covered by a cap. The stigma lies behind the anther.

Pollination

The species is pollinated by naive bumblebees.

Lündstrom, in 1888, was the first to observe the pollination by *Bombus* spec. Gumprecht (1977) describes the pollination by *Bombus pascuorum* as follows: 'When the animals do not find anything at the pseudostamens, they try to find at least nectar. Directed exactly by the sharp-edged column, they penetrate the flowerbase and find themselves deceived again; the spur, lying parallel to the landing-platform is no nectar source. In the search for it the insect-bodies need to bend extremely, and consequently the only bare spot on the back becomes uncovered, on which alone the quadrangular viscidium can be attached. By this double deception it becomes clear why the decisive second visit so rarely takes place. Often the pollinia are removed, but the stigma not supplied.'

Wollin (Wollin & Sundermann 1976) observed in Sweden that of 13 plants which all had the pollinia removed, only two were pollinated. At a second locality, of 13 plants seven had pollinia removed, none was pollinated. He observed pollination by *B.pascuorum* and photographed the animal with pollinia.

Stoutamire (1971) hypothesised that bumblebees are attracted to *Calypso* flowers var. *americana* because of their conspicuousness. In 1981 Stoutamire studied the pollination of var. *occidentalis* and found queens of *Bombus caliginosus*, *B.vosnesens-*

kii, *B.edwardsii*, *B.mixtus* and females of *Psithyrus crawfordi* as pollinators at one place and queens of seven *Bombus* species (*B.melanopygus*, *B.rufocinctus*, *B.centralis*, *B.pleuralis*, *B.bifarius nearcticus*, *B.flavifrons*, *B.frigidus*) at other localities. He mentions *B.californicus* and *B.occidentalis* as known pollinatiors of var. *americana*.

In var. *occidentalis* abrupt rises in the number of flowers pollinated correspond with sudden increases in available pollinators. Var. *occidentalis* proved to be a generalised food-mimic, without a specific model and without nectar.

Boyden (1982) studied pollination in the USA, in Alberta near Banff (var. *americana*).

Of 843 flowers 87% proved to be unvisited, 12% had pollinia removed, 1% was pollinated. He caught bumblebees (mainly queens of *B.melanopygus* and of *B.bifarius nearcticus*, few of *B.occidentalis*) to count the number that had visited the flowers by looking at pollinia or remnants of these on the abdomen. He found 57% as carriers out of 168 individuals.

He found no indications for mimicry of other foodplants. He tested this by comparing bumblebees caught near a *Calypso* population with that of a *Dodecatheon radicatum* population (*D.radiatum* is the only species that from a distance looks superficially similar to *Calypso*). The first group visited other foodplant species. Both collections carried about the same amount of pollinaria or viscidia.

It seems that bumblebees need only one experience to find out that the species is not worth visiting. Only a few animals take a second chance. Short-term memory experiments in the related honeybees have demonstrated that bees show one-trial learning. In long-term memory experiments their accuracy decreased slowly and neared the level of spontaneous choice after six days. They therefore seem to need reinforcement. It can be assumed that this will be the same in bumblebees. Boyden assumes that no laboratory experiments have been done on learning in bumblebees, obviously overlooking the experiments of Manning, which have been popularised by Tinbergen. Bumblebees, after finding new food sources, make orienting circular flights before leaving to their nests and in this way imprint the exact location. When such food sources are removed the bumblebees fly around after return for some time, clearly searching for it. Manning discovered also that in cases of conspicuous flowers (like foxgloves), which can be seen from a distance, the bumblebees don't trouble themselves with learning the location of individual plants. It is obvious then that bumblebees are able to learn even difficult things quickly if it pays. This is highly relevant to the behaviour towards *Calypso bulbosa*.

Boyden suggest that it would be worthwhile to compare seed set in two experimental groups; one with artificial nectar added on a consistent basis and the other a control group. Ackerman tried this in *Calypso*, but probably used too infrequent and too low a concentration of sucrose.

Proctor & Harder (1995) investigated the effect of pollination success on the longevity of flowers in *C.bulbosa*. The life-span of individual flowers is often affected by pollination success. Species differ regarding whether male function (pollen removal), female function (pollen deposition), or both trigger senescence. In *C.bulbosa* pollen deposition resulted in dramatic changes in colour and shape within four days. Unmanipulated flowers and those that had the pollinia removed remained unchanged for 8-11 days. Deceptive (i.e. rewardless) flowers are more likely to satisfy male than female function since the latter requires that a pollinator is fooled twice, first to pick up pollen and second to deposit it. Proctor and Harder found 95% pollen removal versus 66% pollen deposition of visited flowers (41%).

Selection appears to have favoured the female function to trigger senescence and in this way increases pollination chances.

Proctor & Harder (1993) hand-pollinated *C.bulbosa* to investigate the relations between pollen load, capsule weight and seed production. Pollen load affected seed number significantly. The results of the investigation suggest that in *C.bulbosa* the natural unit of pollen deposition, the pollinium flake, of which it has four, contains sufficient pollen to fertilise most *ovules*. Pollinaria are thus able to fertilise more than one flower. Since the stigma normally receives two or more pollinia under natural conditions, seed production in this species is not likely to be pollen-limited.

Alexandersson & Ågren (1996) studied populations in Sweden and the results of their study indicate that life-time seed production may nevertheless be pollen-limited in European populations, and that variation in population size and local abundance of the early-flowering, nectar-producing cooccurring *Salix caprea*, the most important foodplant for the pollinators prior to flowering of *C.bulbosa*, can only partly explain the extensive variation in pollinator visitation among populations of *C.bulbosa* that was observed in consecutive years. So nearby rewarding plants have probably less influence than expected on visitation by pollinators.

Supplemental hand pollination increased cumulative fruit production significantly (1.8 times) without affecting flower production, plant size, or survival. The average fruit set in this study ranged from 21% to 48%, which is relatively high, compared to North American populations.

According to Mosquin (1971) *C.bulbosa* is one of the early flowering species that probably has adapted its flowering time as a result of competition for pollinators with 'cornucopian' species like *Salix* spp. and *Taraxacum officinale*, which flower abundantly in late spring.

Catling & Catling (1991), referring to several authors, mention food-deceptive pollination in var. *occidentalis* by *Bombus* queens and *Psithyrus* females.

Fruit set

According to Catling & Catling (1991) fruit set varies from less than 10% in much of eastern North America, to more than 90% in the Rocky Mountains (Alberta, British Columbia) in both small and large populations.

Genus Corallorhiza

About ten saprophytic species occur in the world, of which six in North America.

C.maculata (var. *maculata* and *occidentalis*), *C.odontorhiza* and *C.trifida* are reported to be autogamous (Catling 1990), by means of rotation of pollinaria (in *C.odonthoriza* there is no rostellum and the flowers keep closed). The first species is also pollinated by predacious dance flies (Empididae) (Catling & Catling 1991b).

Corallorhiza trifida

C.trifida is visited and pollinated by small Diptera, Hymenoptera and Coleoptera, at least in Europe (Löjtnant & Jacobsen 1976). Silén reported *Syrphus cinctellus* for southern Finland, Lang (1980) hoverflies and dungflies (Scatophaga). The entomophilous pollination is, however, incidental, because pollinia are attached to insects only with difficulty and crumble with age. Van der Pijl and Dodson call it a special syrphid-fly flower and Summerhayes (1976) mentions hoverflies and other small insects landing on the labellum. Experiments, however, show that the pollinia do not usually adhere and self-pollination is actually the rule, with the pollinia falling from the stamen on to the stigma below. It is realised by the simultaneous dessication and lifting of the anther cap (Claessens & Kleynen 1998). Nectar is absent and the flowers are weakly scented. The scent is musk-like.

This seems quite an effective method of fertilisation judging by the fact that usually from 85-100% of the flowers set seed (Summerhayes 1976). The ovary and also the sepals remain green and actively produce food for the maturing seeds. The excellent level of seed production enables the plant to colonise any new suitable habitats in the neighbourhood. On the other hand, the means of vegetative multiplication are nothing as effective as in other saprophytic orchids and merely consist of the autonomous separation of branches of the rhizome as the rear parts die away.

Füller too mentions syrphid flies, but records the same phenomena of the column degenerating. Hybrids are not known.

Genus Tipularia

Of this small genus of three species one occurs in the eastern US: the Crane-Fly orchid *Tipularia discolor.*

Tipularia discolor

Pollination

This woodland species is extensively investigated concerning its pollination (Stoutamire 1978, Whigham & McWethy 1980) and ecology (Whigham & O'Neill 1988, Snow & Whigham 1989, Whigham 1990, Proctor & Harder 1994).

This orchid displays modified floral symmetry: the flowers face outward from the central axis with the column oriented either to the right or to the left of the nectary opening. Sepal and petal positions

Plates
28. *Platanthera grandiflora.* Photo J.Mulder, July 1995, Dorcas Bay (p. 59).
29. *P.psycodes.* Photo J.Mulder, July 1995, Sauble Beach (p. 59).
30. *P.psycodes* with *Polites mystic.* Photo W.P.Stoutamire (p. 60).
31. *P.obtusata* with *Aedes* mosquito. Photo by W.P.Stoutamire (p. 61).
32. *P.stricta.* Photo J.Mulder, June 1995, Trail (p. 62).

28	30	32
29	31	

are also modified and this gives the flowers an irregular appearance. According to Stoutamire this could be an adaptation to medium-sized moths (so-called 'millers') that have their bodies, legs and wings covered with scales, to which no pollinia will adhere. This leaves only the tongue and compound eyes free to attach pollinia to. The distance between nectary opening and column tip matches the distance between tongue base and compound eye of these moths.

They locate the dull-coloured racemes presumably by their scent and alight first on the lower flowers, working upwards on the racemes. Schnell and Stoutamire collected *Plusia oxygramma*, *P.precationis* and *Cucullia convexipennis* with pollinaria. Moths carrying pollen on their right eye can only pollinate dextral (right-handed) flowers but not sinistral (left-handed) ones. Each raceme carries both types. Anther caps remain for some time on the pollen masses after removal and this may reduce selfing (see below).

The other known *Tipularia* species from the Himalya (*T.josephii*) and Japan (*T.japonica*) have normal bilateral symmetric flowers.

According to Whigham and McWethy *T.discolor* flowers in mid-summer, when pollinator resources (pollen and nectar) are minimal. They studied pollination ecology by experiments in which pollinia were removed and inflorescences covered to study selfing. Nectar amounts were measured in different experiments. The data suggest that *Tipularia* does not automatically self and is not apomictic, although self-pollination is not excluded and may be mediated by pollinators, since the species is self-compatible. Visiting moths are capable of removing nectar without taking away pollinia. As potential pollinating species they found the small geometrid moths *Protoboarmia porcelaria* and *Xanthorhoe ferrugata*, but these only occasionally visited the orchid flowers and did not appear to remove nectar. The large noctuid moth *Pseudaletia unipunctata*, however, removes nectar and may pollinate, although it was found that pollinia may remain intact after visitation, notably so early in the flowering phase.

It appears that most pollinations are effected when there is less nectar in the spur, so that the moth has to probe deeper to reach it. Pollinators located open flowers within a day or two of the beginning of anthesis. *Pseudaletia unipunctata* seems to exploit the nectar only for a part of the flowering period and ceases foraging after nectar production declines below some minimal level.

Catling & Catling (1991b) point out that the anther cap is retained for 8-30 minutes after pollinarium removal and that this enhances cross-fertilisation, since a pollinarium with a cap is unlikely to adhere to a stigmatic surface because the elastic strength of the pollen stalk is sufficient to pull it away. After the flaps of the cap have opened and it has fallen off, the pollen masses are readily deposited on a stigmatic surface because the elastic threads attaching each of them to the pollinaria are much weaker. They break. The time lapse is long enough to allow the pollinator to move away to a different plant.

Predation

Whigham & O'Neill (1988) found that populations of *T.discolor* are influenced by predators. White-tailed deer are the primary predators and they may remove almost 100% of the leaves. This leads to a significant decrease in plant size and delay in the frequency of flowering in the following years. Experimental defoliation shows that after two successive removals plants do not reproduce sexually (Whigham 1989).

On the other hand greater investment in reproduction also leads to increasingly negative effects on future corm size and leaf area. (Snow & Whigham 1989). The dynamics of populations is most strongly influenced by predation of corms. Combined influences of leaf herbivory, corm predation and lack of seedling establishment have resulted in decreases of populations and individuals (Whigham 1990).

33 | 34
35
36 | 37

Plates
33. *Platanthera dilatata* var. *albiflora*. Photo J.Mulder, June 1995, Manning park (p. 62).
34. *P.hyperborea*. Photo J.Mulder, July 1995, Dorcas Bay (p. 63).
35. *Liparis lilifolia* with mycetophylid fly. Photo by W.P.Stoutamire (p. 66).
36. *Calypso bulbosa* with visiting hoverfly. Photo W.Upton, Mc Leans Creek, Kananaskis Country (p. 66).
37. *Calopogon tuberosus*. Photo J.Mulder, June 1995, Dorcas Bay (p. 72).

CYMBIDIEAE: GOVENIINAE

One genus with 20 species from tropical America.

Genus **Govenia**

The status of this genus is not entirely clear, because details of the column morphology have not been investigated very well. It is the only tropical genus that Dressler formerly placed within the subtribe Corallorhizinae because of the large coralloid masses in the substrate and the initial saprophytic stage. But in his present classification (1993) he considers them as probably a sister group of the Maxillarieae. Schlechter (1993) puts them in the Cymbidiinae.

Govenia utriculata
Govenia utriculata is the only species that ranges into southern Florida. There are no reports about pollination hitherto.

CYMBIDIEAE: EULOPHIINAE

A pantropical subtribe with 264 species, mainly Old World.

Genus **Eulophia**

A large, mainly African genus, with two species in the south-eastern US. In Africa pollination by bees has been observed.

Eulophia alta
This species is reportedly self-pollinated in Puerto Rico and Mexico.

Eulophia ecristata
This species lacks the lamellar crest on the lip. No data found about pollination.

Genus **Oeceoclades**

Oeceoclades maculata shows rain-assisted autogamy in Puerto Rico (see Chapter 5).

CYMBIDIEAE: CYRTOPODIINAE

About 140 species in 12 genera. Pantropical, but mainly Old World.

Genus **Cyrtopodium**

Neotropical genus with two representatives in Florida. Pollination by male *Euglossa* (*hemichlora*), which take fragrant substances from the flowers, is reported, but no data are found about pollination of *C.punctatum* and *C.andersonii* in Florida.

Genus **Galeandra**

A neotropical genus of about 25 epiphytic or terrestrial species of which a terrestrial one occurs in Florida. According to Dressler (1981) the pollinaria of *Galeandra* have been found on *Euglossa* and on large anthophorid bees. He supposes members of the subtribe to be pollinated by bees of appropriate sizes.

MAXILLARIEAE: MAXILLARIINAE

A subtribe from tropical America, with one species in North America.

Genus **Maxillaria**

Large Neotropical genus of epiphytic or terrestrial species, with only one epiphytic species occuring in Florida: *Maxillaria crassifolia.* Some Brazilian species are pollinated by bees that gather wax from the lip. No reports on pollination in Florida about the uncommon and elusive *M.crassifolia.*

MAXILLARIEAE: ONCIDIINAE

A large subtribe with over 1200 species in about 77 genera from tropical America, with some species intruding into North America.

Genus **Brassia**

A neotropical genus with one peripheral species in Florida: *Brassia caudata. Brassia* is known to be wasp-pollinated, but no data are found referring to this species.

Genus **Ionopsis**

About 10 species of this genus are found in Central and South America, only one (*Ionopsis utriculari-*

oides) has intruded into southern Florida. On pollination no details are reported, but some research has been done about fruit production, which will be dealt with in Chapter 5.

Genus **Leochilus**

One rare representative of this small genus of neotropical orchids occurs in Florida. In *Leochilus* wasp pollination is observed, but no data are found on *Leochilus labiatus* of Florida. Chase (1986) mentions pollination by a halictid bee (*Lasioglossum*) and pollinaria attachment between or to the base of the antennae.

Genus **Macradenia**

Out of a dozen epiphytes of this genus of the American tropics and subtropics one species has reached Florida: *Macradenia lutescens*. No data on pollination.

Genus **Oncidium**

In Central and South America this genus is bee-pollinated. There are no reports about the pollination of the four Floridan species.

Genus **Tolumnia**

Tolumnia bahamense, a species of Grand Bahama Island, which also occurs on the coast of Florida, is pollinated by male *Centris versicolor* bees. This is an example of so-called 'pseudoantagonism'.

Nierenberg (1972) observed that in two concurrently flowering syntopic species of *Tolumnia*, the same pollinator species is exploited. The orchid species, however, are effectively isolated, because in *T.bahamense* male *Centris versicolor* bees attack the flowers, mistaking them for moving insects. The other species *T.lucayana* (syn. *Oncidium lucayanum*) is visited by females of the same bee species, which visit the flowers for food (food deception).

EPIDENDROIDEAE: EPIDENDROID PHYLAD

ARETHUSEAE: ARETHUSINAE

A small subtribe from temperate Asia and North America. Three species are known in two genera: *Arethusa* and *Eleorchis*.

Genus **Arethusa**

A monotypic genus of north-eastern America, characterised by bulbous corms and a solitary leaf that develops after the solitary flower opens.

Arethusa bulbosa

As mentioned above, this orchid shows flower convergence with *Pogonia ophioglossoides* and *Calopogon tuberosus*.

Bees are usually attracted to groups of plants only. In visiting the flowers, they land on the fleshy protuberances, slowly crawl down into the flower to seek nectar (which the species does produce in small quantities only), and remain in the flower for about a minute. When backing out they receive the pollinia on the back of the thorax. The stigma is located immediately before the anther case and an insect bearing pollinia pollinates the flower in the process of backing out. Since the distance between lip and anther is about 0.5 cm, only bees the size of *Bombus* queens are capable of pollinating. The workers are not large enough to scrape the anther to remove the pollen.

A.bulbosa is self-compatible, but adapted to outcrossing (Argue 1995).

ARETHUSEAE: BLETIINAE

Almost 400 species in about 21 genera of terrestrial and epiphytic or saprophytic species. Pantropical, extending into temperate areas, among which North America.

Genus **Bletia**

A Central and South American genus with two representatives in southern Florida.

Bletia purpurea and **Bletia patula**

Pollination by mixed groups of bees has been reported for *Bletia purpurata*, including *Euglossa*, *Thygater* and possibly *Melipona*, but Catling mentions self-pollination for *B.purpurea* in Florida, based on personal communication by Luer. This again is an example of autogamy in an otherwise non-autogamous species near the periphery of its distribution range, a phenomenon which is considered as an adaptation to the lack of suitable pollinators (Van der Pijl & Dodson 1966). The rostellum is lacking in the Florida plants. Nothing is reported about the pollination of the second species, the similar *B.patula*, and it is doubtful whether this species still exists in Florida.

Genus Calopogon

All four species of the genus are found in North America, confined to the eastern part. According to Thien (1973) the four species can be found growing intermixed or in adjacent habitats. Van der Pijl & Dodson (1966) have suggested that flower size (and therefore pollinator difference) is a sufficient barrier to prevent crossings between *C.tuberosus*, with its large flower, and the other three species. Thien made intercrossings between the species and found that they all produce good seeds when intercrossed, and each species is self-compatible. He concluded that the main isolating mechanisms between the species are flower size, seasonal isolation and habitat. The attraction by similar colour and ultraviolet patterns on the lip makes the attraction of the flowers to the insect identical.

Calopogon barbatus

The fertilisation of *C.barbatus* (syn. *C.parviflorus*) is described by Robertson (1887). Small bees, Andrenidae, approaching the non-resupinated flower in front, alight upon the crest, when the labellum bends suddenly, so that the dorsal surface of the insect comes down upon the column. He captured the bees *Augochlora sumptuosa* and *A.festiva* carrying pollinia of this orchid. *Bombus separatus* and *Halictus* species were observed as visitors. As early as 1887 Robertson recognised that the nectarless flowers attract pollinators with the stamen-like hairs on the crest.

Calopogon multiflorus

This small-flowered species flowers more or less at the same time as *C.barbatus*, but grows only in pine-palmetto fields that have been burned the preceding winter.

Calopogon pallidus

The flowering season of this species overlaps slightly with that of *C.barbatus*. Evidently there is gene flow between these species because in some populations the flowers are indistinct. But cross-pollination is reduced to a trickle.

Calopogon tuberosus

Pollination

The fowering period of this species (syn. *C.pulchellus*, *C.parviflorus*) overlaps partially with that of the convergent species *Arethusa bulbosa*. The hairs on the lip (Plate 37, p. 69) are believed to imitate stamens (Robertson 1887) and thus to deceive pollinators (pollen imitation). Van der Pijl & Dodson (1966) do not exclude edibility of these hairs, however. Nectar is not present. The flowers are visited by a wide variety of bees, but only those of the proper size and weight are capable of effecting pollination (e.g. *Halictus* species, as observed by Robertson). The flower has a hinged lip and if the visitors are too small they cannot cause this to fall. Insects which are too large and strong react quickly and push the falling lip upright without consistently removing pollinia. The insects seem to be attracted by groups of flowers. They hover briefly before a flower, and then land on the dense, short-knobbed hairs of the lip. Due to the hinge at the base they are quickly lowered onto the column. They then slide off the column on the back and consequently first strike the viscid stigma and then the anther. In this way pollen masses attach to the sticky abdomens. The lip, when released by the bee, immediately springs into its former position, ready for the next visitor. Pollination occurs when a bee with pollen is lowered onto the column and slides past the stigma. *Bombus* queens, although regularly visiting, are too large and only occasionally remove pollen, usually on their legs or the sides of the thorax. Besides workers of bumblebees, the bee *Megachile melanophea* is also capable of pollinating. Dodson (1966) found *Xylocopa micans* to be the pollinator in Florida. Firmage & Cole (1988) observed *Bombus vagans* and *B.fervidus* and several species of Halictidae as pollinators. Bog coppers and several species of Diptera only visited the flowers. Another frequent visitor was the beetle *Trichiotinus affinis*, which uses the flower as a rendezvous location and was often seen mating.

Mimicry

As mentioned before, *C.tuberosus* shows convergence with *Arethusa bulbosa* and with *Pogonia ophioglossoides*. The three sympatric species and the other pink flowers of the same habitat could therefore be considered as a mimicry complex, appealing with a similar 'Gestalt' to a group of naive bees exploring the habitat on the basis of a 'searching image'.

Argue (1995), however, believes that convergence makes little sense in species where none provide any significant incentive in the form of a nectar source. Instead, if this type of mimicry is involved, it is likely that those species have converged upon other more common taxa in the same *biotope* that do offer a rich reward, like *Sarracenia purpurea*, *Epilobium* species, and *Polygala paucifolia*. Pollen transfer may, however, simply depend largely upon the early, tentative explorations of potential new food resources by inexperienced bees (guild mimesis or nonmodel mimicry).

Wal Upton (pers. comm.) mentions mimicry of *Rhexia virginica* in East Texas as observed by D. Goldman.

According to Thien and Marcks a study of the pollination biology of the other species of the genus would be worthwile to reveal if they associate with other plants as do the three convergent species mentioned.

Heinrich (1975) mentions conspicuous variability of flower-colour in the visible part of the spectrum in a population in the Maine bog. Bumblebees were observed to visit the widely scattered flowers and stopped also at *Pogonia ophioglossoides*. The bees passed hurriedly through the area and sampled only a few of the available flowers. Heinrich supposed that, since the non-rewarding flowers were primarily visited by unconditioned pollinators, which had not yet established a specific foraging area and flower preferences, the flower's ability to deceive was increased by its variability in colour (polymorphy).

Reproduction

Firmage & Cole (1988) studied reproductive success of the species in relation to inflorescence size and dispersion pattern. Fruit set was limited by the number of pollinator visits. Reproductive success increased with increasing inflorescence size, although not above theoretical predictions. Sequential flowering of large inflorescences increased the probability of a pollinator visit over the blooming period. Fruits were produced by self-, geitonogamous, and cross-pollination, all through the mediation of insects [but under natural conditions the pollination mechanism probably reduces self-pollination and timing of flower opening reduces geitonogamy (Argue 1995)]. Large inflorescences were not more attractive than smaller ones, because the plant presents only few open flowers at a time. Flowers on plants growing in clumps of 2-8 plants had a higher probability of setting fruit, apparently because of increased pollinator attraction.

Genus Hexalectris

A genus of seven saprophytic species with not less than five representatives in the southern US. The genus has been placed erroneously in the Corallorhizinae, but the flower structure indicates placing them in the Bletiinae, according to Dressler.

Hexalectris nitida is reported to be autogamous (Catling 1990) and autogamy could as well be the mode for the other species, since saprophytic species often reproduce by autogamy. This obviously relates to the lack of pollinators in the habitat. The often reduced column and other rudimentary morphological characters in autogamic species causes special problems to the taxonomist.

Epidendreae I – New World Groups

Epidendreae: Laeliinae

Over 1450 species in 43 genera from tropical America, mostly epiphytes and some terrestrials.

Genus Basiphyllaea

Basiphyllaea corallicola auto-pollinates in Florida, the rostellum being absent.

Genus Encyclia

This very large neotropical genus, with its centre in Mexico and the West Indies, has four representatives in Florida.

These are *Encyclia boothiana*, *E.cochleata*, *E.pygmaea* and *E.tampensis*.

Little is known about the pollination, but Dressler (1981) mentions that most species suggest bee pollination and that this is what he observed, with some wasp pollination in some species.

However, he also mentions self-pollination in *E.cochleata* and *E.boothiana* in Florida. He gives the following explanation: 'Self-pollination has a clear advantage when a plant species is extending itself beyond the geographic range of its usual pollinator. Thus we find a number of self-pollinating forms in the West Indies and Florida that are represented by normal, outcrossing forms on mainland tropical America. (...) Self-pollinating flowers can function quite well without many of the orchid features, and mutations with petal-like lips, distorted columns, and so on, may persist and multiply when they are associated with self-pollination. It is common to find flowers with three fertile anthers in self-pollinating populations (*Encyclia cochleata*, *E.boothiana*, ... for example).'

The self-pollination in *E.cochleata* var. *triandra* is attained by development of two additional anthers that contact the stigma. This is also observed in the West Indies.

E.pygmaea self-pollinates in Puerto Rico.

Genus Epidendrum

Of this very large genus (appr. 600 species) of neotropical American species, seven species are found in the US. Epidendrums are mainly epiphytic.

Epidendrum anceps

This species is widely distributed and extends from Brazil to Mexico and throughout the Antilles into southern Florida. In Florida it is found in cypress swamps and moist hammocks. It grows on many species of trees, often in close association with *E.difforme* and *E.rigidum*. It occurs also sympatrically with the more localized and rare *E.strobiliferum* and *E.acunae* (Adams & Goss 1976).

Peak of flowering is in winter, but in Florida it blooms sporadically throughout the year. The yellow-green to red-brown flowers are long-lived. Flowers in an inflorescence usually undergo synchronous anthesis. They are receptive for 2-3 weeks. The labellum forms a nectar tube that extends into the ovary. There is no free nectar and pollinators have to have long tongues that can pierce the tissue. Fragrance is malodorous, reminiscent of over-ripe vegetables.

Adams and Goss found that four species of moths pollinate successfully at night: *Eucereon carolina*, *Lymire edwardsii*, *Cisseps fulvicollis* and *Oxydia vesulia*. The ball-shaped inflorescence with its aggregated flowers functions as a landing platform. The moths crawl from nectar tube to nectar tube. The phalanges of the rostellum and the cleft that they form guide the proboscis into the nectary. They release the nectar from the thin-walled cells which line the nectary by rasping them with fine spines.

Pollination occurs when the insect visits the next flower and since there are strong self-incompability barriers, fertilisation occurs only when flowers of two different plants are involved in a pollination sequence.

Cross-fertilisation is promoted by anther cap retention. The cap remains attached for a short period after pollinarium removal. The anther cap prevents the pollinarium from touching the stigma.

Severe winters restrict pollinator activity and hence fruit production is low in that case. Since most of the moths appeared to be males, Goss and Adams tested the hypothesis of sexually selective attraction by experiment. Additionally they observed pollinators at work: among 94 individuals of the pollinator *Lymire edwardsii* 91 were male. In total, out of 102 individuals of five species of moths, 98 were male.

From the experiments it was concluded that the flowers produce volatile compounds which specifically attract male moths. This is explained as follows: the volatile attractants may be by-products of pyrrolizidine degradation, chemical mimics thereof or chemical homologues. Pyrrolizidine is required as a source for the biosynthesis of the moths' sex pheromones. The moths associate the volatile attractants with the presence of a pyrrolizidine source. Nectar gathering is an attempt to collect pyrrolizidines, which may be present or not and if present act as reward, and if not the orchid deceives the moths.

Such hypotheses, however, are still quite speculative and need further testing.

Epidendrum difforme

Goss tested cultivated plants for fragrance production, autogamy and self- and cross-compabilities in screened enclosures. The Florida populations do not rely on autogamy and are self-incompatible. Goss also made nocturnal observations for pollinators. The flowers are weakly zygomorphic and have therefore an unorchidlike appearance. They emit a heavily sweet perfume, especially during late afternoon. Pollinators are nocturnal and crepuscular moths. Three species pollinate successfully: *Lymire edwardsii*, *Anticarsia gemmatilis* and *Phyprosopus callitrichoides*.

Diaphania indica and *Pseudoplusia includens* were observed as non-pollinating visitors. Probably also diurnal Lepidoptera act as pollinators. Behaviour of the moths on the inflorescence is similar to that of pollinators on *E.anceps* and the pollination mechanisms are identical. The copious nectar may be exuded from the nectar tube. The distinct fragrance provides pollinator specifity, as was concluded from experiments in which additional captured moths species proved able to pollinate when placed on the inflorescences, but were never attracted naturally.

In Ecuador *E.difforme* is pollinated by the sphingid moth *Amastus acona*. These specimens have morphologically differing sphingophilous flowers and probably have adapted to these hovering pollinators.

Other species
Of the five sympatric congeners of *E.anceps*, three are probably obligate inbreeders. *E.strobiliferum*, *E.rigidum* and *E. nocturnum* are therefore supposed to be completely isolated from *E.anceps*. Cross-pollination between *E.anceps* and *E.difforme*, however, is highly probable. They share a common pollinator, *Lymire edwardsii*, and their pollination mechanisms are similar. But since strong incompability barriers exist between the two species the probability of gene exchange is low. Goss produced only one viable hybrid capsule after numerous experimental crosses.

E.conopseum has the widest distribution in the US and occurs up north into Carolina and west into Louisiana. There are no records about its pollination, but the floral morphology points in the direction of phalaenophily. Reproduction in *E.acunae* has not been observed either.

Some Florida populations of *E.rigidum* autopollinate. The same is mentioned for populations of *E.nocturnum* in southern Florida and elsewhere (Catling 1990).

EPIDENDREAE: PLEUROTHALLIDINAE

Over 3000 species of epiphytic or terrestrial species in 28 genera are found, mostly in tropical America, but some species are found in subtropical North America.

Genus **Pleurothallis**

This large genus is at home in the American tropics. One epiphytic species is recorded from southern Florida: *Pleurothallis gelida*. It looks a bit like a lily of the valley. Basically, the subtribe Pleurothallidinae is a fly-pollinated group. Many of the small flowers within this subtribe are pollinated by small *Drosophila*-like flies. According to Duque (1993) this species is mildly fragrant and is visited by Diptera: Drosophilidae and Empididae and by lady bugs (Coleoptera: Chrysomelidae, Galerucinae, genus *Diabrotica*).

Genus **Lepanthopsis**

Many species of this genus of miniature tropical American epiphytic orchids were discovered in the last decade. In 1991 Luer recognized 37 species. Only one species, *Lepanthopsis melanantha*, is known form North America, from swamps in southern Florida. No records of pollination.

Genus **Restrepiella**

Again a tropical American species with probably one representative in southern Florida: *Restrepiella ophiocephala*. Williams and Williams mention a few found since its discovery in 1963, but Luer (1991), who first reported the find of one plant in southern Florida in 1974, states that this was undoubtedly based on misinformation. No records of pollination.

EPIDENDREAE II – OLD WORLD GROUPS

EPIDENDREAE: POLYSTACHYINAE

Pantropical, primarily African subtribe, with 154 species in four genera, with a number of species in America.

Genus **Polystachya**

This genus of about 100 species is primarily African, with secondary centres in Asia and America. It is known to be pollinated by small bees that gather pseudopollen from the lip.

Polystachya flavescens
It is once again the only representative of the genus in Northern America and is restricted to Florida. Here the species blooms throughout the year, with a peak reproductive period between August and December. Inflorescences are long-lasting and may produce more than 100 flowers. The yellowish flowers are non-resupinate. The median lobe of the

three-lobed lip curls upward and exposes near the apex a mealy crest, extending along the entire length of the disc midline. This crest is composed of multicellular hairs that resemble pollen grains. There is no nectar. The flowers produce a sweet fresh perfume during the day. The species is not autogamous. According to Goss (1977) small female Halictid bees of the species *Dialictus creberrimus* are the only observed pollinators. A bee lands on the lateral sepals and the labellum, orienting itself parallel to the axis of the lip, facing the upturned portion of the midlobe. It collects the pseudopollen for 5-10 seconds, storing the material in the scopae, and then backs out, all the time grasping the labellum with its legs. The weight of the bee on the hinged labellum keeps its thorax in contact with the column. The sticky viscidium is touched and the pollinarium becomes firmly affixed to the dorsal thorax. A bee usually visits from 3-6 flowers and hovers between each entry. While foraging, a pollinia-bearing bee deposits these on the sticky stigmatic surface, guided by the lateral and apical margins. Of 334 flowers Goss observed 82 (24.5%) with pollinia removed and 33 capsules were produced (10% seed-set)

Any anthophilous bee with similar size and morphology can possibly pollinate *P.flavescens*. It may be expected that local floral variations exist relating to differences in local pollinator species. Since few flowers are open at a time, self-pollination will be reduced, the more so because the stipe of the pollinium needs some time to uncurl and take up the right position for deposition on the stigma.

EPIDENDROIDEAE – DENDROBIOID SUBCLADE

DENDROBIEAE: BULBOPHYLLINAE

Genus **Bulbophyllum**

This very large genus of pantropical epiphytes, with its main points in Asia and Africa, is mainly pollinated by flies. Some reports exist of wasp pollination in African species. Some species produce nectar (Dressler 1981). Also, autogamous species are known.

Bulbophyllum pachyrhachis

The only representative in northern America, *B.pachyrhachis* has a *cupula* or *calyculus* at the base of the perianth of which the function is not clear. Observations of pollination are not reported.

VANDEAE: ANGRAECINAE

Over 400 species in 19 genera, Primarily from Africa and Madagascar, with a few of them in tropical America, of which some species extend into Florida.

Genus **Harrisella**

A small neotropical genus of three species, one species in Florida: *Harrisella porrecta*. The moth pollination syndrome is typical for the Angraecinae, but no observations of pollination are known.

Genus **Campylocentrum**

About 40 species in the American tropics. One species in Florida. No data on pollination.

Genus **Polyrrhiza** (*syn.* **Polyradicion**)

A small genus of the West Indies with a representative in Florida: *Polyrrhiza lindenii*. This showy leafless Palm-Polly, Ghost orchid or Jumping Frog-orchid clearly exhibits flowers with the moth-pollination syndrome.

Most likely the pollinator is the nocturnal moth: *Cocytius antaeus* (Schlechter 1996).

CHAPTER 5

Orchids of South and Central America

Orchids are no flowers – they are satanic creatures – beings who only present us the antennae of their figure and counterfeit for us eyes, lips, tongues in an illusive whirlwind of colours, so that we should not suspect the dreadful viper's body which, invisible death bringing, conceals itself in the riches of the shadows

Meyrink: *Bologneser Tränen*, 1905

In my *Atlas of Orchid Pollination: European Orchids* I have used Karl Peter Buttler's field guide as a help for the arrangement of the data. In Chapter 4 I used Williams and Williams's field guide in the same way. The advantage of this approach is undoubtedly that, since most orchid lovers are familiar with such field guides, they can easily find their way and the data fit into their own cognitive structure.

Furthermore, field guides tend to cover all species of the area for which it was designed. Superimposed on the guide I used Dressler's classification (1993) as a canvas.

For South and Central America this approach is not possible. There is no such guide for the whole area and if there was any, there are too many orchids, literally thousands, in this area, and the reader would be submerged in an ocean of names that would drown the facts of pollination.

Secondly, much less is known about the pollination details of all these genera and species. Therefore I have only made use of Robert L. Dressler's classification (1993) to do this job of presenting the data in an orderly way. As a consequence I had to work the other way round than before, now starting from the literature and arranging the data on the given canvas. I will use the same title descriptions as Dressler does.

Orchids with two or three fertile anthers

CYPRIPEDIOIDEAE

This subfamily is represented by species in three of the four genera in tropical America. World-wide there are about 120-130 species in four (or five) genera. The flowers are one-way trap flowers, without any reward for pollinators, which usually are either bees or flies and sometimes beetles.

Genus Cypripedium

This genus reaches the southernmost limits of its range in Mexico (and Guatemala), where it is represented by two species, *C.irapeanum* and *C.dickinsonianum* (Chiapas, Mexico) Pollination: *C.dickinsonianum* is autogamous by development of contact between stigma and pollinia due to curvature changes and changes in length of the stigmatic branch (Catling 1990).

Genus Selenipedium

This genus of about six species occurs in south Central America and northern South America with a disjunct distribution. *S.palmifolium* is reported to have a sweet odour (too sweet to be used in Chanel #5, according to Dickinson, cited by Stoutamire 1967), which probably attracts bees as pollinators. (Van der Pijl & Dodson 1966). Dressler (1993) also mentions the attractiveness of the flowers and supposes visitation by naive bees. Insects learn to avoid the flowers after a few visits, and consequently pollination rates are usually low.

Genus Phragmipedium

This genus ranges from southern Mexico in the Northwest to the Serra da Mantiqueira in southeast Brazil.

Halictine bees and syrphid flies have been observed as pollinators.

Phragmopedium longifolium var. *hartwegii* was observed to be pollinated by Halictine bees of the genera *Chlegorella* and *Caenohalictus* (Dodson 1966) and by two species of syrphid flies in western Ecuador. The syrphid flies behaved in the same way as the bees, i.e. after landing on the extended lip they wandered to the shield formed by the infolded margins of the sides of the lip, apparently attracted by the green wartlike protrusions, and after attempts to feed on these, slipped, falling into the labellum pouch and working their way out through the exits under the stigma and anthers. One bee of the species *Trigona testacea*, several flies and small bees which were caught in the flowers were ineffective as pollinators. The long tails of *Phragmipedium caudatum* seem to have a function in the attraction of flies. In var. *lindenii* of this species a third anther extends from under the column and presses the pollinium directly against the stigma, in this way automatically inducing autogamy. Along with this anomalous morphological trait the lip is developed into a long narrow petal similar to the lateral sepals.

Dodson observed thousands of plants in Ecuador producing seed by autogamy. A low genetic diversity was found in electrophoretic analysis of isozymes. Vogel found that the strong odour of urine produced by the lateral petals of a *Phragmipedium* is associated with the trichomes lining the petals. They may have an important function in orienting an approaching visitor (Stoutamire 1967).

Primitive orchids with a single fertile anther

GASTRODIEAE

The Gastroideae are saprophytes and of uncertain relationship.

Formerly Dressler (1981) placed them as a tribus within the Epidendroid subfamily, but since 1993 he places them separately, as a tribus within a group of primitive orchids with a single fertile anther.

GASTRODIEAE: GASTRODIINAE

A pantropical subtribe of 36 species in six genera. There is only one species of this group known from tropical America. Self-pollination?

GASTRODIEAE: WULLSCHLAEGELIINAE

Saprophytes from tropical America of unclear relationship. Two species of *Wullschlaegelia* are known, which are self-pollinating. The flowers are non-resupinate in one species (which fits the Cranichidinae, wherein the genus was formerly placed) and resupinate in the other.

PALMORCHIDEAE

Genus **Palmorchis**

Tropical American tribe of about 12 species in one genus. Pollination of three species (among which *P.powellii, P.nitida*) by parasitic bees of the genus *Osiris* has been observed in Panama (Dressler 1983).

Male and female bees of these small wasp-like bees visit these species and were observed to take away pollinia in *P.nitida*, whereas in *P.powellii* pollinia were already taken away when visitation of these bees was observed.

A *Trigona* bee was also observed as visitor and is probably small enough to pollinate as well.

P.sylvicola and *P.trilobulata* have flowers with only a small cleft between the tips of the sepals. The flowers keep closed, but when touched may, at least in the first species, suddenly open. Probably pollinators may open these flowers, as is known from the comparable closed flowers of *Calathea* (Marantaceae) and some Loranthaceae. The flowers of *Palmorchis* species are short-lived, lasting only a few hours and show gregarious flowering (see below, *Triphora*)

TRIPHOREAE

Terrestrial or saprophytic species of unclear relationship with other groups. Features of Neottieae, Diurideae, Epidendroideae and Vandoideae are recognised (Dressler 1981). It is distinctly a primitive group, difficult to place (Dressler 1993).

This tropical American group with intruders into North America consists of 28 species in three genera. The Triphoreae appear to be a relic group, supericially similar to the Neottieae. Some species are self-pollinating, but otherwise no data are known about pollination. The flowers suggest pollination by small bees.

The genera *Triphora* and *Psilochilus* show synchronous flowering, probably an adaptation to pollinators, which are thought to pollinate gregarious flowering species more effectively (Dressler 1993) (see also Chapter 4).

Psilochilus physurifolius is autogamous (Catling 1990).

Vanilleae: Vanillinae

A pantropical subtribe of five genera with 119 species.

Genus **Vanilla**

This genus is wellknown for its fragrance and is cultivated for this purpose. In the American tropics *Vanilla* is frequently pollinated by large euglossine bees. Records of *Melipona beechei* are doubtful.

Many authors seem to have relied on the same source and have copied each other.

Roubik & Ackerman (1987) have caught several euglossine bees (*Euglossa*, *Eulaema*, *Eufriesea*) with pollinaria of *Vanilla* species: *V.planifolia*, *V.pompona* and unidentified species. They suggest food-deception flowers instead of fragrance flowers in these orchid species.

V.fragrans is mentioned by Ackerman (1983d) as euglossine bee pollinated on Barro Colorado Island, Panama.

Most Vannillinae seem to be pollinated by large bees. The pollen forms a triangular mass on the bee's scutellum. In some cases, the same bees that pollinate the flowers may be the dispersal agents for the seeds as well. Vanillin strongly attracts the males of *Eulaema cingulata* and other euglossine bee species. This perfume may attract them to the fruits.

The seeds stick to forceps and probably as well to the feet (Dressler 1993).

More advanched orchids with a single fertile anther

SPIRANTHOIDEAE

Tropidieae

A pantropical subtribe of two genera and about 43 species. No data about pollination.

Cranichideae: Goodyerinae

A widespread, mainly tropical subtribe of 35 genera and about 476 species, with few representatives in tropical America. Bumblebee pollination is reported for *Goodyera*. *G.striata* from Mexico is autogamous (Catling 1990).

Cranichideae: Prescottiinae

A subtribe from tropical America with seven genera and 99 species.

Genus **Myrosmodes**

M.cochleare is a terrestrial orchid restricted to high elevations of the northern Andes. It is the highest altitude orchid species in Venezuela, where it occurs between 3800-4200 m. Berry & Calvo (1991) studied its reproduction. The species is self-compatible but not autogamous. The plants have to withstand the stresses of daily freeze-thaw cycles, high solar insolation, and strong solifluction. The species forms a dense spike with non-resupinate nectarous flowers that open from top to bottom and increase in size in that direction. Caged, non-manipulated flowers set significantly fewer fruits (2%) than in observed natural fruit set (14-29%). Flowers at the top were unable to set fruit, as no fruit set occurred after hand pollination.

Fruit set in open-pollinated flowers was highest in the middle position, while in hand-pollinated flowers this was, however, at the base. Insect visits were rare and restricted to sunny periods at midday and temperatures above 12°C. During eight hours of observation, five individuals of the eumenid wasp, *Incodynerus romandinus* and one fly (Calliphoridae spec.) visited a total of 86 flowers. The wasps visited up to 10 flowers/inflorescence and moved rapidly between plants; from 1-6 pollinia stuck to the head or ventrally to the thorax. The fly also carried a pollinarium. Position-related fruit set has been explained in terms of resource competition among developing fruits by different biologists. In this case, however, the observed pattern is likely due to both higher levels of structural female sterility in the smaller, uppermost flowers and to the limited presence and particular foraging behaviour of the pollinators. Floral display is largest towards the middle of the inflorescence. The large wasps outsize the flowers and generally used the emerging lower flowers to land and position themselves to probe the open flowers higher up. Foraging behaviour and increased visibility thus both enhance pollination of the middle flowers.

Genus **Porphyrostachys**

P.pilifera (syn. *Stenoptera pilifera*) is presumably pollinated by hummingbirds (Dressler 1981).

Van der Pijl and Dodson (1966) also suggested bird pollination.

Genus **Prescottia**

P. cf. *stachyodes* is reported as autogamous in Mexico, S. Vera Cruz.

P.densiflora (Singer & Cocucci 1999b) from Paraná, South Brazil is pollinated by halictid bees (*Pseudoaugochloropsis* aff. *cupreola* and *Dialictus* sp.) as the main pollinators, with minor contributions of other insects, such as syrphid flies. The bees can only introduce their mouth parts into the tiny perianth.

The pollinaria are powdery and many flowers can be pollinated with a single pollinarium or a single stigma can receive mixed pollen loads.

CRANICHIDEAE: SPIRANTHINAE

About 409 species are known in 30-40 genera, primarily in the American Tropics, but especially *Spiranthes* is found also outside tropical areas in both the Old and New World. Not in tropical and southern Africa. A revision involving a reduction of the number of genera to 16 was made by Burns-Balogh (1986), but needs evaluation (Dressler 1993).

Most Spiranthinae are probably bee-pollinated and most of them seem well pollinated under natural conditions. The spiral arrangement of the flowers in *Spiranthes* is evidently an adaptation to pollination by bumblebees. *Spiranthes* is pollinated by bumblebees, halictid and megachilid bees.

Stenorrhynchus shows the features typical of hummingbird pollination. According to Light (1998) *Stenorrhynchus* (=*Dichromanthus*) *aurantiacum* and *S.*(=*D.*)*cinnabarinus* are attractive to hummingbirds and have incredibly hard flowers, thwarting the nectar-robbing efforts of birds (and bees).

Pelexia ekmanii was observed in southern Brazil to be pollinated by bumblebees (Dressler 1993).

Euglossine bees with pollinia of *Sarcoglottis* behind their mouthparts have been collected (see also below). *Stenorrhynchos* seems adapted to hummingbirds (with yellow and red flowers). Hummingbird pollination has been observed in *Sacoila lanceolata*, however outside its normal range (Catling 1987, Dressler 1993).

Pollination experiments demonstrated a reliance on pollen vectors in plants of var. *lanceolata* from central Guyana, where a variety of hummingbird pollinators are available, but the plants of the same variety from southern Florida are agamospermic by adventitious embryony. Plants of var. *paludicola* from southern Florida self-pollinate. This phenomenon is associated with pollinator-paucity.

Autogamy is reported in *Cyclopogon prasophyllum* (Chiapas, Mexico) and a *Cyclopogon* species of S. Vera Cruz, Mexico. Also in *C.stenoglossa.*

C.diversifolius from Paraná, South Brazil is exclusively pollinated by halictid bees (Singer, pers. comm.).

A number of species from other genera is also reportedly autogamous: *Eurystylis cotyledon*, *Discyphus scopulariae*, *Hapalorchis lineatus* (Brazil) *Helonoma* (*Spiranthes*) *bifida* (Venezuela), a *Pelexia* sp., a *Sarcoglottis* sp., *Stenorrhynchos* (*Sacoila*) *hassleri*, a *Schiedeella* sp. (Mexico), *Spiranthes prasophylla*. var. *cleistogama* (Guatemala).

Genus **Aracamunia**

In 1987 a peculiar species was found in the Cerro Acaramunia in Venezulela with liguliform (tongue-shaped) glandular structures at the base of the leaves that are supposed to excrete insectivorous substances.

This would be the first described carnivorous orchid ever. It was placed in a separate monotypic genus. Pollination unknown.

Genus **Sarcoglottis**

Pollinaria of *S.acaulis* from Panama, which is presumed to be a nectar flower, were found on *Eulaema cingulata* (Roubik & Ackerman 1987).

Genus **Spiranthes**

Spiranthes speciosa and *S.standleyi* show the bird-pollination syndrome (Van der Pijl & Dodson 1966).

CRANICHIDEAE: CRANICHIDINAE

About 152 species in nine genera, mostly in tropical America. Not much is known about pollination.

Genus **Ponthieva**

Ponthieva racemosa is reportedly visited by halictid bees (Luer 1972), but Dressler (1993) found that the lip produces oil, rather than nectar, so that the pollinators may be anthophorid bees that gather oil as food for the larvae.

Genus **Cranichis**

C.fertilis from Venezuela and *C.muscosa* of Puerto Rico are mentioned by Catling as being autogamous.

ORCHIDOIDEAE

DIURIDAE: CHLORAEINAE

About 80 species in six genera of mostly terrestrial South American species. Also known from New Caladonia.

Gumprecht (1977) describes the pollination of a *Chloraea* species of Chile. The genus *Chloraea*, together with two other endemic genera *Bipinnula* and *Gavilea* have conserved their original primitive status in their isolation. They have pollen without a viscidium. The back of the pollinating insect is smeared with stigmatic fluid and is brought into contact with the completely dry pollen masses. These three genera have developed two devices:

1. Nectar 'canals', produced in two sacs of the fruit, lying adjacent to the placentae and ovules. The two entrances are formed by the column.

2. Stigma and pollen lie very close to each other and partly overlap. The upper part of the stigma disconnects from the column and forms a kind of 'coat-pocket', containing the pollen, and from which the pollen is pulled out as a 'handkerchief'.

The insect lands on the lip which is hinged. The more the insect intrudes into the flower, the more steeply the movable lip with the insect positions itself. As the animal reaches the entrances with its head it stands fully upright and in this position can only put in its tongue into one of the entrances. The back of the insect touches the stigma and if it already carries pollen it dislodges this. In backing out for the first millimeter the movable anther slides back. The pollinia are pulled out and are fully stretched forward in the direction of the insect's back. *Colletes araucariae*, the wasp that pollinated *Chloraea cylindrostachya* has been photographed by Gumprecht.

Catling lists the following autogamous species from Chile and the Falkland Islands: *Bipinnula* sp., *Chloraea falklandica*, *C.fonchii*, *C.inconspicum*, *C.philippi*, *Gavilea* (*Asarca*) *araucana*, *Gavilea* (*Asarca*) *commersonii* and *Codonorchis tetraphylla* (Catling 1990).

ORCHIDEAE: HABENARIINAE

This large subtribe with more than 20 genera is mainly African, but in part pantropical. For most of the subtribe lepidopteran pollination is indicated.

Genus **Habenaria**

Many species are found in South America. Brazil has 200 native species, Argentina has 14 native species and Chile has one native species. Most species can be considered as sphingophilous or phalaenophilous by their flower characteristics.

Habenaria pleiophylla of Brazil is reportedly pollinated by the Passion Vine Butterfly *Heliconius eratophyllis*, which receives pollinia to the base of the eye near the proboscis.

According to Hoehne (in Singer & Cocucci 1997b) species of the *pratensis* group deviate from the sphingophilous syndrome by yellow perianth colour, much expanded labellum lobes and diurnal fragrance. These characteristics indicate butterfly pollination.

Some species are reported as autogamous: *H.mesodactyla* and *H.rodeiensis* from Belize.

Singer & Cocucci (1997b) studied morphological adaptations to sphingophily and phalaenophily in five species from Argentina. *H.gourlieana* and *H.hieronymi* are pollinated respectively by hawkmoths (*Agrius cingulatus*, *Manduca sexta*) and settling moths (*Rachiplusia nu*). Hemipollinaria are attached to the eyes. In two other species *H.paucifolia* and *H.rupicola* pollination by settling moths with hemipollinaria attachment to the eyes can be inferrred by indirect evidence, such as placement of scales and massulae on flowers and by flower structure. For *H.montevidensis* pollination by small moths or mosquitos with hemipollinaria attachment to the proboscis was postulated.

A comparitive study of floral features revealed clear morphological divergence (in spur length and form and position of petal lobes and median rostellar lobe, etc.) between the sphingophilous and phalaenophilous species, in conformity with differences in pollinator behaviour and morphology (hovering vs settling and lengths of probosces). In pollination by hawkmoths, dangling hemipollinaria at both sides of the insect are dragged over the stigmatic arms and pollinia contact and adhere to the sticky stigma. After leaving, the hemipollinaria break away some massulae. Thus the massulae can be distributed over several flowers. The upturned perianth lobes, by guiding the moths' approaching movement, favour frontal strikes of the pollinia against the stigma and prevent lateral strikes. This favours symmetrical pollinations and frontal contact. Most species with long spurs have relatively long petal lobes. In *H.hamata* the flowers lack lateral petal lobes, but have strongly curved spurs. In this species the spur curvature forces the pollinators to adopt a median position.

In four species from the same area Singer and Cocucci compared male efficiency. *Male efficiency factors* proved (unexpectedly) low in all species. Only in one species, *H.hieronymi*, each pollination donation accounted for more than one pollination. More, longterm studies seem necessary.

H.gourlieana is part of a more or less rich sphingophilous flora interacting with the two long-tongued hawkmoth species mentioned. It had the lowest pollination rate of the four species (about 20%). It is predicted that *H.longicauda* with spurs up to 25 cm will be pollinated by *Neococytius cluentius* and *Amphimoeca walkeri* with spur lengths of resp. 25.5 and 28 cm. The Argentine species *H.bractescens* with a spur of 5-7 cm is also expected to be hawkmoth-pollinated.

EPIDENDROIDEAE: CYMBIDIOID PHYLAD

MALAXIDEAE

This is a world-wide distributed tribus of about 960 species in six genera. Pollination by mosquitoes is known from *Liparis* and *Malaxis* in temperate regions. In *Liparis* also hemipterans have been observed. Self-pollination is reported from a *Malaxis* from Mexico (Catling 1990).

Genus **Malaxis**

Malaxis massonii

The nectarless *Malaxis massonii* from the Luquillo Mountains of Puerto Rico was studied by Carrasco & Ackerman (1997). It was used to test the *density dependence hypothesis* in relation to deception pollination. This hypothesis states that deception pollination systems have no fitness cost because rewards do not increase reproductive success at low natural pollination densities. Under manipulated conditions flower and fruit production were measured and pollination removal monitored. The results of the study suggest that density dependent processes are occurring at other scales than those measured, and that density effects are not detected immediately in the current reproduction season, or that reproduction in this orchid species is not density dependent.

CALYPSOEAE

About 35 species in nine genera, distributed in Northern temperate and subtropical America. *Tipularia* is pollinated by noctuid moths. *Calypso* deceives bumblebees by pseudopollen. *Aplectrum* is dependent on halictid bees. *Corallorhiza* is reportedly pollinated by syrphid flies. Species from Mexico are self-pollinating (Catling 1990).

CYMBIDIEAE: GOVENIINAE

Some 20 species in one genus of which the pollination is unknown, but probably by bees.

CYMBIDIEAE: EULOPHIINAE

A pantropical subtribe of 264 species, mainly from the Old World.

Genus **Oeceoclades**

A genus of Africa and Madagascar with some representatives in America.

Oeceoclades maculata

This species is to be found in both the New and the Old World. This peculiar distribution of this terrestrial plant may be explained by seed-dispersal across the ocean and successful colonisation due to its autogamous reproduction and wide range of environmental tolerance (Stern 1988). It even invaded recently into Florida.

González-Díaz & Ackerman (1988) investigated reproduction in Puerto Rico. In this species the anther cap falls off and the two pollinia hang free, attached to the column by the thread-like stipe and viscidium. The stipe bends down and brings the pollinia near the stigma. The pollinia have to be disturbed physically to contact the stigma while hanging. Disturbance is caused by rainfall. Flowers protected from rain produce very few fruits (2%) compared to flowers affected by rain (50-60%). Other factors than pollination limit fruit set, as proved by experiments in which hand pollination yielded the same results as rain-assisted pollination.

Genus **Pteroglossaspis**

A primarily African genus of about seven species, of which two are native in the New World.

Pteroglossaspis ruwenzoriensis

This is the only representative of the genus occurring in South America, but is also known from Africa. It is a terrestrial species, recorded from central Argentina, Southern Brazil and Uruguay, growing in disperse populations of single plants or small plant clusters. Singer & Cocucci (1997) studied its pollination in Argentina. Visitors observed were beetles (Cetoninae), known to feed on ripe fruits, fermented sap and flower nectar. The only pollinator observed was *Euphoria lurida*. Visitors approached the inflorescences in a zigzag upwind flight suggesting long-distance olfactory attraction. After alighting the beetles crawl between flowers, mainly from basal older flowers upward. Flowers are entered from the lip and the insects have to push petals and sepals apart with force to reach the nectar. They gnaw the jelly-like, more or less solid nectary exudation at the column's base. Of 115 flowers (11 plants) 63 fruits (54.8%) were formed. Male function was 96.4%, female 66.1%.

CYMBIDIEAE: CYRTOPODIINAE

Genus **Cyrtopodium**

Dodson and Frymire found pollinaria of a *Cyrtopodium* species, (possibly *C.punctatum*) on the metathorax of *Euglossa hemichlora* in Ecuador. Dodson (1962) however mentions *Eg.viridissima* as pollinator of *C.punctatum*. No nectar is found in the flowers. The scents of *C.andersonii* and *C.punctatum* have not been identified, but do not seem to have a strong euglossine-syndrome character. Pollination by euglossine bees (*Euglossa*) has also been reported by Van der Pijl and Dodson (1966). No details.

Genus **Galeandra**

Pollinaria of *Galeandra* (probably *G.stangeata*) have been found on *Euglossa* and large anthophorid bees (Dressler 1976b, 1981). Gerlach & Schill (1991) found pollinia of *G.devoniana* on male *Euglossa imperialis* in Venezuela. Probably both sexes visit the spurred flowers.

CYMBIDIEAE: CATASETINAE

All species of Catasetinae (about 145 in five genera, Dressler 1981) are euglossine-pollinated. Often flowers are unisexual (Fig. 21).

This is a bit unexpected, since the functional bond between the highly specialised *androecium* and *gynoecium* in the orchid family would seem unbreakable. Superabundance of floral drugs induces male Euglossidae to return repeatedly to the same flower to gather perfume. A self-incompatible bisexual flower would result in a wastage of pollen. The evolution of a differently constructed and positioned flower became advantageous in these cases of overprecise pollination (Van der Pijl 1978).

Genus **Catasetum**

Catasetum, formerly including *Dressleria* and *Clowesia* as sections, is now a homogeneous group, comprising more than 100 species (Williams 1982, Schlechter 1996).

Dimorphism

Male flowers are usually found on small plants and in poor habitats. It is believed that well-nourished plants that are exposed to plenty of sun tend to be female (Gregg 1975), but Warford and Harrell think that factors other than fertiliser and sun may be responsible for the production of exclusively female flowers. Male flowers are more numerous than female ones. Data indicate that approximately

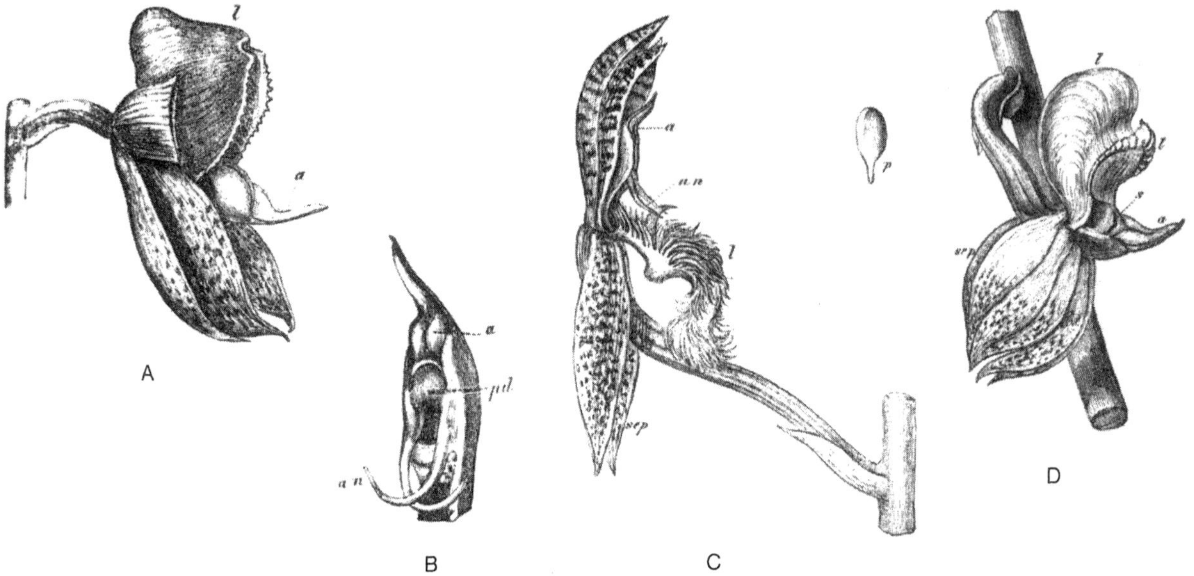

Figure 21. *Catasetum tridentatum*, a species with three different forms of flowers, sometimes even found on the same plant. Initially they were considered as different genera (Figs 30 and 31 from Darwin 1888). A) Side view of a male flower in its natural position, with two of the sepals cut off. B) Front view of a male flower of column, in position reverse of A. C) Side view of *Myanthus barbatus* in its natural position, the sterile hermaphrodite form. Note: According to Oakes Ames (in Cribb 1999) this was in reality the normal form of *Catasetum*. D) Side view of *Monachanthus viridis* in its natural position, the female flower (the shading in C and D has been added from Mr Reiss' drawing in the 'Linnean Transactions'). a = anther, an = antennae, l = labellum, p = pollen-mass, rudimentary, pd = pedicel of pollinium, sep = two lower sepals, s = stigmatic cleft.

25 male flowers occur for each female. Sometimes flowers are found with intermediate polymorphic flowers ranging from male-like to female-like, combining the features of male and female flowers in one sterile entity. Flowers previously thought to be hermaphroditic are in most cases non-functional and appear to be 'errors' (Warford & Harrell 1996). Generally, *Catasetum* plants are found in association with dead trees (Dodson 1978). The flowers produced by different species are quite distinct morphologically. Two or three days after anthesis has begun they begin producing odour. The floral fragrances of many species have been studied. They attract the male euglossine bees which usually land on the lip. Bees may even fight for flowers and try to pull each other away (Janzen 1981).

The column of the male flowers of most Catasetums (not in the four species of section *Pseudocatasetum*) have one or two long, equal or unequal antennae (on the form of which the division into different sections is based) that are very sensitive to touch (Plate 38, p. 84; Fig. 22). Bees that have landed scratch at the fragrance-producing parts of the lip near the ends of the antennae. The anther is under great tension and is released violently at the slightest touch. The viscidium with the pollinarium is cemented to the back of the bee and after about 20 minutes drying of the stipe the anther cap falls

Plates

38. *Catasetum viridiflavum*. Note the antennae. Photo N.A.van der Cingel (pp 84 and 89).
39-41. Inflorescence of *Catasetum fimbriatum* attracting Euglossa cordata males which collect fragrances. Photos R.B.Singer (p. 87).
42. Male *Eufriesia violacea* (Euglossinae) with pollinarium of *Catasetum cernuum*. Photo R.B.Singer, Serra do Japí, mesophytic forest of São Paulo State (p. 88).
43. Male *Eulaema nigrita* hovering at a bait of cineol. This technique for attraction of male euglossine bees is widely used in orchid pollination studies. Photo A.Cocucci (p. 88).
44. Male *Euglossa chalybeata iopoecila* with pollinarium of *Catasetum trulla*. Photo R.B.Singer, Tagaçava, Atlantic rain forest in Paraná (p. 88).
45. The same euglossine species from the same locality with pollinarium of (apparently) *Houlettia brockelhurstiana* (Stanhopeinae). Photo A.Cocucci (pp 88 and 101).

38	40	41
42	44	
39	43	45

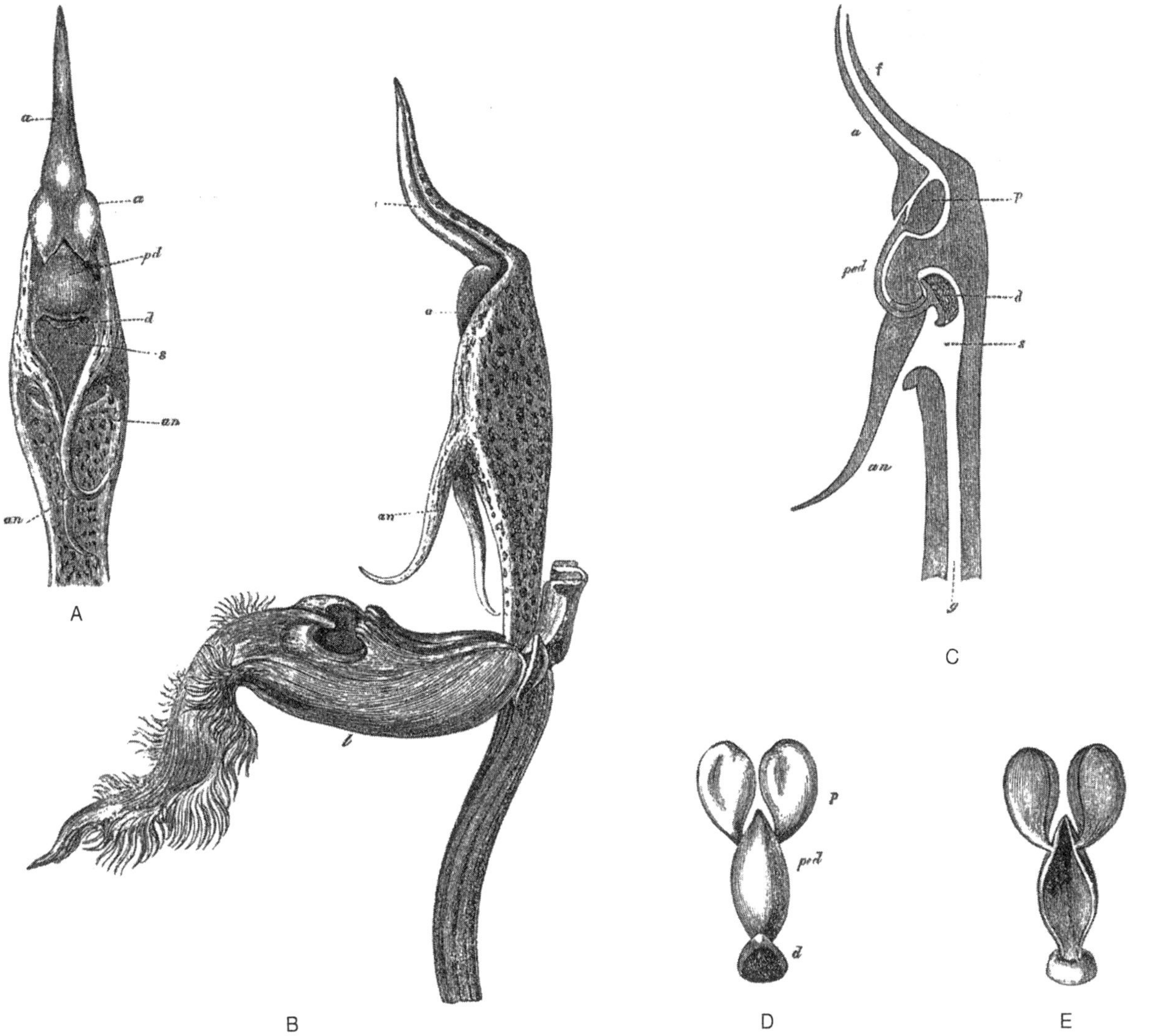

Figure 22. *Catasetum saccatum*. The pollinium is freed explosively after the left-hand antenna is touched (Figs 28 and 29 from Darwin 1888). A) Front view of column. B) Side view of flower, with all the sepals and petals removed except the labellum. C) Diagrammatic section through the column, with all parts a little separated. D) Pollinium, upper surface. E) Pollinium, lower surface, which before removal lies in close contact with the rostellum. a = anther, an = antennae of the rostellum, d = disc of pollination, f = filament of anther, g = germen or ovarium, l = labellum, p = pollen-mass, pd or ped = pedicel of pollinium, s = stigmatic chamber.

46	47	
50	51	
48	49	52

Plates

46-47. *Stanhopea wardii*. Note the construction of the 'fall through' flower. Photo N.A.van der Cingel (p. 104).

48. Noctuid moths (Noctuidae) with pollinaria of *Epidendrum latilabre*. Photo R.B.Singer, Coastal Paraná (p. 121).

49. *Epidendrum radicans*. Mimicry complex of three plants with bicoloured flowers, attracting the same butterfly pollinators. Photo by N.A.van der Cingel (p. 121).

50. *Asclepias curassavica*. Mimicry complex of three plants with bicoloured flowers, attracting the same butterfly pollinators. Photo by N.A.van der Cingel (p. 121).

51. *Lantana camara*. Mimicry complex of three plants with bicoloured flowers, attracting the same butterfly pollinators. Photo by N.A.van der Cingel (p. 121).

52. Butterfly on flowers of *Lantana camara*. Photo N.A.van der Cingel, November 1997, Costa Rica (p. 121).

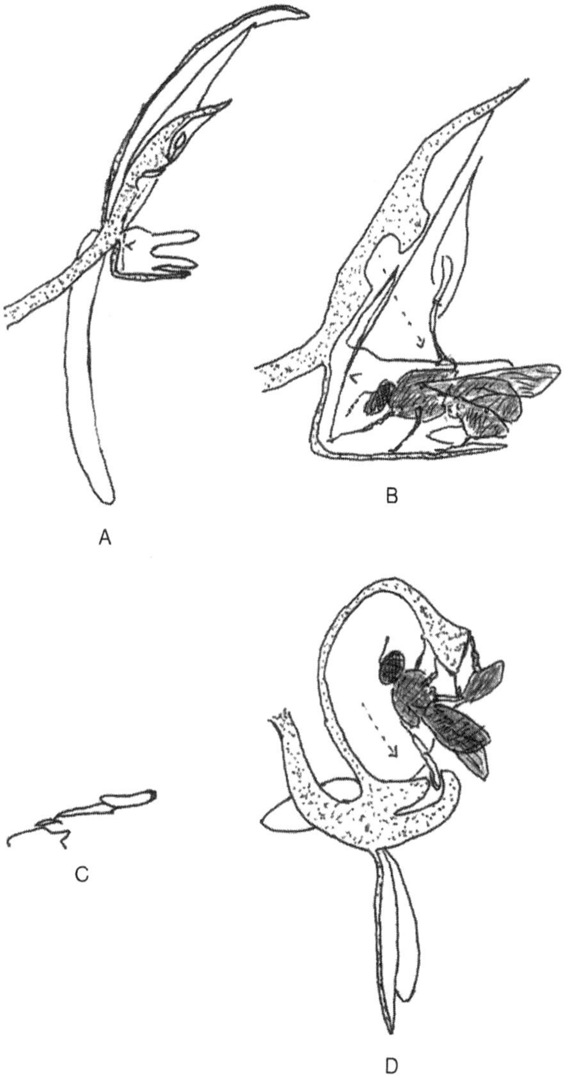

Figure 23. *Catasetum bicolor*. Pollination mechanism. A) Longitudinal section of male flower with pollinarium in place. B) Part of male flower, showing *Euglossa cordata* touching the antenna and receiving the pollinarium on the thorax. C) Placement of pollinarium on thorax. D) Longitudinal section of female flower with bee backing out and inserting a pollinium in the stigma (after Dressler 1968).

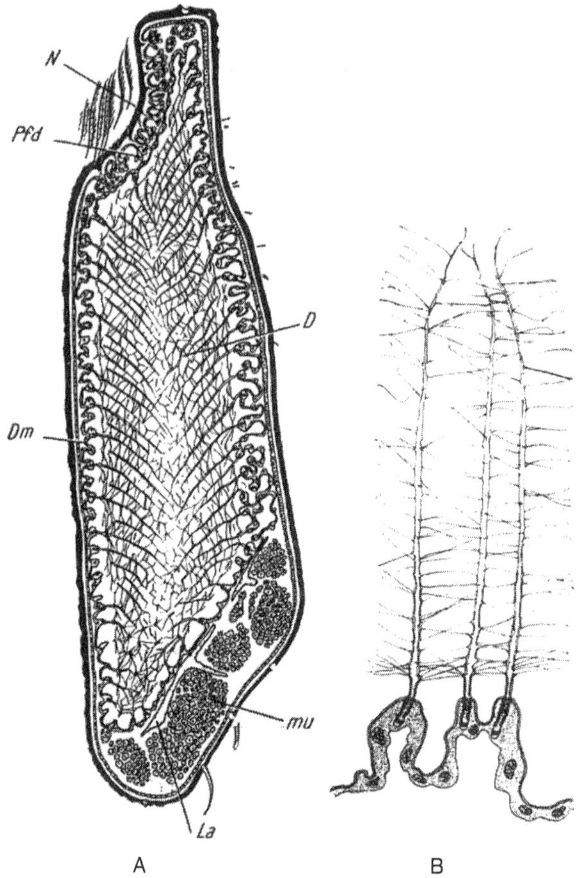

Figure 24. Fragrance container of *Euglossa cordata*. A) Transsection of hind tibia. B) Three separate hairs of the capillary-apparatus. *N* = large container or inflation, *Pfd* = glandular tissue, *D* = cavity, *Dm* = membrane of the cavity, *mu* = muscles of the leg, *La* = blood-vessel(?) (after Vogel 1966).

off and the pollinia are elevated above the back of the bee, taking the position needed for pollination of the female flower (Fig. 23). The production of fragrance in the male flower soon stops. According to Romero & Nelson (1986) *Euglossa* bees avoid male flowers after having forcibly received a pollinarium, but not the female flowers. This behaviour must lead to competition among male flowers. Among *Catasetum* species, the extent of dimorphism is correlated with the degree of apparent aversion caused by the male flower (viz. the weight of the pollinarium). Aversive pollinarium placement is supposed to be responsible for widespread sexual dimorphism in *Catasetum*. The long-lived female flowers are remarkably uniform in different species and not so easily distinguished. They start the production of fragrance later than the male flowers, about 3-4 days after the beginning of anthesis, and can live up to a month if they are not pollinated. Production of fragrance stops soon after pollination. Bees attracted land on the large hooded lip and enter this upside down. Inside, the bee scratches the source of odour-production and on leaving the pollinia are caught by the stigmatic cleft. Usually one pollinium is left behind. The bees store the fragrance in containers in the hind tibia (Fig. 24).

Pollinators

Williams lists the pollinators and visitors as follows:

Orchid	Pollinator/visitor	Note
C.barbatum	*Euglossa cordata*	a
C.bicolor	*Eg.cordata*	
	Eg.tridentata	a
	Eg.cyanaspis	a
C.cernuum	*Euplusia violacea*	a
C.costatum	*Euglossa* sp.	
C.discolor	*Eulaema meriana*	
	El. cingulata	a
	Euglossa chlorosoma	b
C.fimbriatum	*Euplusia auriceps*	
C.hookeri	*Euglossa cordata*	a
C.integerrimum	*Eulaema cingulata*	a
	El.polychroma	a
C.longifolium	*El.bombiformis*	
	El.cingulata	
	El.meriana	
C.luridum	*Euglossa cordata*	a
C.macrocarpum	*Eg.imperialis*	b
	Eulaema cingulata	a
	El.bombiformis	
	El. aff. *luteola*	
	El.meriana	a
	El.benettii	a
	El.nigrita	a
C.macroglossum	*El.bomboides*	a
	El.cingulata	a
	El.polychroma	a
	El.speciosa	
C.maculatum	*El.cingulata*	a
	El.polychroma	a
C.platyglossum	*El.bomboides*	
	El.cingulata	a
	El.polychroma	
C.reichenbachianum	*Euglossa* sp.	
C.saccatum	*Eulaema cingulata*	a
	Euglossa ignita	b
	Eg.augaspis	b
C.tabulare	*Eulaema cingulata*	a
C.thompsonii	*Euglossa augaspis*	
	Eg.cognata	
	Eg.cordata	
	Eg.liopoda	
	Eg.mixta	
	Eg. sp. UM-10	
	Eg. sp. Rd-1215	
	Eulaema cingulata	b
C.viridiflavum	*El.cingulata*	a

a = observed pollinating, b = nonpollinating visitor.

Euplusia is nowadays included in Eufriesea (Dressler 1982).

To this list some other observations can be added:

Dodson (1962) mentions observations of *El.dimidiata* and *Eg.piliventris* as pollinators of *C.macrocarpum* (Ducke, Crüger) and of *El.cingulata* for *C.oerstedii*. (Both Ducke and Crüger thought euglossine bees search for food in the orchid flowers, see Dressler 1968a.) Furthermore *El.cingulata* for *C.* aff. *eburneum* (as *C.fragrans*).

Braga (1977) caught an *Euglossa cordata* in a *C.barbatum* flower that had already lost its pollinia. He was able to attract bees of the same species with a combination of odorous substances known from the flowers of this orchid. The flowers have the strongest scent in the morning.

C.discolor is also pollinated by *Eulaema bombiformis*, *C.expansum* by *El.bomboides*, *El.cingulata*, *El.polychroma*, *C.pileatum* by *El.cingulata* and *El.peruviana*, *C.viridiflavum* (Plate 4, p. 12) also by *El.nigrita* (Whitten et al. 1986).

C.maculatum is also visited by *El.meriana* (Janzen 1981).

The Mexican species *C.pendulum* and *C.laminatum* attract and are pollinated by *El.polychroma* (Warford & Harrell 1996). *C.pendulum* is visited by the larger *El.seabrae* as well, but this bee species was not observed with pollinia.

C.ochraceum is visited by *Euglossa gaianii* and *Eg.modestior* (Romero & Nelson 1986).

Roubik & Ackerman (1987) found pollinaria of *C.bicolor* on *Euglossa allosticta*, *Eg.cognata*, *Eg.cybelia*, *Eg.deceptrix*, *Eg.dissimula*, *Eg.dodsonii*, *Eg.gorgonensis*, *Eg.hemichlora*, *Eg.mixta*, *Eg.tridentata* and *Eg.variabilis*, those of *C.viridiflavum* on *Eulaema cingulata* and *El.nigrita*.

According to Singer (pers. comm.) the legitimate pollinators of *C.fimbriatum* are euglossines of the genus *Eufriesea*. Plates 39-41 (p. 84) illustrate the attraction of *Euglossa cordata* male bees collecting fragrance from *C.fimbriatum* flowers. The whitish, callum-like area seems to be the most important part in this function, although the whole lip is involved in fragrance production. *E.cordata* bees are too small to touch the rostella arms to trigger the pollinarium.

Fragrance composition and specificity

The fragrances of both male and female flowers of a species are the same and rather specific. Hills et al. (1972) investigated the chemical composition of fragrances of *Catasetum* species and identified 14 out of 39 compounds. The results of this investigation demonstrate that each species of orchid sampled which is known to have a species specific pollinator also has a species specific fragrance, and that those species of *Catasetum* which have identical floral fragrances have the same pollinators.

Isolation
The fragrances thus may contribute to the isolation of species from each other. Those orchids which produce distinct fragrances and attract different pollinators may be considered separate biospecies in the sense of Paulus & Gack (1983). By this criterion, *Catasetum dilectum* of Panama and *C.dilectum* of Ecuador are separate species. The same is true for *C.barbatum* of Brazil and *C.barbatum* of Ecuador (Hills et al. 1972). Differential fragrance production is one of the primary barriers to gene exchange among otherwise interfertile entities in the genus *Catasetum* (Williams 1982). In some cases geographical, mechanical, ecological and seasonal isolation seem to be important as well.

Species integrity and specificity of fragrances
(Plate 42, p. 84).
As can be seen from the lists above, some bees are pollinators of several species and some orchids are pollinated by different species of bees. Dodson (1978) e.g. found four species of *Catasetum* coblooming in Guyana. All four were visited by *Eulaema cingulata*. Size relationships precluded this bee as a pollinator of one of the species. Also temporal differences were noted in the visitation of different flowers by different bee species, contributing to some temporal isolation. But occasionally bees (*Eulaema meriana*) were captured which carried the pollinia of three species. Hybrids were found between three of the four species. Isolation therefore does not always seem perfect and may contribute to the formation of new species. For a discussion of the long range effect of such hybridisation, see Dodson (1978). Warford & Harrell (1996) consider the enormous variation in the genus *Catasetum* to be the result of natural hybridising and backcrossing, following different authors in this respect. The promiscuous South American Catasetums, they state, demonstrate a surprising lack of barriers toward interbreeding, with perfume blends attractive to as many of five different euglossine pollinators.

All this raises the questions of the specifity of the fragrances to attract adequate pollinators and of the species integrity and evolution of the orchid species.

Several researchers (Williams 1982, 1985a, 1985b, Williams et al. 1983, Gerlach & Schill 1991) have contributed to answers to these questions by fragrance analysis and field-experiments with bioassays to identify the attractive properties of different fragrances and chemical compounds (especially numerous monoterpenes) found in these fragrances to different bee species. Such field-experiments (Plate 43, p. 84) have been carried out in almost all of the countries where these euglossinophilous orchids and euglossine bees occur. In general, it can be concluded that, with some exceptions, substances which are structurally similar to compounds identified in orchid fragrances are not very good attractants for male euglossine bees and that some chemicals not found in orchids are good attractants. Combinations of compounds with good attracting properties in separate form can be less attractive and some chemicals act as repellents and reduce the effects of attractive substances.

These somewhat embarrassing results indicate that specificity is attained by different mixtures of fragrant compounds, composed of both general and specific repellents and attractants. Some compounds may be modifiers that impart species specifity to the mandibular gland secretions of euglossine bee pollinators (Williams 1980). Small genetic changes in the plants may lead to important changes in the metabolism of fragrance production in the osmophores of the labellum and consequently in attraction of pollinators. In general, the floral fragrances in *Catasetum* attract specific pollinators or groups of pollinators and contribute to the isolation of closely related sympatric species.

Note 1: Ackerman (1983d) states that pollinator specificity occurs in less than half of the euglossine-pollinated orchids, and that host specificity is rare. There is little evidence that the bees are dependent on the fragrance hosts. The orchids tap the majority of euglossine species and individuals for pollinator services during a given season, but most of those bee species that temporarily lack orchid fragrance persist in the area, continually emerge from nests, and seek floral fragrance compounds from alternative fragrance sources (Plates 42, 44 and 45, p. 84). Nearly a third of the local male euglossine bee species censused by him are not pollinators of any fragrance orchids in the area.

Note 2: Obviously there are no fragrance patterns common to all euglossine-pollinated orchids, not even in those attracting perfume-collecting males. The only fragrance character common to most euglossinophilous flowers is that they contain monoterpenes and aromatic esters in higher quantities than other plants (Gerlach & Schill 1991).

Long distance attraction
The production of fragrances is thought to be controlled by hormones and adapted to the activities of the bees. Air temperature plays an important role in the production and release of scents. On cool days, when the bees are not active, no fragrances are produced. The scents attract pollinators over long distances, which is important in relation to the 'trap line' behaviour of the bees in the habitats of tropical epiphytes, where individual plants can be widely dispersed.

A number of plants pollinated by male euglossines have widely dispersed populations, often occurring as single plants per acre, comparable to the thin distribution of a number of rain forest trees. Just as long distance pollination is important for these tree species, so must these orchids depend upon the same phenomenon. Their population structure reflects pollination by insects with long flight ranges (Williams & Dodson 1972).

Mimicry
It should be noted that euglossine bees may mimic each other. (Dressler 1979) describes Müllerian mimicry between *Eulaema bombiformis*, *El.meriana* and *E.seabrae* in the Amazon basin, for example. The same three species are, however, not mimetic in coastal Brazil. He also mentions other examples of mimetic complexes, involving *Euplusia* and *Eulaema*. It will be clear that this makes identification of euglossine bees sometimes very difficult.

Catasetum viridiflavum
Zimmerman et al. (1989) studied flowering patterns of this species in central Panama and related these to data of seasonal abundance of its pollinator *Eulaema cingulata*. *C.viridiflavum* (Plate 38, p. 84) did not flower when pollinators were abundant. *E.cingulata* pollinates in that area at least six different species of orchid and thus its phenology was not expected to be closely tied to that of a single orchid species having a relatively small geographical range. The flowering time of *C.viridiflavum* appears to be maintained by factors that preclude pollinator tracking; bee population and flowering peaks are frequently out of phase by as much as six months. Patterns of fruit loss indicate for counter-selection against plants that flower early, near the time of peak pollinator abundance. Early flowering plants have a considerable higher fruit set (≥50%) than later in the season (4-21%). It is suggested that plants maximise growth following dormancy as a precondition for reproduction later in the growing season by the general increase in flowering towards the later growing season. Flowering time is not simply a function of pollinator abundance, and plants appear not to have responded to selection as predicted by the hypothesis of optimal pollination. It seems that constraints on growth and flowering imposed by drought-induced dormancy best explain the phenology in this species.

Genus **Clowesia**

Clowesia is a genus of five species with small to medium-sized bisexual gullet flowers: *C.rosea*, *C.thylaciochila*, *C.warscewiczii*, *C.russeliana* and *C.glaucoglossa*. The column in these species is placed relatively near the lip, such that the pollinator brushes the anther with the upper thorax and than receives the pollinarium (Dodson 1962), mostly on the scutellum (Dressler 1981).

Clowesia dodsoniana
C.dodsoniana belongs with *C.thylaciochila*, *C.russelliana* and *C.glaucoglossa* to the *C.russelliana* group. *C.thylaciochila* as cited in the older literature is in fact *C.dodsoniana*, and the true *C.thylaciochila* is more restricted in distribution than the latter. It also has a much smaller pollinarium and stigma. In *C.dodsonianal*, a viscidium curtain, which aids in blocking pollination until the pollinarium has been removed, is absent. It is successfully pollinated by *Eufriesea caerulescens*. Warford & Harrell (1996) describe its pollination. The orchid has a powerful sweet scent and attracts bees as soon as the odour is emitted. This fragrance differs from that of *C.thylaciochila*, which species is pollinated by the same insect, according to Romero. The bees enter the sack-like base of the flower and emerge with a pollinarium (sometimes two or three) stuck to the scutellum. The anther is not released from the pollinarium for several hours, a contrivance avoiding self-pollination. The osmophores, which can be demonstrated by a neutral red solution of about 1:1000 (Vogel 1962), are located on the primary stripes of the flowers and by some markings at the column apex and around the stigma. They function as fragrance guides to the bees. The most intense concentrations of dye were taken up by the frilled ornamentation at the tip of the labellum, by

the sac-like inner surface of the base, and by the glands at the base of the column.

Clowesia ruselliana
Dodson (1962) has described the pollination of this species, then still incorporated in the genus *Catasetum* as *C.russellianum*, subgenus *Clowesia.* In the Missouri Botanical Garden Dodson experimented with captive *Eulaema cingulata* bees. The bee, after landing on the lip, in attempting to reach the saccate part, touches the stipe and the pollinarium is released. It swings down into such a position that the bee will push against the viscidium with its metathorax when backing out. After drying of the anther cap it falls off and, on the next visit to another flower, the pollinia (one or both) are left behind in the stigmatic cavity as the bee exits. In Mexico *C.russelliana* is pollinated by *Euplusia mexicana*, attracted by a fragrance that contains 78% cineole. It is the natural pollinator of the species.

Clowesia warscewiczii
Together with *C.rosea* this species forms a different group with fringed labellae.

This species is pollinated by *Eulaema nigrita* and *E.bombiformis*. Here the viscidium becomes attached to the leg just above the joint between the femur and tibia. This curious type of pollination is called 'bees-knees pollination' by Williams (1982). Pollinaria of *C.warscewiczii* were found on *Eulaema meriana* and *El.bombiformis* in Panama.

Other species of **Clowesia**
The pollinators of the other species are unknown, although it is believed that *C.thylaciochila* is pollinated by *Euplusia concava* as is suggested by its geographical range and by the presence of 2-phenylethylacetate in the fragrance, which is a strong attractant for this bee.

Genus Cycnoches

This is taxonomically a complex genus, due to dimorphic species and hybrids. They have unisexual flowers, so the sexes are separated on different plants, although occasionally hermaphroditic flowers are produced.

Allen (Williams 1982) divided the genus into two sections, one (of five species) with male and female flowers of almost equal size and appearance and the other (15 species) with strikingly dissimilar male and female flowers.

In the first section (section *Cycnoches*) pollination is as follows: The male bee approaches the flower and lands inverted or gradually swinging around into an inverted position on the non-resupinate lip, grasping the large callus. In the male flowers the callus is slightly larger and more extended than in the female ones (which appear to have a stronger odour). The bee than manoeuvres itself into the proper position to collect fragrances from the callus. On the male flower the bee must release the grasp of his hind legs to get closer to the source of the odour. The abdomen then swings down, brushing its tip against the apex of the column. The tensed trigger mechanism of the anther discharges the viscidium violently and this sticks to the end of the abdomen on releasing the stipe. This dries and straightens (in 40 minutes in *C.lehmannii*) and after a few hours the anther cap falls off. The female flowers are slightly smaller and this forces the bee, after brushing, to launch into the air to transfer the collected fragrances. The bee falls a short distance and the abdomen swings upward. If it carries a pollinarium, this is caught in the stigmatic cleft and is pulled from the stipe.

In the second section (*Heteranthae*), with large female and small male flowers, the male bees (usually of the genus *Eulaema*, which are smaller than those of the genus *Euglossa*) land on the blade of the flexible lip of the male flowers and crawl to the underside toward the source of the odour. The blade swings down with the weight of the bee. In doing so the insect touches the column-apex with its abdomen and receives the pollinarium (Fig. 25). After drying (in one hour appr.) it can be transferred to the female flowers in the same way as described for the first section.

Arditti (1966) mentions *Centris fasciata* as pollinator of the genus but gives no details.

Cycnoches egertonianum
This species belongs to the Heteranthae section. According to Dressler and Dodson (in Van der Pijl & Dodson, but see also Dressler 1968b) a number of distinct morphological taxa have been lumped into this species. They found four forms in Panama that were species-specific in the attraction of pollinators and mutually exclusive in that sense. The typeform is pollinated by *Euglossa ignita* and *E.flammae*.

Dodson (1962) mentions *Eg.viridissima*, but this is not repeated in the 1966 book..

Pollinators of the other forms are: *Euglossa cyanura*, *E.tridentata* and *E.hansonii*.

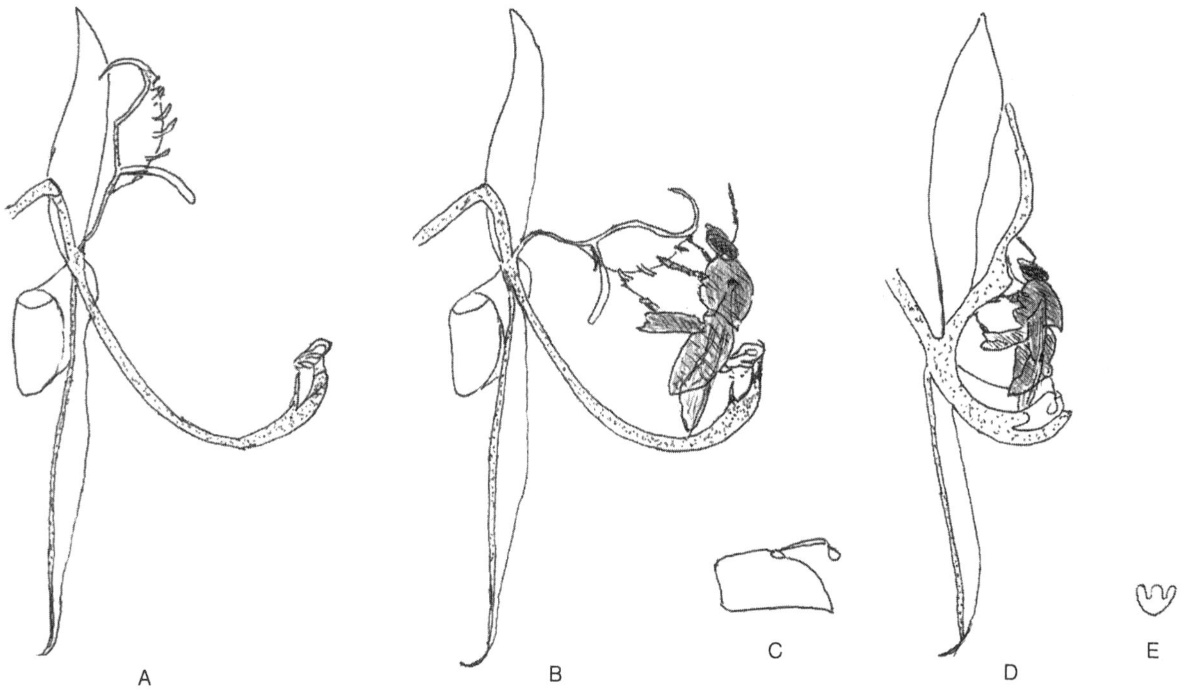

Figure 25. Pollination in *Cycnoches aureum.* A) Longitudinal section of male flower with pollinia in place. B) The weight of the pollinating bee (*Eulaema nigrita*) pulls the lip down and the abdomen of the bee strikes the apex of the column, causing the pollinarium to be expelled and placed near the apex of the abdomen. C) Placement of pollinarium on abdomen. D) Longitudinal section through female flower with the bee inserting a pollinarium in the stigma. E) Apex of the female column, showing the slits which catch the pollinia (after Dressler 1968).

Other species
Williams (following Van der Pijl and Dodson) lists the following species and pollinators/visitors:

C.aureum: The species is pollinated by *Eulaema nigrita.* As nonpollinating visitors *Euglossa tridentata*, *Eg.crasssipunctata* and *Eg.cyanaspis* were observed.

C.lehmanii is pollinated by *Eulaema cingulata.*

C.pentadactylon: *Euplusia superba* was caught with pollinaria.

C.peruviana is pollinated *by Euglossa hemichlora.*

C.ventricosum (including var. *warscewiczii*) is also pollinated by *Eulaema cingulata* (Dodson 1962 mentions Allen).

Roubik & Ackerman (1987) list several species of euglossine bees that were caught with pollinaria of *C.guttulatum*, *C.egertonianum*, *C.aureum* and other unidentified species.

Genus **Dressleria**

Dressleria is the most primitive genus of the Catasetinae. All four species are structurally distinct and each species is pollinated by a different group of euglossine bees. The species, of which some are sympatric, are reproductively isolated on the basis of the pollinators they attract (Williams 1982).

Dressleria eburnea

D.eburnea is pollinated by *Eulaema cingulata.* The male bees land on any open surface of the flower and try to enter the cavity formed by the lip in which the fragrance is to be found. On their way they brush against the anther cap and release the pollinarium, which is under tension. It flings the viscidium which strikes the bee on the legs. The stipe is curled as it is first released and it therefore is impossible for the bee to immediately pollinate the same flower. The viscidium, moreover, covers the stigma and the pollinarium has first to be removed before pollination can take place.

Other **Dressleria** species
D.dilecta is pollinated by *Euglossa hansoni. D.helleri* by *Eg.asarophora*, *Eg.championi*, *Eulema nigrita*, *Euplusia schmidtiana*, *Ep.rufocauda*, *Ep.anisochlora. D.suavis* by *Ep.ornata* and *Euglossa tridentata.*

Genus **Mormodes**

The flowers of *Mormodes* are functionally male until the pollinarium has been removed. There are two types, according to Dressler. One type has functional, short-lived male ('staminoid') flowers with a narrow column, which are more numerous than the functionally female ('pistilloid') flowers (Fig. 26). The other type has flowers that have a wider column and a larger stigma and which become female even if the flower is not visited and the pollinia not disturbed.

The column in *Mormodes* arches over the lip. A long slender appendage nearly touches the lip and is often touched by brushing bees. This causes the pollinarium to be discharged and attached to the back of the bee. It takes about 30 minutes for the stipe of the pollinarium to straighten from the initially coiled position to the erect position that is necessary for successful pollination. The pollinarium is held erect on the thorax and, on flowers in female condition, the bee will deposit the pollinium (or both pollinia) in the stigma.

Although sharing of pollinators by different species is noticed, in general the species have specific pollinators. Hybrids between species are known and cause taxonomic confusion.

Pollinators
Williams (1982) summarises the known pollination data up to 1982 as follows (references left out):

Orchid	Pollinator/Visitor	Note
M.atropurpureum	*Euglossa championi*	a
	Eg.mixta	a
	Eg.cybelia	
M. cf. *variabilis*	*Eg.hemichlora*	a
M.cartonii	*Eg.cordata*	
	Eg.mixta	b
M.colossus	*Eg.mixta*	
	Eg.asarophora	a
	Eg.maculilabris	
	Eulema cingulata	b
	El.meriana	b
	El.nigrita	b
M.flavidum	*Euglossa viridissima*	a
M.igneum	*Eg.igniventris*	a
	Eg.mixta	b
M.lineatum	*Eg.viridissima*	a
M.maculatum	*Eg.viridissima*	a
M.powellii	*Eg.tridentata*	a
M.uncia	*Eg.* sp.	

a = observed pollinating, b = nonpollinating visitor.

In Panama, Roubik & Ackerman (1987) found pollinaria of *M.igneum* on *Euglossa dissimila*, of *M.cartonii* on *Eg.variabilis* and of (an?) unidentified species on *Eg.deceptrix* and *Eg.mixta*.

Dodson (1962) mentions *Mormodes buccinator* as pollinated by an *Eg.* sp. as observed by Oestland and *M.buccinator* 'Quevedo' population as pollinated by *Eg.viridissima*. *Eg.cordata* is recorded for *M.igneum*.

Dressler (1968b) mentions that in *M.colossus* bees from the species *Eulaema cingulata* and *El.meriana* received the viscidia on the head or feet and quickly removed these. The visits therefore seem detrimental. An unidentified species from the Atlantic coast attracted *Euglossa cyanaspis*.

MAXILLARIEAE: CRYPTARRHENINAE

Seven species of *Cryptarrhena* are known, all from tropical America. Pollination unknown.

MAXILLARIEAE: ZYGOPETALINAE

About 330 species, all in Tropical America, are classified in 30 genera in three closely related alliances and four distinctive genera. All records indicate euglossine pollination, in most cases by perfume-gathering males. In the simplest cases we see gullet flowers in which the bees receive the pollinaria on their back or especially behind the head. Some other species have more complex systems. *Chaubardiella* places pollinaria on the basis of the antennae. Some *Cochleanthes* seem to have deceit flowers and attract naive bees.

Zygopetalum alliance

Genus **Batemania**

Vogel (1969) observed pollination in *Batemania colleyi* by an unidentified large Hymenoptere, probably a long-tongued *Eulaema*. It had the pollinarium on the front of the back and was probably sucking nectar from the sham spur.

Genus **Mendoncella (=Galeottia)**

The species *M.grandiflora* is mentioned by Roubik & Ackerman (1987) as probably pollinated by *Eu-*

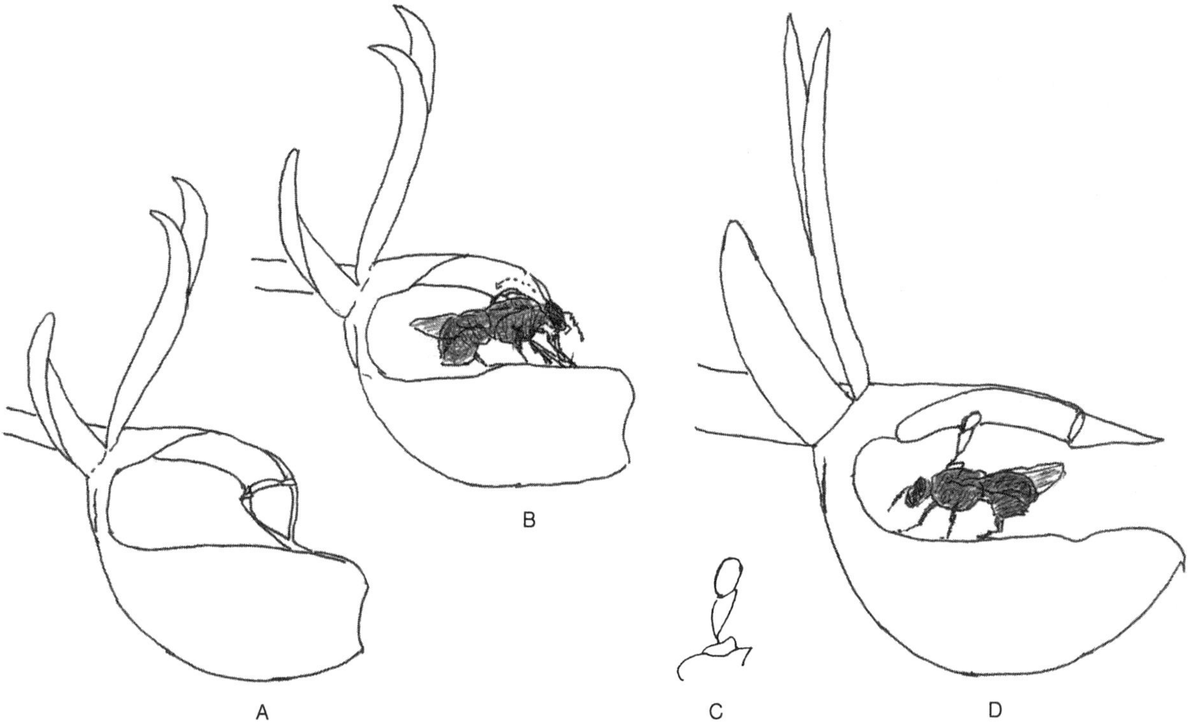

Figure 26. Pollination in *Mormodes igneum.* A) Staminoid flower of the truncate extreme, with the pollinarium in place. B) Same, with *Euglossa igniventris* touching the apex of the column, causing the pollinarium to be expelled. The viscidium usually strikes the thorax of the bee. C) Placement of pollinarium. D) Pistilloid flower of the elliptic extreme, with a bee passing beneath the stigma, effecting pollination (after Dressler 1968).

laema cingulata and *El.meriana*, since pollinaria were found on bees baited in Panama.

Genera Zygopetalum *and* Zygosepalum

Of each genus one species is reported to be pollinated by euglossine bees. A simple mechanism is probable, the pollinarium being placed on the head of the visiting bee.

Chondrorhyncha alliance

Genus Chaubardiella

In this genus the pollinaria are placed on the trochanter of the bee's leg.

Genera Chondroryncha *and* Pescatorea

A specimen of *Euglossa hemichlora* was found by Dodson and Frymire with a pollinarium on the abdomen, possibly of a species belonging to one of these two genera. Van der Pijl and Dodson mention pollination of a *Ch.*species of Panama by *Eulaema speciosa.* (In 1962 Dodson mentions *Eg.viridissima* instead.) They publish a photograph of *Pescatorea wallisii*, pollinated by *Eulaema polychroma.* Dodson (1962) records *El.cingulata.* (Dressler 1961 has one of the same species pollinated by *Eulaema tropica.*) *Eg.polychroma* lands on the lip and works its way to the base of the lip, near the column. The pollinarium is placed on the end of the scutellum as the bee backs out to hover (Williams 1982). But see also pollination of *Cochleanthus lipscombiae* below.

Genera Cochleanthes, Huntleya *and* Kefersteinia

In *Cochleanthus aromatica* the column arches over the head of the bee as it is in the brushing position. On leaving, the pollinarium is consequently deposited on the top of the head.

Van der Pijl and Dodson publish a photo with pollinating *Eulaema seabrae.*

Vogel supposes euglossine pollination in *C.marginata.*

Dressler mentions the collection of female Eulaemas with pollinaria of *Cochleanthus* and sees this as a complicating factor in understanding the pollination in this genus.

A similar system is found in *Huntleya meleagris* (a photograph is published in Van der Pijl and Dodson of a male Eulaema visiting this species). In *Kefersteinia* the bee lands on the lip and makes its way to the base of the lip, where the viscidium and the base of the stipe are wrapped around the basal segment (scape) of the antenna. This deviating method results in mechanical isolation among sympatric species of this group.

Dressler collected bees with pollinaria of *Kefersteinia lactea* on Barro Colorado Island (Dressler 1982). Roubik and Ackerman did the same using baits in Panama on various places and found pollinaria on *Euglossa bursigera*, *Eg.cybelia*, *Eg.dodsonii* and *Eg.tridentata* of *K.lactea*, of *K.costaricenses* on *Eg.mixta* and *Eg.deceptrix*, of *K.parvilabris* on *Eulaema speciosa* and of unidentified species on *Euglossa hansonii* and *Eg.mixta*. Dressler (1976b) mentions pollinaria of an unnamed species on *Euglossa cybelia*. Those of *Huntleya burtii* were found on *Eulaema cingulata*, those of *Cochleanthes lipscombiae* on *Eulaema bombiformis* and *El.meriana*.

Cochleanthus flabelliformis (Puerto Rico) is autogamous (Catling 1990).

Some species of *Kefersteinia* are scentless and may be not euglossinophilous (Gerlach & Schill 1991).

Cochleanthes lipscombiae

Ackerman (1983) studied pollination in this species, which is effected by both male and female euglossine bees. The flowers lack pollinator rewards and deceive bees in search of nectar by mimicking a sympatric legume *Clitoria javacensis*. The female bees were discovered with pollinaria while visiting the foodplant *Cassia fruticosa*. The flower of the orchid is fragrant and was previously thought to attract male pollinators especially, because males with pollinaria of this species were caught at chemical baits. The inflorescences bear a single, relatively large, gullet-type flower.

A female *Eulaema meriana* was observed to enter the flower with extended tongue and was guided by the basal callus of the lip. On leaving, the anther was dislodged and attached to the back of the head. Attempts to get rid of the load resulted in removing the anther cap, the exposed pollinarium remaining in place. Other female bees of the same and other species (*Euglossa cognata*, *Eg.imperialis*) as well as male bees (*Eg.imperialis*) were observed entering the flowers. The latter two species are too small to dislodge pollinaria.

A male *Euglossa tridentata* was observed to collect floral fragrances from the lateral sepals without entering the flower. Pollinarium carriers were regularly noticed, all large bees (five ♂♂ and 51 ♀♀ of *El.meriana*, two ♀♀ and four ♂♂ of *El.bombiformis*, one ♂ *El.nigrita*). The back-swept lateral sepals with their single, groove-forming enrolled margin are unique, and central to the mechanics of pollination. Many *Cochleanthes* species, but not all, share the same sepal morphology, as do several species of *Chondrorhyncha*, and it is suggested that these have a similar pollination mode.

Dressler noted floral similarities of *C.lipscombiae* and the liana *Clitoria javacensis*. This legume has nectariferous flowers visited by large euglossines. It is supposed that the orchid mimics this food plant, but an alternative explanation could be that the bees are young, recently emerged bees, which explore their environment and have not yet developed host preferences, or are bees that tend to explore continually. Mimicry of specific food plants and deception of exploring bees by mimicking a more general image of potential food sources are well-known phenomena in several European orchids (Van der Cingel 1995).

Genus **Dichaea**

This genus comprises about 45 species. As far as is known they are pollinated exclusively by male euglossine bees. No species is known to produce nectar, nor is there any species with characteristics of any other syndrome.

The character of the labellum and column define not only the flowers of this group as different from other orchids, but also constitute important features in distinguishing the many species. According to Folsom (1994) the pollination mechanism has to be understood to appreciate the biology of the flowers.

Dressler mentions a few observations of pollination by euglossine bees, and he frequently saw *Dichaea panamensis* pollinaria on the clypeus of bees. The pollinaria are very delicate and are usually lost in the net or killing jar after catching of the bees.

Roubik and Ackerman found pollinaria of *D.pa-*

namensis on *Euglossa allosticta*, *Eg.cyanaspis*, *Eg.despecta*, *Eg.dissimula*, *Eg.dressleri*, *Eg.heterosticta*, *Eg.mixta*, *Eg.tridentata*, *Eg.varabilis* and *Eufriesea pulchra*, of unidentified *Dichaea* species on *Euglossa allosticta*, *Eg.deceptrix* and *Eg.hansonii* (see also Ackerman 1983d). Folsom (1985) reports pollination by male *Eulaema meriana*.

Pollen flow

Folsom (1985, 1994) studied the pollination of these plants in La Selva Biological Reserve in Costa Rica, at first in 1981, and became convinced that *Dichaea* was unique for pollination studies in which the pollen flow could be followed in the population after dyeing the pollen. The delicate anthers, however, caused a problem till he found *Dichaea potamophila*, a new species with large viscidia, in Columbia. When properly stained the coloured pollinia could be detected through the white translucent anther cap. Plants with freshly stained pollinia of different colour were placed at set points along branches of the main tree in a cluster of chiparo trees and observations were made of pollinator visitation and resulting movement of pollinia from one flower to another. The species shows synchronized flowering on certain dates. They emit a light, musty-sweet fragrance by day and close at night. *Eulema meriana* pollinates the flowers, mostly in the morning. The bee commonly arrived carrying several pollinia attached to the frons of its head. Approach occurred by flying large circles (≥5 m) and hovering 1-2 m downwind at the height of the flower before landing on the apical half of the labellum. It scratched the base of the lip while climbing and positioning over it. After 2-55 seconds, the bee backed off, hovering a few cm away, all of the time facing the flower while apparently transferring the collected fragrant compound to its hind tibia. More landings could follow. Pollinaria need some time (2-5 minutes) before they collapse forward and get the right position for pollination. Out of 88 available flowers 31 had their pollinaria removed by bees and 18 were pollinated. Twelve of the pollinated flowers were among those that had lost their pollinia. Folsom was able to observe which pollinaria were stained or not and which flowers were pollinated by one, two, or more pollinia of one or more parents.

Bees often visited flowers without obvious result. Emasculation did not end the production of fragrance and emasculated flowers were observed being visited and pollinated. Natural self-pollination also occurred twice. On one day two front pollinia were deposited in the flower's own stigmatic cavity, concurrently with an unstained one from outside. Sometimes pollinaria were mishandled by displacement of the viscidium such that it was no longer available to attach properly to a visitor. More details in Folsom (1994). The overall effectiveness of pollination in *Dichaea* was about 50% of that of a comparable study in *Catasetum* by Janzen. From the viewpoint of total pollinia available in a population of 82 perfect flowers, 128 pollinia were circulated (37% of the total present). Only 12 pollinia (8%) actually reached stigmas. Pollen flow is primarily between closely spaced flowers and not as much a matter of long distance dispersal as is often believed for androeuglossophilous systems. Folsom gives many interesting considerations about the pollination of *Dichaea*, based on his results and observations, proving that the methodology used is quite effective to answer the intriguing questions concerning survival of plants dependent on pollinators.

Dichaea hystricina (Puerto Rico) is reportedly autogamous.

Genus **Scuticaria**

Braga (1977) supposes euglossine bees as pollinators of *Scuticaria steelii*. A *Euglossa stilbonata* was attracted by presented odour with a pollinarium of this orchid. It is probably a co-pollinator of larger euglossine bees, for its size is in fact too small for effective pollination. The lip of the flowers in *S.steelii* bears nutritive hairs. The flowers scent most strongly in the morning.

MAXILLARIEAE: LYCASTINAE

A subtribe with about 127 species in eight genera, in three tentative alliances (resp. *Bifrenaria*, *Lycaste*, *Normoorea*-alliances).

Genera **Bifrenaria**, **Rudolfiella** *and* **Xylobium**

Pollinaria of *Bifrenaria* species have been found on male *Eufriesea violacea*. Braga (1977) mentions Vogel (1966) who considers the syndrome of the flowers of the genus to be the euglossine bee syndrome. The flowers of *B.longicornis* scent during the whole day. The mechanism is supposed to be like that in *Bulbophyllum*.

Braga (1977) supposes Hymenopteran pollination in *Rudolfiella aurantiaca*, the pollinators feeding on nutritive hairs from the lip.

There are some observations of *Xylobium* species being pollinated by stingless bees of the genus *Trigona*. *Trigona amalthea* visits and pollinates *X.latilabium* (Peru, photograph in Van der Pijl & Dodson 1966) and smaller *Trigona* bees, identified also as *T.amalthea*, pollinate *X.variegatum* in Peru and Costa Rica, but did not visit *X.latilabium*, although observed flying past this species on their way to *X.variegatum* (Van der Pijl & Dodson 1966). *X.squalens* is pollinated by *Trigona postica*.

Genera Lycaste, Anguloa *and* Neomoorea

Both *Anguloa* and *Lycaste* are euglossine-pollinated. The pollinaria of *Lycaste* are sometimes found on female bees, suggesting some reward other than perfume.

L.barringtoniae, growing in the West Indies, where euglossine-pollinated orchids are rare, as well as euglossine bees, may nevertheless be euglossine-pollinated and on this basis Dressler (1968a) predicts the presence of *Euglossa* in Cuba and Hispaniola. Van der Pijl and Dodson predicted pollination of *L.gigantea*, *L.denningiana* and *L.ciliata* (*L.costata*) to be pollinated by nocturnal bees, because of odour production after sundown. Hummingbirds observed visiting *L.gigantea* in Ecuador by Dodson did not disturb the pollen apparatus and are therefore not considered pollinators. Pollinaria of a *Lycaste* were found on *Euglossa cybelia*, *Eg.deceptrix*, *Eg.dissimula*, *Eg.dressleri*.

Dressler (1961) observed *Euglossa viridissima* visiting *L.xytriphora* in Ecuador. This *Lycaste* was pollinated in a simple manner, the bee entering the large flower by landing on the outstretched lip and receiving the pollinarium at the underside of the dorsal projection of the metathorax while backing out. Furthermore Dressler (1968b) mentions a male *Eg.tridentata* caught with a pollinarium of a small *Lycaste* species.

Dodson (1962) mentions pollination of *L.aromatica* by *Eg.cordata*.

Anguloa only attracts male bees and is clearly andro-euglossophilous (Dodson 1966, Van der Pijl & Dodson 1966). The bees are flipped against the column by the hinged lip. The sepals and petals form a large cup. The lip stands erect and articulates with the base of the column by a thin strap. *Anguloa clowesii* is pollinated by *Eulaema boliviensis* in Columbia. Here the bee tries to enter the flower head first, but finds the source of the odour, the small shelf near the apex of the lip, inaccessible from that position. The bee turns and backs into the flower and scratches in this position, holding the edges of the petals with the middle legs. On leaving, the hold is released and the bee grasps the lip with all six legs. The lip than falls toward the column by the weight of the bee and the bee falls against the column. The viscidium becomes attached to the abdomen. *Eulaema cingulata* was observed pollinating *A. ruckeri* in a garden in Costa Rica.

MAXILLARIEAE: MAXILLARIINAE

About 470 species in eight genera. Tropical America.

Genus Cryptocentrum

According to Vogel (1969) the flowers are adapted to small moths (phaleanophilous) and produce a sweet scent. They possess deeply spurred flowers with dull colours.

Genus Maxillaria

This genus is pollinated by food-seeking bees. Some Brazilian species are pollinated by bees that gather wax from the lip, probably used as a nest-building material.

Dodson (1962) mentions *M.grandiflora* and *M.sanderiana* to be pollinated by *Eulaema cingulata* females, and *M.fuerstenbergae* by an undetermined female bee.

Maxillaria pendens (Braga 1977) is pollinated by a vespid wasp (*Stelopolybia*), which also gathers wax from the lip.

The flowers scent most strongly in the afternoon. In landing on the lip the wasp moves this down slightly and in that way the entrance to the flower becomes free. The pollinator collects the wax from the callus. After collecting it backs out along the same way and picks up the pollinia on the head or eye. The visit is a rapid one. After leaving the flower with pollinia the wasp is first disoriented in flight. The load of the pollinia unbalances it. It lands on leafs or stem before flying away and penetrating a second flower of another plant.

According to Buchmann (1987) 20 out of 300 species of *Maxillaria* produce oil or wax from elaiophores. Dressler (1993) found pollinia on stingless *Trigona* bees and observed *Melipona* bees gathering pseudopollen from the lip of an Ecuadorian species.

M.fulgens may be bird-pollinated, showing cryptic pollinia (Van der Pijl & Dodson 1966).

M.rufescens (including var. *cryptogama*) from Puerto Rico is autogamous (Catling 1990).

M.picta from Curitiba, Paraná, Southern Brazil is pollinated by *Trigona spinipes* (Apidae: Meliponinae) worker bees which are deceived (Singer & Cocucci 1999a). The bees are interested in the flowers for only a few days. After visiting they seem somewhat drowsy, probably due to the strong fragrance.

Genera Trigonidium, Mormolyca *and* Cyrtidiorchis

Trigonidium is a genus of about 13 species, ranging from Mexico to southern Brazil. *Trigonidium obtusum* is pollinated by male bees of the genus *Trigona*, which are reported to copulate with the glandular apices of the petals of the flowers because of their smell which reminds them of *Trigona* queens (Kerr & Lopez 1963, Vogel 1966, Van der Pijl & Dodson 1966). This, however, seems a bit doubtful, and the drawing after Sakigami which they present reminds one of the kind of behaviour of euglossine bees on orchid flowers. More observations are needed.

In *Mormolyca* and *Cyrtidiorchis* (=*Cyrtidium*) the flowers are much more insect-like than in *Trigonidium*, with reflexed sepals and petals and an insect-shaped lip.

Trigonidium egertonianum produces extrafloral nectar.

Genus Pityphyllum

Braga (1977) suggests pollination by Hymenoptera in *Ornithidium* (=*Pityphyllum*) *parviflorum*. The flowers scent strongly in the afternoon. The pollination mechanism is like that of *Maxillaria*.

MAXILLARIEAE: STANHOPEINAE

There are 22 genera within this subtribe comprising about 250 species, all living in tropical America and all pollinated by male euglossine bees. They range from simple gullet flowers (*Acineta*, *Lycomormium*) to very complex ones with most interesting pollination systems. In *Sievekingia* e.g. the gullet-like flower is entered upside down with respect to the column. *Coeliopsis* is a keyhole flower, *Peristeria* a flip-trap. *Kegeliella*, *Paphinia* and *Polycycnis* have hinged lips. In these flowers the weight of the bee causes the column and the lip to come together. *Gongora*, *Lacaena* and *Stanhopea* have flowers through which the bees slide and are caught by the column. Possibly the most intriguing pollination system is found in *Coryanthes* with its bathtub flower.

Genus Acineta

All flowers point downward on a pendant raceme. Sepals and petals form a kind of hood, arching over lip and column, leaving a tunnel for the bee to enter. Bees crawl upward into the flower, and scratch at the inside of the saccate base of the lip. On backing out they touch the rostellum and pick up the viscidium of the pollinarium on the scutellum (Fig. 27). If a bee enters another flower, the pollinia are

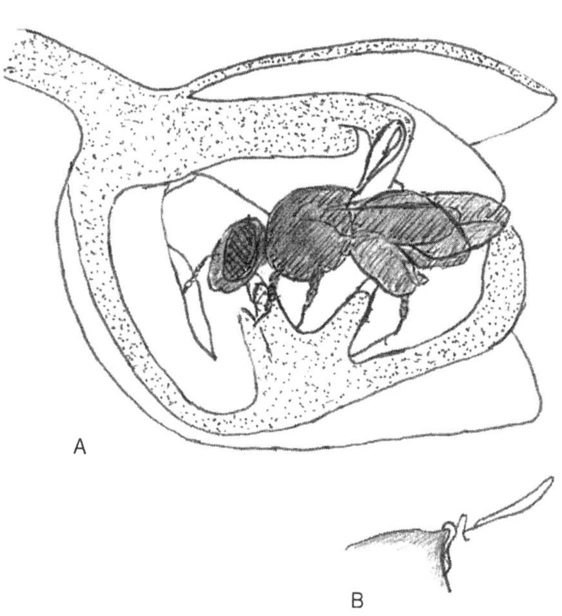

Figure 27. Pollination in *Acineta superba*. A) Longitudinal section of the flower showing *Euplusia* (=*Eufriesea*) *concava* as it backs out, receiving the viscidium beneath the scutellum. B) The placement of the pollinarium on and beneath the scutellum; the thickness of the viscidium is exaggerated (after Dressler 1968).

placed on the stigma when it leaves. The genus appears exclusively pollinated by bees of the genus *Euplusia*. Pollinators are *Euplusia concava* and *Ep.venezolana*. *Ep.concava* is attracted by cineole, a compound that is produced by *Acineta chrysantha*.

Pollinations by *Ep.concava* are observed in *Acineta barkeri*, *A.chrysantha* and *A.superba*.

A.dalessandroi is autogamous in Ecuador.

Genus Cirrhea

Flowers of this genus are erect on a pendant inflorescence. The male bees land on the lip. They touch the extended rostellum. The viscidium is placed on the leg. There are observations of pollination in this genus by *Euplusia violacea* by Hoehne, and six bees of the genus *Euglossa mandibularis* were found in the museum of the University of Kansas with in total 17 pollinaria attached to their legs (Van der Pijl & Dodson 1966).

Genus Coeliopsis

In this genus the pollination system is very simple, probably the simplest within the Stanhopeinae, but very effective (Fig. 28). *Coeliopsis hyacinthosma* is pollinated (in Panama) by male bees of the species *Eulaema cingulata*, *El.meriana*, *Euplusia schmidtii* and *Ep.* cf. *surinamensis*. *Euglossa dodsonii* and *Eg.tridentata* were observed as nonpollinating visitors. Bees land on the inflorescence and reach into the bell-shaped flower, where they scratch at the base of the lip. They force their way in as far as possible and contact the sticky viscidium with the clypeus of the head (which type of placement is unusual among Stanhopeinae (Dressler 1968b). On visiting a second flower the pollinia slide into the stigmatic cleft as the bee backs out.

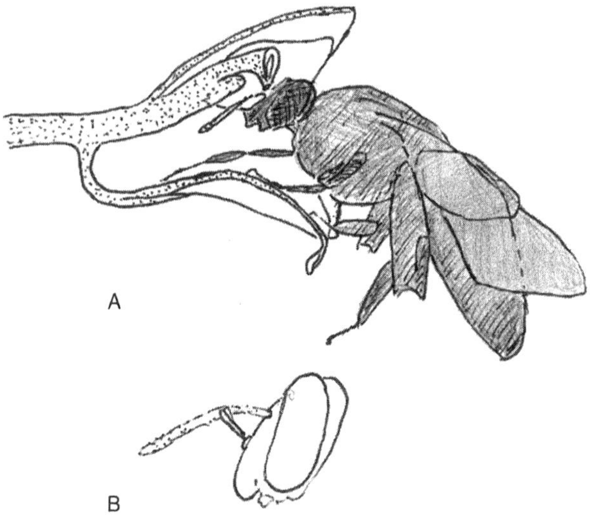

Figure 28. Pollination in *Coeliopsis hyacinthosma*. A) Longitudinal section through the flower showing *Eulaema meriana* receiving a pollinarium on the clypeus. B) The placement of the pollinarium on the clypeus (after Dressler 1968).

Genus Coryanthes

Morphology

The flowers in this genus, which hang down, are very large. Sepals and petals are folded back when the flower opens. The strange, large lip is divided into three parts, the globular or hood-shaped hypochile, the long, tubular *mesochile* and the bucket-shaped epichile. The epichile is partially filled with a watery, slightly sweet fluid during the period just before and after opening, 'tapped' from two glands at the base of the column. The hypochile produces fragrances, which are gathered by euglossine bees.

Bee behaviour

Crüger, who observed the bees in Trinidad (1865) on *Coryanthes macrantha*, stated that the bees may 'be seen in great numbers disputing with each other for a place on the edge of the hypochile. Partly by this contest, partly perhaps intoxicated by the matter they are indulging in, they tumble down into the bucket' (in Darwin 1888). Crüger wrongly interpreted the behaviour of the bees as gnawing, instead of scratching. The bees after slipping off and falling into the 'bath-tub' crawl towards the anterior side, where they find a tunnel-shaped passage between the column and the epichile. They are unable to climb out otherwise because of the steep, waxy sides of the bucket and cannot fly out because of their wetted wings. Forcing their way out through the tunnel, which takes about 15-30 minutes if the anther is not removed, the bees receive the pollinarium upon the abdomen (Fig. 29). In a succeeding visit the pollinia are deposited in the stigma. The bees can, when the anther is already removed, creep through in about one minute. Each species has a distinct pollinator, or group of pollinators, attracted by species-specific fragrances. Sympatric species, which occur in several areas, are isolated by specifity of pollinators.

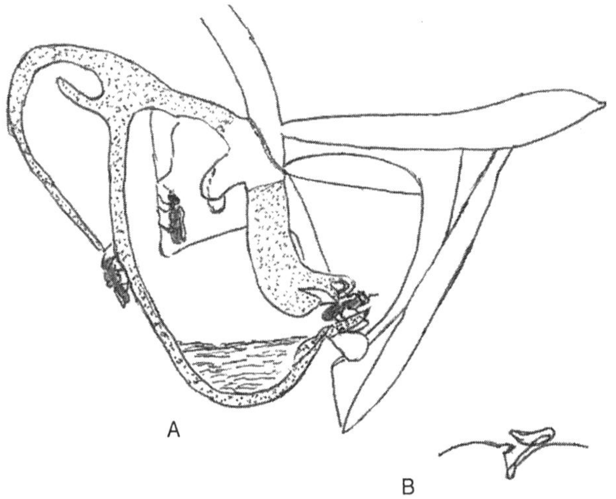

Figure 29. Pollination in *Coryanthes maculata*. A) Longitudinal section through the flower showing two Euglossas brushing on the lip near the hood and a third bee emerging beneath the apex of the column, receiving the pollinarium on the base of the abdomen. B) The placement of the pollinarium between thorax and abdomen (after Dressler 1968).

Pollinators
C. macrantha is pollinated by the really 'promiscuous' *Eulaema cingulata* (observed in Peru and Trinidad) and by *Eulaema basalis* (Trinidad). Dodson (1962) mentions also *Euglossa cordata* as observed by Crüger and Allen. *C.trifoliata* is pollinated by *Euglossa ignita* and *E.dressleri* (but not by *E.augaspis* also attracted, which is too small, according to Dodson 1965), *C.rodriguezii* by *Euplusia superba* (visited by *E.meriana*, *C.maculata* by *Euglossa azureoviridis*. Pollinaria on *Euglossa azureoviridis*, *Eg.tridentata* and *Eg.despecta* were collected by Roubik & Ackerman (1987).

Flowers of *C.leucocorys* (in a garden near Iquitos) were visited by *Euglossa ignita* and *Eulaema meriana*, the first being too small to be an effective pollinator. *C.bicalcarata* was observed to be pollinated by *Euglossa cordata* in the same garden. Both species occur sympatrically in Peru and may be pollinated there by these bees or by other bees of similar size.

C.wolfii (=*C.elegantium*) is pollinated by *Euglossa hemichlora* in Ecuador.

In Iquitos five species (Dodson 1965 mentions *C.macrantha*, *C.trifoliata*, *C.rodriguezii*, *C.maculata* and *C.alborosea*) grow sympatrically but appear to be isolated by attraction of specific pollinators through different fragrances. Hybrids were not found although crosses proved the species interfertile.

Pollinaria of *C.speciosa* were found on *Euglossa dressleri*. Those of other species of *Coryanthes* were found on *Euglossa gorgonensis*, *Eg.imperialis* and *Eg.hemichlora*.

Baker (1963) mentions *Euglossa cordata* as pollinator of *C.speciosa*, as was observed by Allen (1950), see below. Van der Pijl and Dodson mention *Eg.alleni* as pollinator of a *C.* aff. *speciosa*. Dressler observed the same in Panama (Dodson 1965, Dressler 1968b). He observed attraction of pollinia bearing *Eg.azureoviridis* by *C.maculata*.

Coryanthes speciosa
Allen (1950) observed visitation of this species by green metallic bees in northern Honduras. Flowers had just opened, as evidenced by a single drop of moisture and fresh sepals, when a bee was observed with pollinia on its back. Part of the plant was transported to Costa Rica, where a new inflorescence developed. After exposure of the flowers into the open early in the morning, metallic green male bees (*Euglossa cordata*) immediately appeared. They showed no interest in the liquid produced in the glands, nor did they gnaw at any part of the flower. They were obviously attracted by the scent. After contact with one of the poised droplets on the glands for a split second, a bee dropped like a stone to the ground three feet below. Two other bees had the same kind of experience, one of these fell into the 'bucket'. This bee tried to escape by the lateral exit, forcing the head and frontal pair of legs through almost at once. It took, however, the bee three quarters of an hour to get free, emerging with pollinia fixed to the back. It crawled to the lateral sepals where it spent about two minutes in freeing its wings, after which it took a direct, clumsy flight away from the scene. After being caught and put on the flower it refused to take hold and tried to move away. The fragrance disappeared at 9:00 o'clock in the morning, but reappeared the next morning, again attracting bees. Also the droplets had dried up within a few hours after opening. This disappearance of scent and fluid correlated in time with the period the bee needed to free itself and this was interpreted as a contrivance to promote cross-fertilisation or xenogamy. The bees bearing pollinia are thought to be interested again the following day, as was concluded from the observation in Honduras.

Genus Gongora

Pollination
Crüger was again the first who observed the polli-

nation in the genus *Gongora* in Trinidad. He thought the bees were gnawing instead of what we know they do, brushing or scratching the surface of the labellum projections with the tarsal brushes of the front legs. The pollination is, however, not accomplished by the tip of the extremely long pendant tongues of the *Euglossa* bees, as he supposed.

Flowers are pendant on long racemes. The lip is situated above the column. Two small petals are adnate to the column and look like small wings. The bees land on top of the lip and crawl to the underside to collect the fragrance that is produced on the underside of the base of the lip. As a bee releases his hold on the lip and tries to right himself and transfer the collected fragrance to the inflated hind tibia, he slips and falls down the column (Fig. 30). Petals and column form a kind of chute. The bee first passes the stigma and then the anther. In passing the anther, the viscidium is hooked under the scutellum. The stigma opens after the pollinarium is removed. Different species appear to have different pollinators or groups of pollinators. Sympatric species are ethologically isolated.

There has been a lot of confusion about the taxonomy of species, due to wrong identifications, distortion of herbarium material, colour and odour variability, vague descriptions and lack of information about countries of origin in cultivated plants etc. (Dressler 1966).

Williams et al. (1985) found different floral fragrance forms of morphologically identical plants of *Gongora quinquenervis*, each known to attract different pollinators. They call them *sibling species* of the *Gongora quinquenervis* complex. Some of these forms produced the major attractant fragrance in crystalline form. The crystals are only found in the early morning, when production is strong and air temperature relatively low.

Pollinators

Williams lists the following pollination data (adapted):

Orchid	Pollinator/visitor	Note
G.armeniaca	*Euglossa viridissima*	a
G.armeniaca var. *bicornuta* (syn. *G. horichiana*)	*Eg.dodsonii*	
G.bufonia	*Euplusia violacea*	
G.grossa	*Euglossa hemichlora*	a
G.quinquenervis (Costa Rica)	*Eg.cordata*	
G.quinquenervis (Panama)	*Eg.cordata*	
	Eg.hemichlora	a
	Eg.townsendii	a
	Eg.tridentata	a
	Eg.cyanaspis	
G.quinquenervis (Ecuador)	*Eg.* cf. *variabilis*	
	Eg.tridentata	
G.quinquenervis (Peru)	*Eg.ignita*	a
	Eg.augaspis	a
	Eg.decorata	a
	Eg.cordata	
G. sp. (Costa Rica)	*Eg.flammea*	a
	Eg.dodsonii	
G. sp. ('Guanacaste red')	*Eg.viridissima*	a
G. sp. (Dressler 1968a)	*Eg.gorgonensis*	a
	Eg.asarophora	b
	Eg.nigrosignata	
	Eg.villosa	b
G. sp. (Dressler 1968a)	*Eg.hansonii*	a
	Eg.cybelia	a
	Eulaema cingulata	b
	El.nigrifacies	b
	El.polychroma	b
	El.speciosa	b
G.tricolor	*Euglossa cyanura*	a
	Exaerete smaragdina	
G.unicolor	*Euglossa purpurea*	a

a = observed pollinating, b = nonpollinating visitor.

Note 1: Euplusia is now included in Eufriesea (Dressler 1982).

Note 2: Roubik & Ackerman (1987) found pollinaria of *G.gibba* on *Euglossa asarophora*, of *G.quinquenervis* on *Eg.allosticta*, *Eg.bursigera*, *Eg.cognata*, *Eg.deceptrix*, *Eg.despecta*, *Eg.heterosticta*, *Eg.imperialis*, *Eg.mixta*, *Eg.tridentata* and *Eg.variabilis*.

Note 3: Arditti (1966) mentions Allen, who suggested pollination of *G.maculata* (=*G.jeanischii*) by *Euglossa cordata*. The same is reported by Baker (1963) who points out that this is also the pollinator of *Coryanthes speciosa* with an entirely different mechanism. See also Dodson 1965 who mentions Allen (1951) for this last observation.

Note 4: Dodson (1962) mentions *Eg.viridissima* for *Gongora grossa*, but obviously changed this into *Eg.hemichlora* in the 1966 book, although both *Eg.*species are recognised therein. For *G.maculata* the same pollinator was mentioned, but this is not repeated in the book.

Note 5: *Exaerete* is a genus of parasitic euglos-

Figure 30. Pollination in *Gongora tricolor*. A) The position of the bee while brushing is shown by dotted lines. When the bee falls its course is guided by the column and petals, and the pollinarium is placed beneath the scutellum. B) Placement of pollinarium (thickness of viscidium and stipe exaggerated (after Dressler 1968).

sine bees, laying their eggs in the nests of *Eulaema* and perhaps *Euplusia*. Both males and females have reduced hind tibiae and their reaction to orchids is, although present, perhaps rudimentary (Dressler 1968b).

Genus Houlletia

There are no observations of pollination in this genus. Pollination has been studied indirectly. Floral fragrances were analysed and used in chemical baits to observe attractiveness for euglossine bees. Dressler showed that *Houlletia tigrina* was probably pollinated by *Eulaema meriana* by identifying a pollinarium, found on a leg. Roubik and Ackerman found pollinaria of this species on *Eulaema cingulata*, *El.nigrita* and *El.meriana*. Williams (1982) mentions an observation of an *Euglossa* sp. in relation to *H.brocklehurstiana* (see also Plate 45, p. 84).

Genus Kegeliella

Kegeliella atropilosa is pollinated by *Euplusia concava*. Dressler found not only bees with pollinaria on their head, but observed that the flowers of this orchid attract these bee species. Roubik and Ackerman found pollinaria of a *Kegeliella* on *Euglossa deceptrix*.

Genus Lacaena

The pollination mechanism of this genus resembles that of *Gongora*. Bees enter the pendant flowers upside down and brush on the midlobe of the lip. As a bee, on leaving, releases its grip it falls down, the head passing the viscidium. The pollinarium becomes fixed to the back of the head near the thorax (Fig. 31). *Lacaena bicolor* is pollinated by *Euplusia* cf. *caerulescens*, *L.spectabilis* by *Euglossa maculilabris*.

Genera Luedemannia *and* Lycomormium

Luedemannia pescatorei and *Lycomormium squalidum* contain among other compounds cineole and alpha- and beta-pinene in their fragrances. Pollinators are not known as yet.

Genus Paphinia

Paphinia clausula is visited by bees of the genus *Euglossa*. Dressler observed four species, of which

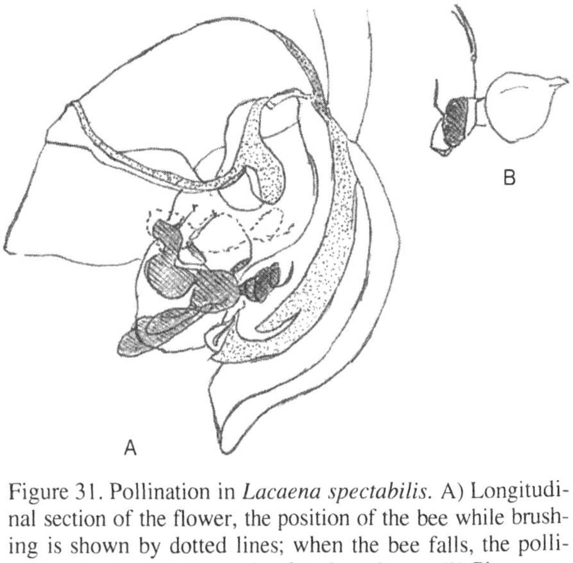

Figure 31. Pollination in *Lacaena spectabilis*. A) Longitudinal section of the flower, the position of the bee while brushing is shown by dotted lines; when the bee falls, the pollinarium is placed between head and prothorax. B) Placement of the pollinarium (after Dressler 1968).

E.gorgonensis was seen to remove pollinia. The pendant semi-closed flower is non-resupinate. Bees enter the inverted flowers, with their papillate lip, right side up. Pollinaria are attached to the middle legs on leaving. Williams suggests this to be probably an atypical system in the genus. But pollinators of other species are unknown and fragrances have not been analysed hitherto.

According to Gerlach & Schill (1991) several *Paphinia* species do not produce any scent perceptible to the human nose. Nevertheless among Euglossini baited with pollinia from French Guiana the majority carried those of the scentless *P.cristata*. Presumably the bees are highly sensitive to compounds which are not detectable by available methods.

Genus **Peristeria**

This genus is allied to *Acineta* but the flowers have a different pollination system. The lip is hinged, and the weight of the pollinator, landing on the lip, causes the hinge to tip (Fig. 32). The bee is thrown against the column. The side lobes of the lip and wings on the column hold the bee, and the pollinarium is attached to the thorax on struggling out. *P.pendula* is pollinated by *Euglossa ignita* and visited by *Eg.mixta* and *Eulaema meriana*, *P.elata*, the national flower of Panama, is pollinated by *Euplusia concava* and visited by *Euglossa crassipunctata*. Pollinia were found on *Eufriesea concava*.

Another, not identified *Peristeria* species was observed in Panama by Dressler to be pollinated by *Euglossa deceptrix* and *Eg.dodsonii* and visited by many different species of other *Euglossa* bees, some *Eulaema* species and an *Euplusia*. It appears that species without close sympatric allies are less specific in their pollination relationships than those with close sympatric allies and attract more species, which in part may be merely visitors instead of pollinators (Dressler 1968b). Pollinaria of unnamed species were found on *Euglossa bursigera*, *Eg.championi*, *Eg.cybelia*, *Eg.deceptrix*, *Eg.dissimula*, *Eg.dressleri*, *Eg.hansoni*, *Eg.maculilabris* (Roubik & Ackerman 1987).

Genus **Polycycnis**

In *Polycycnis* the bees land on the lip and rapidly move to the base of it to start scratching. The weight of the bee pulls the flower down (Fig. 33). When the bee starts hovering to transfer the collected fragrance, the viscidium, which projects from the long slender column, slides under the dorsal thorax and becomes hooked beneath the scutellum. Pollination is effected by inserting the pollinia into the stigma on leaving.

P.barbata is pollinated by *Eulaema speciosa* and *P.gratiosa* by *Euglossa villosa*. (Dressler 1968b, 1977) Pollinaria of this species were found on *Euplusia (=Eufriesea) anisochlora*, *Ef.chrysopyga* and *Ef.rufocauda*. Two undescribed species of *Euglossa* were observed to carry pollinaria of *P.muscifera* in Panama, one of which was later named *E.oleolucens*.

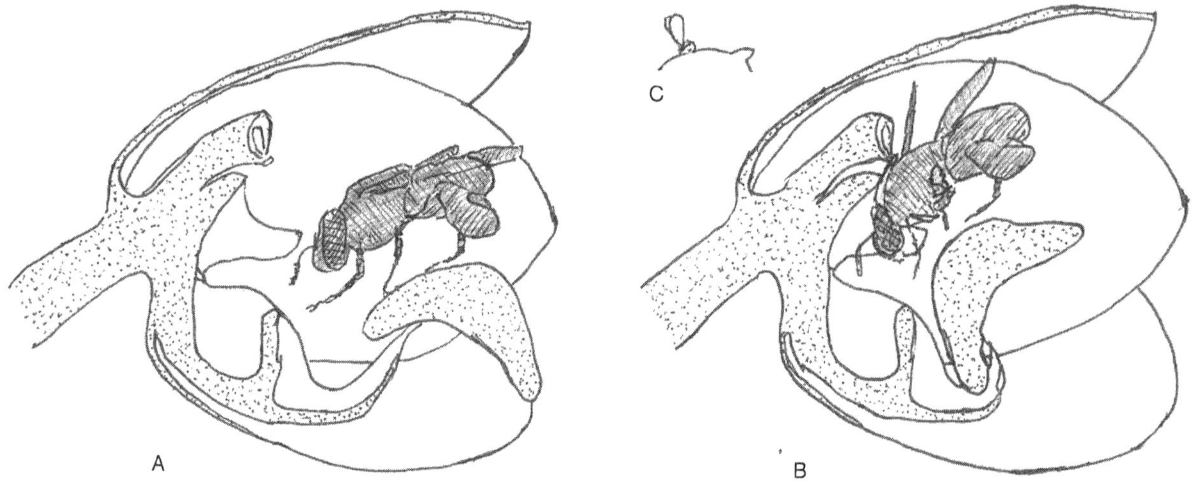

Figure 32. Pollination in *Peristeria* spec. A) Longitudinal section of the flower with *Euglossa dodsoni* on the hinged lip. B) Ditto, showing the lip overbalanced and the bee thrown against the column. C) Placement of pollinarium on thorax (after Dressler 1968).

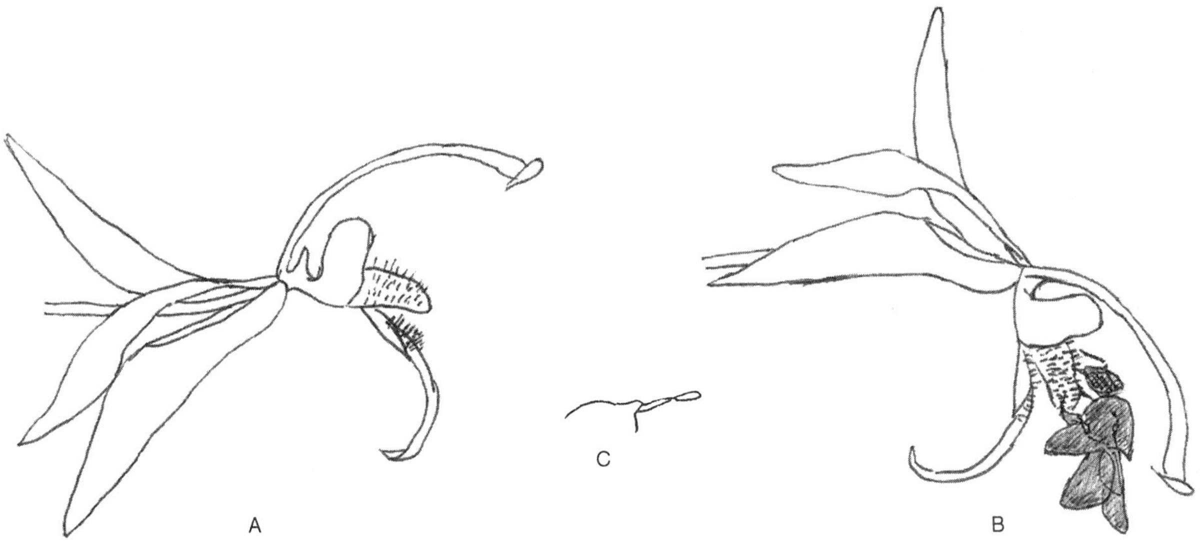

Figure 33. Pollination in *Polycycnis gratiosa.* A) Flower in normal position. B) Flower with *Euglossa villosa* on lip; the weight of the bee causes the flower to bend downward, the pollinia then being hooked beneath the scutellum of the bee. C) Placement of the pollinarium (after Dressler 1968).

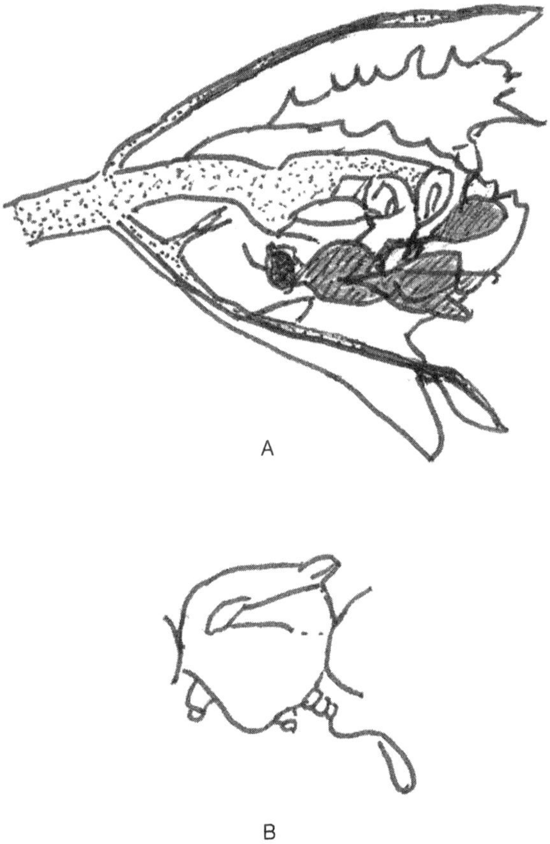

Figure 34. Pollination in *Sievekingia fimbriata.* A) Longitudinal section of the flower with *Euglossa sapphirina* entering the flower upside down. The hooked viscidium catches on the trochanter of the mid or hind leg. B) Thorax of the bee, showing pollinarium placement on trochanter of hind leg (after Dressler 1968).

Genus **Schlimia**

Schlimia, sometimes wrongly written as *Schlimmia*, is a genus of three rare species from the Andean regions, from Venezuela to Ecuador (Sweet 1973).

Schlimia trifida has flowers that are unusual in that the sepals are united and form a hood. The lip is reduced. Dodson observed several male *Euplusia* cf. *purpurata* in Colombia, visiting this species. They entered the flower and scratched at the lip. The pollinarium is apparently attached to the scutellum. There is no information available on fragrance composition.

Genus **Sievekingia**

Pollination

In *Sievekingia* species (Dressler 1968b, 1976) the flowers point downward on pendant spikes (cf. *Acineta*) and are densely clustered in a head-like inflorescence. In *S.jenmannii* male *Euglossa nigropilosa* bees were observed hovering under the inflorescence and crawling upward into the flowers after landing on the petals end entering the flower upside down (Fig. 34). Spine-like processes from the callus inhibit their entrance and guide them into position over the column. The cupped viscidium is attached to the trochanter of the midleg or hind leg. In other species pollination is similar. *S.fimbriata* attracts small Euglossas as well as the larger *Eufriesea duckei*. The larger bees cannot get as far into the flower as the smaller ones. *Ef.duckei* receives the pollinaria on the foreleg, the smaller bees receive them on the midleg or hind leg. Some bees, observed in *S.suavis* and *S.fimbriata*, appear to be of improper size or are wrongly oriented and are ineffective pollinators.

Pollinators

Williams' list of species and their pollinators is as follows:

Orchid	Pollinator/visitor	Note
S.butcheri	*Euglossa cyanura*	
S.fimbriata	*Eg.mixta*	b
	Eg.sapphirina	a
	Eg.cybelia	
	Eg.crassipunctata	b, c
	Eg.dressleri	b
	Euplusia duckei	
	Ep. mussitans	
S.jenmannii	*Euglossa nigropilosa*	c
S.reichenbachiana	*Euplusia surinamensis*	
S.rhonhofiae	*Euglossa hansonii*	
	Eg.tridentata	
	Eg. sp. nov.	
	Eg.heterosticta	
S.suavis	*Eg. dodsonii*	a
	Eg.townsendii	b
	Eg.bursigera	c

a = observed pollinating, b = nonpollinating visitor, c = captured carrying pollinia.

Genus **Stanhopea**

Pollination

The more primitive representatives of this genus have a pollination system as in *Acineta*. The flowers are small and rather closed. When the bees enter the flower, they brush at the base of the lip. The pollinarium is deposited on the underside of the scutellum of the bee as it leaves the flower and passes by the apex of the column. To these species belong among others *Stanhopea ecornuta* (which is pollinated by *Euplusia schmidtiana*), *S.pulla* and *S.annulata*. *S.ecornuta* (Fig. 35) may be approached and left from the side. In most cases, however, the bees backs out with the projections on each side of the apex of the lip keeping them on the centre of the lip, which is the correct position for receiving the pollinarium. The more advanced species have 'fall-through' flowers (Plates 46 and 47, p. 85). The flowers point downward and the bees fall a bit on leaving (Fig. 36). The bees do not land on the flower to crawl to the lip, but fly directly into it, alighting either on the front or on the back of the hypochile. They move towards the source of the fragrance, situated in the base of the lip and then often lose their grip and fall through, abdomen first. The epidermis of the labellum, especially the upper side, is slippery from small oil drops and very flat. Insects lose their grip if they step on this surface after entering or leaving the fragrant tissue (Pohl 1926). If they first succeed in collecting fragrance

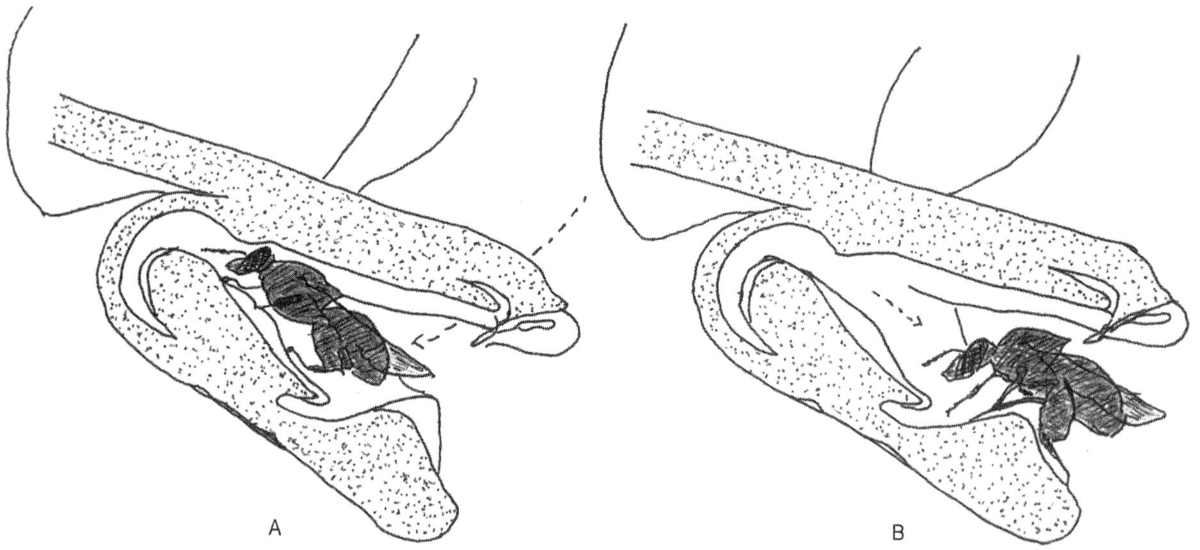

Figure 35. Pollination in *Stanhopea ecornuta*. A) Longitudinal section of flower with the bee brushing near the base of the lip (having entered from the side). B) The bee backing out beneath the apex of the column. The placement of the pollinarium is probably as in Figure 36, though the stipe is much shorter (after Dressler 1968).

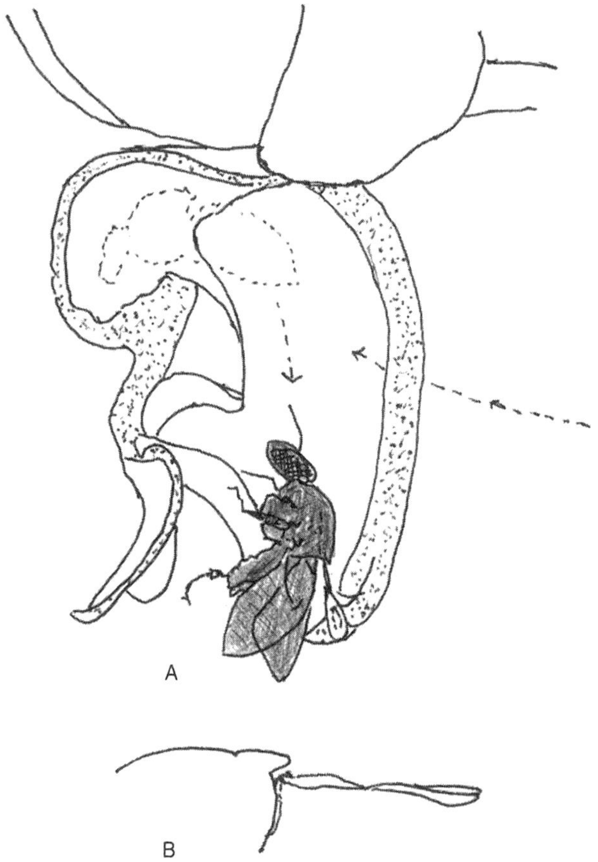

Figure 36. Pollination in *Stanhopea oculata*. A) the bee enters from the side (arrows), brushes in the hypochile (dotted outline), and, on withdrawal, may slip and fall through the flower, the pollinarium being placed beneath the scutellum. B) placement of *Stanhopea costaricensis* pollinarium on *Euplusia* (=*Eufriesea*) *schmidtiana*; the viscidium (thickness exaggerated) is largely placed on the propodeum (after Dressler 1968).

they will fall when they attempt to launch into the air to hover and to transfer the collected fragrance. The horns on the mesochile direct them down through the chute past the apex of the column, where the viscidium slips under the scutellum. As in *Gongora* the pollinarium has to be removed first before the stigma becomes receptive. Usually Stanhopeas are pollinated by large *Eulaema* bees. Some visiting *Euglossa* bees that are too small leave without pollinaria.

Fragrance production

A large number of floral fragrances have been identified and their specificity investigated. As in other Stanhopeinae most species have specific pollinators or groups of pollinators and isolation may be mechanical or ethological in sympatric species. The osmophores, which contain the fragrances, are investigated cytologically and chemically by Stern et al. (1987). [Vogel (1963) was the first to perceive that the labellae of Catasetinae and Stanhopeinae contain osmophores and that the euglossine bees were not looking for food, but he misinterpreted their behaviour by supposing that they were attracted by pheromones (*gamokinese*) from the osmophores and were deceived, the orchids being parasitic.] The external morphology of the osmophores of *Stanhopea* and other euglossinophilous taxa varies considerably. In *Stanhopea* we find papillae in some species and smooth tissue in other ones. The weak odour detected at night suggests that the synthesis of fragrance is stopped then. The persistence of aroma might be due to a gradual volatilisation of fragrance accumulated in the daytime. Curry (1987) studied the metabolic pathways by which the terpenoids and aromatics that compose the fragrance of *Stanhopea* osmophores are synthesised.

Pollinators

Van der Pijl & Dodson (1966), Williams (1982) and Williams & Whitten (1983) list species and their pollinators, which lists I have combined:

Orchid	Pollinator/visitor	Note
S.anfracta	*unknown*	
S.annulata	*Euglossa grantii*	a
S.candida	*Eg.chlorosoma*	
(=*S.randii*)	*Eg.ignita*	a
	Eulaema meriana	b
S.cirrhata	*Euglossa* cf. *flammea*	a
S.connata	*Eulaema speciosa*	a
	Euglossa nigripilosa	b
S.costaricensis	*Eufriesea schmidtiana*	c
	Ef.rufocauda	c
	Eulaema meriana	a
	El.luteola	a
	El.nigrita	a
	El.seabrae	a
S.ecornuta	*Eufriesea schmidtiana*	a
	Eulaema seabrae	
	El.nigrita	b
	Euglossa allosticta	b
	Eg.imperialis	b
	Eg.tridentata	b
S.embreei	*Eulaema bomboides*	c
S.florida	*Euglossa nigropilosa*	a
	Eulaema meriana	b
S.frymerei	*El.bomboides*	
S.gibbosa	*El.meriana*	a

S.grandiflora	*El.meriana*	a
	Euglossa ignita	b
S.impressa	*Eg.grantii*	a
S. aff. *jenischiana*	*Eulaema bomboides*	
S. cf. *oculata*	*Euglossa crassipunctata*	b
	Eg.cyanaspis	b
	Eg.flammea	b
	Eg.hemichlora	b
	Eg.tridentata	b
	Eufriesea carulescens	a
S.panamensis	*Ef.ornata*	c
	Ef.mussitans	?
	Euglossa crassipunctata	b
	Eg.cyanaspis	b
	Eg.deceptrix	b
	Eg.despecta	b
	Eg.hemichlora	b
	Eg.tridentata	b
S.pulla	*Eg.asarophora*	a
S.reichenbachiana	*Eulaema leucopyga*	
S.rodigasiana (=*Embreea rodigasiana*)	unknown	
S.ruckeri	unknown	
S.saccata	*Euglossa viridissima*	a
S.tigrina	*Eg.viridissima*	?
	Eufriesea caerulescens	a
S.tricornis	*Eulaema meriana*	a
S.wardii	*El.polychroma*	
S.warscewicziana	*Eufriesea macroglossa*	a
	Ef.chrysopyga	c
	Ef.concava	a
	Ef.rufocauda	c
	Eulaema polychroma	a

a = observed pollinating, b = nonpollinating visitor, c = captured carrying pollinia, ?= role uncertain.

Note 1: The lists present some contradictory results. In such cases I have chosen the most recent information.

Note 2: Braga (1976) mentions *Eulaema mocsaryi* (near Manaus, Brazil) as the pollinator of *S.candida* and *Euglossa ignita* as an 'odour thief'.

Note 3: *S.rodigasiana* is now removed from *Stanhopea* on the basis of its distinct floral and vegetative morphology and forms a monotypic genus (*Embreea*).

Note 4: Dodson (1962) mentions *S.oculata* 'Naranjapata' as pollinated by *El.bomboides*. Probably this was later changed into *S.* aff. *jenischiana*. He records *E.viridissima* in the same paper as pollinator of *S.connata*.

Note 5: Dressler (1976b) found pollinia of *S.cirrhata* on *Euglossa championi*, *Eg.imperialis*, *Eg.hansonii*, *Eg.gorgonensis*.

Pollinaria of unnamed species were found on *Eulaema cingulata* and *Eufriesea ornata*.

Speciation

Stanhopea has been mentioned as an example of socalled '*leap-frog' speciation*, i.e. very rapid speciation due to adaptation of species to pollinators which differ in size, but are morphologically and ethologically similar. Each adaptation requires rapid, but minor, morphological adjustments in the flower. This can lead to a series of slightly different species based on the same mechanism but different pollinators. The end of the development is an abundance of sympatric 'biospecies'. The rapidness of the speciation is deduced from the lack of change in the vegetative characteristics of the plants. The same situation is found in *Catasetum* and *Cycnoches* (Van der Pijl & Dodson 1966).

Dressler (1968b) in discussing this situation, in which an orchid species, invading a new habitat, may encounter a 'pre-adapted' pollinator, admits that this concept of 'leap-frog' speciation was proposed when they were still very naïve about the euglossine bees (as to number and size). Euglossine pollination, however, would seem to provide a better model for sympatric speciation than flower constancy. Dressler discusses in this respect the probable consequences of mutations in odours produced by the orchids. Euglossinae supply a fertile background for speciation. Euglossine pollination provides effective isolating mechanisms, permitting sympatric speciation.

Several species of *Stanhopea* have produced natural hybrids. *S.annulata* and *S.impressa*, for example, are both pollinated by *Euglossa grantii* and hybrids between these species are found rarely in Ecuador and Columbia (Williams & Whitten 1983). The attractant is phenylethyl acetate. The main isolating mechanism appears to be mechanical. The flowers of the hybrids differ morphologically from both parents and visiting bees are unable to pollinate, which contributes to maintenance of species integrity. Moreover, the species are also geographically isolated by a difference in elevation (100-600 m and 700-1500 m). The same occurs between *S.ecornuta* and *S.costaricensis*, which share three common pollinators (see list above). The common attractant in this case is cineole (Williams & Whitten 1983).

Genus **Trevoria**

The five species of this genus with gullet-like flowers are not very common and nothing is known about their pollination. On the basis of the shape of the viscidium, Dressler suggested a pollination mechanism similar to that of *Schlimia*, probably with a larger pollinator. Lehmann (in Garay 1970) wrote in 1897: 'The position of the column and lip, standing straight upright in an urceolate (=urn-shaped) manner, the thick fleshy process at the base of the lip, being exactly the same length as the column, and standing parallel and closely appressed to that organ, the peculiar shape and insertion of the stigma and other items are wholly novel, and the arrangement of them to each other may set at naught some of the pet views of modern hypocritical speculators with respect to the aid of insects in the fertilisation of orchids, or give rise to quite new conceptions'.

MAXILLARIEAE: TELIPOGONINAE

A tropical American subtribe of 126 species in four genera. The flowers offer no reward, but mimic female flies. The striking similarity has led to the common name 'La Mosca' for the plants throughout their range (Dodson & Escobar 1987). The presence of a distinctive, spiny, insect-like callus or column in most of the species, together with the occurrence of several species growing together in the same locality, suggests highly specific attraction of pollinators. The few reports of pollination observations indicate pollination by pseudocopulation by male tachinid flies that are deceived by the similarity of the column and callus of the flowers to the spiny abdomens of the female flies. The numerous species of tachinid flies are very distinctive in size, colour and spinyness of the abdomen. Visual stimuli seem to be most important in recognition of mating partners. The same kind of stimuli from the flowers provide powerful reproductive isolating mechanisms in the plants. They are also the basis for the present taxonomy of the orchids (Christensen 1994).

Genus **Trichoceros** *and related genera*

Members of this genus and its allies (*Hofmeisterella*, *Telipogon*, *Stellilabium*, *Sodiroella*, *Dipteranthus*, *Cordanthera*, the last four now united into one genus, *Stellilabium*) occur mainly at higher elevations in the Andes but a few extend into Central America.

The lip of the flowers of the six species of the genus (and of some related genera) closely resemble tachinid flies. The simulated insect is commonly so life-like that it appears as though it could easily fly away (Van der Pijl & Dodson 1966). As in the European *Ophrys* species, the similarity leads to pseudocopulatory behaviour of the pollinators. The flowers attract male flies that try to mate with the mimicking structures and have sometimes been observed to do so alternating with female flies, obviously indiscriminately. The shiny stigmatic surface, situated more or less at the apex of the 'false-abdomen' of the flower reflects sunlight, as does the genital orifice of the female fly. Female flies, when ready for copulation, land on leaves or other flat objects and alternately open and close the genital orifice when male flies pass by. This behaviour serves as a signal to the males which then rapidly approach to copulate. If the male flies approach the flowers, misled by the similarity with females, they strike the flower just enough to pick up pollinia. Pollinaria are attached to the legs of deceived males. The long stipe of the pollinarium bends down afterwards and is forced into the stigma during a next flower-visit. The flies feed mainly on the abundant nectar of the flowers of the genus *Mikania* (Compositae), which attract them by sweet fragrances and which grow in the same habitat, usually the banks of streams. *Trichoceros antennifer(a)* (syn. *T.parviflorum*) from Colombia and Peru simulates the female flies of *Paragymnomma* species. Since there is no fragrance, attraction seems based entirely on visual stimuli.

In *Telipogon* the fly-like appearance is not as pronounced as in *Trichoceros*, but there is no doubt about the mode of pollination as based on the same principle. In sympatrically growing species hybridisation seems prevented by the striking differences in the flowers, probably attracting different flies.

Maduro has observed pseudocopulation by a tachinid fly in *Stellilabium* (*Telipogon*) (Dressler 1981).

MAXILLARIEAE: ORNITHOCEPHALINAE

About 76 species in 14 genera, all tropical American.

There are several observations of *Ornithocephalus* being pollinated by bees of the genus *Paratetrapedia*. *O.avicula* and *O.* cf. *patentilobus* are pollinated by *Paratetrapedia testacea* in Peru. *O.bicornis* and *O.powellii* are pollinated by *Paratetrapedia calcarata* in Panama. La Fontaine caught one individual of *Euglossa cybelia* carrying five pollinaria of an *Ornithocephalus* species. Probably the bee was in search of nectar.

Vogel has indicated that the gland on the lip is actually an oil gland. Buchmann (1987) mentions all 50 species of *Ornithocephalus* and four species of *Zygostates* to possess oil-producing elaiophores, and attract anthophorid bees that gather oil as food for their larvae. This is apparently also true for the other members of this group (Dressler 1993).

The closely related *Dipteranthus obliquus* is presumably pollinated by small nectar-seeking bees. Its fragrance components have never been found in euglossophilous flowers (Gerlach & Schill 1991).

MAXILLARIEAE: ONCIDIINAE

This group, comprising 1232 species in about 77 genera, shows a great diversity of pollination systems, including wasp-, bee-, hummingbird- and butterfly pollination. The subtribe as a whole has one of the most diverse types of pollination systems of all (sub)tribes in the Orchidaceae. Only moths and flies are except.

Wasp pollination is reported in *Ada*, *Brassia*, *Leochilus*.

Seven genera are euglossine-pollinated: *Aspasia*, *Lockhartia*, *Macroclinium*, *Notylia*, *Rodriguezia*, *Trichocentrum*, *Thrichopilia*. In *Gomesa* and *Miltioniopsis* non-euglossine bees have been observed. Some *Odontoglossum* species seem to be pollinated by bumblebees and several species of *Oncidium* by anthophorid bees, notably *Centris*, males or females in different relationships (see below). *Sigmatostalix* is pollinated by oil-gathering bees. Hummingbirds have been observed as pollinators in some genera, (see below).

The great diversity of systems has tempted Williams to hypothesise on the evolution of pollination systems within this subtribe. He postulates that the most primitive type is based on the attraction of nectar-feeding bees and wasps to flowers of genera such as *Leochilus*, *Mesopinidium* and the like. From this type various types of spurred nectaries, like found in *Comparettia*, *Ionopsis*, and *Rodriguezia* could have developed. From such basic bee-nectar flowers, pollination by hummingbirds is thought to have evolved, as in *Symphyglossum*, *Rodriguezia*, *Cochlioda*. According to Van der Pijl & Dodson, birds appear to be rather late comers in the pollination of orchids and therefore we may expect isolated cases in groups which are basically pollinated by other agents.

Euglossine bees, males and females, do visit various genera for nectar, e.g. *Cischweinfia*, and for floral fragrances, as in *Notylia*, *Macradenia* and others. The male euglossine pollination could have been derived from bee-nectar flowers, with such genera as *Aspasia*, *Trichocentrum* and *Trichopilia* currently intermediate between nectar-producing flowers and pure euglossine flowers. In the intermediate-stage flowers, a nectary-like structure is present, but no nectar is produced. In the pure euglossine-syndrome flowers, such as *Notylia* and *Macradenia*, not even a nectary-like structure remains (Williams 1982).

Only those genera of which the pollination system is known or has been recorded are mentioned below.

Genus **Ada**

According to Van der Pijl and Dodson the flowers fit the ornithophilous syndrome with connation into a tube 'on the way'.

Dressler, however, states that in *Ada* wasp pollination is found. Of course, both can be true, but no concrete observations are reported. *Ada aurantiaca* has creamy or greyish white pollinia, as is characteristic for several hummingbird-pollinated orchids.

Genus **Aspasia**

A few observations are available on pollination of this genus. Williams lists as follows:

Orchid	Pollinator/visitor	Note
A.principissa	*Eulaema cingulata*	c
	El.meriana	c
	Exaerete frontalis	c
	Ex.smaragdina	a, c
A.psittacina	*Eulaema cingulata*	a
	El.polychroma	a

a = observed pollinating, c = captured carrying pollinia.

The pollination system is still unclear

A.psittacina has a strong odour of eugenol (clove oil), which attracts *Eulaema cingulata*, *Exaerete frontalis*, *Ex.smaragdina* and other bees. Other species, like *A.epidendroides* and *A.principissa* have the same aroma. There is a nectar tube or chamber, but nectar is not produced. Williams has suggested that at one time nectar was produced and both male and female euglossine bees were attracted.

Aspasia variegata is known to produce trans-carvone oxide as an attractant. Although no pollinator is known as yet, five species of *Eulaema* and three species of *Euglossa* are known to be attracted by synthetic trans-carvone oxide and to be pollinators of various species of *Catasetum*. Trans-carvone oxide has been found as a floral fragrance component in nine species of *Catasetum*, one species each of *Aspasia* and *Notylia*. It is also found in a non-orchidaceous species of *Dalechampia* (Euphorbiaceae). The synthesis and secretion of this compound is seen as the result of evolutionary convergence in floral rewards (Whitten et al. 1986).

Dressler (1961) observed pollination by male and female *Eulaema cingulata* and *El.tropica* of *A.epidendroides*.

Aspasia principissa

Zimmerman & Aide (1989) monitored a population of this orchid in central Panama over a period of two years. The species is pollinated by large euglossine bees (*Eulaema*, *Exaerete*). The flowers contain a long nectary-like structure but produce no nectar. They mimic food flowers of the bees. *A.principissa* is self-compatible. Observations and experiments produced evidence that within a season, fruit production was limited by pollinator availability. Fruit set for hand-pollinated flowers was over six times greater than that for naturally pollinated flowers. In plants that produced more than one fruit, however, fruit size declined in subsequent fruits, and plants producing fruits in one year, produced smaller shoots and inflorescences in the next year. Plants of *A.principissa*, therefore, seem resource-limited over their lifetime. Few individuals in the population contribute to reproduction, which fact, according to the authors, may probably lead to increased genetic drift in small populations and thus to rapid diversification of orchid species in evolution. But see also Calvo & Horvitz (1990), who criticise the conclusions about lifetime *resource limitation* on the basis of their transition-matrix demographic approach.

Genus Brassia

The genus is reported to have species with wasp pollination. Van der Pijl and Dodson published a photograph of *B.* aff. *antherotes* visited and pollinated by the wasp *Campsomeris columba*.

Genus Capanemia

C.thereziae is pollinated by female *Polybia fastidiosuscula* wasps which lick nectar on the surface of the lip. The wasps hover in front of the flowers for some seconds and land on the inflorescence or on the surface of the tree on which it grows and then walk around to visit the flowers of the pendulous inflorescence. A single wasp may carry several pollinaria, which seems to disturb the insects, which were sometimes observed trying to clean the pollinaria from the clypeus with their forelimbs (Singer & Cocucci 1999a).

Genus Cischweinfia

Male and female euglossine bees visit *Cischweinfia dasyandra* in search for nectar and pollinate them (Williams 1982).

Genus Cochlioda

C.vulcanica is visited by skippers and butterflies, but pollinia were seen to be left intact on examination of visited flowers. Hummingbirds are effective pollinators.

C.rosea also is bird-pollinated (Van der Pijl & Dodson 1966).

Genus Comparettia

Fourteen species have been described, of which on closer examination about ten remain (Schlechter 1996). According to Vogel (1969) they are psychophilous, at least *C.speciosa* and *C.macroplectum*, whereas *C.falcata* could be ornithophilous, which is indeed the case. The red flowers of *C.coccinea* are also attractive to hummingbirds (Light 1998).

Comparettia falcata

Rodríguez-Robles et al. (1992, 1994) and Meléndez et al. (1997) investigated the effects of display size, flowering phenology and nectar availability on the effective visitation frequency in *Comparettia falcata* in Puerto Rico. This is an allogamous species, pollinated by the endemic hummingbird *Chlorostilbon maegaeus*.

The lip of *C.falcata* has a callus that partially closes the floral tube at the level of anther and stigma, a common arrangement among ornithophilous orchids. It guides the beak of the bird against the column. Pollinaria are consequently attached to the beak and show a cryptic colour (Dressler 1971).

Van der Pijl and Dodson mention another hummingbird, *Amazilia tzacatl*, as pollinator of *C.falcata*. Honeybees, *Apis mellifera*, also visit *C.falcata*, but are unlikely pollinators. Autogamous pollinations are possible, but may result in reduced seed-set. For the 1989 and 1990 flowering seasons, the probability of pollinarium removals and natural pollinations increased with individual inflorescence display size. However, the frequency of effective pollinator visits was independent of flowering phenology in both years. Reproductive success at the two study sites did not differ in either year when adjusted for differences in pollinator activity. Nectar availability was not a good predictor of effective visitation. The small, dilute nectar volumes secreted by *C.falcata* may benefit the plant by increasing interplant movement and pollen dispersal. The species has a higher natural fruit set than tropical deceptive orchids. In subsequent experiments with manipulated flowers, from which the spur tips were clipped at one of the study sites, rendering them nectarless, it became evident that pollinarium removals from intact flowers were significantly different from those with cut spurs, and total visits differed significantly as well. However, the difference in female success was clearly not significant between the two treatments. The number of visits per flower dropped dramatically from previous years despite higher bird activity. The data are consistent with the hypothesis that the pollinator perceives the nectar reward despite its small quantity and low sugar content, suggesting that 'a meagre nectar offering' by this epiphytic orchid is better than no reward. The studies also demonstrated that there is not necessarily a dichotomy between the availability of resources and pollinators as limiting factors in the sexual reproduction of plants.

Genus Fernandezia

About pollination nothing is known, but Dressler (1981) suggests that *Fernandezia* may be hummingbird-pollinated. The pollinaria are rather like those of *Lockhartia*, but other features are different.

Genus Gomesa

The genus is bee pollinated as observed by F. Müller (Dodson 1962, Dressler 1981). No further data are available.

Genus Ionopsis

Ionopsis utricularioides

Montalvo & Ackerman (1987) as well as Echevarría et al. (1997) studied limitations to fruit production in this non-rewarding species in, respectively Puerto Rico and Cuba, with similar results. Natural pollination was infrequent. Hand pollination experiments led to the conclusion that the species is resource limited. Less than half of the manipulated flowers developed mature fruits, independent of whether flowers were selfed or outcrossed. The failure to set fruit is related to plant size, with smaller plants aborting more fruits. High fruit set in one year may result in decreased vegetative growth in the subsequent year. The first named authors speculate on the basis of these results that resource limitation of fruit production may have influenced the evolution of nectar-deception pollination in many orchids.

Calvo & Horvitz (1990) however, think that the available evidence does not support the hypothesis of lifetime resource limitation and suggests that the species may be limited by pollinators.

The kind of pollinator of *I.utricularioides* is not mentioned.

Genus Leochilus

A genus of nine species of which pollination for five species is reported. All pollinators are wasps or halictid bees, which feed on the nectar.

Chase (1986) studied the population ecology of two sympatric, synchronously flowering species of this genus in Costa Rica. *Leochilus scriptus* is pollinated by two polybiine wasps of the genus *Stelopolybia*: *S.areata* and *S.hamiltoni*. *Leochilus labiatus* by a halictid bee of the genus *Lasioglossum*. Both species have a wide distribution over the same general range. They can occupy the same microhabitat, often growing on the same branch, although they have different host preferences. They commonly coexist on *Hibiscus rosa-sinensis*. Hybridisation in nature, however, has not been reported. They are self-compatible species. Many insects, including the pollinators, visited the dully coloured (cream-yellowish green with red-brown to red-purple spots and stripes) flowers and did not pick up pollinaria. The flowers contain very low quantities of a very viscous nectar (72-73% ± 10% sugar). The nine captured wasps pollinating *L.scriptus* were females. Considering time, few pollinators were visiting: 15 in 28 observation days of 8-9 hours each day. Visiting *Trigona* bees, fruit flies and skipper butterflies did not pollinate *L.scriptus*. Of 32 pollinated flowers 81% were selfed (geitonogamy), 85% received only one of the two pollinia present on a pollinarium. For 12 marked flowers, the average was 1.12 visits per flower; 10 were pollinated. None was visited twice. The wasps often succeed in removing the pollinaria before leaving an inflorescence.

On *L.labiatus* seven *Lasioglossum* bees (subgenus *Dialictus*), all females, were captured. They made 64 pollinating visits during 28 days. Of 118 flowers pollinated, 60% were selfed by geitonogamy, but all pollinated flowers received only one of the two pollinia of a pollinarium. Of 12 marked flowers, 11 were pollinated. The average number of visits was 2.8. Often bees try to remove pollinaria and thereby often fall off the flowers. They then sometimes land on a different plant and this behaviour can lead to outcrossing. Effective pollen flow is restricted to 1-2 m in both species of orchids because of the successful removing of pollinaria during or shortly after leaving the plants.

The two orchid species have a different fragrance: *L.scriptus* smelling like ripe watermelon, *L.labiatus* like grated coconut.

The *Lasioglossum* species are too small to pollinate *L.scriptus*. The two wasps, on the other hand, are too large to reach into the nectary of *L.labiatus*. Thus, on a pure mechanical basis the inappropriate sized vectors are excluded from the relevant parts of all but the correct flowers. Visiting non-pollinating insects never landed and were probably deterred or repelled by the fragrance produced.

Genus **Macradenia**

This genus is pollinated by male euglossine bees and has a characteristic 'euglossine' aroma. *M.brassavolae* is pollinated by *Euglossa hemichlora* and *Eg.villosiventris*. The flowers lack a nectary. Dressler (1993b) presents a photograph of the species with its pollinator.

Genus **Mesospinidium**

Nectar-feeding wasps and bees are attracted to the nectar-producing flowers, but no details are reported.

Genus **Miltonia**

The bee *Ptiloglossa ducalis* of the Colletidae was captured by Dodson while pollinating the flowers of *Miltonia endresii*. The flowers were visited in complete darkness and the pollinator proved to be very efficient. Several kinds of bees are known to be nocturnal or crepuscular. In Ecuador *M.vexillaria* appeared to be pollinated before dawn, but by day no visits were recorded during a three-day observation. Also *M.phalaenopsis* and *M.roezlii* are supposedly pollinated in the same way. The white or glowingly pink flowers of these species are easily seen in dim light. Other species have different colours and different construction and are not expected to be pollinated at night (Van der Pijl & Dodson 1966).

Genus **Miltoniopsis**

The genus is bee-pollinated (Dressler 1981). No details are mentioned.

Genus **Neokoehleria**

Vogel (1969) states that the flowers are undoubtedly pollinated by Hymenoptera, which have to force their tongue between the elastic lip and rostellum.

Genus **Notylia**

The 50 (or so) species of this widespread genus can be divided in two groups with different plant and florescence morphology and different flower colours, but similar flower form. One group is 'fan-leaved', the other 'pseudobulbous'.

Pollination

All pollinators observed are male bees of the Euglossini and the syndrome is clearly that of androeuglossinophily. Pollinaria become attached to the frons of the bees that scratch the base of the lip. The pollinia are lowered after some time, caused by differential drying of the two surfaces of the stipe. Only one pollinium can enter the next visited flower, due to the very small size of the stigma. The bees are much larger than the flowers, 4-5 times, which nevertheless withstand the rough-and-tumble treatment by these pollinators. They collect the crystalline liquid from the papillose trichomes of the lip. The scent is indiscernible for humans. Bees of the same species may pollinate more than one species of *Notylia*, but these species are geographically isolated. Sympatric species have different pollinators. Visits of nonpollinating insects of the wrong size are reported. The flowers must be open for at least 24 hours before the fragrance leads to interaction between flower and pollinator. The stigma needs two days to widen enough to accept a pollinium.

Pollinators

Dodson (1967) lists observations about the pollination of seven species in Peru, Brazil, Panama and Ecuador.

N.wulschlaegeliana and *N.xyphorius* of the 'fan-leaved' group are pollinated by large males of *Euplusia surinamensis* (=*Ep.smaragdina*?) Five species of the 'pseudobulbous' group are pollinated by bees of the genus *Euglossa*. *Notylia* aff. *buchtenii* was observed in Iquitos, Peru, to be pollinated by male *Euglossa ignita*, *N.buchtenii* by male *Eg.augaspis*. *Eg.ignita* visited the same species but did not act as pollinator, apparently because of its large size. The two species are clearly kept distinct by the pollinators. *N.panamensis* has been found to be pollinated by *Eg.hemichlora* in Panama and *N.* aff. *barkeri* by another unnamed *Euglossa* (provisionally designated as KU 9). Williams mentions *Eg.erythrochlora* as a pollinator of *N.* cf. *barkeri* (see also Dressler 1968b) in Costa Rica and for a species of the same aff. *barkeri Eg.ignita*, *Eg.hansoni*, *Eg.tridentata* (captured with pollinia), and *Eg.sapphirina* (nonpollinating visitor). Another *Notylia* of the same section was visited by *Eg.ignita*, *Eg.hansoni* and the same unidentified species KU 9. *N.pentachne* is pollinated by *Eulaema cingulata*.

Several species have been observed to have strong self-incompability barriers and seed pods drop off about three weeks after self-pollination. The flowers need very precise behaviour by the specific pollinators. The pollinators have to orient very exactly in reference to the parts of the flower, visually guided by the position of the sepals, petals and lip. This could explain why species of both groups mentioned have such similar flowers. Speciation probably depended on the attraction of specific pollinators, all euglossine bees, along with geographical isolation as the fundamental barrier to hybridisation.

Warford (1992) observed pollination of three species in Mexico, sharing a common pollinator *Euglossa viridissima*: *Notylia trisepala*, *N.tridachne* and *N.barkeri*. The species have small differences in lip angle, column morphology and pollinia orientation on the stipe. These differences seem to be responsible for the maintenance of species integrity.

The orchids in Mexico were ant-guarded; the plants excrete an extrafloral nectar form the base of the flower bracts. The ants also tend the newly developing fruits, which seem to offer some reward.

Roubik and Ackerman mention pollinaria of unidentified species on *Euglossa championi*, *Eg.tridentata* and *Eg.mixta*, of *N.linearis* on *Eg.cybelia* and *Eg.dodsonii*, of *N.barkeri* on *Eg.dissimula* and *Eg.variabilis*, of *N.albida* on *Eg.hemichlora*, of *N.pentachne* on *Eulaema bombiformis*, *El.meriana* and *El.cingulata*.

Genus **Odontoglossum**

This genus is in general bee-pollinated. This formerly large genus has been split into different genera: *Odontoglossum*, *Rossioglossum*, *Lemboglossum*, *Ticoglossum*, *Otoglossum*, *Systeloglossum*. These genera will be treated here together, although some have been placed in different alliances.

Rossioglossum grande is pollinated by (female and possibly male) bees of the genus *Centris*.

Odontoglossum kegeljani from Ecuador is pollinated by male *Bombus hortulanus* var. *robustus*, which try to find nectar in the false nectaries. The teeth of the prominent callus on the lip impedes the advance to the flower and in their struggles the bees detach the viscidium of the pollinarium with their heads. The stipe curves downward, bringing the pollinia to a position in front of the head, the right position to pollinate. Other, smaller bees were observed as well, trying to reach the false nectaries, but these were to small to contact the viscidium. A scarabid beetle, *Macrodactylus* spec. was observed to eat the callus and damaged the flowers considerably.

O.retusum, *O.flavescens*, *O.mystacinum* are, at high elevations in the Andes in southern Ecuador, expected to be hummingbird-pollinated by their superficial similarity to various ericaceous shrubs (like *Gaultheria*), which are constantly in flower in that area. The plants have flowers that are atypical of the genera to which they belong, and may be seen as an adaptation to the hummingbirds, due to the lack of the normal pollinators in an area which is too cold for them. *O.retusum* has cryptic coloured pollinia.

Genus **Oncidium**

In this large genus pollination by the large bees of the genus *Centris* seems to predominate, sometimes motivated by food, sometimes by deceit of one sort or another (Dressler 1981).

Dodson (1962), puzzled by the lack of observations of pollinations, after his first accidental encounter with *Centris* bees pollinating *O.planilabre*, concluded that the bees work very fast and pollinated 17 out of 20 flowers within two minutes.

The glistening 'tabula infrastigmatica' of the flower are associated with food by the bees, which grasp it with their mandibles.

Mimicry and oil-gathering
There is evidence that some *Oncidium* species mimic the flowers of Malpighiaceous vines. These are for example *O.guianense*, *O.haitiense*, *O.quadrilobium*, *O.osmentii* and *O.compressicaule* (Dod 1986). (Note: Most of these species now are renamed as *Tolumnia* species, see below.) Female bees of the genus *Centris* gather oil (lipids) or a waxlike substance from the paired sepaline glands of Malphigiaceae. These are called epithelial elaiophores, small areas of secretory epidermal cells in which lipids accumulate in great quantity, up to 3 mg per flower, beneath a thin protective cuticle. The lipids are 8-acetoxy substituted free fatty acids with a carbon chain length of C_{14}-C_{20}. Energy derived by consumption, digestion, and metabolism of fats is higher per gram than for equivalent amounts of carbohydrates (sugars or starches) or other foods.

It may be that the bee, seeing with ultraviolet light, is attracted visually to the form. The bees scrape the wax with their feet. Odd behaviour by the bees has been noted after such scraping, perhaps indicating some intoxication.

The oil is collected with front and middle legs and transferred to the hind legs, where it is absorbed by capillaries and carried to the brood cells, often mixed with pollen. It is used as food for the larvae (Vogel 1970). *Centris* bees that collect oil from Malpighiaceae flowers have a distinctive oil-harvesting comb formed of a single row of large, flattened, overlapping setae, curved at the tip, along with 2-5 giant spatulate setae (called the 'hood organ' by Vogel) on the ventral surfaces of the anterior and middle basitarsi in opposition to the primary anterior comb. The same bees have been observed to seize *Oncidium* flowers for an instant and then fly away. Buchmann (1987) mentions 50 species out of 350 of the genus to have oil-producing elaiophores. Oil-flowers in general are found most abundantly, but not exclusively, in neotropical savannahs and forests. Buchmann mentions bees of the genus *Tetrapedia* (Tetrapediini, Anthophoridae) as probable floral lipid harvesting bees in Orchidaceae. This is confirmed for a species from Southern Brazil, Paraná by Singer (pers.comm.): *O.paranaense* is pollinated by *Tetrapedia diversipes* gathering oil from two elaiophores on the lip.

Territorial behaviour
Male *Centris* bees use flower clusters of *Oncidium* as the focal point for their territories, where they meet the females. They rest on a twig or leaf where they can survey their territory. The males defend the territories against other flying insects. It is thought that, moving in the wind, the flowers look like an insect and arouse the aggression of the males. The racemes and flowers are arched and the slightest breeze causes the flowers to dance. When the flowers move the bee attacks them. They do not land but merely strike the flower and receive pollinia on their face (frons). The outstretched petals with their coloured bars are thought to act as a flight-pattern guide on the bee's compound eyes. If the breeze continues the bee strikes flower after flower. This phenomenon is called pseudoantagonism, but according to Dressler this could be probably better described as 'pseudotrespassing'. Dressler reports the territoriality to be so pronounced that captured bees returned to the defence of their territory upon release (Van der Pijl & Dodson 1966). The action of the bees is extremely precise.

Examples of orchids with this mode of pollination are *O.hyphaematicum* (by *Centris buchwaldii*) and *O.planilabre* (*Centris geminata*) in the coastal zone of Ecuador and *O.stipitatum* in Panama. Dressler observed the same behaviour of male *Centris* in *O.stipitatum* in Panama and Adams found *O.bahamense* to be pollinated by male *Centris versicolor* (Nierenberg 1972).

Female pollinators
Female *Centris* sp. were observed pollinating *O.och-*

matochilum in Costa Rica by Dodson. The bees acted in a 'frenzied' manner, but behaved different from the way the males displayed.

Observations of female *Xylocopa* cf. *transitoria* pollinating *O.onustum* were made in coastal Ecuador (Van der Pijl & Dodson 1966). Here the plants grow on cacti. The bees had to learn how to deal with the fragrant flowers but were effective pollinators after some time.

Nierenberg (1972) observed female *Centris versicolor* visiting and pollinating *O.lucayanum* (syn. *Tolumnia lucayanum*) on Grand Bahama. These females were hovering in front of the flowers and landed upon the central region to attempt in vain to extract nectar. They grasped the infrastigmatic lobe with their mandibles. In the Dominican Republic he observed the same behaviour of what were believed to be *Centris* females on the flowers of *O.quadrilobium*. These bees behave as on the flowers of malpighs of the genus *Stigmatophyllum* (see above). Ultraviolet photographs of the flowers of these equitant Oncidiums show a central dark region and thick dark lines above the callus and below the stigmatic cavity, whereas the peripheral or petal areas are light (ultraviolet reflecting). The central area corresponds to the callus (and in malpighs where we see comparable reflections and dark UV absorbing spots to the central disk of pistil, stamens and nectaries). On Grand Bahama Island are yellow equitant Oncidiums (now *Tolumnia*), pollinated by female *Centris* bees, and white-flowered ones which elicit pseudoantagonistic behaviour in male *Centris* bees.

Mate mimicry

An interesting case is the pollination of *O.henekenii*, for it is not improbable that this rare species is pollinated in Hispaniola by pseudocopulation. This 'bee' orchid resembles an insect of purple-brown colour (*mate mimicry*).

Dod (1976) observed and photographed visits by a bee and noticed two types of behaviour. One was a 'hit and fly' manoeuvre, the other a definite mounting of the flower, 'caressing' the lip with the hind feet and tapping movements of the abdomen down toward the lip. The latter behaviour was interpreted as a copulation attempt. Bees caught were *Centris* males of a hitherto unknown species, related to *C.versicolor*, probably the unknown male of *C.insularis*.

The place were the pollinator was observed was declared a sanctuary. Nowadays the species is propagated from seed to be reintroduced, as I recently learned at an Orchid Workshop in Soroa, Cuba (1997).

Autogamy

Some *Oncidium* species are autogamous. In certain populations of *O.glossomystax* (which is a natural hybrid between *O.pusillum* and *O.pumilio*) the stipe curls, forcing the pollinia into the stigma. Some clones in a species may be self-incompatible and others are not. Catling mentions as autogamous species: *O.oestlundianum* (W.Mexico) and *O.pollardii* (Mexico).

In some species of the section Heterantha, like *O.heteranthum* and *O.abortivum*, sterile and very small, reduced flowers are found. The significance of this phenomenon is unknown.

Van der Pijl and Dodson mention *O.macranthum* to be pollinated by *Bombus hortulanus* var. *robustus* and a *Centris* species.

Genus **Psygmorchis**

P.gnomus and P.zamorensis are reported as being autogamous.

Genus **Rodriguezia**

According to Vogel (1969) the flowers are typical bee-flowers and probably pollinated by euglossine bees, with the exception of *Rodriguezia secunda* (Plate 9, p. 13), which is observed to be pollinated by hummingbirds at Caracaraí, Brazil (Vogel) and at Iquitos, Peru (Dodson) and has cryptic pollinia (Dressler 1971). However, more species seem to be regularly hummingbird-pollinated.

R.leeana is reportedly euglossine-pollinated. It is, however, not clear if the flowers are true euglossine-syndrome flowers, or produce a small amount of nectar.

In Campina, Brazil, Braga (1977) observed a butterfly *Heliconia hermanthena* as an effective pollinator. He caught two pollinators with pollinaria fixed to the head close to the proboscis and observed a butterfly pollinating a flower while collecting nectar.

Genus **Sigmatostalix**

According to Buchmann (1987) all 16 species have oil-producing elaiophores as a reward for pollinators.

Genus Symphyglossum

Species of this genus show the bird-pollination syndrome according to Van der Pijl & Dodson (1966). *S.sanguineum* has cryptic cream or greyish white pollinia.

Genus Tolumnia

Plants of the genus *Tolumnia* are known as 'equitant oncidiums' and were separated from the genus *Oncidium* in 1986. See above.

Tolumnia variegata

The flowers of this species are self-incompatible and offer no rewards to their pollinators, which are female *Centris versicolor* (Anthophoridae). Fruit production is very low in some populations (<4%). Each capsule contains many thousands of wind dispersed, dust-like seeds.

Ackerman et al. (1996) studied seed establishment in the epiphytic *Tolumnia variegata* in Puerto Rico. They experimentally enhanced fruit production and monitored subsequent seedling establishment over a period of two years. Fruit enhancement increased seed rain and seedling establishment consistently in high-fruit treatment plots. Most recruitment occurred near fruiting plants. It was concluded that if selection for increased pollinator attraction occurs, an increase in seed output should result in population growth.

Sabat & Ackerman (1996) studied fruit set and found low fruit production in natural populations. Flowering time was the most important trait affecting fruit set. The proportion of plants setting at least one fruit was significantly high early and late in the season, but low during the flowering peak. Display size had little effect. Orchid floral abundance per host tree had a significant effect, however. The results show that flowering phenology may be evolutionarily unstable, possibly as a consequence of the deception pollination system. Furthermore, a deception strategy would be relatively unsuccessful in populations where plants are found in either very dense or sparse patches. The results are consistent with what might be expected from flowers that offer no reward. *T.variegata* plants are probably exploiting the exploratory foraging behaviour of recently emerged, immigrant, or forgetful *Centris* bees. More naive pollinators were expected early and late in the season, (i.e. more than during peak flowering), because then the orchid flowers are rare in the community. The peak effect of floral display size is probably the result of the deception strategy, plants with large displays being most likely to attract naive pollinators. The advantage is lost, however, after experience of the pollinator with the flowers.

Ackerman & Galarza-Pérez (1991) studied the patterns and maintenance of the extraordinary variation in this species and put forward as one of their hypotheses that it arose from the pollination system. They suppose that this is based on food deception involving a non-specific mimicry complex of reward flowers. Obvious models do not exist. Pollinators likely abandon the flowers after a few unrewarding visits. The more difficult it is for a pollinator to discriminate the non-rewarding flowers, the more it will visit.

Tolumnia bahamense

Tolumnia bahamense is pollinated by male bees of the species *Centris versicolor*. This is an example of pseudoantagonism.

The flowers imitate enemy insects which are to be driven away by the male bees who defend their territories (see also Chapter 4).

Tolumnia lucayana

The flowers appear as food producing flowers of *Malphigi glabra* to the female bees of *Centris versicolor*. The orchid is sympatric with the previous one but is reproductively isolated by attracting the other sex of the pollinator species (Dodson 1975).

Genus Trichocentrum

The genus is exclusively pollinated by euglossine bees. There is a non-functional nectary. The flowers produce fragrance and are visited by male bees, which enter the flower, brush at the base of the lip, launch into the air to transfer the collected fragrances and receive pollinaria on top of the head or on the frons. In experiments with chemical baits in Panama, bees were caught carrying pollinaria of *Trichocentrum*.

T.panamensis and *T.tigrinum* have distinctive strong fragrances, which have not been identified hitherto. Van der Pijl & Dodson (1966) published a photograph of *Eulaema cingulata* pollinating *T.tigrinum*. *T.panamensis* is pollinated by *Euglossa cordata*.

Pollinaria of *T.capistratum* were found on *Eu-*

glossa allosticta, *Eg.bursigera*, *Eg.crassipunctata*, *Eg.deceptrix*, *Eg.gorgonensis*, *Eg.tridentata* (Roubik & Ackerman 1987).

Genus Trichopilia

T.rostrata is pollinated by *Euglossa hemichlora* (Dodson 1962) and *Eg.viridissima* (Dressler 1961). *Euglossa dressleri* was collected with pollinaria of *T.subulata*, *Eg.heterosticta* with those of *T.* cf. *leucoxantha.* A number of males of several unidentified species of *Euglossa* visited and removed pollinaria from *T.subulata* in Panama (Williams 1982). Pollinaria of *T.maculata* were found on *Euglossa bursigera* and *Eg.tridentata.* Those of *T.subulata* on *Eg.dressleri* and those of an unidentified species on *Euglossa cybelia* (Roubik & Ackerman 1987).

T.fragrans is autogamous (Catling 1990).

Genus Trizeuxis

T.falcata is reportedly self-pollinating. Otherwise not known.

EPIDENDROIDEAE: EPIDENDROID PHYLAD

ARETHUSEAE: BLETIINAE

There are 388 species classified in 21 genera within this pantropical subtribe. Bee pollination is expected for most genera and is observed in several, notably *Bletia* and *Phaius*. Fly pollination is reported for *Plocoglottis*. Lepidopteran pollination is expected in *Calanthe*. Autogamy is reported, e.g. of *Hexalectris nitida* from Mexico, see also *Bletia.*

Genus Bletia

For this genus bee pollination has been reported. Dressler (1961) observed several bee species pollinating *B.purpurea*. *Euglossa viridissima* and two other bee species were caught with pollinia on their dorsal thorax. The species produces abundant nectar.

Dodson (1962) records a female *Xylocopa* sp. as pollinator of *B.catenulata.* Some species are autogamous: *B.rosea* (*B.purpurata*) from Mexico and Guatemala, *B.campanulata*, *B.macristhmochila*, and *B.urbana* from Mexico.

Genus Spathoglottis

Spathoglottis plicata is self-pollinating, according to Dressler (1981) and others, through structural modification (Catling 1990). The species originated in Borneo and is now widely distributed in the Neotropics. It has been introduced into Guadeloupe 15-20 years ago, where Jaffe et al. (1989) investigated the influence of visiting ants. It has extrafloral nectaries on which the ants forage. They suggest that the nectaries prevent the ants from reaching the pollen and keep them foraging on the outer parts of the flower. They presume that this also protects the plants against potential herbivores and may shield the pollinators from potential predators. Not only the orchid but probably also the visiting ant species are exotic to the island. One of the visiting ant species seems to collect pollen and may interfere with the reproduction of the plant. It is, of course, not excluded that self-pollinating species receive pollinators and that they even rely on some outcrossing.

EPIDENDREAE I – NEW WORLD GROUPS

EPIDENDREAE: SOBRALIINAE

A tropical American subtribe with 219 species in four genera.

Genus Sobralia *and related genera*

Most *Sobralia* species are pollinated by various kinds of bees, among which are euglossine bees. They may have ephemeral flowers, lasting sometimes less than one day, and flower gregariously in the whole area of distribution after a period of nonflowering. The flowers appear to contain self-digesting enzymes. Bees tend to immediately visit such gregarious flowering species, whereas they don't visit the rare deviating flowers that bloom on alternate days. In Ecuador, sympatric *S.violacea* and *S.decora* var. *aerata* tend to flower on different days and are pollinated by different bees. According to Williams (1982) it is unclear if the bees visit solely for nectar or if some of them are also attracted by floral fragrances and thus fall within the euglossine-pollination syndrome. Van der Pijl and Dodson mention intoxication symptoms by different bee species (*Eulaema*, *Xylocopa*) after absorbing liquid from *Sobralia violacea* and *S.rosea.* They mention an observation by Darwin of the same kind in *S.macrantha* in a European hothouse. *Sobralia decora* is visited by *Euglossa viridissima*, *S.leucoxantha* is pollinated by *Eulaema speciosa*, *S.rosea* visited by *Euplusia ornata* and pollinated by *Eulaema polyzona* and *Bombus morio* (Ecuador), *S.sessilis* pollinated by *Euglossa cordata*. *S.violacea* is observed to be visited by *Euplusia surinamensis* and to be pollinated by *Eulaema cingulata*, *E.polychroma*, *E.speciosa*. Dressler (1961) mentions also *Euplusia* (=*Eufriesea*) *smaragdina*, *Eulaema tropica* and *Xylocopa frontinalis* as visitors and pollinators in this species, Dodson (1962) male as well as female *Ep.*(*Ef.*)*smaragdina* and ditto *Xylocopa frontinalis*. *S.* aff. *weberbaueriana* is pollinated by *Eulaema polychroma* in Colombia, but also by species of non-euglossine bees, viz. *Xylocopa* species. It appears that *Sobralia* is not, however, specially adapted to *Xylocopa* bees, which are seen as co-pollinators. The same holds for *S.violaceae*, pollinated by male and female *X.transitoria* and *X.frontalis* in Western Ecuador and by the bumblebee *Bombus morio*.

Braga (1977) mentions *Euglossa* cf. *ignita* as a probable pollinator of *S.macrophylla*.

Dressler (1976b) mentions finding *Sobralia* pollinaria on *Eulaema nigrita*.

Roubik & Ackerman (1987) found pollinaria of *Sobralia* species on several euglossine bees (12 species) of the genera *Euglossa*, *Eulaema* and *Exaerete* and suggest that *Sobralia* could be a food-deception flower and not a fragrance flower.

Some *Sobralia* species show adaptations to hummingbird pollination. *Sobralia amabilis* was visited by the hummingbird *Pantrope insignis* (and has cryptic coloured pollinia). This species is self-compatible (Bush & Beach 1995).

In the genera *Elleanthus* and *Sertifera* hummingbird pollination is obvious from the syndrome characteristics shown (including purplish brown or leaden blue cryptic pollinia), and is observed in *Elleanthus* on several occasions (Dressler 1981). In this genus synchronized flowering is the rule in some species, whereas others have a flowerhead with a jelly-like substance through which the flowers emerge in whorls, new whorls opening on consecutive days and thereby eliciting repeated visits. *E.hallii* and *E.rosea* were seen to be visited by unidentified hummingbirds (by Dodson), *E.arpophyllostachys* by *Ocrateus underwoodii*, *E.hymenophorus* by *Amazilia tzacatl*. The light purple flowers of *E.capitatus*, the yellow ones of *E.aurea* and the orange or scarlet ones of *E.aurantiacus* were visited and pollinated by unnamed hummingbirds (Van der Pijl & Dodson 1966). Dressler (1961) mentions hummingbird pollination of *E.capitatus* in Ecuador.

In Lankester Gardens in Costa Rica a picture is sold of a Green Violet-ear (*Colibri thalassinus*) visiting an *Elleanthus* species, probably *E.tonduzii*, photographed by P. Fogden.

EPIDENDREAE: ARPOPHYLLINAE

Five species of *Arpophyllum*, all from tropical America, are probably pollinated by hummingbirds, judged from the syndrome characters, including cryptic pollinia.

EPIDENDREAE: MEIRACYLLIINAE

There is only one genus, *Meiracyllium*, with two, probably three species, (Schlechter 1991). The perfume of methyl cinnamate suggests pollination by male euglossine bees (Dressler 1981).

EPIDENDREAE: COELIINAE

Five epiphytic species are known in the genus *Coelia*, of which the pollination is not known.

EPIDENDREAE: LAELIINAE

This subtribe comprises 43 tropical American genera with together 1466 species, of which relatively much is known concerning pollination. There is a considerable *adaptive radiation*: bees, wasps, butterflies, sphingid moths, hummingbirds and flies are all known as pollinators of the same or different genera. *Cattleya*, *Laelia*, *Schomburgkia* and *Encyclia* are probably bee-pollinated, as judged from only a few observations. Some *Encyclia* (subgenus *Osmophyta*) seem wasp-pollinated. *Myrmecophila*-flowers are without any reward and advertise falsely. The same may be true for more bee-pollinated genera. *Caularthron* is pollinated by carpenter-bees. Sphingid pollination is known from *Brassavola* and *Rhyncholaelia*. Most *Epidendrum* species are pollinated by diurnal or nocturnal Lepidopterans. Hummingbird pollination is probable in *Alemannia*, *Neocogniauxa*, *Sophronites*, *Hexisia* and some species of *Laelia* and *Epidendrum*. In *Scaphyglottis* stingless bees of the genus *Trigona* have been observed as pollinators (Dressler 1993). Braga (1977) suggests fly pollination in *Orleanesia cuneipetala* and *O.yauperensis*. The pollination mechanism is similar to that of *Bulbophyllum*.

Genus Basiphyllaea

Basiphyllaea angustifolia (Puerto Rico) is autogamous.

Genus Brassavola

Brassavola is a moth-pollinated genus of 16 or so species. The sphingophilous flowers, with reduced labellum, are clearly better adapted to hovering pollinators than to landing ones. The flowers are usually creamy white or pale green, produce heavy fragrances at night and have hidden, abundant nectar in cavities embedded in the ovary. The flower position is horizontal or slightly hanging, which aids the moths while feeding.

In several species extrafloral nectaries are present at the base of the sepals. These nectaries attract ants which deter herbivores and other insects that might consume the flowers, and at the same time keep the ants away from the column. Most species produce pollinator-specific fragrances at night. The production is strictly light-controlled (Roebuck & Steinhart 1978).

Braga (1977) observed a hummingbird visiting *B.martiana* in Campina, Central Amazon, but supposes it to be a nectar thief. Theft of nectar by a hummingbird was observed by him in South Brazil in *B.tuberculata*.

Brassavola nodosa

Schemske (1980) examined success in pollinia removal (male success) and fruit production in relation to inflorescence size of this orchid. The species is self-compatible, but not autogamous.

Fruit-production is pollinator-limited. An increase in flower number per inflorescence markedly increases the probability of reproductive success. Failure to produce fruit is very high in one- and two-flowered inflorescences: 92% and 79% resp. In addition, the percentage of inflorescences with male success (pollinaria removal) and female failure (no fruit) was much greater than the other way round, i.e. male failure, female success. There are, however, conflicting selective pressures, because in spite of the low probability of reproductive success for small floral displays, compared to the larger ones, few-flowered displays are dominant in populations. This apparent contradiction arises from the fact that reproductive characteristics have evolved to optimise lifetime reproductive output, not that of a single season. Many-flowered inflorescences require great investment in energy and nutrients. Flower number therefore may be constrained by plant size, which is a function of age and/or habitat. Early reproduction obviously has advantages enough to outweigh the low probability of seed dispersal experienced by small plants.

Many-flowered plants are rare and probably have a low survivorship due to the length of time they need to grow large and may have a lower level of outcrossing because of geitonogamy.

According to Wyatt (1982), who summarises findings on the adaptive significance of inflorescence architecture, Udovic predicts that selection will result in an inflorescence that strikes a balance between resource limitation and pollinator limitation. *B.nodosa* produces extrafloral nectar (Fisher & Zimmerman 1988).

Murren & Ellison (1996) investigated effects of habitat, plant size and floral display on male and female reproductive success in a mangrove habitat in Belize. Pollinium removal was 12%, fruit set 30% in summer flowers, much lower in fall flowers. Resources limit flower production and components of floral display, whereas pollinator availability most likely limits reproductive success.

Brassavola digbyana

This species was observed being visited by sphingid moths in Honduras by F. Fuchs (Van der Pijl & Dodson 1966). The species has caused some difficulties in taxonomy, because of its less distinct morphological sphingophilous character.

Genus Broughtonia

Autogamy is reported of *B.sanguinea*. *Broughtonia sanguinea* has cream or greyish white pollinia which may be an adaptation to hummingbirds (Dressler 1971), but it forms natural hybrids with close allies which appear to be bee-pollinated. Both modes of pollination may be true.

Genus Cattleya

The flowers of *Cattleya* are melittophilous. Fragrances in some species indicate that they might be pollinated by male euglossine bees, although it is known that some species are pollinated by both sexes searching for nectar.

The pollination mechanism was described by Van der Pijl and Dodson, illustrated by Fawcett and by Frymire (see Fig. 9, p. 13).

Cattleya warszewiczii was observed by Dodson being visited by *Xylocopa viridis* and by species of *Eulaema* in Columbia. *Cattleya maxima* was visited by *Eulaema polychroma* (=*tropica*?, Dodson 1962) in Ecuador, *Cattleya mendelii* by *Eulaema cingulata* in Columbia. *Xylocopa*-bees seem better adapted to Cattleyas than *Eulaema* species. *Melipona flavipennis* was encountered on numerous occasions by Dodson pollinating *C.luteola* in Peru. They were visited just before dawn, when the flowers are fragrant. Many hybrids under cultivation in the American tropics trap *Euglossa* bees in the stigma, which consequently die. Bumblebees and honeybees are observed to pollinate Cattleyas in greenhouses in the temperate regions. *C.aurantiaca*, with orange flowers is probably pollinated by hummingbirds. It has also cryptic cream or greyish white pollinia (Dressler 1971). Natural hybrids of this species, however, indicate that also bees may pollinate (Van der Pijl & Dodson 1966).

Scharrenborg (1998) mentions natural hybrids with the large flowering *C.skinneri*, producing C.X *guatemalense* with intermediate sweet scenting flowers. In Guatemala, South Mexico and El Salvador these natural hybrids occur. In El Salvador *C.aurantiaca* flowers don't open fully and here the species is self-pollinating, since all flowers bear fruits.

C.violacea is observed containing crab spiders, which catch visiting insects (Romero 1981). The crab spider seems to mimic in its colours the part of the flower it occupies.

Silva Stort (1973) counted seeds of naturally pollinated *C.loddigesii* and found great variety in fertility, ranging from 20.2% to 90.8% in intraspecific pollinations.

Catling lists as autogamous species *C.aurantiaca*, *C.crispa*, *C.patinii* and two unidentified species, one from Trinidad.

Braga (1977) observed *Eulaema mocsaryi* as pollinator of *C.eldorado*. The flowers scented strongest during the morning. The visits lasted on average 50 seconds. After the visit the pollinator transferred the collected fragrance to the hind legs in about 10 seconds and visited another flower on the same plant.

Genus Caularthron

The genus is pollinated by carpenter bees, according to Dressler (1993).

Braga (1977) observed pollination in *C.bicornutum* by *Xylocopa frontalis*. He describes the arrival of the bee as a kind of attack. After the pollinator lands, the lip moves slightly backward, allowing the bee to enter. After feeding on the callus, the bee moves backward, forced by the auricles of the callus, and presses the lip down again by its weight. The lip suddenly moves backward and the bee's head receives the pollinaria between the eyes. The visit lasts for about 30 seconds and the pollinator usually flies to another flower on the same or another plant.

Caularthron species may also contain ants.

Fisher et al. (1990) studied the relation between ant-plant specificity and the use of host plants as a resource in *C.bilamellatum*. By using isotopes, they determined the portion of the ant's diets derived from the plant. This plant is facultative *myrmecophytic*. Ants enter the pseudobulbs through preformed basal slits and use the chambers inside as nest sites. Six out of 11 inhabiting ant species, which were the most common species in this orchid, were studied, and it was found that these derived nutritional benefit from their host orchids. They used extrafloral nectar on a species-specific level. The relation is supposed to be mutualistic, the orchid receiving protection against herbivores. Autogamy is reported of *C.bilamellatum* (Belize).

Genus Constantia

Matias et al. (1996) studied the reproductive biology of the endemic species *Constantia cipoensis* in Minas Gerais, Brazil. This melittophilous species emits a scent in twilight hours, when the lone pollinator, *Xylocopa* cf. *artifex* returns to its nest, built inside the branches of *Vellozia piresania* and *V.compacta*. The possibility of pollination occurring between distant populations is very remote. Pollen dispersion is limited and there may be reproductive isolation between populations. Pollen is displaced in only a very small number of flowers and consequently there is a low fruit set.

Genus Dilomilis

Hummingbird pollination is reported of *Dilomilis*

montana in Puerto Rico. The bird species observed was *Chlorostilbon maugaeus*, also observed on *Epidendrum* species (Tyrrell & Tyrrell 1990).

Genus Dimerandra

Autogamy is reported in *D.emarginata* (Nicaragua).

Genus Encyclia

The taxonomy of this genus is clearly a complex matter and many changes have been proposed and introduced. There is still widespread confusion, especially concerning the relations with *Epidendrum*, among growers of this genus. Ecologically the two groups differ in their adaptation to different pollinators. *Encyclia* species have been crossed artificially with those of other genera on a wide scale.

The pollination mechanism was described by Hoehne (1949). The pollinator lands on the apical lobe of the labellum and by its weight the lip is put down and the side lobes enter the inner side of the column wings, making the way free to enter. After the visit the pollinator moves backward the same way out. The pollen is fixed to the back or head.

Dressler mentions self-pollination in *E.cochleata* (var. *trianda*) and *E.boothiana* (var. *erythonoides*) at the edge of their geographical range, where the usual pollinators are absent. See, however, Catling (1990), who mentions other areas as well for the first of these species. Catling mentions, moreover, autogamy in *E.bradfordii* (Trinidad), *E.chacaoensis*, *E.gravida*, *E.ochracea* and *E.cretacea* (Mexico), in *E.isochila* var. *tridens* (syn. *Epidendrum belvederense*) in the Dominican Republic, *E.pygmaea*, *E.syntenisii* (Puerto Rico) and in *E.monitcolum* (= *monticolus*) of Jamaica.

Most species suggest bee pollination (especially by carpenter bees), but some are pollinated by wasps. Dodson found *E.pentotis* to be regularly and efficiently pollinated by the scoliid wasp *Campsomeris columba* in Colombia. He observed pollination of *E.crassilabia* (syn. *E.vespa*) in the Andes in Ecuador by *Xylocopa frontalis*. Both pollinations were photographed and published in Van der Pijl & Dodson (1966).

E.vitellina shows the characteristics of hummingbird pollination, including cryptic coloured pollinia.

Encyclia cordigera

Janzen et al. (1980) studied pollination of this species in Costa Rica. The species is pollinated by medium- and large-sized black female carpenter bees (*Xylocopa* sp.) which enter the flower in search of nectar and receive the viscidium on the frons or top of the head. The flowers are very similar to those of *Gliricidia sepium*, a common legume tree visited by the same bees in the same period. Experiments with self-pollinated, single-fathered and multi-fathered flowers suggest a certain self-incapability, because fruit set was highest in the multi-fathered plants, which received pollen from different parents, and lowest in self-pollinated inflorescences. It is possible that the plant is pollinator-limited, but the experiments were not conclusive on this point.

Encyclia krugii

Ackerman (1989) found indications for pollinator limitation in Puerto Rico in this self-incompatible species. Fruit set in a natural setting is low (4%) and is limited by several factors. Many fruits were eaten (e.g. by rats) or aborted. Inflorescences that produced fruits produced fewer flowers next season, suggesting resource limitations to exist. Paucity of pollinators and resource availability are the dominant constraints and combine to limit sexual reproduction. The non-rewarding flowers are thought to act as a 'generalised food-flower mimic' making pollinator visitation rare and brief.

The same or similar reproductive strategy was found in *E.kraenzlinii* and two undescribed *Encyclia* species from Puerto Rico and the Virgin Islands.

Encyclia phoenicea

Vales et al. (1997) studied a population of *E.phoenicea* in Cuba during 1997. More than 50% of the individuals are mature plants, but only part of these complete fruiting. About 20% show evidence of being visited and only 4% set fruit. Predation of flowers is the main cause of the low degree of fructification. Another cause is pollinator activity. Of several visitors only female carpenter bees (*Xylocopa cubincola*) are legitimate pollinators.

Encyclia tarumana

This species is pollinated by *Rubrica nasuta* (Bembicidae, Hymenoptera) according to Braga (1977). The pollinia are fixed to the head. Scent is strongest in the morning. The visit is short.

He also observed a wasp, *Stelopolybia* cf. *pallipes* (Vespidae) which he considers to probably be a nectar thief, in spite of pollinia taken out. Pollinia were not deposited by the wasp on the stigma.

Genus **Epidendrum**

More than 500 species belong to this genus, even with a sharp definition of taxonomic characters needed to belong to it. The taxonomy still needs revision, in spite of several attempts to clarify relations.

According to Dressler this genus is basically Lepidopteran pollinated, with a few species adapted to fly pollination and several to hummingbirds. Ackerman calls it primarily a food-deceptive group, in which most species are pollinated by moths and butterflies (Ackerman 1986b).

Van der Pijl and Dodson mention *E.nocturnum*, *E.arachnoglossum*, *E.ciliare*, *E.purpurescens* and others to exhibit the moth-pollinated (phalaenophilous) type of flowers. The moth *Amastus acona* was found with pollinia of *E.difforme* and also pollinates *E.latilabre* in Costa Rica (see Plate 48, p. 85). The deep tube inside the ovary, often lacking free nectar, however, is typical. *E.secundum*, *E.paniculatum* and *E.radicans* are pollinated by butterflies and are psychophilous. The first one is co-pollinated by a skipper, *Urbanus proteus*, as observed by Dodson, who also observed hummingbirds and the butterfly *Papilio polyxemus* var. *americus* and a second unidentified skipper. The last one was found by him to be pollinated by five kinds of butterflies in Costa Rica. *E.paniculatum* by *Heliconia* spp.

E.falcatum has two fragrance peaks, of jasmine in the morning and lily/narcissus in the late afternoon, suggesting adaptation to both moths and butterflies (Light 1998).

Hummingbird-pollinated (ornithophilous) are *E.pseudepidendrum*, *E.ibaguense*, *E.ardens* and *E.scabrum*. The latter two live high in the Andes, where birds take over from bees in many plant families. The orchids resemble the various ericaceous shrubs of the area, like *Gaultheria* ssp. This type of mimicry of foodplants is often seen in nonrewarding orchids. Also ornithophilous is *E.pfavii*, which orchid is pollinated by the hummingbird *Amazilia tzacatl*.

E.fimbriatum is observed to be visited by tachinid flies (Dodson). The minute flowers offer some nectar and probably attract flies by their white-greenish colour and faint odour. It is likely that other small-flowered species like *E.ramosum*. *E.stiliferum* and *E.strobiliferum* are pollinated by flies as well. Braga (1977), however, supposes a butterfly as pollinator of the last species, the flowers scenting by day and being nectariferous, and states that most of the plants are autogamous and some cleistogamous.

Autogamy is reported from *E.nocturnum* (Florida, Puerto Rico, Brazil), *E.eustirum* (Mexico), *E.funkii* (Ecuador), *E.latifolium* (Puerto Rico), *E.pallens* (Costa Rico), *E.phragmites* (Nicaragua), *E.rigidum* (Florida, Mexico), *E.variegatum* (Trinidad) and some unidentified species from Brazil, Trinidad and Ecuador (Catling 1990).

E.obesum is reportedly self-compatible (Bush & Beach 1995), but not autogamous. It is pollinated by hawkmoths.

Many species have extrafloral nectaries. E.g. *E.cinnabarinum* (Delabie 1995). These nectaries attract ants, which are either prevented from entering flowers and destroying important floral parts and/or keep other herbivores from doing so.

Epidendrum huebneri

Braga (1977) observed a visiting butterfly by day that did not, however, take away pollinia. The strong fragrance in the evening suggests pollination by a moth. The species is terrestrial.

Epidendrum ibaguense, E.radicans and E.secundum

An interesting case of pollination is that of *E.ibaguense*, *E.radicans*, *E.secundum*. These species are similar in that they have different colour forms and have overlapping habitats with similar colour morphs of *Asclepias currassavica* and *Lantana camara*, two pantropical weeds (Plates 49-52, p. 85). The flowers of these orchids do not have nectar, whereas the other two plants do. Whether *E.radicans*, *E.ibaguense* and *E.secundum* are different species is not clear from the literature. Schlechter at least considers *E.ibaguense* and *E. radicans* as different, belonging to different subsections, but does not mention *E.secundum*. Van der Pijl and Dodson mention *E.ibaguense* and *E.secundum* as different species and Bierzychudek does the same with *E.radicans* and *E.secundum*.

Van der Pijl & Dodson (1966) have suggested that in Southern Ecuador, where two colour forms of *E.secundum* grow side by side on the same rocks, these are pollinated by two different hummingbirds. In northern Ecuador occurs a hybrid swarm, that has the two colour forms as parents and is thought to be pollinated by a third hummingbird which does not occur in southern Ecuador.

Different colour forms coexist elsewhere with *Lantana* of the same colour, e.g. yellow in the Cauca Valley, Colombia, cerise near Quito and pink near

Volcan Tunguhura in Ecuador. Gilbert (in Bierzychudek) states the same for other parts of Central and South America.

Dodson (1962) observed visiting skippers on *E.ibaguense*, but believes these to be ineffective pollinators and occasional visitors, whereas hummingbirds are effective. Dodson also noticed the sympatric occurrence of *E.ibaguense* with *Lantana camara* and *Asclepias curassavica*, which species have the same coloured flowers (orange and yellow). They are visited by butterflies, hummingbirds and, to a lesser extent, by bees. It was hypothesised that these species form a mimicry complex. Boyden took up this unpublished hypothesis and tried to test this in Panama. The two weeds were considered to be so-called Müllerian mimics, with the orchid as a Batesian mimic, benefiting in a 'parasitic' way from the similarity.

Boyden saw monarch butterflies *Danaus plexippus* visiting the orchid. These had pollen of *L.camara* and pollinia of *A.currassavica* on body and legs. Pollinia of *E.ibaguense* were found on the proboscis. Bierzychudek tested the hypothesis in Costa Rica with *E.radicans*. She found that the 'Batesian mimic' is not visited more often when interspersed with the model than when alone, which is one of the conditions necessary for the operation of such a mimicry complex. Butterflies were by far the most common visitors to all three plant species (many species have been reported to visit these plants and Boyden summarises the species that were observed visiting at least two of the three species). Occasional visits of bumblebees to *Asclepias*, of meliponinine bees to *Epidendrum* and bees of different kinds and hummingbirds to *Lantana* were noticed. Only the butterflies carried pollinia. The possibly short length of time that the species coexisted and the low number of cross-visitations by individual butterflies casts some doubt on the mimicry hypothesis. The results do not, however, exclude its validity. Other alternative hypotheses exist to explain the colour similarity, e.g. aposomatic colour, to prevent the plants from grazing by herbivores. The conspicuousness of both red and yellow in the same flower could serve to attract both butterflies and hummingbirds, which have different colour preference and perception. Even different butterflies may have a strong preference for some particular colour morphs, as is known for *Lantana camara* for a long time (Kay 1978). Nevertheless, the observations of Dodson and Gilbert support the mimicry hypothesis, although not on a local scale and patch-to-patch basis, but throughout the entire flight range of the pollinators. At high elevations the butterflies could be reminded by *Epidendrum* of the *Asclepias* and *Lantana* of lower elevations. By the way: scientific proof for the existence of this kind of mimicry exists for several European orchids.

Wolfe (1987) studied the influence of inflorescence size on success of pollinaria removal in *Asclepias curassavica* and *E.radicans*. The total number of pollinaria removed per inflorescence was positively correlated with inflorescence size. In *A.currassavica*, 74% had at least one pollinarium removed, in *E.radicans* this was lower, 56%, as could be expected. On the floral level these percentages were 39 and 21. This and other studies suggest that both pollinaria removal (male success) and pod production (female success) are pollinator limited in Costa Rica.

Epidendrum ciliare

Ackerman & Montalvo (1990) found proof for both pollinator and resource limitation in this unrewarding orchid in Puerto Rico. The flowers are self-compatible, but not autogamous. The hawkmoth *Pseudosphinx tetrio* was a pollinator in one of the populations studied. Average natural fruit maturation rate ranged from 5-15% over a four-year period. Only 10-15% flowers of control plants in an investigated population were naturally pollinated over a two-year period. Fruit predation was heavy. Undamaged fruits often were aborted, irrespective of being selfed or cross-pollinated. When fruiting was experimentally augmented, long-term effects of heavy fruit loads became evident by different parameters. Lifetime fitness of *E.ciliare* seems seriously affected by resource constraints. Under such conditions, resources allocated to nectar production, which can be energetically expensive, would be better committed to other functions, thereby providing a basis for the evolution of deceit pollination.

Epidendrum fragrans

Braga (1977) observed a visit of about eight seconds by a large fast-flying digger-wasp (Pompilidae, Hymenoptera), but was unable to observe pollinium removal. Only the head moved into the flower after feeding on nectar and callus. The flowers scent in the morning.

Epidendrum paniculatum

According to De Vries & Stiles (1990), who investigated its pollination in Costa Rica, this species is pollinated by butterflies that seek pyrrolizidine alkaloids (PA), especially ithomiine butterflies. Experiments suggest that the pollinators are attracted by scent to the orchid flowers. Wagner (1973) already discovered the attractiveness of the flowers of *E.paniculatum* for monarch butterflies in greenhouses in Ann Arbor, Michigan, and was informed by Dodson about the possible nature of the chemical involved.

In experiments of De Vries and Stiles with PA-baits over 20 species of butterflies were observed, carrying what were likely to be *E.paniculatum* pollinaria on their probosces. Most butterflies were male individuals (98%). (Wagner, however, found a sex ratio of about 1:1).

The butterflies also visited orchid flowers of the wet season form of the *E.paniculatum* complex (Ithomiinae mostly, and less commonly Danainae and Ctenuchidae) (Plates 53 and 54, p. 124). They obviously detected the flowers by their scent and after approach they landed, uncoiled the proboscis and probed around on an individual flower. They inserted the proboscis into the flower, usually with the head appressed to the opening of the floral spur, to be trapped by the viscidium. To escape, the butterfly pushed vigorously against the petals with its legs until the proboscis was jerked free, with a pollinarium attached near the base. Transport over distances of several kilometers is quite feasible in these long-lived insects.

According to De Vries and Stiles the phenomenon parallels in some respects the reliance upon chemicals obtained from orchids by male euglossine bees.

Genus **Hexisea**

Hexisea aurea is hummingbird-pollinated and has adapted cryptic pollinia (Dressler 1971). *H.imbricata* has also slightly cryptic pollinia of a different colour.

Genus **Isochilus**

Isochilus major shows the hummingbird pollination syndrome with non-yellow pollinia.

Genus **Jacquiniella**

Braga observed a diptere visitor in the flowers of *Jacquiniella globosa*, but no fruit set. The flowers scent most strongly in the afternoon.

Genus **Nageliella**

Nageliella purpurea is hummingbird-pollinated (cryptic pollinia).

Genus **Neocogniauxia**

Neocogniauxia monophylla has features of hummingbird pollination, including cryptic pollinia.

Genus **Oerstedella**

Autogamy is reported in *Oerstedella aberrans* (Guatemala), *O.centradenia* (Costa Rica, Panama) and *O.thurstonorum* (Ecuador).

Genus **Schomburgkia**

Flowers of this genus of about a dozen species are typical bee-flowers. The bee *Xylocopa fimbriata* was caught with pollinia of *Schomburgkia lyonsii* on the Barbados Islands and *S.splendida* is pollinated by *Xylocopa lachnea* in Colombia. In Peru *S.crispa* (syn. *S.moyobambae*) is effectively pollinated by *Trigona* (*Partamona*) *nigrior*. In this genus autogamy seems to occur in certain species after long anthesis.

Schomburgkia tibicinis

Pollination

S.tibicinis is pollinated by the bee *Eulaema polychroma* at dawn, just after sunrise, when floral odour is very perceptible (Rico-Gray & Thien 1987). Visits are very brief, 4-5 seconds. The bee pushes between the lip and the column while making a loud buzzing sound and emerges with the pollinia stuck to the back of the thorax. The bees visit in a sort of random pattern. Flowers missing pollinia or fertilised flowers diminish their attraction to bees, apparently due to physiological changes. *Xylocopa* bees were also observed near the flowers, but did not enter. Honey bees are occasional diurnal visitors and are stunned by the odour. This may cause their death when, during recovery from the intoxication on the lip of the flower, ants kill them.

Fruit set
Low fruit set (0.02 fruit/flower ratio) results from pollinator limitation. The authors associate this limitation with the pollination system, described as mimicry based on naiveté. The flowers do not appear to mimic any other flower in the community, flowering occurs simultaneously with the emergence of the bees, pollination and fruit set decline shortly after the bees emerge and have learned that the flowers are not rewarding. The orchid is allogamous and flower colour varies. All these characters point towards the syndrome mentioned.

Symbiosis with ants
S.tibicinis is well known for its symbiosis with ants. These inhabit the hollow pseudobulbs.

Rico-Gray & Thien (1989) studied the effect of four species of ants on reproductive fitness (number of fruits produced) of this orchid in Mexico. Ants forage day and night for the extrafloral nectar. The results indicate that ant efficiency in disrupting the activities of the main herbivore, a beetle (*Stethobaris* sp.) varies, and that this is related to ant size. The larger the ants, the better the control. The different species of ants are territorial and two species never occur in the same pseudobulb. There are no orchids without ants, but ants may live and feed on other plant species. In general, ant presence should not be associated with benefit to the plants. However, ants pack many of the pseudobulbs with debris and this material moves into various parts of the plant. Direct nutrient uptake was demonstrated by ^{14}C-labeled material (Rico-Gray et al. 1989).

Ant-orchid associations are for that matter not at all rare. Fisher & Zimmerman (1988) investigated these in Panama (Barro Colorado National Monument) and found 16 of 23 orchid species producing extrafloral nectar, the nectaries primarily found on reproductive structures and developing shoots. However, occupation of orchids with resulting nutritional benefit is less likely. Most associations are species-rich and not species-specific.

S.lyonsii, *S.marginata*, *S.undulata* and an unidentified species of Trinidad are reported as self-pollinating.

Genus **Sophronites**

Sophronites cernua has dull blue grey pollinia and shows the syndrome features of hummingbird pollination. The same is true for the red *S.coccinea.*

EPIDENDREAE: PLEUROTHALLIDINAE

This very large subtribe probably comprises as many as 3021 species in 28 or 29 genera (28 in Luer 1993, who misses *Zootrophion*). They are widely distributed from southern Mexico and Florida to southern Brazil. It is basically a fly-pollinated group. Most of the observations concern small *Drosophila*-like flies, and most flowers show the appropriate syndrome of myophily/sapromyophily. Autogamy is recorded of several Pleurothallidinae: *Physosyphon loddigesii*, seven species of *Pleurothallis* (see below), *Restrepiopsis clausa* and *Stelis cleistogama* (Catling 1990).

Some Pleurothallidinae produce nectar (in spp. of *Pleurothallis*, *Myoxanthus*, *Stelis*).

Vogel suggested that the flies in many Pleurothallidinae are guided by osmophores positioned on long appendages or inside the flower. The osmophores are tail-like as in *Restrepia antennifera* (dorsal sepal), *Myoxanthes reymondii* (two petals), *Pleurothallis* (*Dresslerella*) *hirsutissima* (three sepals) or form an area surface on the fused sepals inside the flower, as in *Scaphosepalum ochtodes*, or on the labellum (*Dryadella*). In *Masdevallia* and *Dracula* the elongated tips of the three sepals have osmophoric function. The odours range widely from semenlike, stinkhornlike and musklike to rancid butterlike etc. (Christensen 1994). The append ages, however, may also function as visual attractant, e.g. by moving in a breeze.

Plates

53-54. *Eucereon setosum* (Arctidae:Ctenuchidae) Moth with pollinarium of *Epidendrum paniculatum*. Photos R.B.Singer, The coastal rain forest of Paraná, Southern Brasil (p. 123).

55. Small fungus of the type that is mimicked by *Dracula* species. Photo N.A.van der Cingel, April 1997, Costa Rica (p. 125).

56. *Pleurothallis* spec. With visiting bug. The bug was seen probing flower by flower by the author. Photo P.Boeren, April 1999, Monteverde, Costa Rica (p. 126).

57. *Restrepia guttulata*, clearly showing the myophylous syndrome. Photo N.A.van der Cingel (p. 128).

53	56
54	
55	57

In many Pleurothallidinae a motile, hinged labellum for close-range guidance is present (*Dracula, Masdevallia, Pleurothallis, Dryadella, Zootrophion, Scaphosepalum*). The labellum is hinged at the base, and when a fly moves around the labellum, its weight makes the lip turn at the balancing point and the fly is pressed against the column. During the struggle to free itself, the fly receives the pollinia.

In *Acostaea, Condylago* and *Porroglossum* the labellum is sensitive and springlike. It has evolved independently in these three genera. The spring mechanism works by turgor changes in *Porroglossum*. In *Acostaea* the labellum is under tension with the broad base of the curved column, and a stimulus causes the lip to snap suddenly upward and trap the intruder.

In *Condylago rodrigoi* a pair of armlike extensions from the base of the lip, which rest with some pressure on the column-foot, resemble the condyles of a knee-joint. Unknown stimuli cause these to slip suddenly downward onto the foot, which causes the lip to snap upward, directing the insect to the column (see below).

In some Pleurothallidinae (*Myoxanthes reymondii, Lepanthes calodictyon, Platystele stenostachya*) insectlike floral structures are found, suggesting pseudocopulation, but this has not been confirmed yet by observations (Christensen 1994).

In many species of *Brachonidium, Lepanthopsis, Platystele, Pleurothallis* and *Stelis*, there is a well-demarcated, more or less circular structure on the front surface of the lip just above the base and positioned beneath the stigma. The function of this 'little eye', the glenion, is unproved, but in all likelihood it plays an important role in attracting the pollinator (Luer 1986a). From the diversity in size, colour, and shape of the floral structures in Pleurothallidinae it is likely that a broad array of flies, varying in size and biology, is involved in their pollination.

Genera Dracula *and* Masdevallia

Some of the *Masdevallia* species are pollinated by hummingbirds, others by flies. They live in habitats that lack or are poor in bees, moths and comparable flower-feeding insects.

Vogel (1978a) has suggested that the remarkably fungus-like lip of *Dracula* mimicks toadstools and that the flowers are pollinated by fungus flies. The fungus mimetes, as fungus gnat flowers in general, are dwellers of the dark forest floor, with blossoms close to the bottom (geophily). Their hues are usually dark purple brown, with white or translucent patterns. The odour is mushroom-like and musky and the fungus-like parts show intensive local transpiration (a weakly nectarous fluid in *Dracula*). The gnats habitually lay their eggs in mushrooms and try to do so in these mimicking flowers, misled by olfactory, visual, tactile and humidity stimuli, and in doing so, pollinate the flowers. The eggs are laid in the lamella and pore imitations. Eggs and larvae are incapable of eating the flower tissue and die before anthesis is finished. Mating males also contribute to pollination. Darwin found the eggs of flies inside flowers of *Masdevallia fenestrata* (*Zootrophion atropurpureus*) in a greenhouse in Kew.

The insects belong mainly to the Mycetophylidae and related Sciaridae, which tend to live in the same habitats as the plants (besides orchids, there are also Araceae and other plant families) and have hairy thoraxes that easily pick up pollinia.

Dracula species have first been described as *Masdevallia* species. Today about 100 species are recognised, which are distributed from southern Mexico to Peru (but not in Bolivia, Brazil or Venezuela and adjacent countries). Most species occur in Ecuador and Colombia and grow between 1500 and 2500 m. in altitude. It is the epichile of the lip that imitates the inverted cap of a mushroom, complete with gills (Plate 55, p. 124). In open green-

58	59
61	
62	60

Plates

58. Hawkmoth *Theretra capensis* with pollinaria of *Bonatea speciosa* attached to its eyes. Photo S.Johnson, Knysna, South Africa (p. 133).

59-60. *Meneris tulbaghia* pollinating *Disa ferruginea*. Two different phases of the process. Photos S.Johnson (p. 137).

61. *Disa ferruginea*, Cluster Disa. The orchid is a mimic of some nectar-producing plant species. Photo N.A.van der Cingel, March 1995, Table Mountain (p. 137).

62. *Tritoniopsis triticea*. Mimesis. Model of *D.ferruginea*. Photo N.A.van der Cingel, March 1995, Cape Peninsula (p. 137).

houses in the mountains where potential pollinators have free access, spontaneous hybridisation is a common occurrence and is probably effected by a tiny, promiscuous fruit-fly. Under natural conditions hybridisation is uncommon and species-specific pollinators may be the rule. Some taxa, however, are suspected of being natural hybrids (Luer 1993).

Masdevallia comprises more than 350 species, distributed from southern Mexico to southern Brazil. They are distinguished from the other Pleurothalloids by the calliferous petals and a lip hinged to a free, incurved extension from the apex of the column-foot. *M.rosea* and *M.coccinea* are ornithophilous (Van der Pijl & Dodson 1966). *M.fractiflexa* is pollinated by (blow-)flies, which are attracted by the foul odour (Dodson 1962). Light (1998) has observed houseflies and blowflies attracted by the disagreeable smell of the flowers of *M.peristeria* in Columbia in her shadehouse. The flies were caught at their thorax and trapped. They escaped only by intervention of the human observer, suggesting a larger fly as legitimate pollinator.

Dracula erythrocaete is pollinated by *Drosophila*-like flies.

Some artificial hybrids between *Dracula* and *Masdevallia* are known.

Genera Lepanthopsis, Lepanthes, Pleurothallis *and* Stelis

Dod (1986) mentions the pollination of small Pleurothallidinae of the genera *Lepanthopsis*, *Lepanthes*, *Pleurothallis* and *Stelis* by aphids and thrips in the National Botanical Garden of Santa Domingo, Dominican Republic. In these four genera he found indications that out of 126 species 98% were endemics.

In none of the species of these four genera he found a typical nectary, i.e. a spur. They nevertheless produce nectar in the lip, which appears as a drop of water at first sight. This fluid near the stigma attracts insects (Plate 56, p. 124). In a *Stelis* species very small thrips were attracted by nectar drops from the top of the petals, near the column and appeared able to pollinate the flowers. In the other three genera aphids were observed to back out of the flowers with a pollinium.

Van der Pijl and Dodson mention *Stelis aemulae* to be visited by several kinds of flies and mosquitoes in Panama. A type of fly *Bradysia* sp. (Sciaridae) was well guided and moved from flower to flower taking nectar from the lip. Pollinia were attached under the mouthparts and these flies proved effective pollinators. Dod (1986) noted that some species of *Stelis* have attractive aromas. Most species offer nectar but Chase and Peacor found that some species instead have crystals of calcium oxalate hydrate on the lip and the petals, which may function as 'pseudonectar' (Christensen 1992). Similar crystals are found in *Lepanthes* and *Lepanthopsis*. Soto Arenas observed small gnats visiting a species of *Stelis* and noted that many Mexican *Stelis* species open during the night. This behaviour is confirmed by Dix & Dix (Guatemala), who observed mosquitoes (Cilicidae) and in one case a small moth visiting a *Stelis* sp. (Christensen 1994).

Christensen himself observed a small weevil receiving pollinia on its head after taking nectar from a species of *Stelis* in the montane forests of southern Ecuador.

Stelis cleistogama is autogamous.

Christensen (1994) found larvae in flowers of *Lepanthes* sp. The flowers were not destroyed, which suggests that they do not provide digestible food.

Calvo (1990) found significant differences in fruit set among inflorescence size classes in *Lepanthes wendlandii*, a self-compatible and probably fly-pollinated orchid, but there was no clear tendency. Plants with larger inflorescences, however, do have a higher probability to set at least one fruit.

Ackerman and Zimmerman recorded 9.1% fruit set in *Lepanthes woodburyana* and 5.9% in *L.sanginea* (Christensen 1992).

Salazar Chàvez & Soto Arenas (1996) found similar proportions in Mexican *Lepanthes* species.

In Colombia, near Medellin, Duque (1993) studied pollination in *Pleurothallis* with the following results:

- *P.secunda* was persistently visited by many Bibionidae (Diptera), which are to be seen as the primary pollinators. *P.secunda* was also visited by Drosophilidea;
- *P.cardiophylla* was frequently and persistently visited by tachinid flies. In this and other species of the section Macrophyllae-Fasciculatae the flowers open permanently during day and night. Pollination is in general mediocre, indicating pollinator-limitation;
- *P.cardiothallis* is reportedly self-compatible (Bush & Beach 1995);
- In *P.monocardia* a fungivore fly species (Diptera: Mycetophylidae) is an infrequent visitor.

Fruiting appears to be low (Van der Pijl and Dodson mention pollination of *P.monocardia* by *Lycoria* sp., Sciaridae);

- A not defined *P.* species was infrequently visited by a chrysomelid beetle, sometimes observed with pollinia;
- *P.alvaroi* was visited infrequently by Bibionidae (Diptera), a *Drosophila* species and a small snout-beetle species (Curculionidae). Pollinia were rarely observed on these insects;
- *P.apoxys* is frequently visited by various *Drosophila*;
- *P.phalangifera* is very attractive for *Drosophila* species, and sometimes also visited by Bibionidae;
- *P.aves-seriales* attracts a variety of insects, including Bibionidae, Drosophilidae, Sciaridae (Diptera), Braconidae and Vespidae (Hymenoptera);
- *P.tetragona*. Many insects are frequently found inside the flowers and seem to like staying there for some time: Drosophilidae, and a species of snout-beetle (Curculionidae, Coleoptera);
- *P.gelida* is visited by Drosophilidae and Empididae (Diptera) and also by small beetles (lady-bugs, chrysomelids);
- *P.divaricans*, visited by many and various insects: Bibionidae, Sciaridae, Drosophilidae (Diptera) some Vespidae and Braconidae (Hymenoptera) and a snout-beetle (Curculionidae);
- *P.imraei*. Most common visitors are Drosophilas and a snout-beetle;
- *P.sclerophylla*. The flowers attract many insects: Drosophilas, small wasps (Draconidae) and larger ones (Vespidae), Sciaridae (Diptera) and a snout-beetle;
- The terrestrial orchids *P.floribunda*, *P.galeata*, *P.pluriracemosa* attract a variety of dipteran insects and small wasps with as a result effective pollination;
- *P.chloroleuca* and the similar *P.peduncularis* are pollinated by Drosophilidae and Empididae (Diptera);
- Hybridisation was not observed in this colony of Pleurothallidinae and species integrity seems guaranteed by mechanical barriers. Small spiders caught visiting insects, but this was not of great importance.

Singer (pers. comm.) reports pollination of *P.luteola* from Paraná, Southern Brazil by *Megasellia* flies (Phoridae) which apparently lick nectar on the surface of the lip. For details see Singer & Cocucci (1999a).

Dodson (Van der Pijl & Dodson 1966) observed visitation of *Drosophila*-like flies in *P.xanthochlora* and of flies of the genus *Lacodrosophila* in *P.ruscifolia* and *P.eumecocaulon*. Dodson observed sciarid flies (Lycoriidae) on *P.monocardia*, licking nectar from the lip (Vogel 1978).

Vogel (in Christensen 1994) found fly eggs in the flowers of *Cryptophoranthes* (=*Pleurothallis*) *endresianus* in Costa Rica.

Autogamy is reported of several species: *P.carioi* from Belize, *P.corniculata* from Nicaragua, *P.ghiesbrechtiana* var. *cleistogama* from Guatemala, *P.quadrifada* (Nicaragua), *P.ruscifolia*, and *P.sempergemmata*.

Pleurothallis endotrachys

Chase (1985) studied pollination in this widespread and variable species in Costa Rica. *Drosophila immigrans*, a species with world-wide distribution was observed visiting flowers. The flowers produce a faint rotten-fruit odour early in the day, but do not offer nectar. The flies only visit in the mornings. The flies are initially attracted by the rough surfaces of the sepals. After landing on a sepal and investigating this with its mouthparts, a fly would then move into the centre of the flower and climb onto the hinged lip. Passing the balance point, the fly is thrown lengthwise into the column by the lip. The struggle to free itself leads to removal of pollinia from the anther. They are attached to the base of the scutellum on the thorax. Five flies caught were all males, but sexual motivation seems not to be involved.

Stelis argentata

Stelis is the third largest genus in the Pleurothallidinae and comprises about 500 species. This species, of which the reproductive biology was investigated by Christensen (1992), has nectarless flowers with reduced lip and petals surrounded by a flat dish formed by the dull yellowish-green sepals and covered by reflective crystals. New flowers opened primarily in the morning, in the late afternoon and during rainy weather, coinciding with pollinator activity, as judged by pollinaria removal. (Often flowers close during sunny and dry periods but are able to open again with increasing humidity. Such floral behaviour is common in Pleurothallidinae and was observed by Syme (1880) in Jamaican *S.ophioglossoides* and *S.micrantha*). No insects were actually observed pollinating the flowers but numerous small flies were observed in the immediate surroundings. Some *Drosophila* flies were observed and collected resting on the edge of the sepals. Other tiny flies and thrips were observed in the plants but did not seem to be interested in the flowers. The darker nerves on the sepals are thought to function as a false nectar guide for small flies. The reflective crystals on the lip and petals may help to direct the attention of the insect to the centre of the flower. Flies are known to be attracted to shining surfaces. The crystals may seem similar to free nectar found on the labellum of other species, or to nectar in open nectaries. There was a dramatic increase in pollen removal during cloudy and rainy periods. It is likely that the release of scent in *S.argentata* is determined by high atmospheric humidity. Pollinators are primarily attracted by fresh, maximally fragrant flowers. 67% of the flowers had pollinaria removed and 9.6% of the flowers were pollinated. Fruit set was only 2.2%, so it appears that the orchid is partly self-incompatible. Hand cross-pollination resulted in high fruit set. In hand self-pollinated flowers fruit set was comparable to natural fruit set. It may indicate that the pollinators mostly carry out selfings. Fruit set is in general low in small pleurothallid epiphytes.

Genus **Porroglossum**

Oliver (1888) describes the irritability of *Porroglossum muscosa* (than still named *Masedevallia muscosa*), based on an investigation in Kew Gardens. The movement is a sudden and rapid folding up of the labellum on its band-like 'neck' due to changes in turgor, so that the distal part is brought near the column. The stimulation can be called forth by the gentlest touch of a hair or by the foot of an insect and takes less than two seconds. The insect will be caught in a temporary closed space. Insects have to make their escape after being captured through the aperture left between the two lateral sepals. Small flies were supposed by Oliver to pollinate in this way.

Genus **Restrepia**

In *Restrepia* osmophores are present at the petal apices and dorsal sepal apex (Plate 57, p. 124). Often these have the form of tail-like appendages, e.g. in *R.antennifera.* Secretion of the dense exudate is a dual-stage process. The firstly produced fatty oil may be used as an energy source for synthesising aminoids and terpenes, which then volatilise. In the Pleurothallidinae the aminoid fragrances are probably only advertisements (and not rewards), functioning over long distances to attract myophilous insects (Pridgeon & Stern 1983).

EPIDENDREAE II – OLD WORLD GROUPS

EPIDENDREAE: POLYSTACHYINAE

There are 154 species known, grouped into four genera, in this pantropical tribe which is primarily African but has a second centre in tropical America.

Genus **Polystachya**

This genus is known to be pollinated by small bees that gather pseudopollen. According to Dressler this is probably the pollination system of most species. For the pollination of the widespread *Polystachya flavescens*, see North America. This epiphytic species is in the tropics (Surinam) often found on *Curatella americana*. *P.concreta* (=*P.luteola*, =*P.flavescens*) is autogamous in the West Indies. *P.foliosa* of Puerto Rico is also self-pollinating (see also Chapter 6).

EPIDENDROIDEAE – DENDROBIOID SUBCLADE

DENDROBIEAE: BULBOPHYLLINAE

This subtribe is pantropical and consists of more than 1100 species in 10-15 genera, mainly found in the Old world tropics.

Genus **Bulbophyllum**

Most *Bulbophyllum* species show the fly-pollination syndrome, but in Africa also wasps and bees are observed as pollinators.

Bulbophyllum correae
Braga (1977) observed pollination by milichiid flies in Campina, Brazil. The flowers scent most strongly in the afternoon. The pollinator, after landing, moves the labellum backwards and searches for nectar. The pollinium is attached to the back of the thorax. The pollinator stays for quite a time in

the lip where it feeds on small hairs. It then may walk by the sepals from one flower to the other.

Bulbophyllum warmingianum (=**B.involutum**)
Sazima (1978) describes the pollination of *B.warmingianum* by flies in Minas Gerais, Brazil. This species is pollinated by myrmecophytic female milichiid flies (*Pholeomyia* sp.). The flowers are greenish with purple spots and stripes, which is a typical coloration for myophilous flowers. No scent was perceptible for the human nose. The labellum is very mobile, enabling the weakest air current to keep it constantly vibrating. Flies generally alight on the flower and go directly to the groove at the base of the labellum, where nectar is produced, and lick from it. The weight and movements of the fly are not sufficient to pitch the labellum to the column. Due to movements of the lip in air currents the fly is passively thrown against the column and receives the pollinia on the back. After some struggling it usually flies away. The flowers are self-compatible, but self-pollination is rare. Sazima claims this to be the first known case of wind-assisted pollination of an entomophilous plant.

Borba & Semir (1998) investigated the pollination of this species (*B.involutum*) in campo rupestre in the same area and found also two species of *Pholeomyia* to pollinate as well as one species of *Desmometopa*, trapped by assists of wind. They noticed a carrion odour on the first day of anthesis, which lasts three days. Most visits occurred on the first day. There is only one flower open at a time. The dependency on wind-assistance for successful pollination is seen as an adaptation to the open habitat where this species lives. Fruit set is low.

Bulbophyllum weddellii and **B.ipanemense**
According to Borba & Semir (1998) these two species are pollinated in a comparable way to the pollination of the previous species: wind-assisted fly pollination. In *B.weddellii* two unidentified species of *Pholeomyia* were visiting. The entire day they were observed approaching flowers without landing. Visits were most frequent during a morning period in which the flowers scent and with a second peak in the afternoon, when odour emission is lower. Visits occur with erratic flight approaches and sometimes the insect flies over the flower for some seconds, especially when air velocity is high. The scent is sweet and in younger flowers, which are visited more frequently, slightly acid. The flies land on the white tip of the lateral sepals and may stay there for several minutes, then walk to the sepal base and labellum callus and finally head in the direction of the labellum base, feeding on the small amount of nectar present in the channel formed by the lateral lobes. Winds with a velocity of 1.0-1.5 m/s are necessary to trap the fly against the column, where it is glued to the viscidium. Escape by struggling backward needs about 10-60 minutes. As the fly is finally freed from the column the labellum tips downward and throws the insect out.

More than half of the pollinaria were removed. Deposition of pollinia was observed once and the fly was in that case trapped for more than an hour.

B.ipanemense was visited more frequently than that of the other two species mentioned, whereas approaches without landing were less frequent. Only one species of *Pholeomyia* was the only visitor, with visits of varying duration.

Smaller visitors are ineffective. Wind-assisted fly pollination seems an adaptation to the open campus rupestre and seems to be dependent on small amounts of nectar, keeping these facultative anthophilous insect visitors long enough on the flowers to allow for the wind-velocities that are needed to occur.

VANDEAE: AERIDINAE

Vanda does not belong to this area, but as an imported genus with several cultivated species has been observed to be visited by euglossine bees. Dressler (1976b) mentions capture of *Eulaema nigrita* with *Vanda* pollinaria. This shows that euglossine bees may be attracted by exotic orchids.

VANDEAE: ANGRAECINAE

About 400 species in 19 genera from tropical Africa and Madagascar, with a few genera in tropical America.

In this subtribe the moth-pollination syndrome predominates. Pollination of angraecoid species from Madagascar is reported by Nilsson (1985, 1986) and long before as indirect observation by Schröter (1934). These observations concern the genera *Angraecum*, *Neobathiea* and *Jumellea*. The American genera are *Campylocentrum*, *Dendrophylax*, *Harrisella* and *Polyradicion*. They have probably invaded America from Africa and seem to derive from a leafy, small-flowered plant. The large-flowered species from the West Indies have supposedly evolved separately. *Campylocentrum aromaticum* from Southern Brazil, Paraná, is pollinated by halictid bees and to a minor extent by other insects, such as syrphid flies. Heliconius butterflies acted as nectar thieves.

The halictid bees *Pseudoaugochloropsis grammea*, *Augochlorella* cf. *eusticta*, *Augochlora amphititre* and a second *Augochlora* sp. were the main flower visitors. The bees enter the flowers from their upper side and only introduce their mouth parts into the flower tube to reach the nectar in the curled spur. Pollinaria adhere to mentum and stipes on the ventral or lateral side. On a second visit they are deposited in the markedly concave stigma. *Campylocentrum burchellii* is pollinated by the bees *Trigona spinipes* and *Plebeia emerina* (Apidae, Meloponiae) (Singer & Cocucci 1999b).

Problematic species

Genus **Eriopsis**

The tropical American species *Eriopsis* may, according to Dressler, deserve a separate subtribe. It has traits of the Cyrtopodinae (Cymbidieae) as well as of Maxillarieae. No details about pollination are reported.

CHAPTER 6

Orchids of Africa and Madagascar

Who could believe an ant in theory?
A giraffe in blueprint?
Ten thousand doctors of what's possible
Could reason half of the jungle out of being
John Ciardi

For obvious reasons the pollination of African orchids has been poorly studied, with the exception of South Africa, and more especially the terrestrials of South Africa.

For the terrestrial species of southern Africa I have used the same taxonomic ordering as Stewart et al. (1982) on the generic level. Note: The successor of this book: Linder & Kurzweil (1999) appeared too late to be included. As usual I have used Dressler's taxonomy (1981, 1993) as the superimposed system.

Madagascar is a disjunct geographical entity with a different flora and fauna as a consequence of a separate evolution, but I have not treated it separately as I did for Japan in Chapter 7 because the distinctions are clear and no confusion need be feared, for the resemblance and occurrence of related genera is great enough to allow for a combined treatment.

Primitive orchids with a single fertile anther

GASTRODIEAE: GASTRODIINAE

A few species are found in Africa. The group is primarily Asian. Some species are reportedly autogamous. Pollination of African species unknown.

GASTRODIEAE: EPIPOGIEAE

This small subtribe of three genera and five species is found in Eurasia, tropical Africa, Australasia and tropical Asia.

Epipogium aphyllum, a saprophytic species also of Europe is visited by bumblebees, but is probably cleistogamous, according to Vöth (Van der Cingel 1995). It seems that all species are autogamous.

NEOTTIEAE: LIMODORINAE

No data about pollination from African species. For the taxonomical problems of this group I refer to the note on the Neottieae in Chapter 4.

NERVILIEAE

Since 1993 Dressler places *Nervilia* in a separate tribus within a large group of primitive orchids.

Petterson (1989, 1991) studied the taxonomy of the genus *Nervilia* in Africa and the Arabian Peninsula. He distinguishes 16 species in four sections. (The total of species within the genus is 65 according to Dressler 1993.) Most species are single-flowered. He reports several facts about pollination in this genus (see below).

Dressler (1981) suggested the flowers are adapted to pollination by small bees, whereas in 1993 he also mentions self-pollination for some forms.

In the section Linervia gregarious flowering seems to be the rule. This is triggered by the first rains of the season. Anthesis usually follows three days after the first heavy rains. The main portion of plants flower on the same day and last one full day only. New mass flowering may follow some weeks later. Flowering may be absent for many consecutive years.

Gregarious flowering, according to Van der Pijl & Dodson (1966), may indicate the development of constancy by mass visits of social bees.

When a suitable pollinator enters the flower of a *Nervilia*, its thorax just reaches as far as the stigma and, if already carrying pollinia, massulae become attached to the sticky stigmatic surface. On withdrawal, the insect first gets smeared by the viscidium on the thorax and then the rear rim of the thorax lifts the anther by its front rim, which is somewhat protruding, so that the pollinia are stuck along the thorax (Petterson 1989).

Petterson observed small halictid bees transporting pollinia of *N.crociformis* and *N.petraea* in Malawi and Madagascar. The site of odour production of these species was investigated with the neutral red staining method, indicating probable fragrance-producing tissue on petals, sepals and the midline of the lip. UV absorption is strong in the apical part of the lip and along the midline, indicating false nectar guides invisible to man.

N.petreae was visited by a male scoliid wasp, a bee belonging to the Apidae and by a lepidopteran (Acraeidae), all without pollinia. Pollinium transport was observed of *Lasioglossum emirnense* and *Ceylalictus aldabranus* (Halictidae), *Braunsapis* cf. *madecasella* and *Ceratina nyassensis* (Anthophoridae). The orchid appears attractive to both males and females, although males were in the majority.

N.crociformis was observed in different countries (Malawi and Malagasy) to be visited by bees carrying pollinia from the genus *Lasioglossum* and *Ceratina*. Other bee visitors did not carry pollinia. *Lasioglossum* bees had the pollinia glued to the thorax. The yellow part in the lip throat of this species did not release collecting behaviour and therefore is not considered to mimic pollen-bearing organs. It is probably a generalised signal. Fruit set was over 50%.

The single-flowered *N.stolziana* (section Kyimbilaea) is visited by a scoliid wasp, *Micromeriella rhodesiae*, that fits the flower in size and shape and therefore is presumed to be an important pollinator. A visiting vespoid wasp, *Pardopsis punctatissima*, did not transport pollinia, nor did a lepidopteran (Acraeidae) and a syrphid fly. Pollination success was 50%.

Nervilia flowers are nectarless, which indicates pollination by deceit of insects, probably directed towards solitary wasps (Vespidae). Floral structure, pattern and coloration in this section are very uniform and the species mainly differ in number and size of the flowers. Flower colours are green, yellow and purple and similar to those found in typical wasp flowers. The lip displays false nectar guides of red and purple veins. The correspondence of the lip morphology with this wasp makes it very likely that it is the real pollinator. Wasps of the same group (a too large *Synagris aestuans* and a totally black medium-sized wasp, probably a *Delta* sp., the latter transporting pollinia) visited *N.bicarinata*, but the observations are less conclusive. Fruit set in this species was 67%. On this species, scales of a lepidopteran (Amatidae) were also found.

N.shirensis, the species with the largest flowers in the genus, was visited by specimens of the wasp species *Synagris mirabilis*, (Eumenidae) with one wasp picking up pollinia, and another one caught carrying pollinia. Fruit set was about 10%.

Uniformity of flower structure and coloration suggest adaptation to a group of closely related pollinators with a range of species of different sizes fitting the flowers of the different species in this section.

Autogamy (and/or apogamy) and cleistogamy were found in *N.gassneri* and *N.pectinata*.

N.gassneri has both scent and nectar, but exclusion experiments proves production of seeds independently of insects. Rain-assisted autogamous pollination seems possible.

N.pectinata was seen to be visited by an Ichneumoid and an apoid bee.

N.adolphii is suspected to be autogamous because every flower seems to set seed, as Mrs La Croix reported.

PALMORCHIDEAE

This tribus was also placed by Dressler (1993) in the group of primitive orchids with one fertile anther. Two genera with about 12 species in tropical America and tropical Africa.

Pollination probably by small bees (Dressler 1981, 1983).

VANILLEAE: VANILLINAE

Pantropical subtribe. Five genera with about 165 species are known world-wide.

In the New World *Vanilla* is pollinated by large bees of the genus *Eulaema* (Euglossini).

From Africa no records are available. Euglossini do not occur in Africa, but other large bees may be pollinators here.

Bee pollination is indicated by the flower structure (Dressler 1981).

More advanched orchids with a single fertile anther

SPIRANTHOIDEAE

DICERATOSTELEAE

This tribus contains only one species, *Diceratostele gabonense*, which is probably a primitive element in the Spiranthoideae. Pollination unknown.

TROPIDIEAE

Two genera, *Tropidia* and *Corymborkis*, with 43 species together.

The genus *Corymborkis* has two representatives in Africa. The structure of the long pale flowers in this genus suggest sphingid moth pollinations (Rasmussen 1977, Dressler 1981), but no data are found.

CRANICHIDEAE: MANNIELLINAE

A monotypic subtribus from West tropical Africa. The only species is found in Cameroun. Pollination unknown.

ORCHIDOIDEAE

ORCHIDEAE: ORCHIDINAE

World-wide there are about 600 species in 36 genera. Two African genera are *Bartholina* and *Holothrix*. Both genera are supposedly phalaenophilous (Vogel 1954).

Genus **Holothrix**

Vogel (1954) studied *Holothrix secunda*, a yellow-white flowered species with a 10 mm spur and with petals and lip divided into eight small filaments, which give the flower an actinomorph appearance. The flowers scent at night. *H.burchellii* even has a 'star' with 17 'rays'. *H.scopularia* has, apart from divided petals and lip, a tube formed by three petals. *H.longiflora* from central Africa has labellar fringes of 35 mm.

Comparable fringes are found in *H.grandiflora* of South Africa. No data are reported.

Genus **Bartholina**

The species in this genus have a divided, fringed or fimbriate white lip, giving the flower a starry spider-like appearance, fitting the phalaenophilous syndrome. No observations of pollination are reported.

ORCHIDEAE: HABENARIINAE

About 1100 species in 21 genera in two tentative alliances. Most species are terrestrial. Many African species. Pantropical, ranging into Eurasia.

Genus **Bonatea**

Vogel (1954) mentions Trimen writing about pollination biology of *Bonatea* in 1864, without reporting sphingophily. Periodic emanation of a musk-like scent in the evening and green flowers with white labellum lobes fit the syndrome. Pollinating sphingids have been observed in the Eastern Cape.

Pollinia are attached laterally to the head (pleurotriby).

Darwin (1888) also refers to Trimen, and to Weale, who both stated that the nectary does not contain free nectar. Mr Weale believed that the tissue of which the nectary is composed tastes sweet, so that moths probably penetrate it for the sake of the intercellular fluid. Darwin reports the pollinia to be of astonishing length, hanging down from the weight of the pollen-masses after removal. Some other species is reported by Weale to contain nectar. He found a small butterfly, *Pyrgus elmo*, 'perfectly embarrassed by the number of pollinia of this *Bonatea* attached to its sternum'.

Garside (1922) reports observations by Weale of *Pieris gidica* and *Pieris charina* with pollinia of some unknown species that turned out to be a new *Bonatea* species. This species was (then) named *Bonatea darwinii* in honour of Charles Darwin.

Bonatea speciosa

Johnson & Liltved (1996) studied pollination of this species in a South African coastal forest. Hawkmoths (*Theretra capensis* and *Hyles lineata livornica* were the pollinators, especially the first (Plate 58, p. 125). The white orchid flowers scent strongly in the evening. Hawkmoth activity was confined to a short period of 40 minutes after dusk. The pollinaria are exceptionally elongated (circa 20 mm) and are attached to the eyes of the hawkmoths. Proboscis length is around 30 mm, average length of labellum spurs 34 mm. Pollinaria attached to feeding hawkmoths are brushed over the long stigmatic arms projecting in front of the flowers. The flowers have a tooth-like process at the mouth of the spur which forces hawkmoths to enter from either side (this observation was already made by Darwin 1877). Consequently only one pollinarium is removed at a time. A moth entering the left side will have the left pollinarium fixed to its right eye and vice versa. This will eventually contact the left stigmatic arm. The two functional units of the flower are unusual in Habenariinae and presumably maximise pollen export to other stigmas. Pollination success was 98.7% (306 flowers).

Genus **Habenaria**

There are about 400 terrestrial species world-wide in this genus. Vogel (1954) studied several species from South Africa and typified these as sphingophilous and phalaenophilous. The following species are mentioned.

Habenaria polyphylla, (=*H.epipactidea*), with light green flowers and a white, sweetly scenting elongated lip. The scent becomes intensive in the evening. Vogel observed *Hippotion celerio*, a sphingid, with pollinaria on the tarsae of the front legs.

H.tetrapetala (=*H.falciformis* var. *caffra*) is thought to cause sternotriby (pollinia attached to ventral thorax). The spur is curved and adapted to the proboscides of sphingids. The side lobes are pierced and the lip also has three long teeth as is typical for night-flowers.

H.dregaena, *H.galpinii*, *H.polypodantha* (=*H.malacophylla* ?), *H.dives*, *H.clavata* and *H.arenaria* are sphingophilous and/or phalaenophilous with greenish/white flowers, split up side lobes (petals) and lip, a long curved spur and evening scent.

H.kraenzliniana has very long threadlike appendages of side lobes and lip, and *H.praestans* from East-Africa and *H.engleria* from Togo have the same, and have moreover a spur of 15 cm.

On the same typological basis many more species could be added to this list, but observations are lacking for most of these. According to Dressler (1981) we have several records of moth pollination in *Habenaria.*

Habenaria decaryana

This large species from Madagascar, up to 1 m, occurs in moist habitats along watercourses in primary forest. The greenish white flowers with a tripartite lip have a 13-20 mm long spur, one-third filled with nectar. By means of a light-trap near the population a pollen vector was captured. It was a male noctuid moth *Remigiodes remigina.* It was deduced from floral and moth morphology that the pollinator alights and inserts the proboscis while sitting head-up on the flowers, probably supported by the lateral parts of the labellum. In draining nectar the moth is stimulated to force the entire proboscis into the spur, press the head against the column and receive the viscidia onto the ventral surface of the eyes. Due to hygroscopic action the pollinia rotate so that they point downwards, a mechanism that promotes xenogamy. Several incompletely extracted pollinaria, with viscidia exhibiting scales and hairs, suggest that the mechanism does not always work perfectly. According to Nilsson & Jonsson (1985), who studied the pollination at issue, many African and Malagasy species of *Habenaria* have very long spurs indicative of sphingophily. The relatively short spur of *H.decaryana* reflects phalaenophily by noctuid and perhaps geometrid moths.

Genus **Cynorkis**

This primarily African genus has 143 species, of which 57% are Madagascar endemics. One species, *C.lilacina*, is known to be apomictic. It is thought that the xenogamous species are dependent on bees or butterflies for pollination.

Cynorkis uncata

Cribb & Gasson (1981) report unusual asexual reproduction by this species in Tanzania. The inflorescence of each plant was consistently replaced by an axillary bulbil in a colony of the orchid. The orchid normally produces a spike with up to 20 spurred, pink flowers. The bulbils have been successfully grown at Kew.

The bulbils are spherical, with some indication of at least one leaf. Anatomical similarities and differences between the bulbil and two normal tubers were observed and summarised. The bulbils have two or more layers of cells apparently without contents, but without root hairs as in tubers.

Cynorkis uniflora

Nilsson et al. (1992) studied pollination of *C.uniflora* in Madagascar. This plant is pollinated by hawkmoths. Nectar is produced in a spur and is accessible to hawkmoths with medium-long and long tongues. Maximum emission of the sweet and delicate fragrance was at dusk. Nectar was already produced in the bud stage and accumulated from the tip of the spur. Floral traits and visits indicated pollination association with endemic hawkmoths of the genera *Nephele* and *Hippotion*. Seven species of nocturnal hawkmoths and two species of diurnal butterflies were recorded as visitors. Dominant was *Panogena lingens*, but this long-tongued species, as well as three other long-tongued hawkmoth species, acted principally as nectar thieving parasites and did not contact the floral sexual organs. *Hippotion geryon* and *Nephele densoi*, two endemic and frequent species with moderately long tongues, were the legitimate pollinators.

The hawkmoths were apparently visually guided during visiting, although at a short distance odour may have an important function in the attraction. The probability of only one of the two pollinaria being removed is much higher than for both to be removed simultaneously. The orchid pollinia had extraordinarily long caudicles and were carried attached to the eyes of the hawkmoths, protruding far in front of the face of the pollinator. A relatively low number of male functions (i.e. removed pollinaria) effected many more female functions (pollinations), especially with high flower densities. Evidently, the evolution of extreme male structures in *C.uniflora*, such as the remarkably elongated pollinia, has resulted from a sexual selection process in which those few plant male functions with a progressively extreme specialisation have acquired a disproportionate number of pollen donations.

Since *C.uniflora* is an obligate outbreeder, entirely dependent on certain hawkmoths, its fitness is indirectly coupled to the presence in the surrounding forest areas of foodplants for the larvae of the pollinators (*Impatiens* and asclepiads), as well as of nectar sources for its potential nectar thieves.

Some floral traits like pink coloration, the presence of a spot-pattern on the sepals, a strong dark pigmentation on the base of the labellum and the light, delicate fragrance, conform more to a day pollination syndrome (bee or butterfly pollination) and suggest that *C.uniflora* displays unusual specialization in relation to hawkmoths. It is known that hawkmoths have a high quality of colour vision and discrimination in dim light and it is supposed that early in the evening the pollinator can still perceive the 'shortly-after-dusk' flowers sufficiently. Column morphologies suggest that eye attachment is widespread in *Cynorkis*. In *C.flexuosa*, a yellow-flowered species from Madagascar, the pollinia attach to the eyes of pierid butterflies.

Genus **Stenoglottis**

Stenoglottis fimbriata and *S.woodii* are autogamous species of southern Africa (Schelpe 1970).

Orchideae: Huttonaeinae

Five species, all from Southern Africa. Pollination unknown.

Diseae: Disinae

According to Dressler there are five genera and 143 species. Pollination data are reported from only three of the genera.

Vogel (1954, 1959) has studied flower structure and function and hypothesised about evolution within the Disinae. These studies suggested a radiation to a number of pollination systems within this group. According to Proctor (1978) adaptive radiation is a consequence of a complex of selection pressures within any particular evolutionary line towards diversity, operating in relation to a limited number of groups of specialised pollinators. Such selection pressures can also lead to convergent evolution of members of diverse plant families or groups. This is what we see in orchids as well. Compare, for example, convergent traits in Disinae and Satyriinae (spur lengths, scents, colours) or Oncidiinae and Coryciinae (oil-production). The group Disinae (Fig. 37) has indeed radiated to a number of interesting and different pollination systems (Johnson et al. 1998). They found 19 different specialised pollination systems in 27 species used for a cladistic analysis. Broadly similar pollination systems have repeatedly evolved in unrelated clades. For example, butterfly-pollinated flowers evolved twice; showy deceptive flowers pollinated by carpenter bees ditto; and long-spurred flowers pollinated by long-tongued flies four times; night-scented moth-flowers three times, self-pollination ditto. This suggests that a few dominant pollinator species in a region may be sufficient to generate diversification in plants through repeated floral shifts that never retrace the same pathways.

The taxonomy of the Disinae is expected to be

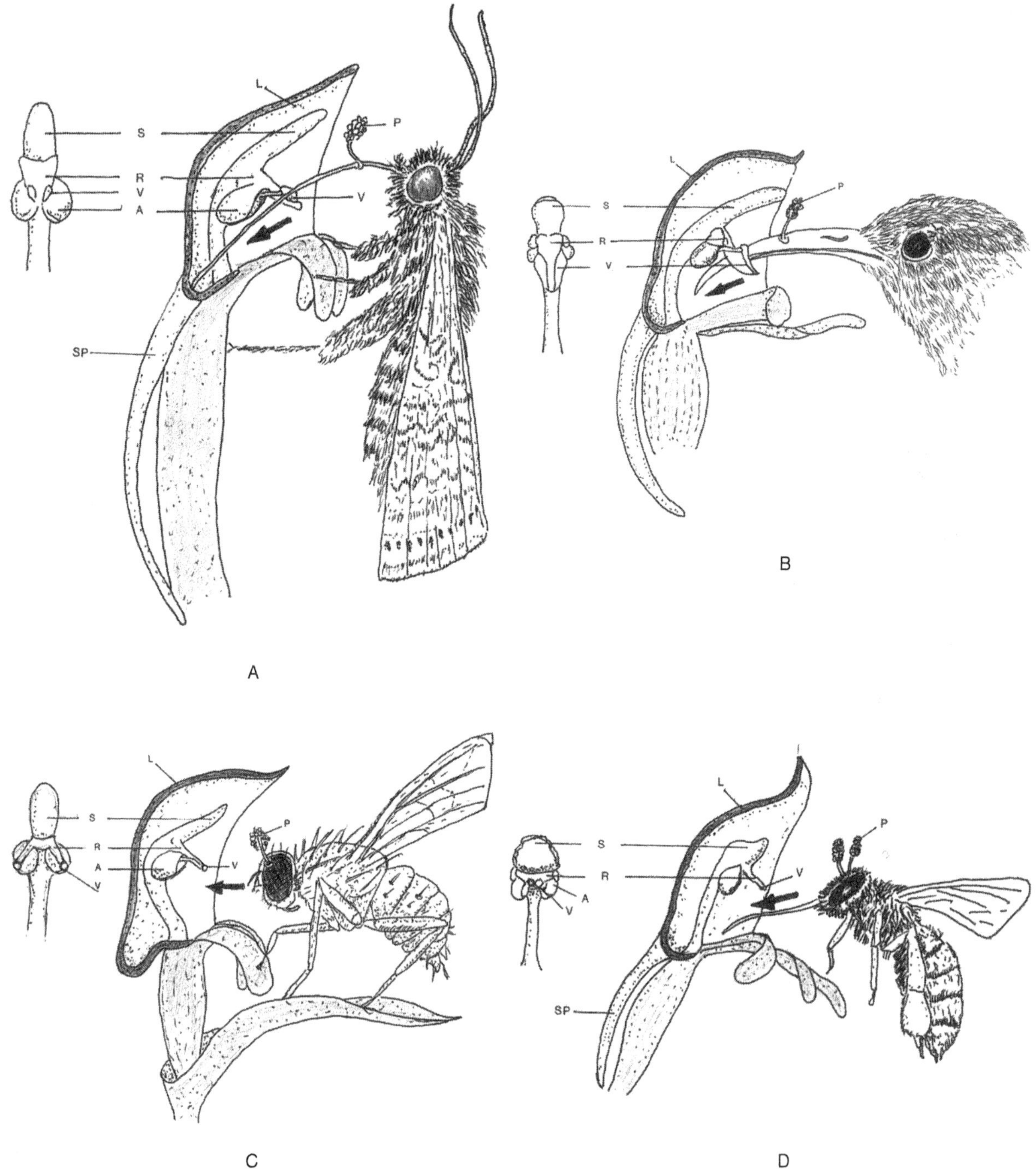

Figure 37. Floral mechanisms in *Satyrium* species (after Johnson 1997). The four different mechanisms are illustrated by lateral views of the flowers with one side of the labellum cut away and a front view of each column. A) Moth-pollinated species (*S.bicorne*). The moth's proboscis is inserted past the lateral notches in the rostellum and into one of the two floral spurs. Pollinaria become fixed when the flexible, plate-like viscidia, seated in lateral rostellum notches, fold around the proboscis. B) Bird-pollinated species (*S.carneum*). Large plate-like viscidia in lateral rostellum notches become attached to the bill of sunbirds when they probe the floral spurs. C) Carrion-fly-pollinated species (*S.bracteatum*). Small globose viscidia are seated in apical notches on the elongated rostellum arms. Pollinaria become fixed to the eyes of a fly when it enters the labellum chamber and presses its eyes against the rostellum arms. D) Bee-pollinated species (*S.erectum*). Small globose viscidia are seated at the apex of a triangular pointed rostellum. A bee, probing the spurs for nectar presses against the rostellum and pollinaria become attached to the frons between the eyes. A = anther, L = labellum, P = pollinarium, R = rostellum, S = stigma, SP = spur, V = viscidium.

changed, like *Disa*, to include some of the minor genera.

Genus **Disa**

This genus of over 120 species is divided into 14 sections (Stewart et al. 1982). Concerning pollination this is one of the best studied genera in southern Africa, although for the majority of species no data are available. In *Disa* the functions of the perianth-parts are interchanged in comparison to the normal orchid flower (Vogel 1959, Van der Pijl & Dodson 1966). The labellum is reduced, the column turns toward the spur and away from the lip. The median, dorsal sepal took over the role of the labellum in regard to nectar secretion and produced the spur. The ventral sepals took over the function of landing platform. The petals became nectar-guides. It demonstrates plasticity in adaptation towards pollinators. New types of flowers, like the tube-type (*Disa ferruginea*), and myophilous *corymbs* of spurless flowers have developed.

Section Stenocarpa

Disa porrecta

The species is mentioned as an almost certain case of bird pollination, but this could be mistaken as with the next species, *D.ferruginea*.

Disa ferruginea

Pollination of this species was already observed by Trimen in 1887 and was observed and described by Marloth in 1895 or 1896 (sources give conflicting information). Pollination biology was studied by Johnson (1993b, 1994a).

This species is entirely dependent on the butterfly *Meneris tulbaghia* (Nymphalidae: Satyrinae) for its pollination (Plates 59 and 60, p. 125). The flowers are without any food reward and attract pollinators by Batesian mimicry of nectar-producing plant species (Plate 61, p. 125). There are two colour forms: a red one which was studied by Johnson in the south-western Cape and an orange one, studied in the Langeberg Mountains. The red-flowered one imitates the flowers of *Tritoniopsis triticea* (Iridaceae; Plate 62, p. 125) and the orange one mimics *Kniphofia uvaria* (Asphodelaceae; Plate 63, p. 140). *D.ferruginea* satisfies the criteria for Batesian mimicry as formulated by several authors, such as Dafni (1984, 1986) and Little (1983):

1. The putative models are sympatric, co-blooming and are visited by the same butterfly, which depends on them for nectar.
2. *D.ferruginea* relies solely on this butterfly and is a widely dispersed species that does not form dense populations, whereas both models are common species that occur in relatively dense populations, so the mimic occurs at lower frequency than the model.
3. The colour and shape of the flowers of the orchid and its models are similar and geographically concordant. It was shown that the butterfly does not discriminate between *D.ferruginea* and *T.triticea* in a mixed stand.
4. The fecundity (fitness) of the orchid is higher when growing sympatrically with the model (*T.triticea*).

The spur depth in flowers of *D.ferruginea* is similar to the mean corolla depth in flowers of *T.triticea*. In the orange form in the Langeberg the spurs of the orchids are shorter than in the red-flowered form, as is the proboscis length of the butterfly in that area. The mean corolla depth of *K.uvaria*, however, is even larger than the corolla length of *T.triticea*.

The red preference of *M.tulbaghia* is rather unusual. It seems that it occupies a very peculiar niche as a 'robber' of bird-pollinated plants with red or orange flowers. Both models *T.triticea* and *K.uvaria* are primarily bird-pollinated. Vogel (1954) thought *D.ferruginea* (as well as the closely related *D.porrecta*) to be an unambiguous ornithophilous flower. Bird visits, though, were not observed in the orchid. Sunbirds have greater visual acuity and can distinguish the orchid from the models (Johnson 1992). Porsch in 1926 (in Vogel 1954) was more on the right track than Vogel when he called it a butterfly flower on its way towards ornithophily.

M.tulbaghia is also the near-exclusive pollinator of about 19 other montane plant species with scentless red flowers which are morphologically adapted to the butterfly, rather than to birds.

The orchid seems only partially self-compatible, but is not autogamous, as was shown by bagging experiments. Fruit set is pollinator-limited. Populations of *D.ferruginea* which are sympatric with *T.triticea* have relatively high levels of pollination and fruit production, compared with populations where the orchid grows alone. The evolution of a pollination system based on Batesian mimicry is in fact difficult to explain for pollinator-limited species, since resources in such a case do not seem to limit fitness and could probably have been invested in floral rewards. There are however indications that in the case of *D.ferruginea* the loss of a reward, combined with effective mimicry, could have resulted in energy savings without reducing fecundity to a lower level than in a nectar-producing species (*Disa uniflora*) of the same type of habitat in the same area.

Disa oreophila

Johnson & Steiner (1995) studied the pollination of *D.oreophila* subsp. *erecta* at Naudesnek in the Drakensberg mountains, eastern Cape province in South Africa. This nectarless orchid is pollinated by long-proboscid flies. The pink, unscented flowers with their long spurs (10-25 mm) were suspected to be pollinated by such flies on the basis of syndrome characteristics: long perianth tubes, bright colours and absence of scent. In South Africa two dipteran families, Tabanidae and Nemestrinidae, with mouthparts of up to 40 mm and 80 mm respectively, have evolved as nectar feeders. The large tanglewing fly *Prosoeca ganglbaueri* (Nemestridinae) was observed to pollinate *D.oreophila* subsp. *erecta* and to carry pollinaria of a sympatric orchid species *Brownleea macroceras*. The two species flower sequentially and live in the short montane grasslands of Naudesnek at high elevations (ca 2500 m). *D.oreophila* ssp. *erecta* flowering first (January) and *B.macroceras* thereafter (February) on the study site. The fly is abundant there and of 10 individuals caught in 1993, nine were carrying (1-10) pollinaria of *D.oreophila* ssp. *erecta* on the underside of the proboscis. Four flies caught in 1988, while *B.macroceras* was flowering, carried 1-13 pollinaria of this species, also attached to the proboscis underside. The flies also visited nectar plants. The proboscis of tanglewing flies is tucked beneath the body during flight, but is brought forwards when a flower is approached. The fly hovers while inserting its proboscis into the horizontal spurred flowers of *D.oreophila*, and then places its front legs on the lateral sepals for support once the proboscis is fully inserted via a small opening between the petals and the rostellum. Hybrids have not been found between the two orchids, probably because of the more or less distinct flowering periods. The same fly has been reported as pollinator of a pink-flowered *Satyrium* species in the Transvaal Drakensberg mountains.

Since taxa related to *D.oreophila* subsp. *erecta* have relatively short spurs it is conceivable that this subspecies has adapted to the long-proboscid fly. This is not likely to be the product of a coevolutionary process, however, since the flowers of the orchid are nectarless and the fly is deceived.

The flies appear to be attracted by the general appearance of the flowers of both orchid species. The orchid seems to fit the category of a general food-source mimic. The Cape orchid *Disa draconis*, which is pollinated by long-proboscid horseflies, falls into the same category of deception (Johnson & Johnson 1993). The long-proboscid flies visit mainly blue, white, or pink flowers, whereas butterflies visiting flowers with the same kind of morphology are attracted to a wider range of colours, including yellow, orange and red. Furthermore, butterflies don't hover in front of visited flowers, whereas long-proboscid flies do. Butterflies tend to settle on flowers with vertically oriented perianth tubes. The flowers visited by long-proboscid flies generally have horizontally oriented tubes. Christensen (1994) mentions Steiner's expectation that at least 13 species of *Disa* may be pollinated by these flies, as well as four species of *Monadenia* and three species of *Brownleea.*

Section Amphigena

Disa salteri

A moth-pollinated species (Johnson et al. 1998).

Section Coryphaea

Disa vaginata and **Disa glandulosa**

These orchid species from the Cape floral region are very similar species, distinguishable by spur length (5-6 mm and 2-3 mm resp.) and by the conspicuous glandular hairs on stems and leaves of *D.glandulosa.* They often grow near each other and flower concurrently. Bagging experiments (Johnson et al. 1994) clearly proved auto-pollination in both species. The pollinia flip onto the stigma while the viscidia remain in their original position. A gentle tap at the base of the plant suffices to bring the pollinia down. *D.vaginata* even shows some form of cleistogamy, auto-pollination sometimes occurring before the flowers open. No insects were observed to visit flowers and none had pollinia removed or pollen from outside deposited on the stigma. Both species show several features that facilitate autogamy, like friable pollinia, flexible caudicles and reduced rostellar lobes. Outcrossing, however, cannot be excluded, since both species have also retained features associated with outcrossing. In general auto-pollination is prevalent in climates unsuitable for insect activity. These orchids occur on shady, south facing rock ledges where rain and wind have free play.

Disa sagittalis

Pollinated by a (short-tongued) hoverfly (Johnson et al. 1998).

Disa draconis

Johnson & Linder (1995) revised the systematic status of *Disa draconis* after field studies in the Cape province, South Africa, in which population variation of different taxa was determined by phenetic (including biometrical) and cladistic methods of analysis. *D.draconis* is split up by them into three species: *D.harveiana* (with subspecies *harveiana* and *longicalcarata*, the latter a new taxon), *D.karooica* (new species), and *D.draconis.* The taxa from different habitats (sand plains, semi-arid regions and montane areas) are distinguished on the basis of floral characteristics, like length of spur, form and measures of petals, flowering period and flower colour. It therefore appears that differentiation into these taxa resulted from adaptation to pollinators. The first observations showed that the non-rewarding, deceptive species were pollinated by a specialised group of long-proboscid (horse) flies (Johnson & Johnson 1993). Later observations (Johnson & Steiner 1997) include long-proboscid tanglewing flies (Nemestrinidae) as well. The orchids show an overall similarity (in colour and flower shapes: e.g. narrow corolla tubes) with the more abundant coexisting nectariferous plants.

According to Johnson & Steiner (1997) *D.draconis* growing in the sand plains of Silverstroomstrand are pollinated by Nemestrinidae (*Moegistorynchus longirostris*) with proboscides of 57 mm on average, well in excess of the spur length of the flower (mean 48 mm), resulting in the pollinaria being attached at various sites along the proboscidae, rather than at the base (Plate 64, p. 140). Horseflies, in this case *Philoliche rostrata*, have shorter probosces (varying from 22-35 mm along sites) than tanglewing flies and flowers having a spur length longer than the proboscis ensure that the flies insert the tongue to its full length and that the pollinaria attach to the base of the proboscis. Short-spurred plants (average spur length 32-38 mm) in several southern mountain populations (Table Mountain, Franschhoek, Du Toitskloof), as well as long-spurred plants in one northern mountain population (Winterhoek), were pollinated by this horsefly. In the northern mountains the spur lengths vary from 57-72 mm.

It seems that variation in spur length among populations in the complex is a result of pollinator-mediated selection, most probably through female function. Artificial shortening of spurs led to a significant decrease in pollination level, but not so in pollinaria removal.

The flies attracted hover in front of the flower and aim the tip of the proboscis into the tiny opening between the nectar guides on the petal and dorsal sepal above and the white rostellum below. As insertion proceeds, the fly may place its front legs on the lateral sepals for support. After extraction of the proboscis the pollinaria dangle below in a position to deposit pollen on the stigma of the next visited flower.

Plants in the *D.draconis* complex are clearly dependent on pollinators for fruit set. Autogamy has not been observed. The plants are only partially self-compatible. Natural fruit set is generally low and varies in different localities. It was highest in Winterhoek where *P.rostrata* is abundant.

Vogelpoel (1997) suggests that tanglewing flies rather than horseflies are the pollinators of the long-spurred subspecies of *D.harveiana* in the northern mountains (Tulbagh and Skurweberg Pass) where they have not yet been observed pollinating.

Johnson and Steiner predict a fly with a proboscis of approximately 70 mm, like *M.longirostris* from the sandplains.

Vogel (1954) classified the flowers as typical psychophilous. He mentions nevertheless that Mrs du Plessis observed a visit by a long-proboscid fly (*Pangonia* spec.) which she caught. It had pollinia on the base of the proboscis. From this it can be argued that one should be careful with assuming certain pollinators on the basis of syndromes.

Section Hircicornes

Disa crassicornis

Weale reported auto-pollination of this species (as *D.macrantha*) in 1871 (Johnson et al. 1994). Vogel (1954) considers it to be a typical psychophilous flower in view of its colour, form and musk-like scent.

Disa cooperi

Disa cooperi has long-spurred white flowers which are strongly fragrant in the evening. Johnson (1995b) observed hawkmoths as frequent visitors and pollinators at dusk in the mountains of Natal, South Africa. The species is found in montane grasslands and bears up to 50 flowers on a tall inflorescence, up to 70 cm in height.

A long nectar-producing spur (about 42 mm) projects from the back of the hoodlike dorsal sepal. The species has a broadly spathulate green lip and pinkish tinges on the sepals and spur.

The petals are modified to form a tunnel-like entrance to the spur.

The hawkmoth *Basiothia schenki*, with a medium length proboscis of on average 4.3 cm, proved an effective pollinator. The proboscis can be fully inserted into the spur and the pollinaria become attached ventrally to the basal portion of the proboscis. The other visiting hawkmoth, *Agrius convolvuli*, has a proboscis of 8-12 cm and was never observed with pollinia. The entrance of the flower of *D.cooperi* is mostly blocked by two erect sepals which leave only a small aperture directly above the column. These petals are curved inwards so that the proboscis is guided into the opening. This contrivance is essential for attachment of the pollinaria as it ensures the proboscis passes over the rostellum which bears the viscidia. The proboscis of *A.convolvuli* is probably too long and although proboscis-viscidium contact could conceivably be made at any point along the proboscis as it is inserted, moths were observed to carry these only at the base. It seems, therefore, that firm contact is made only if the proboscis is fully inserted and the head of the moth presses the petals. Pollinaria are, moreover, most likely to adhere to the thickened basal portion of the proboscis instead of to the slender distal portion. *A.convolvuli*, making about 50% of the visits observed, therefore, is a nectar-thief. This hawkmoth is a pollinator of some other long-spurred orchids on Madagascar (Nilsson & Rabakonandrianina 1988) and a nectar thief of *Cynorkis uniflora* on the same island. Johnson suggests on the basis of the species-specific, broadly spathulate, scent-producing lip being an *autapomorphic* trait within *Disa*, that the species has evolved from a long-proboscid fly-pollinated ancestor. Closely related species like *D.scullyi* and *D.rhodantha* are probably pollinated in that way, share the same floral structure with a long spur, but have thin linear lips without a detectable scent. Probably the ancestor of *D.cooperi* was pre-adapted to hawkmoth pollination and maybe little more was required than a modification of the lip, by which it became able to produce scent.

Disa rhodantha

Johnson et al. (1998) mention a tanglewing fly as pollinator (Plate 65, p. 140).

Disa versicolor

This species is pollinated solely by large female anthophorid bees of the species *Amegilla natalensis* (Plate 66, p. 140), both in South Africa and Lesotho (Johnson 1994b). The orchid has a sharply decurved nectar-containing spur which neatly accommodates the long curved tongue of this bee. Pollinaria are ventrally adhered to the mouthparts. Flowers of this species undergo a colour change from pink to brown during anthesis. This change in colour appears to function as an orientation cue for bees to locate newly opened flowers. The bees always land on the lower portion of the inflorescence with its brown opened flowers. While the pink colour of the buds on the upper portion of the inflorescence probably acts as a long distance attractant, the bees appear to use the transition zone from pink to brown as a short range orientation cue for locating newly opened flowers. Bees accumulated large loads of pollinaria on the ventral surface of the mouthparts. They hang from the dorsal sepal by their front legs when feeding. The sharply decurved spur is an unusual trait in South African orchids,

Plates

63. *Kniphofia uvaria.* Orange form of the mimicry complex of *D.ferruginea.* Photo N.A.van der Cingel, March 1995, Knysna (p. 137).
64. Long-spurred *Disa draconis* form from the sandplains near Cape Town with pollinator, a tanglewing fly *Moegistorynchus longirostris.* Photo S.Johnson (p. 139).
65. Long tongued fly (*Prosoeca ganglebaueri*) on flowers of *Disa rhodantha.* Photo S.Johnson, Mpumulanga province, South Africa (p. 140).
66. Anthophorid bee *Amegilla natalensis* visiting flower of *Disa versicolor.* Photo S.Johnson, Drakensberg, South Africa (p. 140).
67. Male thread-waisted wasp *Podalonia canescens* visiting flowers of the sexually deceptive *Disa atricapilla.* Photo S.Johnson, Ceres, South Africa (p. 142).

63 | 65
64 |
67 | 66

and is probably an adaptation to this long-tongued bee, similar to the adaptation of several European *Orchis* species to bumblebees.

Disa extinctoria

Species with a slender cylinder with numerous small bright red flowers, thought to be pollinated by bees (Vogel 1954) on the basis of syndrome characteristics.

Section Disella

Disa obtusa

Johnson & Steiner (1994) observed pollination of *Disa obtusa* by nematoceran flies of the species *Bibio turneri* (Diptera: Bibionidae) in the Silvermine Nature Reserve, Cape Province, South Africa. Flowering in this orchid occurs in the first season after a fire and consequently very infrequently (5-40 year intervals). The white flowers emit a pungent scent and secrete no nectar. It is almost exclusively pollinated by small flies of the species *Bibio turneri.* Large numbers of flies were active during the observations. They crawl in and out of many flowers before departing. Pollinaria placement differs between the sexes, which visit in equal numbers, as male and female flies show sexual dimorphism in leg and head characters. In females the pollinaria are usually fixed to the swollen front legs, in males (which have slender forelegs) they become mostly attached to the ventral surface of the thorax just behind the head. The flies were regularly seen mating on the inflorescences. Male flies engaged in what appeared to be a form of ritual fighting. Heavy predation on the flies occurred by the ant *Camponotus niveosetosus.* Pollination was efficient and occurred in 84% of 301 flowers examined on 10 randomly chosen inflorescences, whereas about 93% had at least one pollinarium removed. Fruit set was 91.3% (1096 flowers) Other flies, also common in the marsh habitat, were not seen pollinating.

The same fly *B.turneri* was also found on the ssp. *hottentotica* in two (different) places.

Section Intermediae

Disa cornuta

Vogel (1954) believes it to be myophilous, but not unquestionable so. Darwin (1888), quoting Mr Weale, states that *D.macrantha,* which is a synonym (Stewart et al. 1982), often fertilises itself. Weale (1873) himself, however, compares the abundance of seed of *D.macrantha* with that of *D.cornuta,* concluding a difference in this respect. It is not clear for the author whether this refers to different varieties of the same species or to confusion about names.

Section Disa

Disa filicornis

Johnson (1992a, c) caught the pollinator of this species in the Bainskloof mountains of South Africa. *D.filicornis* is pollinated by the mason bee *Chalicodoma karooensis.* It carries the pollinia on the underside of the thorax. The pollinia bend in the middle by more than 90° in a forward direction after being removed from the flower, and thus take the right position to contact the stigma of the next visited flower. One of the two bees that were caught carried 13 pollinia. The bees are deceived, since the orchid does not provide any nectar. It is supposedly 'food-fraud without a model' in Nilsson's definition (Nilsson 1992). Another possibility, however, is that the bees confuse the flowers of *D.filicornis* with the similarly shaped and coloured flowers of *Polygala* species. Since all the bees found carrying pollinia were male, a third possibility is attraction by a female sexual attractant produced by the flowers.

Disa tenuifolia

The mass-flowering of this orchid (syn. *Disa patens*) is stimulated by wild-fires of the preceding year. This species, with bright yellow flowers, is pollinated by various carder and leafcutter bees of the Megachilidae, as well as occasionally by monkey beetles (Scarabaeidae: Hopliini). The flowers have

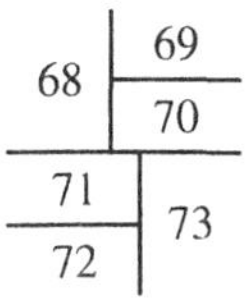

Plates

68-69. Habitat *Disa uniflora.* Photo N.A.van der Cingel, March 1995, Table Mountain, South Africa (p. 144).

70. *D.uniflora.* Note the label indicating pollination research. Photo N.A.van der Cingel, March 1995, Table Mountain, South Africa (p. 144).

71. *D.uniflora.* Photo S.Johnson, Table Mountain, South Africa (p. 144).

72-73. Mountain pride butterfly *Meneris tulbaghia* pollinating *Disa uniflora.* Pollinaria are attached to the legs of the butterfly. Photos S.Johnson, Table Mountain, South Africa (p. 144).

no ultraviolet patterns. Johnson & Steiner (1994) caught five male carder bees of the species *Immanthidium immaculatum*, carrying 4-14 pollinia, two male leafcutter bees *Megachile albohirsuta* with three pollinia each and one male *Megachile semiflava* with five pollinia. The four beetles carried 1-2 pollinia. The strongly elongated pollinia with a flexible caudicle are attached in two distinctive clumps, 3-4 mm apart, on the ventral side of the thorax, such that they dangle loosely onto the stigma during a subsequent visit. The bees merely inspect the non-rewarding flowers and sometimes settle briefly. They show a distinct territorial behaviour, flying repeatedly the same circuits around a patch of approximately 5-20 flowers, returning regularly to the same resting point, which is usually a small rock. This behaviour reminds one of the patrolling flights of bees as observed in the European *Orchis papilionacea* (Vogel 1972), which are called rendezvous flowers, and which are believed to be visited by the male bees to meet females. It is remarkably similar to that of the European species *Immanthidium manicatum*, of which the males set up territories at flowers frequented by females. Although Johnson and Steiner did not observe female bees it is known that these visit yellow legume flowers for food. This may probably explain the preference of the males for the yellow orchid flowers.

The beetles are more or less destructive by eating the pollen and are thought to be inefficient pollinators.

Hand pollination augmented fruit set experimentally by a factor of two compared to natural pollination (30-40%), indicating pollinator limitation.

Disa atricapilla and **Disa bivalvata**

These two closely related 'sister' species have overlapping distributional ranges, occupy similar habitats (seasonal wet sweeps or swamps), flower at the same time (Oct-Feb), and often occur sympatrically (Steiner et al. 1994). The two species differ primarily in shape, orientation and colour of the lateral sepals, being uniformly white or white and partly brown or maroon in *D.bivalvata* and multicoloured, red or maroon plus white (or greenish white) in the upper surface and bluish-black plus greenish-white in the basal half in *D.atricapilla*. The winglike lateral sepals also differ in colour. Furthermore, there are other minor differences in size of the pollinaria and rostellum.

Closely related species that differ in floral features are fairly common in the genus. These two species are terrestrial geophytic orchids endemic to fijnbos vegetation in the Cape region of South Africa. Flowering is strongly stimulated by fires in both species, a common feature in terrestrial orchids in Africa (and Australia). Neither species produces a food reward, both have flowers that are scented, but the scents differ, *D.atricapilla* smelling weakly spicy with a hint of peppermint and *D.bivalvata* being very fragrant with the odour of ripening apples.

D.atricapilla (Plate 67, p. 140) is pollinated by the wasp species *Podalonia canescens* (Sphecidae) and *D.bivalvata* by a different wasp species *Hemipepsis hilaris* (Pompilidae). Both orchid species strongly and uniformly absorb ultraviolet, but *D.atricapilla* has a shiny spectral reflectance on the back of the sepals in visible and UV light and is thought to mimic the shine from a folded pair of insect wings, analogous to similar reflections in wasp-pollinated *Ophrys* species in Europe.

Flowers remain open and fresh for 2-3 weeks. Although *D.atricapilla* is fully self-compatible, the species is not autogamous.

Natural levels of pollination and fruit set vary in different locations, but are relatively high, usually over 50% in both species.

Podalonia canescens, the pollinator of *D.atricapilla*, exhibits typical mate-seeking behaviour when visiting the orchid flowers. This consists of patrolling plants repeatedly and in many cases inspecting inflorescences by hovering or circling briefly without landing. When they land in the centre of the inflorescence the wasps look around very briefly before flying off or move towards the tip of an open flower. In the last case they would probe the point where the two petals come together over the anthers (the site of maximum scent production). Such behaviour is thought to lead to the eventual placement of pollinaria on the ventral surfaces of the wasps. Copulation attempts, however, were not observed. The contrast with normal food-seeking behaviour is obvious.

Of 20 wasp males caught on or around the orchid 18 carried pollinia and an additional 12 males were caught on nearby nectar-secreting flowers (the average number per wasp was 11.2 !).

The wasp *Hemipepsis hilaris* was observed visiting or patrolling the flowers of *D.bivalvata* in different localities. In one place 14 male wasps were caught on the flowers, half of which carried pollinia. A second much larger species, *Hemipepsis capensis*, was commonly observed patrolling in one place and occasionally seen carrying pollinaria. The behaviour of the wasps on the orchid was very similar to that of the pollinators of *D.atricapilla*. Experiments with flower extracts can probably cast more light on the motivation of the wasps in visiting the orchid flowers.

On *D.atricapilla* two beetles, *Peritrichia* sp., were observed carrying pollinia and one was observed pollinating. On *D.bivalvata* hopliniid species of the same genus, but of different species, were captured with pollinaria. The beetle visits were of longer duration and contact with pollinia was more haphazard. Bolus (1893) also found a beetle on *D.bivalvata*, but Van der Pijl & Dodson (1966), who mention this observation, thought this orchid to be fly-pollinated.

Hybrids between the orchid species are known since 1918 but are rare and probably are caused by beetles. The wasps fit the pattern of many aculeate Hymenoptera, especially those involved in sexual deception, with emergence of males in advance of the emergence of females. It is believed that in spite of the lack of resemblance of the flowers to female wasps the male wasps are attracted by sexual deception. Probably certain key optical features such as shine, colour and lip markings function to release mate-seeking behaviour. This is the first case of sexual deception reported from orchids of Africa.

A third related orchid species, *D.elegans*, is reported to be visited also by a hopliinid beetle (*Peritrichia* sp.). Vogel (1954) supposed *D.bivalvata* to be more *allotropous* (attracting insects of different orders) than pure myophilous, demonstrating the weakness of the syndrome concept.

Disa fasciata

Pollinated by a tanglewing fly (Johnson et al. 1998). The pollination system has evolved independently from other long-tongued fly-pollinated species in the Disinae, as concluded from a cladistic analysis. Bolus noted a striking similarity between this rare orchid and *Adenandra venosa* (Rutaceae) and suggested that these species share a common pollinator (Johnson 1992c).

Disa rosea

This species with pink spurless flowers is autogamous (Johnson et al. 1994).

Disa pillansii

Pollinated by a halictid bee (Johnson et al. 1998).

Disa caulescens

Pollinated by a muscid fly (Johnson et al. 1998).

Disa tripetaloides

This species complex shows colour polymorphy with yellow, white and red forms, suggesting adaptation to different pollinators (but see also Nilsson 1980). Intraspecific variation in floral characters indicates an active role of pollinators in the diversification of taxa (Johnson 1996a).

Disa venosa

This species has a secondary chamber formed from the petals as is also known from several other species (*D.racemosa*, *D.filicornis*, *D.tenuifolia*, *D.atricapilla*, *D.bivalvata*). All these species are pollinated by bees or wasps, which forcibly grasp the petals and insert their heads into this petal chamber. These petals are thickened, presumably to prevent being damaged by the insects (Johnson et al. 1998). For *D.venosa* they mention carpenter bees. The species contains no nectar and relies on food deception mimicry (see also Johnson 1993e). The species often is sympatric and co-flowering with the more common *D.racemosa* (see above).

Disa racemosa

According to Vogel (1954) the species is melittophilous and the dish-shaped sepalum contains nectar in a sag. The insects can reach this only by creeping through this mouthlike sepalous 'sham flower'. Pollinia were thought to be placed on the ventral surface of the insect's abdomen (sternotriby). Johnson et al. (1998), however, observed neither nectar, nor sternotriby. They mention xylocopine (*Xylocopa*, carpenter) and anthophorine (*Amegilla*) female pollen-seeking bees, deceived by food deception mimicry (see also Johnson 1993e). Observed pollinators were *Xylocopa rufitarsis*, *Amegilla niveata* and *A.spilostoma*. *X.caffra* was observed visiting, but is probably too large to act as pollinator. Only one male bee (*X.rufitarsis*) carried a pollinarium, all others were females.

A precise floral mechanism ensures that pollinaria are always attached to the middle part of legs on the pollinator. While settled on the flower, the bees clasp the petals with their front legs, while the middle legs rest across the rostellum and the back legs are placed on the lateral sepals. Pollinaria become fixed to the basal segment of the middle legs by means of a large sticky viscidium, and despite variations in bee size.

Levels of pollination and fruiting varies from 4-48% among eight study populations, being generally low.

D.racemosa occurs co-flowering and sympatrically with *D.venosa*, a species which is very similar, but much rarer. Natural hybrids are, however, never found in spite of the fact that both species appear to share pollinators. Species integrity seems to be maintained by a sterility barrier, as indicated by experimental crossing.

Disa uniflora
The relation between this species and its butterfly pollinator *Meneris tulbaghia* (Satyridae) was first reported in 1895 by Marloth and indirectly observed by him on Table Mountain (Plates 68-73, p. 141) where he caught the butterfly with a pollinium attached to one of its legs. Trimen suggested butterfly pollination as early as 1864. The orchid is nectariferous and the only species within the section *Disa* with nectar, which is believed to be an autapomorphic character. The butterfly inserts its proboscis into the spur, guided perhaps by the vein-like markings on the dorsal sepal. The butterfly aligns its body with remarkable precision over the midline of the flower with its head towards the dorsal sepal and the petals within, positioning its long legs on either side. Very soon the long proboscis is unfurled with the tip being inserted between the petal lobes and the interior surface of the dorsal sepal. The proboscis tip soon locates the nectar in the downward directed spur, whereupon a leisurely drink may take a few seconds (Vogelpoel 1994). Pollinaria are glued to the legs or thorax and are withdrawn in a horizontal fashion from the anther on leaving (Johnson 1993b, Johnson & Bond 1992). The pollinaria dangle from the legs and easily come into contact with the stigma of a visited flower. Johnson observed several butterflies carrying about four pollinia. One pollinator even carried 13 pollinia. The affinity of the pollinator for rocky habitats increases pollination significantly in such places, compared to more open habitats along streams in the valleys. Johnson observed ±50% in a gorge population, compared to ±25% in a valley. Marloth wrote in 1915: 'Although fruits (of *Disa uniflora*) are rare along streams, they occur often on cliffs, where the large mountain butterfly *Meneris tulbaghia* is free to visit its flowers, this beautiful but shy insect not readily venturing in between the banks of streams.'

An interesting observation was made by Vogelpoel (1994) who saw a butterfly on a flower of *D.uniflora*, carrying pollinia of both *D.uniflora* and *D. ferruginea*, the first on a leg, the second on the proboscis, thus making obvious that cross-pollination and eventual hybridisation would be most unlikely because of the distance between the stigma and the 'foreign' pollinium.

Section Micranthae

Disa polygonoides
Considered psychophilous due to its syndrome characteristics by Vogel (1954). Darwin (1888), however, mentions a Mrs Barber, who had seen more than once a large fly, allied to *Bombylius*, with the pollinia attached to the base of the proboscis.

Disa woodii
This roadside orchid, colonising road margins like a common weed (Johnson 1997) is capable of self-pollination (Plate 74, p. 148). This seems the consequence of the possession of a vertically held anther like in *Disa ornithantha.*

Disa fragrans
This species is thought to be perhaps moth-pollinated (Van der Pijl & Dodson 1966).

Disa erubescens
The ssp. *D.erubescens* ssp. *carsonii*, formerly named *D.stolzii* was observed to self-pollinate in Malawi on the Zomba plateau by Gassner (1982), by dropping the pollinia from the anther loculi onto the sticky stigmatic surface. Cross-pollination by butterflies (probably) may not be excluded, but has not yet been observed. Petterson (1986) observed the same in a photograph of S. Hellqvist taken in southern Tanzania, but failed to find signs of auto-pollination in his own pictures from Mozambique, taken between 1978 and 1982. This may be explained, according to Petterson, by local variations in the frequency of autogamy or by temporal factors.

Disa ornithantha
Self-pollination by the same mechanism as observed in *D.erubescens* ssp. *carsonii* was observed by the author on the Nyika plateau in Malawi in *D.ornithantha* (Plate 75, p. 148.) (Van der Cingel 1994). The same observation was made by Petterson (1986) in a specimen found near Njombe, southern Malawi.

Genus Herschelianthe (= Disa)

There are 16 species in the genus *Herschelianthe*, formerly *Herschelia*, of which 13 are found in

South Africa. They show a wide variation in floral morphology, suggesting a corresponding wide diversity of pollinators. But little is known. The occurrence of a natural hybrid between *Disa ferruginea* and *Herschelianthe graminifolia* casts some doubt on the generic distinction between *Disa* and *Herschelianthe* (Linder 1985). Note: As a consequence Linder & Kurzweil (1999) have renamed the species now as *Disa graminifolia.*

Herschelianthe (= **Disa**) **graminifolia**
This species (Plate 76, p. 148) with its bright sky-blue, faintly sweet scented flowers, with a purple lip and greenish petals, is pollinated by carpenter bees of the genus *Xylocopa* (Anthophoridae), as was recently discovered. Marloth once observed, at the beginning of the century, a lycaenid butterfly and a small fly as visitors, both without removing or carrying pollinia.

The flowers have a very short spur and offer no nectar (Johnson 1992, 1993).

The orchid is thought to be a food-source mimic. Male and female bees in search of new nectar sources in a period when there is a shortage of food plants are attracted to the flowers. This is sometimes called *naiveté exploration*, which is a better name than nonmodel mimicry (which is a contradictio in terminis). Bees land on the lip and try to crawl into the flower, but their entrance is blocked by the robust column. Bees entering the flowers thus push against the column and receive the two large pollinia, joined by fused viscidia on the ventral thorax. The whole operation lasts about two seconds. Observations during 30 hours produced only 15 pollinator visits. Twelve bees, (10 female and one male *Xylocopa rufitarsis*, one female *X.caffra*) were caught with 2-6 pollinia adhering to the thorax. Pollen deposition (or female success) appeared lower (31%) than pollen removal (male success), which varied from 46-58% in two different locations. The obvious wastage of a part of the removed pollinaria is thought to be due to the sparse population structure and lack of constancy of the bees to the nectarless flowers (which may be the result of a learning process).

A natural hybrid between *H.graminifolia* and the red co-blooming and sympatric *Disa ferruginea* has been found once (*Herscheliodisa* x *vogelpoelii*). Hybrids are obviously rare, due to the different pollinators.

Herschelianthe (= **Disa**) **lugens**
Vogel (1954) writes: 'Typical fly-flower!' The labellum has green beardlike appendages, functioning as an osmophore, with a typical urinary scent. The middle sepal is formed as a spurred hood, green with translucent windows. The flowers are nectarless.

Herschelianthe (= **Disa**) **multifida**
This species, formerly *Disa charpentieri*, according to Bolus (Vogel 1954) has a very remarkable long claw-like lip which 'as it sways in the wind suggests a possible means of attraction for insects'. Vogel therefore supposes this to be a 'Flimmerkörper', i.e. oscillator, a flexible thread-like structure vibrating in the wind, attracting flies.

Observations to confirm this are necessary.

Genus Monadenia

This genus of 16 species is characterised by a single viscidium. All species are known from the Cape floral region of South Africa. Flowering is strongly stimulated by fire in most species.

Monadenia bracteata and **M.densiflora**
This common species is also known from Australia where it has been accidentally introduced and has obviously maintained itself as a consequence of its pollination system (Plate 77, p. 148). It has unattractive, small flowers and readily sets seed by auto-pollination (Kurzweil & Johnson 1993). Bagging experiments proved autogamy. This is attained by means of the pollinia breaking up in the bud stage. The stigma lies directly below the anther and the pollen massulae fall down on it before the flowers open. Some nectar, however, is present. *M.densiflora*, a species that is closely related and can only be distinguished from *M.bracteata* by the shape of its spur, has well developed pollinia and is probably pollinator dependent.

Monadenia ophrydea and other species
M.ophrydea is pollinated by noctuid moths. This species flowers only in the first season after fire. The species shares long spurs (26 mm) with most other species of the genus. The flowers emit a strong cinnamon-like scent towards dusk. As frequent visitors, mostly arriving after 18:00 hours, Johnson (1995) observed noctuid moths. These insects, after landing, crawled rapidly over the inflorescence, stroking the flowers with their antennae and probing open flowers with their proboscides. They usually placed their front legs on the lateral

sepals for support during feeding. The pollinaria become attached to the ventral surface of the proboscis. They hang down from the base of the proboscis and may thus contact the stigma of the next visited flower.

Johnson caught four moth of the species *Syngrapha circumflexa* with 1-6 pairs of pollinaria. A *Cucullia minuta* was captured with as many as 10 pairs of pollinaria.

Male and female pollination success are both high (94% and 93% were measured resp.). Self-pollination does not occur.

The flowering spikes of *M.ophrydea* are very inconspicuous among the blackened and charred branches of recently burnt vegetation. The dark coloration of the flowers seems an adaptation to render the plants cryptic. The plants are obviously palatable as was seen from the many chewed off inflorescences found. Animals like rock hyraxes are known to graze orchids.

The cost of *crypsis* of flowers is reduced conspicuousness to pollinators. Therefore these plants rely almost entirely on scent to attract pollinators. Noctuid moths are capable of locating scented flowers without visual cues, as was demonstrated by Brantjes (1978).

Other species in the genus that have elongate spurs and evening scent are:

Monadenia atrorubens (also inconspicuous due to dark coloration).

M.comosa, *M.bolusiana*, *M.reticulata* (all three species with lime-green flowers) and *M.cernua* (cream-green flowers).

Pollination by medium-sized settling noctuid and possibly geometrid moths is likely in these species. Functional morphology of the flowers points in the same direction. Pollination by long-tongued flies is also possible and in fact falls in the same syndrome.

Genus Schizodium

Schizodium inflexum fits the melittophilous syndrome (Vogel 1954), and possesses small pink flowers on a thin filiform stem.

DISEAE: SATYRIINAE

Three genera (*Pachitis*, *Satyridium* and *Satyrium*) with about 110 species. Primarily African. *Satyrium* species are found also in Asia. Non-resupinate flowers with two spurs or depressions. The only data about pollination refer to *Satyrium*.

Genus Satyrium

Not much is known about pollination of the 100 or so African *Satyrium* species, although recently some data have become available. Johnson (1995) summarised pollination data within this genus.

The morphologically diverse flowers reflect adaptations to a wide range of pollinators, including solitary bees, carrion flies, butterflies, birds and moths (Fig. 37, p. 136). Two basic types of floral mechanism are recognised: long floral spurs and plate-like viscidia seated in lateral rostellum notches (adapted to birds and lepidoptera) and short spurs and globose viscidia seated in terminal rostellum notches (adapted to flies and bees) (Johnson 1997).

Vogel (1954) studied *Satyrium* species for syndrome characteristics. This kind of typology by 'Bautypen' was justly criticised by Van der Pijl (1978) for ignoring ecological influences in the structural classes of flowers, relying too much on *orthogenesis* (i.e. internally directed evolution trends) to explain floral morphology.

The non-resupinate flowers have two spurs formed by the labellum, which is, quite unusually, situated behind the column.

Satyrium longicauda

This is one of the many species that has long spurs and scents sweetly in the evening, clearly showing the phalaenophilous syndrome characteristics (Plates 78 and 79, p. 148).

The white flowers of this species are pollinated by noctuid moths. These feed from the nectar after settling on the flowers, grasping the petals with their legs and inserting the proboscis into the long spur. Pollinaria become fixed to the proboscis of the pollinating moth. The viscidia form small plates, which clasp around the proboscis.

Satyrium odorum

Rebelo (1987) three times observed an opportunistic sunbird (*Nectarinia chalybea* male) probe and remove nectar from the narrow spurs of this normally moth-pollinated species with evening-scented flowers. After landing the bird worked its way upward on the inflorescences. The last few flowers deposited their pollinia on the upper surface of the bird's mandible. The bird, after perching, brushed off the pollinia. Pollen transfer by birds is therefore probably largely restricted to discrete clumps of flowers. Rock Hyraxes ate most of the flowers in the population involved.

Satyrium ligulatum
This creamy white or yellow-green (tinged with purple) flowered species with a sweet scent is pollinated by butterflies by day and by noctuid moths in the evening. There is no obvious periodicity in scent production. Johnson (1997) observed as pollinating Nymphalidae (Lepidoptera): *Acraea horta* and *Vanessa cardui*; Further a noctuid species: *Agrotis segetum*. Mean spur length was 9.23 mm, closely matched by the proboscides lengths of the pollinators. Pollination success was high, up to 98%.

Satyrium stenopetalum
Again a moth-pollinated flower with white evening-scented flowers tinged with cream and traces of pink and with a soapy and pungent scent. Johnson (1997) observed as the main visitor *Agrotis segetum*, a noctuid moth. They settled in upright position on the inflorescence and probed each of the spurs, while grasping the flower with their front legs. Pollinaria were fixed to the dorsal surface of the proboscis. Out of 71 flowers from six inflorescences 65% had been pollinated and 70% had at least one pollinarium removed. Spur length was on average 13.7 mm, slightly longer than the proboscis of *A.segetum*. Self-pollination, not unusual in *Satyrium* as a consequence of multiple probing of both spurs, is avoided in *S.stenopetalum* by bending the caudicles. Pollinaria contact the stigma only when the proboscis is inserted into the same side of the flower as the pollinaria came from.

Satyrium bicallosum
Garside (1922) described pollination by fungus gnats (*Sciara* sp.). Fungivore flies (Plate 80, p. 149) usually lay their eggs on mushrooms and are known as pollinators of several orchid species from Europe and America. The dull greenish-yellow flowers of this African species smell like fungi and the odour attracts the insects.

Purple spots ('eyes'), one above each spur, may act as insect guides. Pollinia are removed one by one, as a consequence of the two apertures, each entrance having its own alighting place formed of a lateral sepal and petal. Garside found the pollinators caught by the leaves of *Drosera*, when he deliberately sought them. Each fly, stuck to the tentacles of the insectivorous plants, carried 1-3 pollinia. They were identified as *Sciara* species. Garside observed flies running rapidly over the ground, overburdened with pollinia and running up a spike and entering one of the openings. Most pollinations probably are effected at dusk, when the flies are most active. Garside supposed that the pollinator would lick the extra-cellular oily secretion of the surface of the short unicellular hairs at the base and side of the spurs, comparable to what Darwin had suggested for European species like *Orchis morio*.

Satyrium bracteatum
This species has fleshy brown to black flowers, which fit the sapromyophilous syndrome. They imitate the colour of the substrate on which carrion flies lay their eggs, viz. faeces and rotting animal corpses, and have a scent faintly recalling putrid meat. Carrion flies (Plate 81, p. 149) are indeed observed as pollinators. They land near the plant and walk into the 'gaping mouth' of the flower. Pollinaria are deposited on the eyes of the flies and these may gradually become blinded by the accumulated mass of pollinaria after several visits.

A form with green bracts from the Table Mountain was visited by small carrion flies: *Helina* sp. (Muscidae) and dungflies: *Scatophaga stercoraria* (Scatophagidae). They land on the extended bract and walk into the flower. Pollinaria are attached to the eyes and upper thorax. Pollination success was 62%, pollinaria removal 64%. A black-flowered form was pollinated by the same dungfly.

No nectar was present in the spurs.

Satyrium pumilum
This species (and *S.saxicolum*) also shows the syndrome of sapromyophily and is, according to Dressler (1981), suspected of being pollinated by carrion flies. The rosette-like plants with leaves and flowers close to the ground have quite large flowers, dull green outside and maroon inside, sometimes banded. They smell strongly (sometimes faintly) of putrid meat. The syndrome characteristics are even more obvious in this species than in the previous one, and from this it may safely be inferred that carrion flies will be the pollinators, although confirming observations are still lacking.

Satyrium hallackii
Two subspecies are known, ssp. *hallackii* and ssp. *ocellatum*.

The subspecies *ocellatum* is clearly adapted to insects with long probosces (Plate 82, p. 149). The spur is 12-31 mm long. Van Ede observed in 1990 a long-tongued fly (Nemestridinae) with pollinaria of *S. hallackii* attached to its proboscis in the wetlands of Verloren Valei Nature Reserve. In February 1993 another species of the Nemestridinae, *Prosoeca ganglbaueri* (known as pollinator of *Disa oreophila*), was observed pollinating the orchid. The same day at 16:00 hours large numbers of hawkmoths appeared and pollinated the flowers. The moths moved rapidly, visiting about 38(!) flowers per minute. Each moth carried many pollinaria on the proboscis. They belonged to four species: *Basiothia schenki*, *Hippotion celerio*, *Nephele comma* and *Agrius convolvuli*. The pollinators require tongues at least 3.5 cm long in order to reach the nectar at the base of the spur. *Prosoeca ganglbaueri* has a proboscis length of 3.5 cm, *Agrius convolvuli* has 9 cm. Over 98% of the orchids appeared pollinated on investigation.

The subspecies *hallackii* of the Western Cape has short spurs (7-8 mm). Hawkmoths are rare in the area. This subspecies is pollinated by carpenter bees (Plate 83, p. 149). According to Johnson (1997) the intraspecific variation in floral traits (in this case: spur length) may reflect an adaptive shift in the pollination system of species. This may in this case have been caused by an ecological gradient in South Africa: carpenter bees are common in coastal habitats, but rare in grassland habitats where there are few available nesting sites. Hawkmoths, on the other hand, are common in grasslands, but rare in fijnbos where there are few palatable host-plants.

Satyrium macrophyllum

The species is considered psychophilous by Vogel (1954) due to the long spur (25 mm). Confirmation by observation needed. See e.g. *S.coriifolium*, which was also supposed psychophilous by Vogel, although suspected of ornithophily by Porsch (in Vogel 1954, who doubted this strongly), was even typified for pollination by Pangonidae or Nemestridinae, but eventually proved ornithophilous. Christensen (1994), however, recently mentioned the possibility that some nectar-producing *Satyrium* species with pink flowers and long columns and spurs of over 20 mm, like *S.macrophyllum*, *S.candidum* and *S.coriifolium* might be pollinated by hovering long-tongued Nemestrini and Tanabidae.

Satyrium bicorne

The species, with yellow-greenish or dull yellow flowers with weak to strong tinges of purplish brown, is pollinated by noctuid and geometrid moths. Spurs are 10-22 mm long (mean length 15.2 mm). Johnson (1997) captured eight moths representing four species, all having pollinaria at the dorsal surface of the probosces, which were about the length of the spur. The species were: Noctuidae: *Syngrapha circumflexa*, *Heliothus armigera* and an unidentified species; Geometridae: *Drepanogynis* sp.

Satyrium coriifolium, S.princeps and S.carneum

These three species are pollinated by sunbirds (Johnson 1996).

Garside (1922) wondered about the lack of observations of pollination in *S.coriifolium*, a species 'with such attractive colour and abundant nectar', and states that the visits of pollinators (he supposed diurnal insects) are by no means frequent, although examination of the anther usually gives evidence of (insect) visits. Marloth 'figured' the pollinator incorrectly to be a bee *Lasioproposa bigoti*. *S.carneum* has red flowers, *S.coriifolium* has orange to bright yellow and *S.princeps* rose pink to carmine red. The orchids have nectar, but apparently lack scent, unlike other *Satyrium* species. They are regularly probed by several species of sunbirds which remove the pollinia, which become firmly fixed to the bill with the help of large viscidia. The plants show

Plates

74. *Disa woodii* showing self-pollination of the flowers when pollinia fall onto the stigma. Photo S.Johnson (p. 144).
75. *D.ornithantha*. Self-pollination. Photo N.A.van der Cingel, April 1994, Nyika-plateau, Malawi (p. 144).
76. *Herschelianthe graminifolia* with predating crab spider. Photo N.A.van der Cingel, March 1995, Table Mountain (p. 145).
77. *Monadenia bracteata*. Self-pollinating orchid, note the early swelling of the ovaria. Photo N.A.van der Cingel, November 1995, Table Mountain (p. 145).
78. *Satyrium longicauda*. Phalaenophilous flowers. Photo N.A.van der Cingel, March 1993, Natal (p. 146).
79. Noctuid moth pollinating *Satyrium longicauda*. Photo S.Johnson, Kwazulu, Natal, South Africa (p. 146).

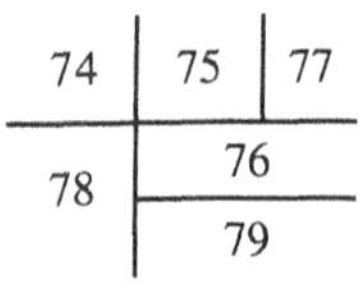

several adaptations to these visitors: apart from the colours and the copious nectar (in *S.carneum* at least, but less so in the other species) there are the sturdy inflorescence stems which can support the weight of perching birds and the large viscidia.

The birds often rub their bills against branches to get rid of the pollinaria that obviously irritate them. Sometimes they succeed, but usually pollinaria remain fixed on the beaks. Natural hybrids occur between the sympatric *S.carneum* and *S.coriifolium*, which are undoubtedly effected by sunbirds like the Malachite (*Nectarinia famosa*), Orange-breasted (*Nectarinia violaceae*), Lesser Double-collared (*Nectarinia chalybea*) or Black sunbird (*Nectarinia amethystina*), which have been observed as pollinators, some of them on both species.

Pollinaria are affixed to the bills. Most of the birds observed to visit *S.carneum* had dense clumps of pollinaria attached to the upper mandible, about 10 mm from the tip. Pollinaria of *S.coriifolium* were attached nearer to the tip of the mandible and those of *S.princeps* about 15 mm from the tip (Plates 84 and 85, p. 149).

Female birds appeared to be more effective pollinators because they spend less time defending territories. Only half of the males carry pollinaria at any one time, but nearly all the females did. Sunbirds probe the flowers while grasping the inflorescence with their feet in a typical feeding posture.

Insects (honey bees and anthophorid bees in *S.coriifolium*, carpenter bees and honey bees in *S.carneum*) occasionally visited the species under study, but did not remove pollinaria because of their too small size, although they may obtain some nectar.

Satyrium erectum

S.erectum from the Western Cape dry inland valleys is pollinated by solitary bees, mostly *Anthophora* species and by Cape honeybees. Bees are relatively rare among the pollinators of *Satyrium* species (Plate 86, p. 164). The flowers are pink, have spurs of circa 9.1 mm and produce a pungent scent during the day. This attracts solitary bees, which probe for nectar and receive pollinaria on their frons. As a 'day-flower' *S.erectum* has conspicuous nectar-guides, streaky patterns which help the bees find the spur entrances. The mean volume of nectar per flower was 0.81 µl and the mean sugar concentration 34.8%.

Satyrium buchananii

According to Vogel (1954) this species should be sphingophilous with a spur up to 70 mm and white colour. Sweet scented.

Satyrium volkensii

East-African species. Phalaenophilous, according to Vogel (1954). Confirmation needed.

Genus Satyridium

Satyridium rostratum, the only South African species, is mentioned by Vogel (1954) as an example of a species with melittophilous syndrome flowers, viz. a medium length spur (5 mm), pink-purple colour of the flowers with dark red spots, weak scent and well developed lower lip. Observations needed.

DISEAE: CORYCIINAE

Five genera with about 110 species. Vogel (1959) suggested pollination by Hymenoptera, probably bees, which since has been proved for *Disperis* and *Ceratandra* and is very likely for the other genera (*Corycium*, *Evotella*, *Pterygodium*) as well because

<table>
<tr><td rowspan="2">80</td><td>81</td></tr>
<tr><td>82</td></tr>
<tr><td rowspan="2">83</td><td>84</td></tr>
<tr><td>85</td></tr>
</table>

Plates

80. Fungus gnat on flowers of the mushroom scented *Satyrium bicallosum*. Photo S.Johnson, Table Mountain, South Africa (p. 147).
81. Carrion fly (*Helina* sp.) visiting the unpleasantly-scented flowers of *Satyrium bracteatum*. Photo S.Johnson, Table Mountain, South Africa (p. 147).
82. Hawkmoth on flowers of *Satyrium hallackii* spp *ocellatum*. Photo S.Johnson, Mpumulanga province, South Africa (p. 147).
83. Carpenter bee (*Xylocopa caffra*) pollinating *Satyrium hallackii* spp *hallackii*. Photo S.Johnson, Kogelberg, South Africa (p. 148).
84. Female lesser double-collared sunbird with pollinarium of *Satyrium carneum* attached to the beak. Photo S.Johnson, Kogelberg, South Africa (p. 149).
85. Male malachite sunbird with pollinaria of *Satyrium coriifolium* attached to its beak. Photo S.Johnson, Ceres, South Africa (p. 149).

of the flowers possessing oil-producing *elaiophores* (Buchmann 1987). The presumption of Van der Pijl & Dodson (1966) about tabanid and nemestrid flies as pollinators of *Disperis*, based on Vogel (1954) who mentioned this for *D.polygonoides*, is not confirmed by any observation, although in *Pterygodium alatum* and *Disperis villosa* they have been observed as visitors. In *Ceratandra* beetle pollination as well as bee pollination has been observed (Steiner 1998).

Genus **Disperis**

About 75 (82) species (different numbers are mentioned by different sources) are recognised of which 26 (28) species are found in southern Africa. The genus even intruded into tropical Asia (Taiwan and New Guinea).

In southern Africa most species are found on open grassland on moist to wet south-facing slopes, most abundantly after fires. Some species are found in shaded forest or bush habitat. Most of the southern African species (17 or 18 out of 28) secrete oil from trichomal elaiophores (Buchmann 1987) and are pollinated by special oil-collecting bees of the genus *Rediviva* (Steiner 1989, 1992). Only two species outside of southern Africa probably also secrete oil. Overall ca 24% of all species do so. Steiner and Whitehead observed *Rediviva* bees carrying pollinia of *Disperis* species during investigations on the pollination biology of the oil producing species of *Diascia* (Scrophulariaceae).

Diascia flowers have two spurs or sacs containing non-volatile oil, which is collected by the solitary female bees with their forelegs. They mop it up with the help of special absorptive hairs. The oil is transferred to the hind legs via the middle legs and transported to the nest where it is mixed with pollen to feed the larvae or to make the nest cells waterproof.

Disperis flowers have no oil in their spurs but hidden deep within the hood of the flower at the tip of the lip appendage, where it is secreted from trichomes. The tip is accessible only through openings on either side of the erect lip. The bees place their heads against the apical lobe (a part of the upper lip) or the tip of the lateral petals in the centre of the flower and insert both forelegs on either side of the lip-base and rub the oil-secreting part with the feet. The viscidium adheres to the inner side of the forelegs and on withdrawal the pollinaria are taken away. In different species pollinaria may become fixed to either coxa (*D.waelei*, *D.oxyglossa* e.g.), tibia (*D.cardiophora*) or femur (*D.tysonii*) of the forelegs, depending on the lengths of the caudicle and viscidia of the species involved. Note: Weale (1873) hypothesised the place of attachment on the basis of the morphology of the flowers and supposed the tarsi to be 'the agents of fertilisation'.

In the Western Cape area (South Africa) *Rediviva longimanus* has been captured with the pollinarium of *Disperis circumflexa*. The female bees of this species have extremely long forelegs, 'dipsticks' (19 mm), exceeding the body length (14 mm).

Rediviva peringueyi pollinates *D.villosa* and *D.bolusiana*.

In the Drakensberg area (South Africa) *R.neliana* pollinates at least five and probably eight species. *R.*'bru' (an undescribed species) visits at least four species (a.o. *D.waelei*, *D.oxyglossa*, *D.stenoplectron*). *R.colorata*, a forest species pollinates at least one species (*D.fanniniae*, Manning & Brothers 1986) and probably two more forest species (*D.lindleyana*, *D.thorncroftii*).

Many species occur sympatrically, are co-blooming and share pollinators. Hybridisation is prevented by differing placement of pollinaria. Oil secretion in the Coryciinae has developed independently of the same phenomenon in *Oncidium* and other American species. Some species may be self-pollinating as is suggested by Steiner.

Disperis capensis

Bolus (1918) made observations on the frequency of removal of pollina in this species (and in *D.villosa*), but did not observe insect visits.

Of the nine non oil-secreting species of southern Africa like *D.capensis*, *D.micrantha*, *D.johnstonii*, *D.virginalis* and others only *D.capensis* has been investigated in more detail (Johnson 1993). This species is thought to mimic the flowers of *Polygala bracteolata* (Polygalaceae), a plant that occurs together with it on the Cape Peninsula and which produces nectar. Especially the purple-pink colour is similar, more than the shape.

Two carpenter bees, *Xylocopa rufitarsis* species, caught on *Polygala bracteolata*, carried pollinaria of *D.capensis* when sought on the basis of the supposed mimicry relationship. The pollinaria were attached to the ventral thorax, just behind the head. In two study sites 20% and 53% of the flowers of 30 and 13 plants appeared pollinated. A white form of *D.capensis*, with smaller flowers which appear a month later, is likely to have a different pollination system and in that case should be considered a different species.

Disperis dicerochila
The author (Van der Cingel 1994), when visiting the Nyika plateau, Malawi in April 1994, was struck by the similarity of the flowers of *Disperis dicerochila* (Plate 87, p. 164), growing in shady forest edges and occurring in several colour variants, from almost white to deep purple-pink, and the flowers of *Diascia* species (Plate 88, p. 164), growing dispersed nearby in the same edges, although more in sunny places, and also occurring in white and purple-pink variants. This could be either Batesian or Müllerian mimicry, depending on the production of oil in the orchid, or not.

Genus Ceratandra

The genus consists of six terrestrial species endemic to South Africa. Several species have an oil-secreting lip callus. These calli are secretory and produce a nonvolatile oil, containing a complex set of lipids, instead of nectar.

Ceratandra atrata and **C. bicolor**
Recent observations have shown these species to be pollinated by the oil-collecting bee *Rediviva gigas* (in Steiner 1998).

Ceratandra globosa
According to Steiner (1998) the species is occasionally pollinated by beetles, but is predominantly (up to 65%) self-pollinating by mechanical stimulation. It lacks a callus.

Ceratandra grandiflora
In *C.grandiflora* a varying proportion of the flowers (33-69%) lack a lip callus and is visited and pollinated by beetles. This may represent a shift from one specialised pollinator (oil-collecting bee) to another (beetle). Callus-containing plants secrete oil.

Steiner (1998) investigated pollination of the species and found two species of beetles to be capable of pollinating: *Lepithrix hilarus* and *Heterochelus podagricus* (Scarabaeidae: Hopliini, socalled monkey beetles). The latter species is sexually dimorphic and mainly males were observed on the plants, suggesting rendezvous pollination. At one of the study sites one female was seen to land on an inflorescence and it mated with the males present. The males set up and defended territories on the inflorescences. They sat for long periods on the flowers without much movement early in the morning whereas in late morning they patrolled with or without landing.

Both species showed no interest in the oil secreted by the lip callus and visited flowers with or without calli indiscriminately. The beetles can be attracted by colour only, preferring some colours above others. The flowers are self-compatible, but normally not self-pollinating.

A shift from bee pollination to beetle pollination may have been caused by a separation of plant and bee pollinator as a result either of dispersal and establishment of plants to areas outside the bee's range (colonisation) or as a result of contraction in the geographic range of the bee.

Genus Corycium

Several species of this genus have flowers with dark colours, and some even have black flowers. Combined with the fact that these species flower after bush fires they probably fall within the same kind of syndrome as was mentioned for Monadenia species which show crypsis (Plate 89, p. 164) to prevent grazing by herbivores.

Manning & Brothers (1986) mention pollination of *Corycium* or *Pterygodium* by *Rediviva* species. They observed a pollinarium on a female of *Rediviva politissima* and on a ditto larger unidentified species.

Genus Pterygodium

P.newdigatae is autogamous (Catling 1990). This obviously refers to var. *cleistogamum.*

Manning et al. (1986) mention possible pollination by *Rediviva* species (see above). Some of the species with greenish-brown flowers are myophilous (Vogel 1954, 1959) and long-tongued flies were observed visiting *P.alatum.*

EPIDENDROIDEAE: CYMBIDIOID PHYLAD

MALAXIDEAE

World-wide tribus with 960 species in six genera, some of which are widespread in Africa. There are almost no observations of pollination, except in

Malaxis (Europe and North America, mainly mosquitoes and other Diptera) and *Liparis loeselii* from Europe, which is autogamous, although pollination by mosquitoes is not excluded. Further unknown.

CYMBIDIEAE: CYRTOPODIINAE

Pantropical, mainly Old World. About 140 species in 12 genera.

Genus Cymbidiella

Cymbidiella flabellata

Nilsson et al. (1986) studied the interactions between the pseudo-terrestrial (roots anchored in fern rhizome), endemic *C.flabellata* and flower-visiting insects in a marsh on the east coast of Madagascar.

The large plant (1-1.5 m) has 10-15 showy, yellowish green flowers with contrasting dark and reddish markings. The habitat was rich in Hymenopterans. UV-absorption and reflection were measured and it was concluded that the brownish black and reddish spots and wavy rim of the lip epichile contrast with the light coloured floral structure in the insect visual system. The flowers have a vanilla-like fragrance during the day. The flowers are nectarless and act by deceit.

The large black sphecid wasps *Sceliphron fuscum* and *Sphex torridus* were the only two species that regularly visited the flowers. Males of *S.torridus* were most common and contributed probably most to the pollination. On both wasps viscidia and/or pollinia were found. They become attached to the anterodorsal surface of the wasp' head. The wasps flew at a very high speed in and above the stands of the orchid and now and then approached one or a few of the inflorescences nearest to their flight path and visited some flowers (low constancy). The wasps depress the hinge of the labellum during visits. The labellum regains its position when the wasp stretches itself into the hypochile. On retreating the hinge is pressed down again. It seems that the flower morphology best fits the females of *S.fuscum*, which seem more efficient pollinators than male *S.torridus*. These wasps are 'mud daubers' and collect their nest material from places were the orchid lives.

Of the flowers (n = 816) 52.7% had their pollinaria removed, but only 8.6% were pollinated. Removal of pollinaria and pollination frequently did not occur during the same flower-visit by a wasp. Data suggest that the orchid is a generalised food-flower mimic.

Male *Sphex torridus* displayed territorial behaviour, patrolling landmarks, including inflorescences of the orchid. They alighted on dead stumps of trees and chased *conspecifics* and other passing insects. A female *Campsomeris pfeifferae* (Scoliidae) was observed visiting a flower, but the extracted pollinarium fell off. Other Hymenopterans were not attracted, although on two occasions females of *Ropalida fraterna* (Vespidae) made exploratory visits to labella and possibly exploited secretions from aphids.

Fruit set is very low (4%), but fruits contain an enormous amount of seeds. The 'excess' of flowers produced reflects maximalisation of the male function (pollen donation), which is not uncommon in a system based on deceit.

Cymbidiella humblotii

Nilsson et al. (1986) have observed visits by patrolling males of *Tachytes indifferens* (Sphecidae) on *C.humblotii* in Madagascar, but the wasps were too small to pollinate the large flowers. This orchid has almost the same fragrance as *C.flabellata*, and these facts suggest that large sphecid wasps may also pollinate this species.

EPIDENDROIDEAE: EPIDENDROID PHYLAD

EPIDENDREAE II – OLD WORLD GROUPS

EPIDENDREAE: GLOMERINAE

Agrostophyllum is the only genus represented in Africa, in the Seychelles. Pollination unknown.

EPIDENDREAE: POLYSTACHYINAE

This is a pantropical, but primarily African tribus with about 150 species in four genera.

Genus Polystachya

A large genus with over 200 species. On Madagascar 21 species are found. The flowers in *Polystachya* are nectarless and are known to be pollinated by small bees that gather pseudopollen, formed as foodhairs. Few observations are reported on pollination (see also Chapter 5).

In West Africa a small stingless *Trigona* bee visited *P.p(r)obeguinii* in cultivation (Johansson), but was not observed to carry pollinia and the pseudopollen in this species seems to fit larger insects.

According to Vogel (1978b) the pseudopollen of *P.pobeguinii* is empty and inedible.

Facultative autogamy is reported in *P.concreta* (as *P.zeylandica* from Mascarene) and in *P.fuliformis* (South Africa). Cleistogamy is observed in *P.golungensis* (Ethiopia).

Polystachya rosea

This species was studied by Pettersson & Nilsson (1993) on an inselberg of the central plateau of Madagascar. The species is terrestrial or epilithic. The flowers are nectarless and odourless. Flowers are non-resupinate, the lip forming a hood over stigma and anther. On the lip sparse soft trichomes are found. The colour varies from white to yellow, orange and different hues of purple, always with purplish markings and a yellow raised area adaxially on the lip.

Male and female bees, *Lasioglossum emirnense* and *L.nitididorsatum* (Halictidae) act as pollen vectors. The pollinating insect is thought to enter upside down, attracted to the UV-absorbing yellow centre, which acts as a false food-guide. *P.rosea* does not provide pseudopollen, contrary to many species in the genus, nor does it mimic any foodplant of the bees. The interior of the flower is narrow and when the bee retreats, it squeezes itself against the column. Viscidia adhere dorsally to the thorax.

Both sexes visit (note: pseudopollen only attracts females). Male bees use the plants as landmarks during mate-seeking in the same way as they do with real foodplants. *P.rosea* thus contributes to mating success by being attractive as a potential foodplant for visiting female bees. This behaviour makes the interaction more mutual. Flowers are explored for nectar or shelter. The bees are polylectic foragers and have a great learning ability that allows individuals to specialise in a wide range of different flowers. Pollination is based on non-model deception, probably facilitated by extensive flower-colour polymorphism, which prevents recognition. Odourlessness also inhibits repeated recognition of the rewardless flowers.

Plants in different places showed a considerable variation in pollination success, depending on nearby availability of foodplants.

Fruit set was low, on average 18 ± 14% for the individuals. Some plants (15%) bore no fruits at all.

EPIDENDROIDEAE – DENDROBIOID SUBCLADE

PODOCHILEAE: ERIINAE

Stolzia is the only African genus (out of eight genera). Pollination unknown.

DENDROBIEAE: BULBOPHYLLINAE

Genera **Bulbophyllum** *and* **Chaseëlla**

According to Dressler (1981, 1993) most *Bulbophyllum* species seem to conform to the fly-pollination syndrome, and records of Asiatic species all agree with this supposition. The perfume of some species leaves no doubt of carrion-fly pollination, but there are also species which produce nectar (Jones & Gray 1976a).

However, some West African species are reportedly pollinated by Hymenoptera (wasps and bees).

Johansson observed wasps, stingless bees and ctenuchid wasps as visitors of three *Bulbophyllum* species in West Africa (1974). Vermeulen (1987) who revised the taxonomy of the African continental species (68 species described, including the only *Chaseëlla* species) mentions only the pollination data of two species: *B.ballii* is thought to be autogamous on the basis of the facts that pollinia easily fall off and the plants form fruits abundantly. *B.curvimentatum* is supposedly cleistogamous, because of abundant fruits being produced in spite of the perianth being persistent on the mature fruits.

Furthermore, the varieties *lucifugum* and *gram inifolium* of the species *B.calyptratum* are kept separate on the basis of apparently differing pollination systems, based on the differing position of the rachis.

Photographs of species like *B.schinzianum* var. *schinzianum* and *phaopogon* show lips that seem to imitate hairy insect bodies and this could indicate pseudocopulation by pollinators (see also Jongejan 1994). Species like *B.imbricatum*, *B.maximum*, *B.purpureorhachis*, *B.resupinatum* and *B.magnibracteatum* show flowers that in their colours and form do indicate pollination by sapromyophilous insects.

VANDEAE: AERIDINAE

A few species of *Acampe* and *Taeniophyllum* are

known from Africa. The subtribe, however, is mainly Asian. Pollination unknown.

VANDEAE: ANGRAECINAE

There are over 400 species in 19 genera. In this subtribe the moth-pollination syndrome predominates. The group is primarily Madagascan, with some representatives on the mainland of Africa and outliers in the Americas (Dressler 1981). At least 70 species of orchids with a long spur of 8 cm or more occur in Madagascar, all probably pollinated by hawkmoths.

Observations of pollination are reported of orchids in three genera only.

Genus **Angraecum**

Angraecum arachnites

This long-spurred (10.1-15.4 cm, nectar level at 4.8 cm) species was studied in a primary forest in central Madagascar (Nilsson et al. 1985, 1987). It was found to be pollinated by and adapted to a single species of hawkmoth, *Panogena lingens*, despite abundance of many concurrent Sphingidae (26 are listed) of which several were also long-tongued (four with lengths over 10 cm). The sphingid proved dimorphic in length and breadth of its proboscis and only one of the two morphs, that with the longest and most slender proboscis, was recorded pollinating the orchid. It therefore seems that the orchid has exclusively and very precisely adapted to this pollinator.

P.lingens was the only sphinx out of many species caught by light-trap near the orchid pollination that carried pollinia of this species. The only other pollen vector was *Xanthopan morgani* (see below under *A.sesquipedale*), but this species carried different pollinaria. In *P.lingens* the pollinaria were attached to the ventral side of the proboscis (basal portion), which mode of attachment differs from that of all the other flowering orchids. About half of the hawkmoths had mixed loads, so visited other orchids as well. One male *P.lingens* hovered at a flower of *Angraecum compactum* (see below). The flowers of this species are much larger. Pollinaria of this species were also found on *P.lingens*. The same holds for pollinaria from two *Aerangis* sp. Probably *Aer.monantha* (syn. *Aer.fuscata*) and *Aer. stylosa* (in the 1987 publication *Aer.articulata* is mentioned as a co-flowering species, whereas *Aer. stylosa* is not mentioned), from *Jumellea teretifolia* and *Neobathiea grandidierana* (syn. *N.filicornu*).

A total number of (approximately) 105 viscidia of *A.arachnites* were found on *P.lingens*. Most of these (two-third) ready lost pollinia to stigmas, obviously to those of *A.arachnites*, as concluded from the unique ventral position of the stigma in this species. The distance between spur entrance and nectar surface in relation to proboscis length is believed to be a crucial factor in the maintenance of loyalty and outcrossing by hawkmoths in *A.arachnites*. The peculiar spur curvature is also crucial for the pollination process.

About 78% of the flowers were pollinated and pollinaria removal (male function) was 87%. Fruit set was 41% and probably higher later in the season.

The flowers had a distinct periodicity in the emission of fragrance and began to scent at dusk. The fragrance contains nitrogen compounds, which may be an important dimension in plant-sphingid relationships (cf. *Epidendrum*, Van der Pijl & Dodson 1966).

The hawkmoth fauna of Madagascar is highly endemic and especially rich in long-tongued forms (Nilsson et al. 1985). It appears that these Sphingidae have persisted on the island, since all long-tongued species belong to relatively archaic genera, according to Rothschild and Jordan. A long tongue represents a plesiomorph character. The monophily (relation with a single pollinator species) in *A.arachnites* is interpreted as a result of a refined long-term specialization developed in a relatively stable environment. The extraordinary number of long-spurred Orchidaceae in Madagascar seems the product of coevolution between these orchids and long-tongued sphingids.

Angraecum compactum

Nilsson et al. (1987) observed a male sphinx *Panogena lingens* visiting a flower of *A.compactum* at 18:15 hours. It approached in a zigzag flight upwind that suggested long-distance attraction by the strong aromatic fragrance. The large viscidia of this orchid were often found to attach dorsally on the proboscis, 2.5-3 mm from the base, in light-trapped moths. In total 26 pollinaria were found. Most *P.lingens* individuals carried mixed loads of pollinia from (in total) five out of six flowering species in the area. *P.lingens* therefore appears to be a generalist forager in a series of specialised pollination systems (see also above under *A.arachnites*). Two sympatric morphs were present within *P.lin-*

gens, differing only in proboscis size (on average 4.4 cm!). The longer-tongued morphs dominated as an orchid pollinator (95.3%).

It is supposed that a *P.lingens*-angraecoid relationship may have reached a specialised state before progenitors of the other long-tongued Sphingidae colonised or evolved in central Madagascar.

Angraecum sesquipedale

There is a famous story about the (probable) pollinator of *A.sesquipedale* (Plate 2, p. 12). It is possibly one of the best known and most famous stories about pollination of orchids. Nevertheless it will be told here again.

The English missionary Ellis, who visited Madagascar several times, found this orchid species there and transported it to England. In 1859 he wrote about his travels to Madagascar and told about the epiphytic growth habits and ecological demands of this and other angraecoid orchids. He even wrote about pollination and observed sunbirds sucking nectar from *A.eburneum* and taking away a pollinium, but suspected sphingids to be the real pollinators, because of the impressive, almost incredible spur-length (max. 42 cm, on average 28-32 cm).

A.sesquipedale flowered for the first time in England in 1857 and became known as 'Christmas orchid', because of the white star-form of the flowers.

Darwin (1888) carefully studied the flower and concluded that only a sphingid moth with a long proboscis of about 25-28 cm, as yet unknown to science, could pollinate it. He wrote: 'In Madagascar *Angraecum sesquipedale* must depend on some gigantic moth'. His conclusion (and prediction!) was met with disbelief ('ridiculed' as he wrote), although some biologists accepted this idea and Wallis asserted in 1891 that science could prognosticate the existence of such a lepidopteran with the same certainty as astronomers predicted the discovery of the planet Neptune.

Darwin described what he believed to have happened in evolution as follows:

'As certain moths of Madagascar became larger through natural selection in relation to their general conditions of life, either in the larval or mature state, or as the proboscis alone was lengthened to obtain honey from the *Angraecum* and other tubular flowers, those individual plants of the Angraecum which had the longest nectaries (and the nectary varies much in length in some orchids), and which, consequently, compelled the moths to insert their probosces up to the very base, would be best fertilised. Those plants would yield most seed, and the seedlings would generally inherit long nectaries; and so it would be in successive generations of the plant and the moth. Thus it would appear that there has been a race in gaining length between the nectary of the *Angraecum* and the proboscis of certain moths; but the *Angraecum* has triumphed, for it flourishes and abounds in the forests of Madagascar, and still troubles each moth to insert its proboscis as deeply as possible in order to drain the last drop of nectar.'

This hypothesis, described as a process of coevolution between the orchid and its pollinator, was only recently tested.

Otherwise than Darwin supposed, the relation is more one-sided and less a matter of mutually directed coevolution. Other sphingophilous plants probably evolved before the orchids came on the scene. But Darwin's conclusion provided the basic idea to a general principle about interactional evolution between animals and flowering plants. The results of research on the evolution of flowers with deep corolla tubes, as done by Nilsson (1983, 1988) in *Platanthera* support Darwin's hypothesis to explain the evolution of flower depth.

Fritz Müller found indeed a sphingid with such an extremely long proboscis, 30-33 cm, but in the wrong place (Brazil). His brother Hermann depicted it in 1873 in Nature. Thirty years later, in 1903, a sphingid from Madagascar with a proboscis of about sufficient length was described as *Xanthopan morgani praedicta*, the last epithet meaning predicted, honouring Darwin's prophetic view. It demonstrated the powerful scientific basis of evolutionary thought.

Although Mr Perrier de la Bâthie observed fruit set in *A.sesquipedale* and pollination in other long-spurred angraecoid orchids by sphingids, pollination of this orchid has not yet been observed.

Genus Jumellea

Jumellea teretifolia was found to be pollinated by the hawkmoth *Panogena lingens* on Madagascar (Nilsson et al. 1987). The plant has smaller flowers than some other orchids pollinated by the same vector. The spur is curved in the proximal part and is pendulous. The column constitutes the upper part of the spur-opening. Viscidia are attached dorsally on the proboscis, mostly 1.5-2 cm from the base. In total 13 pollinaria were found as part of mixed loads of pollinia from different orchids.

Genus Neobathiea

Pollinia from *N.grandidierina* (syn. *N.filicornu*) were found on the hawkmoth *Panogena lingens* in central Madagascar by Nilsson et al. (1987). The small flowers have curved and pendulous spurs. The pollination is similar to that of *Jumellea teretifolia* (see above).

VANDEAE: AERANGIDINAE

About 300 species in 36 epiphytic genera. All tropical African (including Madagascar, where about 20 of the 60 known species of the genus *Aerangis* occur). Many genera indicate moth pollination. Observations of chrysomelid beetles in *Cyrtorchis* and *Tridactyle* as frequent visitors are reported, but these may not be pollinators (Dressler 1981).

Genus Aerangis

Pollination data of two species are reported from Madagascar.

Aerangis ellisii
Nilsson et al. (1988, 1992) studied the pollination of *A.ellisii* by new techniques.

The species was found on an inselberg in central Madagascar.

Since the nectar-drinking hawkmoths that pollinate hover against the flower, the pointed and protruding viscidium apparently penetrates deeply into the hair or scale clothing and adheres among the hairs or scales. Consequently scales and hairs may be found on the stigmas of the next visited plants.

Scale and hair analysis of pollinated stigmas thus can give information about the pollinators, because scales and hairs are species-specific. Stigmatic content of flowers was dissected under a binocular microscope and scales and hairs present were transferred to slides to be compared with prepared reference slides. Out of 144 open flowers 51% had pollinaria removed and 24% had pollinated stigmas. Among 74 flowers that showed evidence of visitation 47% had been pollinated and 60% had scales attached. Of 63 dissected flowers 44% had scales, mostly few, but 28% had more than 20. By this method of analysis it was found that two species of hawkmoths, viz. *Agrius convolvuli* and *Panogena lingens* were visitors of *A.ellisii.* Scales similar to those of *Panogena jasmini* were also found in three cases but could not be identified beyond doubt. The scale traces found mostly came from tips of palpi and frons of the hawkmoths. At least three-fourth of the traces were definitely identified as derived from the two hawkmoth species mentioned. These represent a small fraction of the 21 species of hawkmoth caught at the place of study. *Xanthopan morgani praedicta* and *Coelonia solani* have probosces that obviously are too long for *A.ellisii.* They would deplete the nectar if they were regular visitors and disturb the legitimate interaction. Perhaps the specific fragrance composition is not attractive to these moths, although this seems of a general hawkmoth attracting type. Most other species have probosces that are too short to reach the nectar, which needs at least a length of 6 cm. The proboscis-nectar-spur relation indicates that only the two hawkmoth species mentioned have frequent access to the nectar while not reaching the tip of the spurs.

Fruit set was low. Many flowers aborted, irrespective of the origin of the pollen (self-, geitonogamous or cross-pollination). Fruit production was probably resource-limited.

In 1992 a population of *A.ellisii* was studied by a technique of marking pollinia. Microtags with letter and number combinations, cut from microfilm, were made self-adhesive and placed on pollinia after opening of the anther cap. After reclosure the pollinia thus were individually marked and, if deposited on stigmas by pollinating hawkmoths, fully monitored. Pollen transfers were in this way found to be infrequent, to involve single pollen parents as donors and to occur mostly within 5 m of the flowers from which they were removed. The results suggest that the reproductive success, which is disproportional for individual plants, arises because of variation in phenotypic traits of the plants and probably are the result of preferences of the pollinator, i.e. they result from pollinator-mediated mate choice between male and female parts of the flowers, just as in mutual sexual selection as proposed by Darwin (1871).

Aerangis fuscata
Nilsson et al. (1987) observed pollinaria of this orchid on *Panogena lingens* in central Madagascar. Three pollinaria were found. The pollinator usually carried mixed loads of pollinia from different orchid species. In *Aerangis* the mechanism of pollination is different from that in *Angraecum* and other angraecoid orchids of which *P.lingens* is the vector. The column is positioned above and in front

of the spur-mouth. The hawkmoth hovers against the spur-mouth and the viscidium is attached to the hairy covering of frons or palps of the head. A female *Xanthopan morgana praedicta* (see *Angraecum sesquipedale*) was found with stigmatic mucilage, presumably from *Aer.fuscata*, attached to the proboscis.

The pollination of another long-spurred co-flowering species in the same area, *Aer.articulata*, remains unknown because no vector was found.

Genus **Diaphananthe**

Johansson has reported a moth (*Euchromia* sp.) as frequent visitor of *Diaphananthe*.

Genus **Mystacidium**

According to Vogel (1954) a *Mystacidium* spec., probably *M.chilochistae* should be melittophilous (after Engler 1908). The species is leafless, like many species in the genus. The flowers are small, white and non-resupinate. Nectar is found in a short, round spur, at the entrance of which lies the yellow signalling anther. Fragrance is absent.

Many species are probably psychophilous, phalaenophilous or sphingophilous with white flowers and spurs of differing lengths, up to 50 mm and scenting in the evening or at night. Observations are lacking.

CYMBIDIEAE: EULOPHIINAE

Pantropical subtribe with about 260 species in six genera. Probably most members of this subtribe are pollinated by bees of appropriate sizes (Dressler 1981).

Genus **Eulophia**

This is a large genus with over 200 species found in tropical and subtropical regions around the world, but with the majority in Africa.

Eulophia species are found in many different habitats, but mostly in *Brachystegia* woodland (miombo) and in wet grasslands or bogs (dambo). Woodland species tend to flower early in the season, with the onset of rain, and may produce flowers before leaves are formed.

Autogamous species

Williamson studied self-pollination in eight species of *Eulophia* in Zambia. The species all possess rather small inconspicuous flowers, two species being saprophytic with greenish flowers (*E.ecalcarata*, *E.farcta*). Some of these species occur on rock islands within the miombo and most species flower in the period of maximum firing of the miombo. Due to fire, intensive cultivation and general degradation of the habitat by anthropogenic factors, all the self-pollinating species are rare and, except one species, occur as single plants. Species retricted to rock outcrop islands seem to have 're-treated' to places where they are less damaged by cultivation and fire. Self-pollination as a method of survival thus seems to be linked to anthropogenic factors.

The evolution of self-pollinating systems in *Eulophia* therefore may be a recent development. In these *Eulophia* species the pollinia, at a certain level of maturity of the flowers, become active and outgrowths begin to appear from the lower border of the pollinia. The rostellum usually is a narrow shelf separating the base of the *androclinium* from the anterior border of the stigma. It is possible that the central portion of the rostellum contains some stigmatic tissue. The outgrowths of the pollinia, which are located loosely within the base of the anther cap, take on various species-specific forms:

E.farcta: here the outgrowths are cylindrical and grow towards the stigma, covering the viscidium and rostellum completely.

E.saxicola: large outgrowths lift the anther cap and become fused to the cap base.

E.nyassae: narrow outgrowth penetrating the rostellum on their way to the stigma.

E.ecalcarata: very large outgrowths tend to bypass the viscidium and then curve backwards over the rostellum.

E.katangensis: same pattern as in *E.farcta*, but viscidium obsolete in many flowers. Pollinia joined by the caudicles.

E.penduliflora: pollinia without viscidium, stipe or caudicles, below androclinium. Rostellum shelf deeply cleft and tends to curve around the pollinia. Part of stigmatic tissue projecting into the base of the androclinium. Pollinia become fused to the base of the anther cap. The small pollen outgrowths almost immediately touch the stigma.

E.acutilabra: pollinia below androclinium, attached to viscidium by a reduced stipe in normal flowers. Viscidium and stipe completely absent in flowers where self-pollination takes place. Short distance between stigma and pollinia, so that outgrowths touch easily.

E.macaulayae: anther cap becomes fused to the androclinium. Pollinia immediately overflow on to stigma when growing out.

On most of the species studied, capsules were present but were not verified for seed fertility.

Self-pollination has not been observed in *Eulophia* species growing under higher rainfall conditions or where effective humidity is high (wet grasslands and dambo).

Eulophia cristata

This scentless, nectarless species is common in West Africa. It is pollinated by carpenter bees (*Xylocopa*). Lock & Profita (1975) observed pollination on the Acra Plains in southern Ghana, where it flowers abundantly after the first seasonal rains, when still leafless. It has pinkish flowers, which on average bloomed for about 10 days. Out of about 200 flowers, about 30% lost pollinia. The flowers fall after anthesis has ended. Only four flowers appeared pollinated during the observation period. The species is not autogamous. Hand pollination by selfing, geitonogamy or xenogamy gave fruit set in all flowers treated. Cross-pollination gave better results (96%) than self-pollination. Selfed flowers showed fertility reduced by one-third, seeds having no embryos. As visitors some butterflies (*Danaus chrysippus*, *Pyrrhochalcia iphis*) entered occasionally and probed, but could not pollinate. Various wasps also visited, but were probably hunting for insect prey.

Some bees (2 × *Anthophora*, 1 × *Megachile*) also visited without pollinating. Two *Xylocopa* species, however, removed pollinia. A large unidentified species was seen only once. *Xylocopa olivacea* was the most regular visitor, and appeared to be the real pollinator. It used to land on the lip, entering the spur and staying for 4-5 seconds, backing out with pollinia on the back of the head. The bees tried to get rid of the pollinia, usually without success. Epidermal cells in the empty spur project as papillae. Cells below the papillae contain granules of unknown content, but were not observed to be eaten. It is considered improbable that *E.cristata* is a food-source mimic, but possibly naiveté exploration is the case.

Eulophia horsfallii

Kullenberg (1961) observed carpenter bees, (*Xylocopa* sp.) on the ‘powerfully built flower with no superficial nectar’, tearing holes in the labellum base with their powerful mandibles, and licking the exposed tissue. He refers to *E.horsfallii* in discussing the lack of nectar in *Ophrys* and in *Orchis* species, the latter of which may show holes in the spur epidermis made by insects which then lick the sacchariferous assimilation products.

Eulophia speciosa

Van der Cingel observed carpenter bees (*Xylocopa* sp.) visiting and pollinating this species on Inhaca Island, Mozambique in 1987 (Plate 90, p. 164). The large, ‘noisy’ bees flew very fast and stayed only for 1-2 seconds in the flowers.

Eulophia viridis

Darwin mentions the species, because the depression of the pollinium after removal is peculiar by its small extent. This at least indicates pollination by some insect, probably a bee.

CHAPTER 7

Orchids of Asia

*Words by friends with one and the same heart are just as sweet as the aroma of lan**
Confucius. *The book of changes*

Tropical Asia and mainland of Asia

Asia is a huge and botanically complex continent. It includes the boreal and temperate regions of Siberia and China, the lowlands of e.g. India and the high Himalaya, the islands of Southeast Asia and even New Guinea and other Australasian islands, which belong to a totally different *biogeographical* world ('East of Wallace's line'). Many genera and species are shared with Europe in the northern and Mediterranean areas. Many others are shared with Australia, e.g. in Indonesia and some even with North America. The orchid flora of China for example, is poorly known by western biologists, although matters are improving, but nevertheless the list of genera represented there contains 158 with together about 1000 species (Chen Singh-chi & Tang Tsin 1982). But few studies in that large country seem to be dedicated to the pollination biology of Chinese orchids and although there may be lots of details known, the knowledge available in China is hardly accessible because of language and reading problems. The same lack of knowledge is characteristic of many other parts of Asia, due in part to the same but also to other reasons, the tropical forests being not too well suited for field research of the kind needed to gain insight in pollination matters.

For practical reasons I have included all parts of what is traditionally called Asia in this chapter and only separated Japan as a distinct part. In order to prevent unnecessary duplication in areas of overlap with other chapters I have several times simply referred to these.

Orchids with two or three fertile anthers

APOSTASIOIDEAE

This small subfamily, containing only two genera *Apostasia* and *Neuwiedia*, together less than 20 species, has rather primitive features, especially the genus *Neuwiedia*. They are limited to tropical Asia and Australasia and are probably relics of a once wider distribution. They form a link between the orchids and lilies (Liliaceae). The lip is similar to the petals and the lateral anthers are fertile, while the median anther may be fertile (*Neuwiedia*), sterile, or lacking (Dressler 1993).

Some of the apostasioids are self-pollinating. According to Vogel (1981) both genera may be pollinated by *buzzing* bees, i.e. bees that vibrate on the flower to release the pollen from the anthers. The flower of *Apostasia* resembles known vibration flowers and seems to have lost the third anther in evolving a more efficient buzz-flower. The two remaining anthers form a tube with the style, analogous to the tubular anthers of *Solanum*, which genus has typical vibration flowers.

Seed, pollen and anthers of the apostasioids are orchidlike and therefore this group should not be separated from the Orchidaceae as some taxonomists have proposed, nor should it be placed with the Cypripedioideae, from which it differs significantly.

Genus **Apostasia**

Apostasia wallichii
Preliminary observations of the pollination system in *Apostasia wallichii* in West Sumatra by Okada et al. (1996) suggest that its floral characteristics are different from those in *Neuwiedia veratrifolia* (see below). The flowers are *protogynous* and stigma and anthers are separated. This species probably is not self-pollinating, but dependent on pollen vectors.

*lan = orchid

Genus Neuwiedia

Neuwiedia borneensis

Inoue et al. (1995) mention pollination of this species by *Trigona melanocephala.*

Neuwiedia veratrifolia

Okada et al. (1996) investigated the pollination system of this species at Padang, Sumatra, Indonesia. Many individuals within a population and in neighbouring areas bloomed synchronously. Indeterminate inflorescences of 30-75 flowers daily supplied one open flower on average, with seven as a maximum. Bagging experiments suggested self-compatibility, and high seed set from bagged plants as well as the close proximity of stigma and anthers indicate that this species may be mostly self-pollinating. In addition stingless bees, *Trigona laeviceps* and another *Trigona* species (Apidae, Hymenoptera), seem to act partly as pollen vectors. They were the only flower visitors observed that were dusted with pollen grains and were visiting the flowers. Bees visited only newly opened flowers and walked on the inside. It was concluded that the pollination system differs from that in other Orchidoideae which are outcrossing in breeding system and often have exclusive and complex relationships with pollinators.

CYPRIPEDIOIDEAE

The lady's slippers, with two fertile anthers, are closely allied to the monandrous orchids. They have one-way trap-flowers, with entrance into the lip easiest from the front and an easy exit only from the rear. Insects trapped leaving the flower pass the stigma first and then one of the anthers.

In Asia the genera *Cypripedium* (including *Criosanthes arietina*) and *Paphiopedilum* are widespread, the first genus in the temperate zone, the second in the tropical area. In both genera the gloss on the flowers is caused by fatty oil that functions as a slide for insects (Knoll 1922). The pouch usually has transparent windows that are thought to guide flies to the entrance or exit (window-theory, cf. the behaviour of flies against windows).

Genus Cypripedium

Over 40 species are known and recently new species have been discovered in China. They are found in the temperate zone of North America and Eurasia. Many of them are hardy and quite popular among orchid growers and for that reason severely threatened in their natural habitats.

Most species (over 40) of the genus are supposedly bee traps, depending on naive bees which after some visits learn to avoid the flowers. The bees are attracted by chemicals that may mimic important substances in the life-cycle of the bees, but they offer no reward. The smaller flowers are pollinated especially by Andrenidae, Halictidae and Megachilidae, whereas the larger ones are pollinated by queen bumblebees. Some species have lip structures that mimic mushrooms and are probably pollinated by fungivorous insects. These fungus mimetes are dwellers of the dark forest floor, with blossoms close to the bottom (*geophily*) of dark purple brown hues as a rule. The fungus gnats approach the deceptive flowers, attracted by mushroom odours to lay their eggs into the lamella imitations and then pollinate these flowers when trapped. They are misled by olfactory, visual, tactile and humidity perception. The different stimuli may act successively and lead to a chain of actions with temporal summation into a fixed action pattern (distance attraction by scent, nearby attraction and landing by olfactory and/or visual stimuli and egg-laying induced by tactile stimuli from fungose structures and humidity). Eggs and larvae are not able to make use of the flower tissue for uptake of food and will die before anthesis is ended.

The effectiveness of the pouch as a trapping device may be destroyed if rainwater collects in it prior to successful pollination. Potential pollinators may drown, or the pouch may rot, leaving an alternative escape route for the insect. To prevent accumulation of rainwater, the flower is held in such a position that the dorsal sepal overarches the pouch (Hennesy & Hedge 1989).

In general little is known about the pollination of Asian species of *Cypripedium.*

Cypripedium arietinum (=Criosanthes arietina)

This species, according to Atwood (1984), could be better separated as *Criosanthes arietina* on the basis of the antherlike staminode and other supposedly primitive features. It is found in two remarkably separate areas, in North America and in central China (often denoted as *C.plectrochila*). This disjunct distribution may be seen as an indication for the relic character of a once widespread taxon. In North America it is pollinated by *Lasioglossum* bees. No details are known about its pollination in China.

Cypripedium calceolus

The pollination of this species is extensively described in my *Atlas of Orchid Pollination: European Orchids*. It is also found in temperate Asia. Darwin (1888) studied the morphological aspects of its pollination. Nilsson (1978) its anthecology.

The hairs on the labellum are thought to probably retain pheromones from visiting bees or they may, by light refraction, stimulate photoactive behaviour of pollinating bees under the stigma in addition to the light windows in the labellum. Apart from false nectar guides the flowers emit scents that attract the bees. Nilsson found more than nine fragrant substances that work at close range. Of these, acetates are also found in the cephalic pheromone secretions of visiting *Andrena* bees. These are used by males to odour-mark objects in order to attract females and probably also attract the males themselves, causing them to aggregate. Alphafarnasene is also present in the Dufour glands of the bees and serves as a lining material in nests. Its volatile part acts as a signal on the nest site, releasing landing responses that are probably inherited. Linalool may be associated with food.

Nilsson and others found over 50 visiting insect species alighting on floral parts, among them beetles, butterflies, flies and hoverflies, bumblebees and many kinds of solitary bees. Pollinators are mainly medium-sized bees and especially female *Andrena* bees.

Cypripedium section **Trigonopetalum**

According to Christensen (1994) who cites Cribb, species of this section (i.e. *C.bardolphianum*, *C.fargesii*, *C.margaritaceum*, *C.micranthum*, *C.nutans* and *C.palangshanense* and two new species) are all likely to be myophilous. These species have dull colours, often yellow or cream-speckled with maroon or purple, and the mouth of the saccate lip is quite small. In some species flowers appear remarkably similar to those of the myophilous *Aristolochia* species. Cribb observed unidentified black flies visiting one of the new species.

Cribb (1997) has a different name for the section (Trigonopedia) and puts *C.palangshanense* in the section *Retinervia*. *C.nutans* is synonymous here with *C.bardolphianum*. Three new species are *C.forrestii*, *C.lichiangense*, and *C.wumengense*.

Cypripedium lichiangense

This species from the needlewoods of Yunnan, China, is one of the species which obviously imitates mushrooms in form and colour. On the basis of its appearance I suppose pollination by fungivore insects. Cribb (1997) states that coloration and fragrance suggest fly pollination in this species and in *C.margaritaceum* (see above) and that he has seen flies apparently attracted to the flowers and approaching the lip of both species.

Genus Paphiopedilum

This genus of about 60 species is restricted to Southeast Asia. Many species seem to be pollinated by flies or mosquitoes. Syrphid flies that lay their eggs on the staminode seem to pollinate species of the section *Coryopedilum*. These flies normally lay their eggs on aphids. This phenomenon is also known from *Epipactis consimilis* from Israel, in which structures on the lip mimic aphids. In *Paphiopedilum* foul scents (the often cited 'odore spermatico-urinoso' of Delpino in *P.villosum* and *P.purpuratum* is an example), lurid coloration and 'furry warts' and spots on the floral parts may be adaptations to attract flies. The dark spots and glandular structures may also have mimetic properties. Delpino found Calliphoridae in *P.barbatum*.

Atwood (1982) observed attraction of syrphid flies (*Baccha fuscipennis*) to flowers of several species (*P.phillipinense*, *P.hennisianum*, *P.argus*) in a greenhouse in Florida, far from the native habitats. The observations suggest low pollinator specificity. Kramer (1990) also mentions attraction of syrphid flies, e.g. in *P.phillipinense* and *P.praestans* (=*P.glanduliferum*).

Species of the section *Parvisepalum* are suggested to be bee-pollinated (Cribb 1987) because their flowers resemble those of bee-pollinated species of *Cypripedium* and *Phragmipedium*. Those of the section *Brachypetalum* are thought to be adapted to myophily as judged by their putrid odours (Atwood 1982). The spot patterns on the flowers point in the same direction.

P.lowii and *P.mastersianum* show autogamy by structural modification in which pollinia and stigma develop in contact.

Paphiopedilum charlesworthii

According to Bänziger (1994) the pinkish violet colour mimics that of rewarding flowers like *Camchaya loloana*, *Boea glabrisepala*, *Dichiloboea birmanica* and *Clerodendrum urticifolium* which grow in the vicinity. *Apis cerana* bees were repeatedly seen to fly to the orchid after visiting such flowers. They always flew away, however, without alighting. One hoverfly has been observed acquiring pollinia, attracted at close range by the white staminode and its yellow protuberance to which the pollinator flies. No fruit set was observed (20 flowers).

Paphiopedilum concolor and **P.ciliolare**
Atwood (1982) observed 'ordinary' flies trapped in the flowers of these species, but they died, being unable to find the way out. The observations obviously were not made in the natural habitat, but in a greenhouse in Florida.

Paphiopedilum fairreanum
One of the adaptations mentioned for myophilous/sapromyophilous flowers to attract or guide pollinators is the presence of transparent windows. These windows guide a fly to certain places inside or outside the flower and relate to the behaviour of flies towards light in dark places. Such windows are usually mentioned as guidance for trapped insects inside the lip of Cypripedioideae. Their significance has often been discussed. In *P.fairreanum* there are 'auxiliary transparent windows' in the flag (dorsal sepal), 'serving as a backboard as in basketball', which are therefore supposed to play a role in the attraction of flies (Van der Pijl & Dodson 1966).

Paphiopedilum parishii
Bänziger (1994) observed an *Allographa robinsonii* caught in the pouch, and emerging from the exit, smearing off pollen in passing the stigma.

Paphiopedilum rothschildianum
Atwood (1985) studied the pollination of this species on Mount Kinabalu. The syrphid fly *Dideopsis aegrota* was found as pollinator. The flowers emit a peppery or spicy fragrance and are considered to mimic brood sites of the female flies with their staminode, the glandular hairs on it looking like a colony of aphids, the normal brood sites. Aphids usually are found on nectariferous plants in the same habitat. Atwood found as many as 76 eggs in one single staminode. The flies fall into the lip after alighting on the slippery staminode and can only escape by passing the anthers and stigma. Pollination is effected when the fly repeats the visit in a second flower. The floral fragrance emanates from the centre of the flowers. Atwood was unable to find the pollinators of the sympatrically occurring *P.javanicum* var. *virens* and *P.hookerae* var. *volonteanum*, although he found a single syrphid egg on the staminode of the last species. This plant however grew very near to plants of *P.rothschildianum.*

Paphiopedilum sanderianum
This species is the only one which secretes nectar, as far as is known. The nectar is secreted from glands within the pouch wall and probably aids in attraction of pollinators (Kramer 1990). In this species the light intensity of the nectariferous area is the lowest of any of the *Paphiopedilum* species, so that visual attraction by 'windows' may play a very minor role, but is compensated by attraction through nectar. The nectar is only produced on the upper side portion of the pouch and only 3-4 days after the onset of anthesis. It is still not clear why this species has such long petals of up to about 1 m. Some people think that they may function in the attraction of pollinating insects.

Paphiopediulum villosum
In this species the long-distance lure is apparently by colour contrast. Hoverfly pollinators are attracted by the bright yellow staminode in the centre of the flower, surrounded by less attractive, brownish red petals, sepal and upper part of the pouch. The odour is faintly urine-like, suggesting mammalian body fluids which contain salt and are sought after by many insect species, including hoverflies. Close-range attraction is by the glittering staminode, which mimics droplets of moisture and/or honeydew. It has a slippery wart at its centre. Six species of hoverflies, mainly *Episyrphus alternans*, *Syrphus fulvifacies* and *Betasyrphus serarius* fly to this wart, alight and lose their grip and consequently fall into the pouch. They can only escape up the hind wall and pass the stigma on their way to the exit above the anther. Pollen is attached to the thorax (Bänziger 1994).

About 50% of the flowers which were near wilting after three months of flowering had pollen removed. Fruit set was very low (9%).

Primitive orchids with a single fertile anther

Gastrodieae: Gastrodiinae

A pantropical subtribe of saprophytes, primarily from Asia and Australasia, with 36 species in six genera: *Auxopus*, *Didymoplexiella*, *Didymoplexis*, *Gastrodia*, *Neoclemensia* and *Uleiorchis*. In *Gastrodia* we have a report of pollination in the Australian *G.sesamoides* by a small xylocopid bee, *Exoneura* spec., that gathers pseudopollen from the lip.

Holttum (1953) has mentioned fly pollination in

Gastrodia javanica as observed by Mr Carr. The visiting flies take a powdery substance from the upper surface of the lip. Owing to the presence of the column-foot, there is a gap between column and lip at the base. This the fly enters, and crawls from base to apex of the lip; the apex is pressed close to the anther and so the fly removes the pollinia on leaving. On entering a second flower the pollen is deposited on the stigma at the base of the column.

GASTRODIEAE: EPIPOGIINAE

There are five species known in three genera: *Epipogium*, *Silvorchis* and *Stereosandra*. All species are saprophytic and are distributed in Eurasia, tropical Africa, tropical Asia and Australasia. *Epipogium aphyllum* from Europe is thought to be pollinated by bumblebees, but there is some doubt about the observation. Vöth (1994) believes the species is cleistogamous/autogamous and he observed visiting bumblebees that left without pollinia. Probably most if not all species are self-pollinating. This is observed in *E.mutans* of Sikkim, India and in *E.roseum* (Catling 1990) of Java.

Epipogium roseum

The biology of this species, growing in dark places on the forest floor, was studied by Docters van Leeuwen (1937) in Java. The flowers usually open, but before they do so the anther moves towards the stigma and finally touches this, two days before anthesis. The underside of the pollinia bursts and pollen grains begin to germinate on the moist hairs of the stigma. When the flowers expand the pollen tubes have penetrated into the cavity of the ovary and the ovules have for the greater part been fertilised. The development of the seed is uncommonly rapid, the seeds become scattered on the third or fourth day after the expansion of the flowers. The capsules are pendulous and close to the soil, but the seeds are very light and are carried away by rising currents of air. The wide distribution (from West Africa into Southeast Asia and up to Australia) shows the effectiveness of the dispersal.

NEOTTIEAE: LIMODORINAE

More than 50 species (terrestrials) in four genera: *Aphyllorchis*, *Cephalanthera*, *Epipactis*, *Limodorum*. Part of the species is saprophytic. They are found in the northern hemisphere and in the Old World Tropics (Africa and Asia). Some of the northern Eurasian species extend into the Middle East.

The pollination of *Limodorum*, *Epipactis* and *Cephalanthera* species is extensively described in my *Atlas of Orchid Pollination: European Orchids.*

In general *Cephalanthera* species mimic food-plants of bees. Insect-dependent *Epipactis* species usually produce nectar, whereas self-pollinating species don't. *Epipactis veratrifolia* (=*E.consimilis*) is pollinated by hover-flies in a pseudoparasitic relationship (see below). *Limodorum* species are either cleistogamous/autogamous or bee-pollinated (with production of nectar). There are no reports on pollination in tropical Asian species.

Genus Epipactis

Epipactis veratrifolia

The species was studied in Israel by Ivri & Dafni (1977).

Apart from hover-flies which are the legitimate pollinators, some honey-bees and small Halictus-bees were observed. Eight species from five genera of hover-flies were caught and proved to be effective pollinators. The adult flies are nectar- and pollen-feeders and the larvae eat aphids. The species are: *Sphaerophoria scripta* and *S.rueppellii*, *Metasyrphus corollae*, *Paragus bicolor*, *P.aegyptius* and *P.tibialis*, *Ischioden aegyptius* and *Episyrphus balteatus*.

Sphaerophoria rueppellii

The males of this species show territorial behaviour around plants of *E.veratrifolia*. They sit on or near the orchids and occasionally hover for some minutes. When other males approach they are attacked and chased away. When females approach, efforts are made to copulate. The males visit the labellum from time to time, lick nectar and pick up pollinia or pollinate, the flies being pushed against the column by the springing epichile. Females try to lay eggs on the labellum or on other parts of the flower. They may also pollinate afterwards or before the egg-laying. Females may also lay eggs on other parts of the plant that are infested with aphids. It is assumed that the female is attracted by the pattern of black calli-like swellings on the epichile, which mimic aphids in shape and colour (Plates 91 and 92, p. 165; see also Plate 14, p. 13). This visual deception stimulus may, according to the flight behaviour of approaching animals, be followed by olfactory stimuli from the plant and/or aphids. In the final phase tactile stimuli may come into play from the protuberances on the epichile.

Other species
Out of 13 species of Syrphidae in the area studied of which the larvae feed on aphids, eight were found to act as pollinators of *E.veratrifolia*, and seven of these were observed to lay eggs in the flowers of this orchid. Species with vegetarian larvae were not found as visitors. The species of the genus *Paragus* are too small to be effective pollinators, although they were found to carry pollinia and sometimes stick to the stigma, unable to escape.

This exploitation of reproductive behaviour by orchids by imitation of the substrate for egg-laying is wellknown from *Masdevallia* species which mimic mushrooms and are therefore visited by mosquitoes and other mycophilous insects. The phenomenon of imitating prey, in this case for the larvae of the pollinator, is called pseudoparasitism and forms a special class of this syndrome.

NEOTTIEAE: LISTERINAE

About 30 species of terrestrials in two genera (*Listera, Neottia*), some of which are saprophytes, are distributed in the northern hemisphere. The genus *Listera* is pollinated primarily by nectar-seeking wasps and flies (for details see Chapter 4 and my *Atlas of Orchid Pollination: European Orchids*).

Genus Listera

Listera ovata
Much is known about the pollination of this widespread Eurasian species. It ranges from the whole of Europe into Central Asia to the Himalaya. It has been found recently on an island in Lake Huron, Ontario as well. It is mentioned here because of the interesting picture I took (Plates 92-95, p. 165) in 1997 in which not only different insects were observed as visitors and pollinators, but also a crab spider, preying on the visiting insects. As a rewarding species it is attractive for many species of insects from different families. Pollinators are also manifold, the most effective being relatively strong insects of about 8-12 mm length, with short mouthparts and hairless medium-sized heads (Nilsson 1981a, b).

Genus Neottia

Neottia nidus-avis
This species occupies a large part of Europe and temperate Asia into Korea and Japan. It is a saprophyte from dark woods.

Its pollination is poorly known. Darwin found many pollinia removed and concluded that the flowers must be freely visited by insects. Müller mentioned sucking Diptera to Darwin. If pollinators stay away the pollen becomes incoherent and some falls on the stigma, into which the pollen tubes then penetrate. Thrips and other small insects may spread the pollen by crawling over the stigma. Prochazka and Velísek mention flies as pollinators (in Burns-Balogh et al. 1987). Reinhard et al. mention flies as probable regular pollinators and ants as pollen carriers that may cause geitonogamy. The picture of a 'daddy-long-legs' (*Tipula* spec.: Tipulidae) was taken by me in the Teutoburgerwald, Germany in 1997 and confirms fly pollination (Plate 96, p. 172).

NERVILIEAE

Genus Nervilia

The genus contains 65 species and is distributed in tropical Asia, Australasia and Africa. Some forms are self-pollinating. Pollination by wasps and small

Plates
86. *Satyrium erectum* one of the few bee-pollinated *Satyrium* species. Photo S.Johnson, Worcester, South Africa (p. 149).
87. *Disperis dicerochila.* Photo N.A.van der Cingel, April 1994, Nyika Plateau, Malawi (p. 151).
88. *Diascia* sp. Possible model of *Disperis* sp. Photo N.A.van der Cingel, April 1994, Nyika Plateau, Malawi (p. 151).
89. *Corycium microglossum.* Crypsis. Photo N.A.van der Cingel, December 1995, Tsitsikamma (p. 151).
90. *Eulophia speciosa.* Melittophilous flowers. Photo N.A.van der Cingel, 1989, Inhaca Island, Mozambique (p. 159).

86	87 / 88	89
	90	

bees is reported from Africa (Petterson 1989). Some species produce nectar, others deceive pollinators by false advertisement. No details are known on the pollination of Asian species.

Vanilleae: Galeolinae

The Galeolinae are distributed in tropical and subtropical Asia (and extend into Australia). Saprophytic vines or herbs with fleshy stems and rudimentary or absent leaves. There are 27 species known in four genera. Pollination unknown.

Vanilleae: Vanillinae

A pantropical subtribe with about 120 species in five genera. In tropical America *Vanilla* species are pollinated by male euglossine bees, attracted by vanillin. No reports on the pollination of Asian species.

Vanilleae: Lecanorchidae

About 20 saprophytic, leafless species in one genus *Lecanorchis*, distributed in tropical Asia. Pollination unknown, except that *Lecanorchis javanica* is self-pollinating.

Primitive orchids of uncertain classification

Pogoniinae

There are 64 species known in five genera. Some of these are found in eastern Asia. Species from North America attract bees by deceit. Some species simulate pollen with hairy lips.

More advanched orchids with a single fertile anther

SPIRANTHOIDEAE

Tropidieae

A pantropical tribus with about 43 species in two genera: *Corymborkis*, *Tropidia*. Nothing is known about the pollination, but several species have white or pale green flowers, suggesting moth pollination. Rasmussen (1977) has suggested moth pollination in *Corymborkis*. Comber (pers.comm.) observed moth pollination in Java (Plate 97, p. 172).

Cranichideae: Goodyerinae

A large subtribe with about 476 species in 35 genera (two alliances). Distribution widespread, but especially in tropical Asia. Little is known about the pollination of the many species.

Pollination by bumblebees and *Lasioglossum* bees is reported for *Goodyera* (Europe, North America). Butterfly pollination is reported for *Ludisia* (*Haemaria*) and is probable for *Macodes* as well. In *Myrmechis* autogamy is observed by Docters van Leeuwen (Van der Pijl & Dodson 1966). Catling mentions as autogamous: *Anoectochilus longicalcarata* from Sumatra, *Cyrtorchis aphylla* from Java (and Borneo, Chan et al. 1994 also mention self-fertilisation), *Goodyera carnea* from Sikkim, India, *Goodyera triandra*, *Kuhlhasseltia papuana* and a *Lepidogyne* from New Guinea.

91	94
92	
93	95

Plates

91. *Epipactis veratrifolia.* Eurasian orchid with aphid-imitating calli on the lip, imitating prey for egg-laying hoverflies. Photo N.A.van der Cingel, 1995, Cyprus (p. 163).
92. *E.helleborine* with hoverflies and aphids. It is imaginable that the pseudoparasitic relation in *E.veratrifolia* developed from this kind of tripartite relationship. Photo N.A.van der Cingel (pp 163 and 164).
93. *Listera ovata.* With visiting beetle (Cantharidae). Photo N.A.van der Cingel, May 1997, Germany (p. 164).
94. *L.ovata.* Visitors: Diptera, probably Mycetophylidae, a pollinator with pollinia and a predating spider, catching a pollinator. Photo N.A.van der Cingel, May 1997, Germany (p. 164).
95. *L.ovata.* With visiting daddy-long-legs (Tupilidae). Photo N.A.van der Cingel, May 1997, Germany (p. 164).

Genus **Ludisia**

Ludisia discolor

This species is pollinated by the butterfly *Pleisconeura asmara* as observed by Ridley in 1896. Pollinaria are deposited on the legs. The column is twisted to one side. Anther and viscidium form a hook at the apex to catch the legs of the visiting butterfly (Plate 98, p. 172). In *Macodes* the column is even more twisted, the anther facing the dorsal sepal.

CRANICHIDEAE: SPIRANTHINAE

Over 400 species in 30-40 genera. Distribution primarily in America, but also represented in Asia and other continents.

In *Spiranthes* the flowers are arranged in spirals and this is seen as an adaptation to pollination by bees and especially bumblebees, which habitually visit the inflorescences in upward spiral flight. The combination of protandrous flowers and anthesis starting downward guarantees cross-pollination. No details known of the pollination of Asian species.

ORCHIDOIDEAE

DIURIDEAE: CALADENIINAE

Distribution: Australasia (for details see Chapter 8).

DIURIDEAE: DRAKAEINAE

As for Caladeniinae (see Chapter 8).

DIURIDEAE: PTEROSTYLIDINAE

As for Caladeniinae (see Chapter 8).

DIURIDEAE: ACIANTHINAE

About 150 species in five genera (*Acianthus*, *Corybas*, *Cyrtostylis*, *Stigmatodactylus*, *Townsonia*), distributed in Australasia and tropical Asia.

The genera *Acianthus* and *Corybas* are reportedly pollinated by fungus gnats, in spite of their much differing flowers (see Chapter 8 for details).

Stigmatodactylus paradoxa from Sikkim, India is autogamous.

Corybas himalaica has gullet flowers. The gullet is produced from the median sepal and the labellum.

DIURIDEAE: CRYPTOSTYLIDINAE

About 16 species in two genera: *Coilochilus*, *Cryptostylis*. Apart from Australasia and the Pacific Islands widespread in tropical Asia.

Cryptostylis is pollinated by pseudocopulating ichneumon wasps (see Chapter 8 for details).

Cryptostylis arachnites and *C.fulva* from New Guinea are autogamous species.

DIURIDEAE: DIURIDINAE

Distribution Australasia (for details see Chapter 8).

DIURIDEAE: THELYMITRINAE

Distribution Australasia and into Malaysia. About 75 species in two genera: *Calochilus*, *Thelymitra* (see Chapter 8 for details).

DIURIDEAE: PRASOPHYLLINAE

About 100 species in three genera: *Genoplesium*, *Microtis*, *Prasophyllum* (for detailed information see Chapter 8).

ORCHIDEAE: ORCHIDINAE

About 370 species in 34 genera in four tentative alliances. Distribution primarily in the northern hemisphere, especially Europe, and ranging into tropical Asia (and into Africa).

Pollination is pluriform: nectariferous rewarders are rare except in genera like *Platanthera*. Mimicry of all kinds is well developed and extensively investigated. A complete survey of the pollination of European species is published in my *Atlas of Orchid Pollination: European Orchids*. Some Asian (Near East) species from Israel are included in the descriptions and these are not repeated here. This refers e.g. to *Orchis israelitica*, a supposed foodplant (Liliaceae) mimic, *Orchis caspia*, thought profit from the vicinity of nectariferous flowers in the same habitat by parasitic mimesis and *Orchis galilaea*, which is believed to mimic in scent a female bee to attract males of the same species (*Lasioglossum marginatus*). Furthermore *Serapias* species, of which the flowers imitate sleeping nests for small bees.

Genus Dactylorhiza

Dactylorhiza incarnata

This is a widespread species from Eurasia and ranges from Ireland in Europe far into Siberia and Turkestan.

Lammi & Kuitunen (1995) used the species in Finland (Europe) to test the so-called *magnet species hypothesis*. It is supposed that in a deceptive pollination system a nectarless plant benefits from growing near nectar-containing species. Another hypothesis is that such a system functions best in remote habitats such as marshes, where flowering plants attractive to pollinators are rare (*remote habitats hypothesis*). By adding nectar-containing violets (*Viola* x *wittrockiana*) near populations of *D.incarnata* they found that fruit set in the orchid species was adversely affected by the violets, probably because interspecific exploitation competition for pollinators took place in favour of the violets at the expense of orchids. It was therefore concluded that the result gives support to the remote habitats hypothesis.

Another aspect of *D.incarnata* is *polymorphism*. This phenomenon is not unusual in several *Dactylorhiza* species (*D.romana*, *D.sambucina*). In these species different colour morphs exist and may grow together in the same population. In *D.incarnata* a pale yellow morph exists, which in some populations is the only morph, but in others is mixed up with the incarnate morph, as I observed in Poland.

It is believed (Nilsson 1980) that in these mixed populations of deceptive orchids naive pollinators may alternate between the different morphs (*disassortative pollination*) and in this way considerably increase the chances of pollination. It is possible in such populations to maintain a stable polymorphism as a consequence (Kay 1978).

There is also evidence that in some polymorphic species lepidopteran and dipteran pollinators show strong preferences for particular colour morphs (e.g. *Lantana camara*). This *assortative pollination* (Kay 1978), however, only maintains a stable polymorphism if certain criteria for choice are met by the pollinators. In deceptive orchids these criteria are not easily fulfilled.

Genus Platanthera

Platanthera blumei

Platanthera blumei from the mountainous area of West Java, Indonesia was observed on Mt. Panggrango Gedeh to be pollinated by a noctoid moth *Agrotis nigrum* (Docters van Leeuwen 1933).

Genus Steveniella

Steveniella satyroides

Nazarov (1995) studied this species in the Russian Crimea.

This study was not incorporated in in my *Atlas of Orchid Pollination*: *European Orchids*. Since the species is Asian as well as European, ranging from Anatolia into northern Iran, the results of the study are mentioned here.

This endangered species is pollinated by wasps. All nine visiting wasps at the first site studied were workers of the wasp species *Paravespula vulgaris*.

One worker of the wasp species *Dolichovespula sylvestris* was observed as visitor at a second site.

Most wasps observed had pollinia (hemipollinaria) on the clypeus (forehead). Mouthparts of the wasps and length of the orchid spur fit each other well.

The brownish green colour of the flowers is similar to that of other wasp pollinated orchid species. The flowers however do not contain free nectar. Nevertheless at the end of the flowering season 93% of all flowers had been visited in the first population and 69% bore fruits.

The wasps are not only interested in the spur, but also in the special lip. The wasps after alighting first inspect the papillae at the base of the lip, before plunging their mandibles into the spur. It is supposed that the predatory wasps see the reddish-brown centre of the flower as a piece of animal food. The flat papillae at the base imitate the texture of a fleshy substratum. The wasps attempt to tear off a section of the tissue. The reddish coloured inner cavity of the spur may resemble a customary feeding site for the wasps.

Scratches from the mandibles are found on the inner surface of the spur in 78% of the flowers. In pollination the hemipollinaria are pressed onto the stigma with great force and significant portions of the massulae remain therefore on the stigma, due to this 'aggressive' behaviour.

Nazarov describes the mode of pollination as *'false prey' syndrome*.

ORCHIDEAE: HABENARIINAE

A pantropical (and African) subtribe, ranging into Eurasia (and North America). Almost 1000 species in 23 genera in two tentative alliances.

Pollination is pluriform, but as far as is known mostly done by moths and butterflies. *Peristylis grandis* from Java is an autogamous species.

The European species are mentioned in my *Atlas of Orchid Pollination: European Orchids*. The American and African ones are discussed in the relevant chapters of this book.

There are no detailed studies of Asian genera or species concerning pollination (Plate 99, p. 172).

DISEAE: SATYRIINAE

This primarily African subtribe has one genus, *Satyrium*, which extends into Asia.

In this genus fly pollination and pollination by Lepidoptera, birds and bees has been reported. No details are known on Asian species.

DISEAE: CORYCIINAE

A primarily African subtribe that also has one genus *Disperis* extending into tropical Asia.

Several African *Disperis* species produce oil instead of nectar and attract oil-collecting bees of the genus *Rediviva* (Melittidae). Whether oil-production is found in the Asian *Disperis* species is not known.

EPIDENDROIDEAE: CYMBIDIOID PHYLAD

MALAXIDEAE

A subtribe with world-wide distribution and 960 species in six genera. To give an example of the distribution in Asia: In China the genus *Liparis* is represented by 45 species (out of about 250 worldwide), *Malaxis* by 15 (of 300), *Oberonia* 20 (300), *Hippeophyllum* 1 (5), *Risleya* 1 (1) and *Orestias* 0 (3) (Chen & Tang 1982).

Little is known about pollination of the Malaxideae and nothing of what has been published refers to Asian species except species found also elsewhere (see for details Chapter 4 and my *Atlas of Orchid Pollination: European Orchids*).

Oberonia thwaitsii from Ceylon is comparable with *Bulbophyllum*, section *Megaclinium*, which possesses a vibrating lip and flat rachis that may insure the first landing in the gradual approach for flies.

As autogamous species Catling (1990) mentions: *Liparis caespitosa*, *L.cleistogama* (Ambon), *L.longipes*, *Malaxis (Microstylis) histonianthe*, *M.maximowicziana* (Sikkim), *M.soleiformis* (Java), *M.neproglossa* (New Guinea) and *Oberonia linearis* (New Guinea).

Percival (in Van der Pijl & Dodson 1966) has cited *Liparis caespitosa* as cleistogamous in misty rainforests, but *chasmogamous* in drier areas outside the forests.

CALYPSOEAE

A Northern temperate and subtropical subtribe of 235 species in nine genera with few representatives in Asia.

There are no details known about the pollination of Asian species, except those species found also elsewhere.

For *Corallorhiza* and *Calypso bulbosa* see Chapter 4 and my *Atlas of Orchid Pollination: European Orchids*. For *Tipularia* and *Aplectrum* see Chapter 4.

CYMBIDIEAE: BROMHEADIINAE

A tropical Asian subtribe of one genus with 12 species. Pollination probably by bees, as is supposed on the basis of a morphological trait, i.e. a lunate viscidium in *Bromheadia truncata*, which resembles that of *Galeandra* or *Maxillaria* and could be attached to a bee's scutellum (Dressler 1993).

The anthesis is of short duration ('Malaysian one day flowers'), which is compensated by gregarious flowering. The gregarious flowering is triggered by a fall in temperature as in *Dendrobium crumenatum*. *B.alticola* usually flowers one day before *D.crumenatum*.

Ridley (1890) mentioned pollination by *Xylocopa* species in *B.alticola*, *B.aporoides* and *B.palustris*.

CYMBIDIEAE: EULOPHIINAE

Pantropical subtribe: 264 species in six genera.

In *Eulophia* bee pollination is reported (Africa). In *Dipodium* a female megachilid bee has been observed as pollinator (Australia).

CYMBIDIEAE: THECOSTELINAE

A small subtribe of four tropical Asian species in two genera: *Thecostele*, *Thecopus*, of which the pollination is not known.

CYMBIDIEAE: CYRTOPODIINAE

A pantropical, mainly Old World subtribe of 139 species in 12 genera.

Genus **Cymbidium**

Pollination of *Cymbidium* by honey bees, stingless bees, carpenter bees, and bumblebees has been reported (see also Chapter 8 and Japan of this chapter).

Hybrids in greenhouses in Holland have been observed to be pollinated by bumblebees.

Arditti (1985) mentions that the calli in the flowers are thought either to mimic food or to guide bees when entering the flower. Post-pollination phenomena like accelerated senescence and colour changes, save the energy of pollinators by conserving pollinator-power, and are due to hormone controlled production of ethylene and anthocyanin synthesis induced by pollination.

Cymbidium aloifolium
According to Ridley (Van der Pijl & Dodson 1966) this species should be pollinated by *Xylocopa* bees.

Cymbidium atropurpureum
Chan et al. (1994) mention bumblebees as visitors and probable pollinators.

Cymbidium devonianum
In northern India bees were drugged by flowers in this species, according to Ghose, who gave no details (Van der Pijl & Dodson 1966).

Cymbidium finlaysonianum
In Malaysia this species is, according to Burkill (1919), efficiently but rarely pollinated by the giant Indian honeybee *Apis dorsata* (Van der Pijl & Dodson 1966).

In Thailand I saw this species cultivated near houses in its natural habitat and these plants had several large developed fruits (Plate 100, p. 172).

Chan et al. (1994) mention bumblebees as likely pollinators.

Cymbidium insigne
Kjellson et al. (1985, also in Du Puy & Cribb 1988) observed pollination of this species by a bumblebee *Bombus eximius* and of the orchid *Dendrobium infundibulum* in north-eastern Thailand (Fig. 38). The bumblebee is a regular pollinator of *Rhodondrendron lyi* in the regarding area. The two species of orchids are suggested mimics of *R.lyi* and the three species constitute a floral mimicry system.

The flowers of the three species resemble each other in shape, size and main colour. *C.insigne* occurs in two colour variants in the area, of which the unstriped form, which resembles the other species best, is involved in the observed mimicry system. The flowering of *R.lyi* coincides to a great extent with that of the orchids. The flowers of the orchids are long-lasting, about one month if not pollinated.

The workers of *Bombus eximius* are only found in early spring when the three species flower. The two orchids were pollinated mainly by workers and to a lesser extent by queen bees. A second *Bombus* species, *B.rotundiceps*, was seen a few times to carry pollinaria of *D.infundibulum*. Other visiting insects, observed without pollinaria, were the bees *Apis dorsata* and *Anthophora himalayensis*, two species of hawkmoths and a beetle (*Clinteria* sp.). The flowers were visited most regularly during morning hours, especially from 6:00-7:00 hours. The orchids do not produce nectar, nor any other reward. The pollinaria are attached to different parts of the bee for the two orchid species as a result of their differing column morphology. This prevents interference between the reproductive systems. The frequency of visits in *D.infundibulum* was much lower than in *C.insigne*. About 20% of all observed bumblebee individuals (of which 60% were worker bees and the rest older queens) were carrying pollinaria of *D.infundibulum*, but only a few (five out of 97) carried pollinaria or their remnants from *C.insigne*. This is explained by the much higher total number of flowers of *D.infundibulum* and probably also by better fixation of *Dendrobium* pollinaria. From 329 *D.infundibulum* flowers 18 (5.5%) had pollinaria removed and from 106 *C.insigne* flowers 30 (28.3%). Fruit set was low in *D.infundibulum* (3.4%). On fruit set in *C.insigne* no data are available, but much higher numbers are to be expected.

Cymbidium pumilum
Cultivated plants in Japan, where it was introduced around 1850 from China, are pollinated by drones and workers of *Apis cerana japonica* (Sasaki et al.

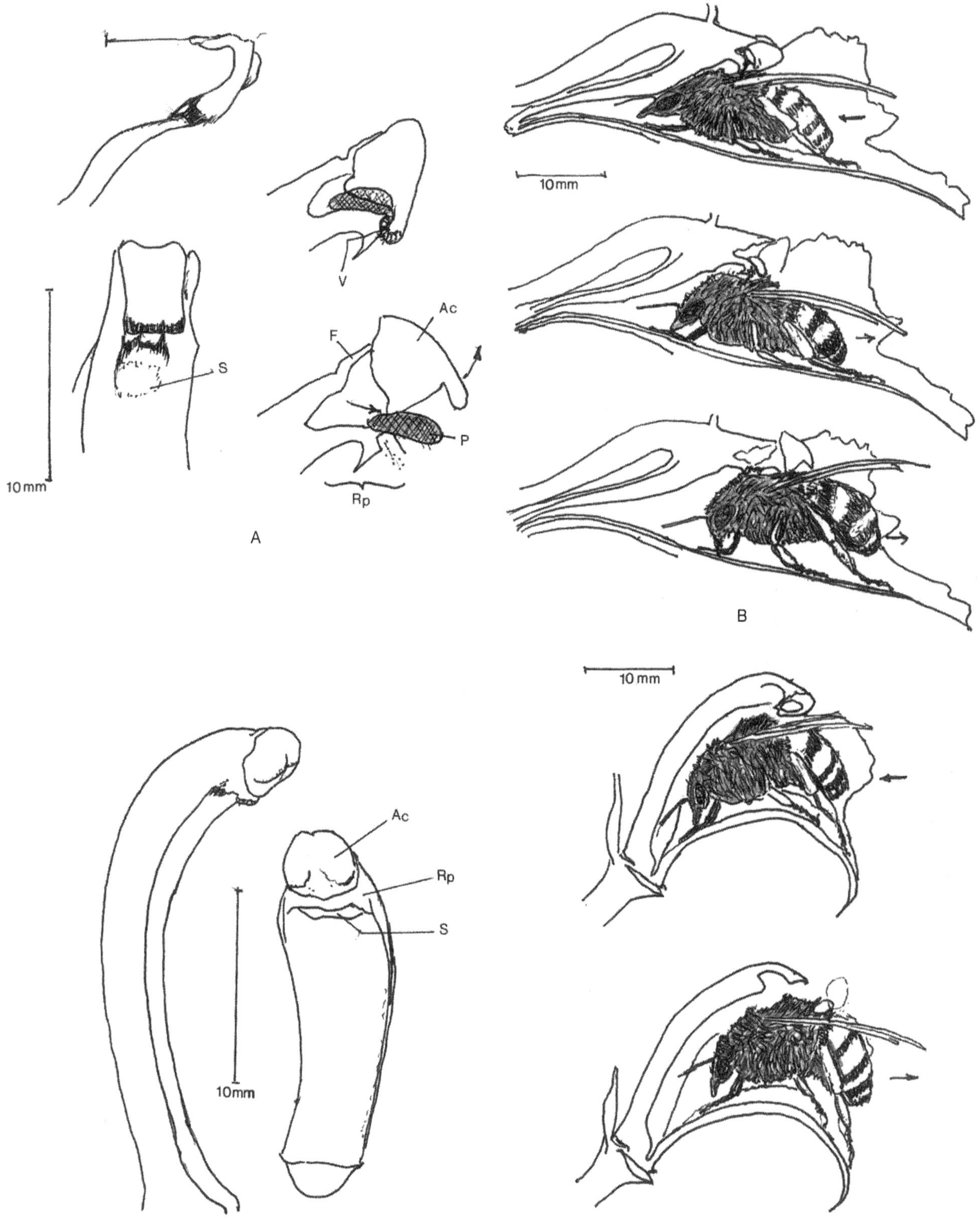

Figure 38. Floral mimicry in *Dendrobium infundibulum* and *Cymbidium insigne* (after Kjellsson et al. 1985). A) Gynostemium in *D.infundibulum.* B) Pollination of *D.infundibulum.* C) Gynostemium of *C.insigne.* D) Pollination of *C.insigne.* Ac = anther cap, F = filament, P = pollinia, Rp = rostellar projection, S = stigma, V = viscidium.

1991, 1993). The attraction of drones is unique. It is caused by a volatile substance from the fragrance of the orchid that is, however, not related to a pheromone. It is thought that the relation between the bees and the orchid evolved in China in the ancestral bee species *Apis cerana cerana*, but this hypothesis needs testing. *Apis cerana indica* and other bee species are not attracted. The pollination by drones is at a frequency lower than that caused by workers (see also Japan of this chapter).

Genus Grammatophyllum

Grammatophyllum multiflorum

The plants are famous for the longevity of their flowers, which are said to last for nine months if they are not pollinated. These are the longest-lasting flowers known. The trait is seen as an adaptation to insure pollination.

Grammatophyllum speciosum

In this species we find heteranthy, with large sterile flowers on the lower parts of the inflorescence, with the lateral sepals fused and the labellum absent. The species is sometimes visited by *Xylocopa* bees, but these large bees are too heavy and depress the lip so much that the pollinarium cannot be deposited (Ridley 1905, Burkill 1919). *Apis dorsata* is an effective pollinator (Ridley 1905). In Florida the species is pollinated by *Augochlora pura* under cultivated circumstances.

According to Burkill (1919) Ridley considered the large *Vespa cincta* as pollinator. It goes to the flowers at times for nectar, and at times to prey on bees.

Genus Porphyroglottis

Porphyroglottis maxwelliae

Van der Pijl & Dodson (1966) have suggested pollination by pseudocopulation in this Malaysian species on the basis of the appearance of its flowers. Hitherto this has not been confirmed by observations.

Cymbidieae: Acriopsidinae

This monogeneric subtribe comprises six epiphytic species with unknown pollination, all from tropical Asia.

EPIDENDROIDEAE: EPIDENDROID PHYLAD

Arethuseae: Arethusinae

The Arethusinae – two genera: *Arethusa* and *Eleorchis*, with together three species – are distributed in temperate Asia and North America. The pollination of the two Asian species *Eleorchis japonica* (Japan) and *E.conformis* (Korea) is not known.

Arethuseae: Bletiinae

Almost 400 (388) species in about 21 (-26) genera. Pantropical and extending into temperate Asia. Terrestrials, epiphytes or saprophytes.

Pollination by bees is reported for *Bletia* (*B.catenulata*) of Peru by *Xylocopa tricuspidifera* and mixed groups of bees in *B.purpurea* in North America and for *Phaius* (see below) and is probable for many of the genera.

In *Calanthe* Lepidopterans are to be expected as pollinators because of the syndrome characteristics of the flowers (see also Japan).

In the Javanese mountains *Calanthe* species show a trace of autogamy. Catling mentions as autogamous: *C.aureifolia* (Sumatra), *C.inaperta*, *C.manii* (Sikkim, Himalaya), *C.papuana* and *C.veratrifolia* var. *cleistogama* (New Guinea), *C.vestita* (Sumatra).

In *Phaius tankervilliae* autogamous (and cleistogamous) forms are known of which the flowers do not open fully and the pollen drops while the flowers are still in bud. It is supposed that naturalisation of this species in areas to which it was brought as a horticultural plant may be due to the introduction of such autogamous forms (Gandawidjaja & Arditti 1982). Allogamous forms of *P.tankervilliae* are pollinated by a species of *Xylocopa* bee (Van der Pijl & Dodson 1966). Autogamy in the genus *Phaius* is also mentioned for: *P.maculata*, *P.amboinensis*, *P.albescens*.

In *Plocoglottis* most species have a motile lip as in *Bulbophyllum* species, indicating fly pollination. *P.glaucescens* is autogamous.

Nephelaphyllum shows a non-resupinate white lip, whereas the other petals are curved back and are greenish. This similarity to the unrelated American *Epidendrum* tube flowers, with the free part of the lip advertising and the other petals inconspicuous and greenish, points in the direction of pollination by moths or sphinxes (Van der Pijl & Dodson 1966).

In *Spathoglottis* autogamy has been observed in

S.microchilina (Java), *S.paulina*, *S.plicata* (Malaysia) and a species from New Guinea. *Tainia penagiana* of Ambon and Java is also autogamous.

Genus **Plocoglottis**

The labellum in *Plocoglottis* throws visitors actively against the column by tension released by the visitor. The mechanism is described by Ridley as follows: 'In the bud, the lip lies flat against the column, but, as the flower opens, it is drawn down in such a manner as to lie flat at some distance below it. This is effected by the lateral sepals the thickened inner edges of which develop the prolonged angles of the lip, and, as they are deflexed, draw down the lip. When a fly alights on the lip in search of nectar, it sets it free by its weight or by slightly separating the lateral sepals with its feet, and the lip springs up suddenly (its flexible claw acting as a spring) and strikes the face of the column, where it remains permanently, pressing the insect against the pollen masses so that it cannot escape from the trap without withdrawing the pollen'.

Some species are purple spotted and foetid, indicating sapromyophily. *P.foetida* is observed to be fly-pollinated by Ridley in 1896 (Van der Pijl & Dodson 1966). For *P.lowii* the mechanism with the motile lip was described by Burkill (1913) and Holttum (1953). For *P.atroviridis* by Schneckenberger (1990). The article is illustrated with elucidating pictures.

According to Van der Pijl (1941) the labellum of *P.javanica* is triggered only by pressure on two sensitive stretches of tissue in the basal angles between the column and the lip. When an insect alights, it touches these spots and the labellum springs up against the column. The active tissue can be overstretched by repeated bends downwards but this can be 'repaired' by pressure on the folded tissue backwards beyond the deadpoint.

The petals have purple spots but the flower is scentless. In this respect it differs from *P.gigantea*, and is therefore believed by Van der Pijl to be pollinated probably by different flies. A New Guinean species is pollinated by *Xylocopa* bees.

COELOGYNEAE: THUNIINAE

A monogeneric subtribe with five species. The pollination of *Thunia* is unknown, but the flower structure suggests bee pollination (Dressler 1993).

COELOGYNEAE: COELOGYNINAE

Orchids from Australasia, Tropical Asia and into China. Terrestrial or epiphytic. About 285 species in 20 genera. The flowers of *Coelogyne* suggest bee pollination. Wasp pollination is reported for *Coelogyne fragrans* (Plate 101, p. 172). Cultivated Asian *C.lawrenciana* in Costa Rica were pollinated by *Trigona amalthea*. It is supposed that *Xylocopa* bees could be pollinators of several species. *C.fuliginosa* scents of mushrooms (Vogel 1978) and is probably sapromyophilous. *Pleione* is almost certainly pollinated by bumblebees. In cultivation bumblebees are observed as pollinators. The callus-ridges on the lip guide the pollinators so that they are exactly positioned to pick up pollinia or to place them in the stigmatic cavity. This callus structure is species-specific (Cribb & Butterfield 1988).

Autogamy has been reported for *Entomophobia kinabaluensis* (Borneo), *Otochilus* sp. and *Pholidota* sp. both from India.

Plates

96. *Neottia nidus-avis* with pollinator (Tipulidae). Photo N.A.van der Cingel, May 1997, Germany (p. 164).
97. A sphingid moth visiting a Cucurbitacae in Taiwan. The same type of pollinator was observed on *Corymborkis veratrifolia* in East Java by the photographer. Photo by J.Comber (p. 165).
98. *Ludisia discolor.* Note the twisted column, a contrivance to catch the legs of pollinating butterflies. Photo N.A.van der Cingel (p. 166).
99. *Habenaria* spec. With visiting mosquito. Photo E.van Wijngaarden, Yunnan-province, China (p. 168).
100. *Cymbidium finlaysonianum* with large fruits. Photo N.A.van der Cingel, January 1997, Thale Noi, Thailand (p. 169).
101. *Coelogyne flaccida* with visiting wasp. Photo N.A.van der Cingel, January 1996, Khao Yai, Thailand (p. 172).

<table>
<tr><td colspan="2" rowspan="2">96</td><td>97</td></tr>
<tr><td>98</td></tr>
<tr><td>99</td><td>100</td><td>101</td></tr>
</table>

Genus **Dendrochilum**

In this genus anthesis may be synchronised, which has the effect of visual collaboration of the small flowers in the attraction of pollinators.

Dendrochilum longibracteatum

Dendrochilum longibracteatum was studied by Pedersen (1995) for its phenology, flower biology, flower visitors and seed set in Sumatra. On 20 inflorescenses as many as 1236 animal visits were recorded, of insects and spiders of great taxonomic diversity. The visits, mostly paid in the morning, were correlated with the number of open flowers. The flowers are fragrant, emitting a fresh sweet scent. Pedersen caught 762 visiting insects and spiders, of which only 14 were pollen vectors. These were all facultatively nectarivorous Diptera and Hymenoptera. The main pollinators supposedly are Diptera (viz. Ceratopogonidae, Chironomidae, Sciaridae) and Hymenoptera (viz. Braconidae, Diapriidae, Dryinidae, Ichneumonidae), none of which are strictly anthophilous. *D.longibracteatum* is genetically self-incompatible. Fruit set consequently is low, not to say very low (0.2%). Certain features shared by *D.longibracteatum* (and many other species of this genus!) and the European *Listera ovata* can be characterised as 'a syndrome for pollination by a variously specific selection of facultative anthophilous insects' (according to Nilsson 1981). This syndrome combines elements of the cantharophilous and myophilous syndromes. Both species grow in dense clones and produce long racemes with numerous small flowers emitting a distinct scent. The flowers offer superficial nectar secreted from a linear shallow nectary placed longitudinally on the flat labellum. It is likely that this syndrome is widespread among orchids, although some specialisation by attraction of a relatively small group of insects of similar size may be released by e.g. specific scents as in *Listera cordata* (fungus gnats) and *Herminium monorchis* (parasitic wasps).

Dendrochilum longifolium

Ridley observed beetles (*Rhyncophorus* sp.) creeping about among the nectarless flowers with pollen masses on their heads.

Dendrochilum pterogyne and **D.haslamii**

These Bornean species show such a constantly rich fruit set that they are thought to be autogamous (J. Wood in Pedersen 1995).

EPIDENDREAE: GLOMERINAE

Epiphytic or terrestrial species of tropical Asia and Australasia. There are 230 species in seven genera. The pollination mode is unknown, but the flower structure suggests moth pollination in many cases. Autogamy has been reported for: *Agrostophyllum crassicaule*, *A.compressum*, *A.dichorense*, *A.graminifolium*, *A.montanum*, *A.pelorioides*, *A.stenophyllum*, *A.superpositum* and *A.verriciforum*, all from New Guinea, *A.denbergii* from Krakatau, Indonesia.

EPIDENDREAE: ADRORHIZINAE

Three species in two genera from tropical Asia (southern India, Ceylon). Pollination unknown.

EPIDENDREAE: POLYSTACHYINAE

Pantropical, primarily African subtribe, with over 150 species in four genera. The flowers of some *Polystachya* species are pollinated by small bees that gather mealy pseudopollen from the lip. The cosmopolitan *Polystachya flavescens* is known to

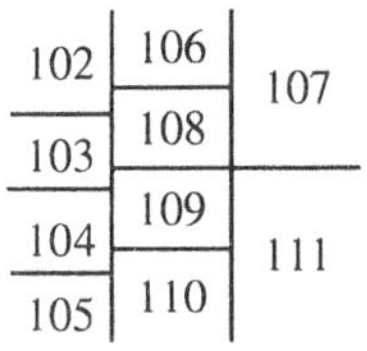

Plates
102. *Epigeneium nakahari*. Photo J.Indsto (p. 175 and Photograph 1, opposite p. 188).
103. *Dendrobium lindleyi*. Photo J.Indsto (p. 175 and Photograph 2, opposite p. 188).
104. *D.chrysothoxum*. Photo J.Indsto (p. 175 and Photograph 3, opposite p. 188).
105. *D.bellatulum*. Photo J.Indsto (p. 175 and Photograph 5, opposite p. 188).
106. *D.primulinum*. Photo J.Indsto (p. 175 and Photograph 7, opposite p. 188).
107. *D.rhodostictum*. Photo J.Indsto (p. 175 and Photograph 9, opposite p. 189).
108. *D.heterocarpum*. Photo J.Indsto (p. 175 and Photograph 8, opposite p. 188).
109. *D.cucumerinum*. Photo J.Indsto (p. 175 and Photograph 12, opposite p. 189).
110. *D.pentapterum*. Photo J.Indsto (p. 175 and Photograph 13, opposite p. 189).
111. *D.obtusisepalum*. Photo J.Indsto (p. 175 and Photograph 14, opposite p. 189).

be pollinated by small female Halictid pseudo-pollen-gathering bees in Florida (US), and it is thought that any anthophilous bee of the same size will be able to pollinate this species. There are however no reports on its pollination elsewhere, e.g. in Asiatic countries. *Polystachya singaporensis* is autogamous.

EPIDENDROIDEAE – DENDROBIOID SUBCLADE

PODOCHILEAE: ERIINAE

A subtribe of mostly epiphytic species, about 700 in 10 genera, from tropical Asia and Australasia (one genus occurs in Africa).

Cryptochilus (e.g. *C.sanguinea*) and *Mediocalcar* have flowers that show features of the bird-pollination syndrome (Dressler 1993), but reports on observed pollination by birds are lacking. In my view *Mediocalcar* flowers are rather small for bird pollination and the flowers seem to mimic Ericaceae with a saccate spur. Since many species are partly terrestrial this could indicate bee pollination by mimicking food flowers.

In *Cryptochilus* there is connation of sepals.

Many *Eria* species (*E.paniculata*, *E.monostachya*, *E.ridleyi*, *E.brevicaulus* e.g.) have pseudopollen on the lip and are most probably pollinated by bees. *E.loheriana* has a mobile lip (Goebel 1920). The callus of the nectarless *E.vulpina* secretes an oily viscid substance. *E.ignea* and *E.ornata* have increased their showiness to attract birds by coloured bracts. Several *Eria* species show autogamy, especially species from the rainforests of New Guinea. Catling mentions: *E.albotomentosa*, *E.bifalcis* (NG), *E.clausa* (Sikkim), *E.cleistogama* (NG), *E.excavata* (Sikkim), E.*inamoena* (NG), *E.isochila*, *E.* aff. *javensis*, *E.stenophylla* (NG) and *E.teysmannii* (=*Trochotosia dajakorum*) (Borneo).

Christensen (1994) finds it likely that the 11 species of *Porpax* are pollinated by small flies. For example, *P.lanii* has small, slightly hairy, meat-coloured flowers close to the substrate, the sepals and petal being connate almost to the apex and forming a tubelike structure. The tiny labellum seems to be motile.

PODOCHILEAE: PODOCHILINAE

About 140 species (epiphytic or terrestrial) in six genera from tropical Asia and Australasia. Pollination not known, except that autogamy is reported for many *Appendicula* species from New Guinea (*A.biumbonata*, *A.bracteata*, *A.carinifera*, *A.cleistogama*, *A.concava*, *A.cryptostigma*, *A.diamuensis*, *A.flaccida*, *A.isologlossa*, *A.kaniensis*, *A.lobogyne*, *A.lutea*, *A.oblonga*, *A.pseudopendula*, *A.torricelliana*) and for a *Cyphochilus* sp., also from NG.

PODOCHILEAE: THELASIINAE

Epiphytes from tropical Asia and Australasia, 234 species in six genera with unknown pollination as far as allogamy is concerned. Autogamy is reported in: *Oxyanthera abbreviata*, *O.papuana*, *Phraetia coelonychia*, *P.densiflora*, *P.gracilis*, *P.microphyton*, *P.oreogena* and *P.trilobulata* (all from New Guinea), *Thelasis capitata* (Padang, Sumatra), *Th.compacta* and *Th.papuana* (both from NG).

PODOCHILEAE: RIDLEYELLINAE

Ridleyella is a monotypical genus from Australasia (Papua New Guinea) of which hardly anything is known, except that it is closely related to the Thelasiinae and has unusual flower colour (blue).

DENDROBIEAE: DENDROBIINAE

About 1150 epiphytic (occasionally terrestrial) species in six genera from tropical Asia and Australasia. The centre of the distribution of *Dendrobium* is in the Asian area.

There are many and good records about bee pollination of Australian species (see Chapter 8). Much less is known about the pollination of Asian species, but there are good reasons to expect bee pollination in many *Dendrobium* species. Probably also pollination by syrphid flies, thynnid wasps and other Hymenoptera or even birds may be found outside Australia. Many species flower gregariously and are very fragrant, but nectarless. They attract native bees of adequate size. Several from higher altitudes show the bird-pollination syndrome.

Autogamy is reported for not less than 34 species, mostly from the Indonesian islands and one from Sikkim, one from the Malay Peninsula. For autogamous species names, see Catling (1990). Many species in the genus *Dendrobium* which are pollinated by bees, possess a dramatic bullseye pattern in the near-UV (370-390 nm). Many other species within *Dendrobium* and related Dendrobiinae also have modified or reduced bullseye near-UV patterns. But this is not the case in all taxa of the subtribe. Sometimes other phenomena, including overall strong *near-UV absorbance*, and in other cases moderate to high *near-UV reflectance* without a contrasting pattern to the human visible range, are found. These different patterns may be different adaptations to different pollinators (Indsto & Weston 1999).

Among bird-pollinated species are many in the section Pedilonum which are rather dull and nondescript in the near-UV, a feature which appears to be the norm for bird-pollinated flowers. However, most, if not all birds are known to have near-UV vision, so this may be expected at least sometimes to be important in their foraging behaviour. In this respect it is interesting that a number of species have quite bright near-UV reflectance and this possibly may have a role in recognition of these flowers by birds. Also, it is not uncommon for fly-pollinated flowers to have moderate, or even bright near-UV reflectance. Broadly based observations suggest distinct near-UV reflectance features in such flowers. Some species in *Dendrobium* would appear, on the basis of dullish, redbrown colouring and near-UV reflectance properties, to be fly-pollinated. More observations are needed (J. Indsto, pers. comm.).

Genus **Dendrobium**

Plates 102-111 (p. 175) and Photographs 1-14 (opposite pp 188 and 189) on near-UV photography.

Dendrobium antennatum

In a Time-Life book on New Guinea a photograph is published with a flower spider catching a moth (Plate 112, p. 180). It was sent to me by Grahame Muller who states that a 'honey spider' in Australia which looks exactly the same is often found in orchid flowers and that often the flowers in which spiders are observed end up as seed pods. It is possible that pollination takes place unintentionally during a struggle between the spider and its prey.

Dendrobium apendiculatum

This is reportedly the shortest lived flower known. It opens for five minutes only. Short anthesis is known from many orchids and is often compensated by mass flowering.

Dendrobium crumenatum

This species, famous for its gregarious flowering, nine days after a sudden fall in temperature has triggered anthesis, is pollinated by *Apis dorsata*, the Indian honeybee, which may do so by sudden mass visits.

Dendrobium hasseltii

Was observed by Docters van Leeuwen on a mountain summit on Java to be visited by birds that tried many plants.

Dendrobium infundibulum

As mentioned before (see *Cymbidium insigne*) this species was observed in northern Thailand to be pollinated by bumblebees (Fig. 38, p. 170). The non-rewarding orchid mimics the more frequently occurring rewarding *Rhodondendron lyi* (Cribb & Butterfield 1988).

Dendrobium lawesii and other species from New Guinea

D.lawesii was observed by Slade (1962) to be visited by birds at high elevations (above 3000 m) in New Guinea. Here it grows together with species like *D.sophronites*, *D.roseum* and *D.flammula*, which are all similar and brilliantly coloured (as their names already suggest). Bird pollination therefore is to be expected in all these species, which have a narrowly-clawed lip which forms a tube with abundant nectar production (Van der Pijl & Dodson 1966).

D.lawesii blooms almost continuously. The bell-shaped flower is variable in colour, scarlet, cerise, rose, bright purple, orange or bicoloured with white or yellow-tipped flowers. The front lobe of the labellum is turned upwards, making a door protecting the column and nectar-bearing elongated spur; this effectively bars insects, but is easily opened by a nectar-eating bird's beak. *D.lawesii* grows in areas where there are epiphytic *Rhodondendron* species which have somewhat similar shaped and coloured flowers. It is therefore supposed that the orchid could mimic the rhodondrendons to deceive the pollinator (Banks & Smedley 1997).

Dendrobium nobile

This species has a latency period of flowering after a sudden fall in temperature, usually caused by a shower after a dry spell (cf. *D.crumenatum*).

Dendrobium secundum

This species is reportedly pollinated by a sunbird

Cyrtostomis pectoralis, as observed in Singapore by Burkill in 1919.

Dendrobium superbum

Visited by *Apis dorsata* and *Apis indica* (Burkill 1919). Cultivated specimens in Japan are visited by male melon flies, *Dacus cucurbitae*, which lick the flowers and sequester the attractant 4-(4-hydroxyphenyl)-2-butanone (or raspberry ketone) in their rectal glands, where it is retained for up to six days. The males produce a smokelike substance from the gland during courtship, but its role is not clear. Flies seldom visit inside the lip and are unlikely to be pollinators. (Nishida et al. 1993).

Dendrobium unicum

Kjellsson & Rasmussen (1987) suggest that this species in northern Thailand attracts pollinators by pseudopollen on the lip. It has a longitudinal callus on the lip that is covered with multicellular trichomes, which are grazed probably by bees or wasps. In flowers in which pollinaria were found to be removed, 17.5% of flowers in the field, the trichomes were removed as well. Therefore the hairs are believed to function as food-bodies, attracting pollen-seeking insects. The species is self-incompatible and must rely on anthophilous insects for its pollination. It is not clear whether the pseudopollen acts by deceit or that the trichomes (callus granulae) have food-value.

Genus Epigeneium

Epigeneium treacherianum

According to Chan et al. (1994) the flowers of this species smell of dessicated coconut. Colonies of plants along the branches of dipterocarp trees are cloaked in masses of pink flowers, which are visited by the bee *Apis dorsata*.

DENDROBIEAE: BULBOPHYLLINAE

Pantropical subtribe with over 1100 epiphytic species in 10-15 genera. Most *Bulbophyllum* species seem to conform to fly-pollination syndromes. The available records from Asian species indicate sapromyophily (carrion flies) and myophily (nectar seeking flies). Some West African species are pollinated by wasps or bees.

Christensen (1994) studied drawings and photographs of *Trias* and *Drymoda* species and hypothesises that these are pollinated by flies, because size, shape and colours of the flowers of *Trias* show similarities to those of *Stelis* (Pleurothallidinae). *Trias* usually has a motile labellum, unlike these *Stelis* species they are similar to (see also Chapter 8).

Genus Bulbophyllum

Bulbophyllinae can be seen as the Old World counterpart to the neotropical Pleurothallidinae with respect to floral morphology and biology, probably due to a parallel evolution of adaptation to fly pollination (Christensen 1994). Contrivances for attraction of flies are similar to those in the Pleurothallidinae: foul odours, spots, glands, slime, hairs, motile lips or appendages, and dull cream or yellowish-green to brownish-purple colours of the flowers, and in several species osmophores, often produced in 'antennae' (*B.cerambyx*) or appendages.

The myophilous or sapromyophilous syndrome in this genus expresses itself in different ways. Some species like *B.medusae*, *B.ornatissimum* and *B.arachnites* have tails that probably help to form a large landing surface. Other species have closed flowers with side-slits, e.g. *B.singaporeanum* and *B.grandiflorum*. Fringe-hairs on the lip of *B.barbigerum* cause it to flutter in air-currents. In *B.psittacoides* the small dipteran pollinators settle on pendant sepals, from where they climb up until they reach the lip and, after passing the balancing point, are pitched onto the column. In *B.johannulis* the petals form several long hairs, in *B.macrorhophalum* each petal has three motile appendages, etc.

According to Pohl (1935) species like *B.macranthum* produce oil and sugar on the base of the lip and adjacent flower parts. He supposed the oil to be food for visiting insects and to make the flower a slippery slide-trap and mentions observations by Ridley, who saw flies cleaning the tips of the sepalum completely by licking the fatty oil.

The mechanism by which many species of *Bulbophyllum* (e.g. *B.lobbii*) trap flies, with a motile lip, a balanced see-saw, hinged by a narrow springy claw which tips over when the equilibrium is disturbed by flies landing on it, is described by several authors (e.g. Ridley and Pohl). Darwin did not understand the extreme flexibility and liability to movement of the lip and thought it probably served to attract the notice of insects, which indeed may be the case.

The flower has developed the hinged labellum as

a means of throwing poorly directed flies against the column precisely, so that pollination is effected.

Many species have a purple flower or a purplish lip and smell foul.

In the *Cirrhopetalum* section the flowers aggregate in a fan-like whorl, a peculiar kind of return to radiality. The flexible labellae are a kind of oscillators. In typical *Cirrhopetalum* flowers the lateral sepals function as an attractive alighting place, a function usually performed by the lip (Ridley 1890).

In the section *Micromonanthe* there is also some return to radiality and the flowers form a flat disk with all tepals becoming alike. Such a return to radiality may have mimetic character and function to attract insects by similarity with radial flowers. Some species have a kind of lantern-type structure with slits or holes in the side (*B.ipanemensis*, *B.fritillariiflorum*), which is supposed to correspond to some innate behaviour of flies.

In the section *Monosepalum* from New Guinea there are clublike antennae and a motile lip, whereas the fused sepals form a tube. See also *Monosepalum torricellense* (below).

Jongejan (1994) speaks of hide-away flowers. The flowers of an undescribed species of the section *Polyblepharon* from Papua New Guinea possess flowers that are hidden between fibrous remnants of the rhizome scales. Jongejan suggests mimicry of prey-insects like caterpillars, seeking a safe place against parasitic wasps. This kind of hypothetical system, if it exists could probably only work if the parasitic wasp is attracted by a scent that mimics the odour of the prey and if the close range visual cues help in positioning the parasitic insect in the right way to bring about pollination. More proof is needed.

Autogamy has been reported for: *B.apahanopetalum*, *B.cleistogamum* (Riau, Indonesia), *B.dasyphyllum* (*New Guinea*), *B.dischidiifolium*, *B.nieuwenhuisii* (*Borneo*), *B.scobiculilabre*, *B.trisandrum* (both from NG), *B.verruciferum* var. *carinatisepalum*.

Bulbophyllum articola
This species is typical for the section *Hapalochilus*, a group with usually bright-coloured species. The column is transparent, thin and bent in the middle. It produces glistening droplets of sticky fluid on the lip a few hours after opening. An insect attracted by these droplets is caught by the sticky fluid. The only way to escape is to try to drag themselves away and grip the column. In this way pollination is effected (Jongejan 1994).

Bulbophyllum auratum
A species from the section *Cirrhopetalum* with vertical petals forming a bell shape. The flowers produce a huge amount of fluid, covering the lower parts, assembling on the underside and sometimes falling off. Small insects may be drowned. Larger insects can regain freedom by climbing up and pulling down the lip. If the resilience of the lip exceeds the force of the glue the insect frees itself and is thrown against the column and will receive pollen (Jongejan 1994).

Bulbophyllum beccari
The species has a putrid smell, suggesting pollination by carrion flies (Chan et al. 1994). The odour has been likened to a herd of dead elephants (Pridgeon 1994).

Bulbophyllum cimicinum
The species from Papua New Guinea has a strange appearance, looking similar to a spider. It is suggested that the striking similarity could lure large spider-preying flies or wasps to visit and pollinate the flowers (Christensen 1994). These could also be trying to lay eggs in spiders, so that the larvae can live as parasites on them. If this appears to be true we can speak of pseudoparasitism or *pseudocarnivory* as the pollination syndrome. In *B.kanburiense*, where slight movements cause the small, hairy, purplish spotted lip to move sideways (!), insect larvae crawling in the flower could be mimicked, also luring potential predators or parasitic insects.

Jongejan (1994) speaks of a beetle- or flealike shape of lip and part of the petals of *B.cimicinum*, with the elongated cells lining the 'body' giving the impression of paws and hairy appendages on both sides (lower parts of the petals) imitating insect wings.

Bulbophyllum cupreum and **B.campanulatum**
B.cupreum has a nectar-groove on the lip, *B.campanulatum* has a thick layer of slime on the upturned undersides of the lateral sepals. *B.lobbii* has oil, as mentioned above. This production of food for visitors is seen as a secondary trait (Van der Pijl & Dodson 1966).

Bulbophyllum caputgnomonis
The insect-trapping mechanism by which this species is pollinated is described by Jongejan (1994) and resembles that of other species of the section *Vesicisepalum* (see below: *B. folliculiferum*), although more complicated. Whether the pollinators

look for a place to lay eggs, as supposed, remains to be investigated.

Bulbophyllum echinolabium

Jongejan (1994) supposes carrion foraging beetles as pollinators of the large flowers with a spiny, meat-like coloured basal part of the lip. An insect crawling inside will push its head underneath the column, kicking the lip backwards. The lip has a stick-like extension pointing downwards and will cause the lip to turn inwards again, pressing the insect in- and upwards. The flowers develop one by one, promoting cross-pollination.

Bulbophyllum ecornutum

Pollinated by syrphid flies (Plate 113, p. 180).

Bulbophyllum folliculiferum

In this species from the section *Vesicisepalum* the sepals have a bladder-like shape and the flowers hang down from a long rhizome like little dark purple eggs. They hardly open except between the lateral sepals close to the base, where a small slit of 1 cm length opens about 3 mm. Insects attracted by the potato-like smell have to crawl inside this hole to find what they are looking for. The lip is used as a kind of spring which, as soon as the insect tries to climb on it, swings against the column, producing some extra pressure, needed to fix the pollinia to the head or, if already present, to press them against the stigma.

Jongejan (1994) supposes egg-laying as the motive to enter.

Bulbophyllum graveolens

This species from New Guinea is pollinated by large flies (Plate 114, p. 180) attracted to the foul odour of its flowers. During the day the smell is strongest, the flies become very agitated and stumble over the lip, which catapults them forcefully upside down onto the column (Howcroft 1976). Flies often become stuck and perish because they are unable to free themselves from the sticky stigma.

Jongejan (1994) observed flies (Calliphoridae) entering the flowers in his glass-house, attracted by the foul smell, and sometimes trapped fatally.

Bulbophyllum lobbii

This species has the smell of fresh cucumbers and is not putrid as with so many other species. It catches flies.

Pohl (1935) found oil and sugar on the lip base. The yellow colour with purple stripes is a bit unusual in *Bulbophyllum*, but fits the myophile syndrome.

Bulbophyllum macphersonii

This species from the mangroves of Papua New Guinea is also known from Queensland, Australia, where it is pollinated by midge flies (*Forcipomyia santesi*). It offers no reward (Bartareau 1994).

Bulbophyllum macranthum

The flower is non-resupinate. This position helps in making the flower a sliding trap-device instead of a see-saw. The same mechanism is probably valid for *B.emiliorum*.

According to Ridley (1890), who observed pollination in the Singapore Botanical Garden, the pollinator is a small and active fly. Ridley saw this fly always close by when the plant was in bloom and it is also attracted by *Dendrobium superbum* (not native to the area!) and another small flowered *Bulbophyllum*. The three plants have a certain amount of deep red colour in common and share a strong but different scent. The insect usually commences by licking with its short tongue the upper part of the sepals. It eventually settles down to the front of the lateral sepals, especially on the dark coloured grooved apices. As long as it is at work in the narrow part of the sepals, it can hold on to their edges, but when it gets to the broad part, it cannot reach across and its feet slip from the glassy surface. It then clutches wildly at the lip. When its weight falls upon the lip this drops back, pitching the insect stern first into the column between the *stelidia*, which have enough springiness in them to separate a little and then close tightly on to the abdomen of the captive. The fly strikes the disc of the pollinia with its abdomen where these become fixed upon the first segment. The lip, released by the fly, instantly returns to its original position, and the insect is left struggling on its back, but soon extricates itself and flies away. In a second visit the fly falls off the lip more rapidly because of the extra weight of the pollinia and thus, falling a little lower, strikes the sticky stigma with the attached pollinia.

Millar (1978) presents a series of photographs of a fly pollination of *B.macranthum.*

Jongejan (1994) observed insect eggs deposited around the anther cap.

Bulbophyllum makoyanum and **B.lepidum**

B.makoyanum, a species from Singapore, has flowers that are placed in a circle and the whole appearance of the flower can be interpreted as mimicking other nectar-offering radial flowers (Knerr 1981).

B.lepidum has a comparable appearance, looking like a (normally yellow) flower of *Gazania uniflora* (Asteraceae), especially in UV light, as seen by many insects (Koehler & Davenport 1983).

Bulbophyllum ornatissimum
This species from Sikkim Himalaya and Assam has two different groups of osmophores. Tail osmophores on the distal part of the lateral sepals produce a smell of cod-liver oil. Labellum osmophores produce a trimethylamine-like odour.

Vogel (1963a) believes each odour to possess a distinct function. The tail osmophores, far from the gynostemium, are not directly involved in the pollination mechanism but may serve for long-distance attraction and provide a landing place, whereas the motile labellum with its aminoid smell and nectar serves for close-range attraction and as a trapping device.

Van der Pijl & Dodson (1966) observed two kinds of blowflies (Calliphoridae) visiting the flowers of a cultivated plant in a greenhouse, removing the pollinia. The flies landed on the sepals, crawled up the labellum and were thrown against the column when the lip became overbalanced. The osmophore tails, petal and dorsal sepal bear clublike motile trichomes, functioning as visual attractants (*vibrissae*).

Bulbophyllum pahudii
This species of Java attracts similar flies as does *B.weinthalii* in Queensland. The last orchid is pollinated by Calliphoridae. These blowflies are attracted by the smell of fish meal and collect nectar from a groove near the base of the labellum. In *B.pahudii* (Plates 115-117, p. 180) the smell is one of rotten fish. (Comber, in Christensen 1994).

Bulbophyllum polygaliflorum
The lip of the small flowers of this species looks like an insect, waiting for mating (Jongejan 1994). The lateral margins of the lip have paw-like elongated cells (cf. *B.cimicinum*) and the tip of the lip ends in a wart-like extension, probably mimicking the female genitalia.

Bulbophyllum striatellum
Ridley (1890) saw a minute fly entering the flower of *B.striatellum* between the sepals, presumably in search of nectar, climbing upon the balanced lip and when overbalancing being thrown headfirst into the arms of the column. The erect petals guide the insect, preventing it flying away, and the pollinia are planted between the eyes of the fly. The delicate balance of the lip is believed by Ridley to be absolutely essential in such smallflowered species on account of the very light weight of the pollinator, this being the reason for the extreme mobility of the lip. According to Ridley the mobility does not play a role in the attraction of the insect in this case.

Bulbophyllum ustusfortiter
The flowers of this species (section *Vecisisepalum*), with a bladder-like shape, hardly open and the buds which form a cluster of pseudobulbs, resemble some black contaminating fungus. The triangular opening at the end of the flower gives access to small creeping insects. Once a visitor has entered and climbs upon the lip this suddenly turns inward, throwing the animal head over heels against the column.

Bulbophyllum virescens
Okamoto (1986) observed visitation of this large-flowered species by flies in a montane forest on Seram Island (Indonesia). A medium-sized fly was entrapped in the space between column and the overturned lip as soon as it alighted on the lip. Pollinia were not removed and it is supposed that flies able to carry away pollinia have to be larger.

Comber photographed pollinating flies in North-Sumatra (Plate 118, p. 180).

Genus Monosepalum

Monosepalum torricellense
Jongejan (1994) describes the bell-shaped fused sepals of the flower (as is typical for the genus) as being able to move independently to a certain extent from the other parts of the flower, as it is suspended on a thin, elastic tissue, comprising the base of the sepals. As soon as an insect touches the elongated points of the sepals lower down, it starts to swing back and forth. Petals, lip and column stay immobile inside the bell shape. The petals, which are divided into three movable worm-shaped appendages at their tips, touch the inner side of the sepals. The appendages are rubbed against the rough inner side of the sepals and start to move as well, offering the very strange view of worms wriggling above the insect, inviting it to climb upstairs to pollinate the flower. It is, however, not

mentioned on what kind of innate behaviour this reaction is based. It reminds one of the movements of the lip of *Bulbophyllum kanburiense*, which seem to mimic crawling larvae that probably lure parasitic insects (see above under *B.cimicinum*).

Genus **Pedilochilus**

In this genus of plants that live low down in dark places near the ground, visiting insects can enter the balloon-shaped lip via its opening near the tip and are trapped because the inward-curled edges around the opening prevent escape. Thus the insect is forced to search for another way to escape, helped by nearly transparent parts in the petals which cover the lip. It has to pass underneath the column, pollinating the flower during the passage.

Jongejan (1994), who describes this mechanism, supposes that insects enter the sac to lay eggs, however without giving further arguments or proof from observations.

Vandeae: Aeridinae

There are over 1250 species in about 103 genera arranged in series. They are mainly found in tropical Asia and Australasia (with two species in Africa). There are few records on pollination in this large group.

In *Phalaenopsis* and *Vanda* bee pollination by carpenter bees (Xylocopinae) is recorded.

In *Ascocentrum* and *Renanthera* bird pollination is to be expected. Slade (1980) mentions bird pollination of *Ascocentrum* 'Sagarik Gold' in his garden in the New Hebrides (see Chapter 8).

Australian *Sarcochilus* and *Pomatocalpa* are reportedly pollinated by *Carbonaria* bees.

Beetle pollination is described in *Vanda* by Pradhan (1983), see below and in *Peristeranthus* (see Chapter 8) (Wallace 1980).

An interesting case is the pollination of a *Cottonia* species by pseudocopulation, probably by a beetle.

In *Arachnis*, *Arachnante* (*Arachnis* x *Renanthera*) and *Renanthera* there are two kinds of fertile flowers. The lower ones are odourless, the upper ones have a different colour and emit a strong scent. In contrast to the lower ones they remain fresh and odorous for a long time, even after pollination. The significance is not clear, but probably relates to male function.

Thrixspermum arachnites has gregarious flowering.

In *Arachnis flos-aeris* special odour-glands on the tip of the median sepal, far from the column, produce a musky odour.

Autogamy is reported for: *Cleisostoma parishii*, *Papilionanthe longicornu* (Sikkim, wind-assisted), *Schoenorchis paniculata* (Malaysia) *Taeniophyllum hasseltii* (Java), *Thrixspermum* sp. (Malaysia).

Genus **Neofinetia**

Light (1998) mentions *N.falcata* as having syndrome characteristics of a night-flying moth-pollinated flower, with fragrance of honeysuckle and vanilla, scented at night.

Genus **Phalaenopsis**

Phalaenopsis amabilis

This species, although not adapted to *Xylocopa*, is sometimes pollinated by bees of this genus in nature. A form from New Guinea has a strong sweetish odour, whereas forms from other places have not.

Plates

112. *Dendrobium antennatum*. A crab spider has caught a moth, while hiding in a flower. With courtesy of Time-Life International (Netherlands) BV (p. 175).
113. A syrphid fly pollinating *Bulbophyllum ecornutum*. Photo by J.Comber, East Java at 2500 ft (p. 178).
114. Cultivated *Bulbophyllum graveolens* with fly entering a flower. Photo W.Upton, Australia (p. 178).
115-117. Flies pollinating *Bulbophyllum pahudii*. Photos by J.Comber, West-Java at 3000 ft (p. 179).
118. Flies visiting and pollinating *Bulbophyllum virescens*. Note the fly with pollinia. Photo by J.Comber, North Sumatra at 2500 ft (p. 179).
119. Extrafloral nectaries of *Arundina graminifolia*, visited by different insects, wasps and flies. Photo N.A.van der Cingel, 1998, Costa Rica (p. 182).

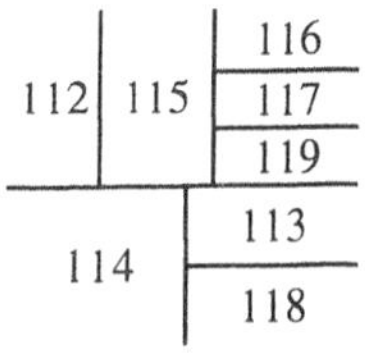

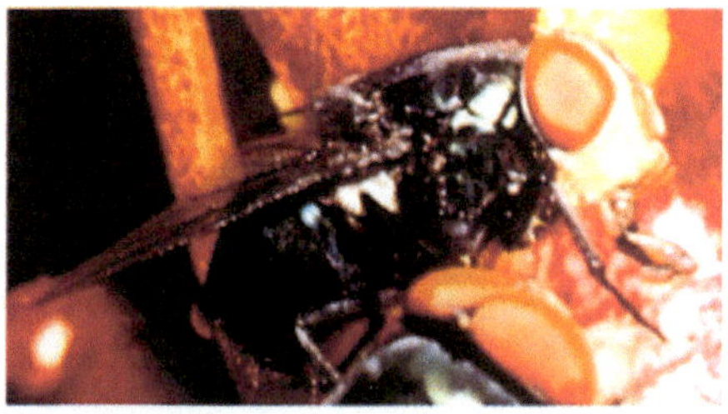

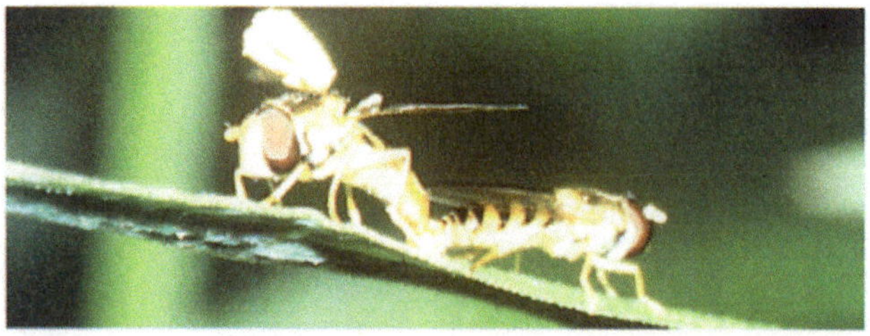

METRIC 1

Genus **Vanda**

Vanda cristata (= Trudelia cristata)
This species from Nepal, Buthan, Sikkim and India inhabits the outskirts of forests as an epiphyte. It has green flowers which are long-lasting and occur in two distinct flower forms, the larger forms with 1-2 flowers per spike growing at elevations of around 1400 m. The other form with up to five flowers is found at 800-1000 m. The apical portion of the lip produces two horns. The lip is white with brick-red to purple-black longitudinal lines. The plant produces several spikes. The faint fragrance attracts beetles of one species. The beetles land on the host tree or on the vegetative part of the orchid and then fly or crawl onto the lip. There it dips the head into the spur for a period up to about four minutes. It does not struggle and reverses slowly out of the flower, carrying pollinia on the back. It usually visits a second flower without pollinating, but then carries another set of pollinia. After visiting several flowers these pollinia clump on the back as a small hump. With a quantity of pollinia the pressure exerted by the beetle is sufficient to deposit one or more pollinia on the stigma. Usually plants in the wild have a great number of seed pods on them.

Vanda teres
In flowers of this species dead specimens of *Xylocopa latipes* bees have been found.

Vanda pollinia are frequently found on captured bees.

The odours of *V.teres* and *V.hookeriana* agree with those of the flowers of foodplants of *Xylocopa* bees.

Genus **Cottonia**

Cottonia peduncularis
The only species of this genus is an epiphyte, native to south-western India and Sri Lanka.

Yadav (1995) observed pseudocopulation of an unidentified insect, probably a beetle, on a labellum of a flower hanging in a wire-house in the Botanical Garden, Shivaji University, in Kolhapur. The glossy, maroon-red labellum mimics the (female?) insect in odour, shape, texture and colour.

VANDEAE: ANGRAECINAE

One species in Ceylon. Most angraecoids are moth-pollinated.

Problematic species

Dressler (1993) groups some subtribes and species that are insufficiently known for proper classification in a chapter with the name *Misfits, Parallelisms* and *Miscellaneous Problems*.

The misfits and leftovers of Tropical Asia are treated below.

ARUNDINAE

There are seven species in two genera. Pollination is probably by bees, as is suggested by the form of the flowers.

In *Arundina speciosa* autogamy in the bud is ob-

120	124
121	125
122	126
123	127

Plates
120. Copulating *Sphaerophoria macrogaster*. The male bearing a pollinarium of *Epipactis thunbergii* on the thorax. Photo N.Sugiura (p. 183).
121. *Sphaerophoria macrogaster* visiting a flower of *Epipactis thunbergii* and taking up pollinia on the dorsal surface of the thorax. Photo N.Sugiura (p. 183).
122. *Episyrphus balteatus* on a flower of *Epipactis thunbergii*, bearing a pollinarium on the head. Photo N.Sugiura (p. 183).
123. Inflorescences of *Goodyera foliosa* var. *maximowicziana*. Photo N.Sugiura (p. 183).
124. A worker of *Bombus diversus diversus* with six pollinaria of *Goodyera foliosa* var. *maximowicziana* on its proboscis. Photo N.Sugiura (p. 183).
125. A geometrid moth *Odontopera arida* pollinates *Platanthera mandarinorum* ssp. *hachijoensis*. Photo K.Inoue (p. 184).
126. Pollinaria of *Platanthera mandarinorum* ssp. *hachijoensis* on the compound eyes of *Odontopera arida*. Photo K.Inoue (p. 184).
127. Pollinaria of *Platanthera metabifolia* on the tongue of *Deilephila askoldensis* (Sphingidae). Photo K.Inoue (p. 185).

served, but the species is not cleistogamous. *Xylocopa*-bees have a preference for flowers with a secluded ante-room, as provided by the upturned side lobes of the labellum of species like *A.speciosa* (and e.g. *Phaius tankervilleae* or *Vanda*).

A.graminifolia has spread in tropical areas all over the world by escaping from cultivation. It has no nectar in the flower, but extrafloral nectaries are visited by many insects. Sugiura & Maeta (1995) mention six species of Lepidoptera and two small bee species visiting the flowers in south-west Japan, however, without pollinating. The extrafloral nectaries were visited by six species of hunting wasps and an ant species (*Anoplolepis longipes* workers), which probably keep away phytophagous pests. The author (Plate 119, p. 180) observed visiting wasps on the extrafloral nectaries in Costa Rica.

Dilochia wallichii from Java and Sumatra is also autogamous.

COLLABIINAE

Some 28 terrestrial species in three genera from tropical Asia of which the mode of pollination is not known.

Genus **Chrysoglossum**

A species of this Asian genus is reported to be self-pollinating.

Japan

In the geographical sense Japan is on the intersection of several continents and therefore shares genera and even species with Eurasia, America and Australasia. Besides, it has several endemics.

Not much is known about pollination of orchids in Japan itself, apart from the work of Inoue on e.g. *Platanthera* and the work of Sugiura et al. (1997), but the interest in pollination biology is clearly growing, so that it may be supposed that in the near future more detailed knowledge will be gained.

I will only deal with those species of which the pollination has been studied or reported in Japan itself and those species which Japan shares with other areas and of which the pollination is known from elsewhere.

Orchids with two or three fertile anthers

CYPRIPEDIOIDEAE

Genus **Cypripedium**

See pages 160 and 161. Japan has four species of *Cypripedium*.

Cypripedium debile

The labellum structures of this species seem to imitate mushrooms and the flowers are supposedly pollinated by fungus gnats.

Stoutamire (1967) was the first to point out this phenomenon. Vogel analysed the syndrome of myophily and described the example of *C.debile*.

The species inhabits dark montane needlewoods in Japan, but is also known from China and Taiwan.

The dwarf species (10-15 cm), with greenwhite flowers of 1 cm with white labella, grows terrestrially and the flowers are close to the bottom by the downward bending of the stem (geophily). The flowers emit a strong odour of 'supermarket mushrooms' and the pouch with its circular entrance at the underside resembles a small leaflike mushroom with its many additional black-brown markings mimicking gills.

Primitive orchids with a single fertile anther

GASTRODIEAE: EPIPOGIINAE

Genus **Epipogium**

Both known species of this genus are found in Japan.

Epipogium aphyllum is also known from Europe, where bumblebees and wasps are recorded as pollinators, although there is serious doubt about the role of bumblebees. Nectar is present. Seed set, however, is very low and this is probably due to its growing in dark places in woods where insects are rare.

Autogamy (cleistogamy) may work in individuals with upright, instead of hanging pollinia (see also p. 163).

NEOTTIEAE: LIMODORINAE

Genus **Epipactis**

Epipactis thunbergii
Four species of syrphid flies legitimately pollinate the flowers of this species, among which *Sphaerophoria macrogaster* is the most effective (Plates 120-122, p. 181).

Episyrphus balteatus, *Sphaerophoria philanthes* and *Melanostoma sclare* are also able to withdraw and deposit pollinaria. The same was found once for the bee *Lasioglossum scitulum*. (Sugiura 1996).

The movable epichile, possessing a furrow at its base, played a critical role. It threw the pollinator onto the stigmatic surface when both sides of the basal slanting surface of the furrow were presumably pressed in the direction of the hypochile by the fore and middle legs of the retreating fly. The pollinia adhered to the thorax (the normal place of adherence) at that moment. Ovipositions of three out of four species of syrphid flies were observed on the epichile, the floral bud and the bract, in spite of aphids being rarely found around these places. This probably suggests chemical stimulation by olfactory stimuli from the flower, since other stimuli are less important than olfactory ones in inducing egg laying. The orange-yellow raised markings on the inner surface of the hypochile may, however, visually mimic aphids (cf. Ivri & Dafni 1977 on *E.consimilis* and Van der Cingel 1997 on *E.helleborine*).

Mean fruit set observed by Sugiura was 54.7% per rachis.

Ohno (1968, in Sugiura 1996) reported the syrphid fly *Baccha maculata* receiving a pollinarium of *E.thunbergii* on its thorax.

NEOTTIEAE: LISTERINAE

Genus **Listera**

Of five species of the genus *Listera*, two are endemic.

Listera cordata
L.cordata is represented with three subspecies, one of which is the circumboreal ssp. *cordata*. whose pollination is described in Chapter 4. The other subspecies, according to Maekawa (1971), differ in lip morphology.

Genus **Neottia**

Neottia nidus-avis
Neottia nidus-avis is represented by a var. *mandshurica*, which supposedly is fly-pollinated like the Eurasian variety. Also, ants are reported as geitonogamous pollinators (Van der Cingel 1995). If pollinators stay away, the pollen becomes incoherent and may fall down on the stigma, in this way effecting autogamy.

More advanched orchids with a single fertile anther

SPIRANTHOIDEAE

CRANICHIDEAE: GOODYERINAE

Genus **Goodyera**

Several species of this usually bee-pollinated genus (especially by bumblebees) exist in Japan. *Goodyera repens* is a boreal species, of which the pollination is long known from Europe. The mechanism is practically the same as in *Spiranthes* (see below) and was described extensively by Darwin (1866) (see also Chapter 4 for more details).

Goodyera foliosa var. **maximowicziana**
Sugiura & Yamaguchi (1997) studied the pollination of this species in Matsue. The flowers were rarely visited by five species of insects, among which two species of ants, thrips, a syrphid fly and a bumblebee, of which only the latter, *Bombus diversus diversus*, mostly workers, pollinated legitimately (Plates 123 and 124, p. 181). They received the pollinarium on the galea of the proboscis. Nectar-robbing ants frequently were 'caught' by the viscidium of the pollinarium and such pollinaria were never dislodged by the pollinator. Fruit set was on average 57.1% per rachis. While the percentage of pollinated flowers was 60.1%, only 17.5% of flowers lost a pollinarium. Single pollinaria were used repeatedly to pollinate several flowers. In the southernmost regions of Japan, where the species is also distributed, no bumblebees are found, suggesting a different breeding system.

CRANICHIDEAE: SPIRANTHINAE

Genus **Spiranthes**

Spiranthes sinensis

In Japan the species is represented by var. *amoena*. For pollination of the species see Chapter 8.

ORCHIDOIDEAE

ORCHIDEAE: ORCHIDINAE

Japan has several genera (Maekawa 1971), some with species represented that are also known from Europe, like *Gymnadenia conopsea*, *Neottianthe cucullata*, *Coeloglossum viride*. These species have been dealt with in my *Atlas of Orchid Pollination: European Orchids*. Therefore the reader interested in the pollination of Japanese orchids is advised to check up this if possible in cases thought to be relevant.

Genus **Platanthera**

Plantanthera is the largest genus in Japan with 21 species. (Inoue 1983). It is also one of the largest genera of Orchidaceae in the whole temperate zone with about 100 species. As in the USA and Europe, the flowers of the Japanese species are adapted to pollination by lepidopterans, offering nectar in a usually long spur. Most species are phalaenophilous i.e. adapted to moths. Some are sphingophilous. Exceptions to adaptation to lepidopterans are *P.chorisiana*, which is pollinated by a small beetle and *P.brevicalcarata*, an autogamous species.

Inoue, who studied the genus extensively, recognises three types of pollen deposition on lepidopterans which are correlated with column features. Pollinaria are attached to either the compound eye, the basal part of the proboscis, or its middle part. Adaptation to the different pollinators has led to different anthecotypes.

Platanthera mandarinorum

Platanthera mandarinorum ssp. *hachijoensis* var. *hachijoensis* is pollinated by several moths, belonging to the Geometridae, Noctuidae and Pyralidae. Nine vector species were observed with pollinaria attached to the compound eyes. *Odontopera arida* (Geometridae) was dominant among these and transported about 87% of the pollinaria (Plates 125 and 126, p. 181). The moths alight on the flowers and usually work at first on one of the lowermost flowers, whereafter they move upwards on the inflorescence. *Odontopera* is much faster than other vectors and shows typical behaviour by walking with vibrating wings. During nectar-feeding, the moths usually clasp the lateral sepals with their forelegs, pressing the thorax on the basal part of the lip and the head into the cavity formed by the column.

In a separate study on Hachijo Island Inoue (1985) observed that the peak of flowering coincides with the massive emergence of *Odontopera arida*. Due to anthesis characteristics (flowers open one by one), and the pollinating geometrids not being restricted by the covers of surrounding herbs in finding the flowers by scent, the visits occur irrespective of spike size. The species appears pollinator-unlimited, since nearly all the flowers are pollinated.

In field experiments (Inoue 1986), in which spur length was manipulated or pollination excluded by bagging, the effects of variation of spur length and of pollinator activity on the reproductive success (RS) of individual plants was investigated. Plants with a natural spur length range show nearly equal RS, whereas plants with artificially shortened spurs had decreased RS proportional to spur length reduction. Pollinaria attachment, which is a measure of male RS, seems to demand stricter morphological correspondence than pollen receipt (=female RS).

Results of bagging experiments indicated that female RS was greater than male RS in cases of high pollinator activity and consequent high pollen carryover. *P.hachijoensis* has the potential for high pollen carryover; i.e. more than ten flowers can be pollinated by one pollinarium, a part of which is used in one pollination event. In cases of high carryover a relatively small number of pollinaria may pollinate a large number of stigmas in the early phase of pollination.

The facts suggest that *fecundity selection* operates more intensely on the male function in the case of high pollen carryover and on both female and male functions in the case of low pollinator activity. Individuals with different spur lengths in natural populations have nearly equal probabilities of RS after excluding the extreme variants through reduction of male RS (cases of high pollen carryover), i.e. individuals with spur lengths within certain ranges, having similar fitness values, are selected. Selection operates against individuals poorly fitted for pollination, more strongly in cases of low pollinator activity (*stabilizing selection*).

P. mandarinorum ssp. *ophrydioides* var. *ophrydioides*, has similar, but smaller flowers and is correspondingly pollinated by smaller moths, which also bear pollinaria on the eyes. Two vectors were captured of which *Apamea hampsoni*, with 11 individuals, transported 76% of the pollinia and *Deileptenia ribeata* (two ind.) 24%. *P.mandarinorum* ssp. *mandarinorum* var. *mandarinorum* has much larger flowers with spurs of 30 mm length and is pollinated by only one species of sphingid: *Theretra japonica*.

This also carries the pollinaria on the eyes. The range of proboscis lengths is distinct from those of sympatrically existing noctuids and geometrids.

P.mandarinorum ssp. *mandarinorum* var. *brachycentron* possesses flowers almost similar to those of *ophrydioides*. In the small population investigated only one noctuid, *Acanthoplusia agnata*, out of four conspecific individuals, was captured with pollinaria.

Platanthera okuboi

This orchid is pollinated by a small sphingid, *Rhagastis trilineata*. Since the distance between the viscidia is small the pollinaria are attached to the basal part of the proboscis. The spurlength is 3.5 cm. The number of individuals of the sphingid increases rapidly in parallel to the flowering of the orchid. The sphingid feeds also on other plants, e.g. on *Lonicera japonica*, and only one out of 10 individuals caught by a light trap carried pollinia of *P.okuboi*.

Inoue (1985) investigated predation and influence of inflorescence size. The species appears pollinator-limited. In closed habitats pollination degree diminishes and predation increases. Large inflorescences are better pollinated.

Platanthera metabifolia

The flowers are slightly smaller than those of *P.okuboi*. The pollinaria are deposited on the basal part of the probosces. Five pollen-vectors were identified, of which four are sphingids and one is a noctuid. The most important pollinators were *Deilephila askoldensis* (Plate 127, p. 181) (46%) and *D.elpenor* (42%). The latter is known as a pollinator of *P.bifolia* and *P.chlorantha* from Europe.

Inoue (1986b) investigated the fecundity of this species in relation to different types of behaviour and morphology in both noctuid and sphingid pollinator moths in three orchid populations. Sphingids selected spikes with large numbers of flowers in the early stages of flowering (flowers with spur lengths of 13-33 mm were equally pollinated), whereas noctuids selected spikes with short (13-23 mm) and medium-sized (23-25 mm) spurs in the late stage of flowering. The proboscis length of sphingids, varying from 15-26 mm, allows them to reach the nectar which accumulates up to half the depth of the spur, whereas on the other hand noctuids with a proboscis length of ca 12 mm seem to avoid variants with long spurs, the nectar of which they cannot extract.

Different patterns of capsule sequences were found. Patterns due to sphingids are sporadic and generally on one side of a spike. Those due to noctuid moths are formed in a continuous fashion. Large spikes seem more attractive to sphingid moths than small ones. The size and density of spikes, however, scarcely affected the pollination activity of noctuids.

The selection of early or late-blooming flowers by sphingids and noctuids is correlated with the periods of emergence of the moths.

Sphingids, which fly long distances, seem to be instrumental in maintaining uniformity among the three populations, whereas noctuids, which operate within smaller ranges, may exert disruptive selection pressure. At present, the pollination characteristics of *P.metabifolia*, coincide best with those of sphingophily in the populations studied.

Platanthera boninensis

The anther connective of this orchid is narrow and the viscidia lie above the orifice of the slender spur. Therefore the pollinaria are attached to the middle of the proboscis. They are consequently broken during foraging trips and only smears of the viscidia remain on the proboscis.

Some Noctuidae and Geometridae (four and three species resp.) were caught with pollinaria.

Platanthera tipuloides

The subspecies *tipuloides* var. *sororia* from Ukishima Moore, where it is sympatric with *P.mandarinorum* ssp. *ophrydioides* and with *P.tipuloides* ssp. *nipponica*, has the pollinaria deposited on the lower surface of the probosces and they are also broken during subsequent foraging, leaving only smears. Var. *sororia* flowers before *nipponica*. Putative hybrids of *ophrydioides* and *sororia* are sometimes found and this is explained by their sharing of the pollinators *Apamea hampsoni* and *Deileptenia ribeata*. The pollinators of the subspecies *nipponica* have smears of the pollinaria deposited on the upper surface of the proboscis. The spur of the orchid is upturned, suggesting upside down foraging on these flowers.

Platanthera hologlottis

The white flowers of *P.hologlottis* have a small ridge on the midrib of the lip in front of the spur which partially obstructs the centre of the flower. Together with a callus that projects downwards below the lateral stigma lobes, this ridge produces a double nectary entrance. Two furrows between the median and marginal ridges provide a tongue guide to it. The American species *P.flava* and the Japanese species *P.ussuriensis* have a similar construction. Pollinaria are attached to the probosces of the moths. The Noctuid *Trichoplusia intermixta* seems to play a dominant role in one of the populations investigated, but a few other pollinators and visitors are known from another population.

Platanthera ussuriensis

As in *P.hologlottis* the spur entrance is doubled by twin obstructions formed by a callus and a ridge. The lateral lobes of the lip may be utilised as a guide for the proboscis to the nectary entrance under each viscidium.

The flowers are yellowish green with a spur of 5-6 mm long, resembling that of the mosquito-pollinated *P.flava*. *P.ussuriensis* however, living in a different habitat, is pollinated by several species of Pyralidae which receive the pollinaria on their probosces. Inoue lists 45 individuals of seven species, which together transported 74 pollinaria.

Platanthera sachalinensis

The lip of this species has one prominent callus at the centre of the basal portion which contacts the stigma. This obstruction guides the proboscis to the nectary entrance below each viscidium. The pollinaria are attached to the proboscis and are broken during subsequent foraging, leaving smears only. Three species of Noctuidae with smears of viscidia on the proboscis were captured.

Platanthera chorisiana

This orchid is pollinated in the daytime by a small beetle, *Oedemeronia lucidicollis* (Inoue 1981). The flower morphology is in accordance with this fact. A short spur less than 2 mm long allows the short-tongued beetle to reach the nectar. Pollinaria become attached to the head between the antennae and around the mouth. They obviously visit the flowers for both nectar and pollen. Other small insects are not excluded as potential pollinators.

Platanthera brevicalcarata

This is an autogamous species, as was observed in both wild and cultivated plants. Stigma and anther cells contact each other during development to full size of the stigma and its lateral lobes.

Typology

As said above, Inoue recognises three types of pollen deposition (Figs 39 and 40):

1. *Type e*. If the anther connective is broad and the distance between the viscidia is wide, the pollinaria are deposited on the compound eyes. Near the orifice the spur is always wider than the proboscis, which can be easily inserted.

2. *Type p-1*. The viscidia are placed just above or below the orifice of the spur and the pollinaria are deposited on the proboscis. There is no obstruction for the insect near the centre of the orifice. The pollinaria or pollen smears are deposited on the upper surfaces of the probosces, except in *P.tipuloides*, where the viscidia are situated below the orifice instead of above them, with consequent attachment of the smears on the lower surface of the proboscis.

3. *Type p-2*. Obstructions at the centre of the spur entrance, derived from the callus at the middle of the lip base and/or a projection below the stigma, produce a pair of nectar entrances, above which the viscidia are situated and into which the insects insert their probosces.

In general the pollinators of *Platanthera*-species are not dependent on the orchids for nectar. There are no distinct differences in gross morphology between moths that forage on *Platanthera* and those foraging on other flowers. Consequently there is no specific adaptation of the moths to a particular species of orchid. This is reflected in the fact that usually several species of moths are associated with pollination in each species of *Platanthera*. Whereever only one species of pollinator is found, as in *P.okuboi* and *P.mandarinorum*, this seems to have evolved as a result of adaptation to the flowering period of the species.

Adaptation to pollinators and habitats may lead to slight differences in flower morphology like length, width and position of the spur; posture and shape of the perianth; distance between viscidia; colour; odour and blooming season. This may be a continuous process and may lead to socalled anthecotypes (Nilsson 1978, 1979). Adaptation to the different

1a 1b 2a 2b
3a 3b 4a 4b
5a 5b 6a 6b
7a 7b 8a 8b

Paired black and white photographs of orchid flowers (Dendrobiinae) to compare visible light exposures (i.e. in the HVR = Human Visible Range) to near-UV exposures (370-390 nanometre range). The relative brightness of near-UV reflectance is variable and is indicated for each species: VB = very bright, B = bright, M = moderate, L = low, and A = near-UV absorbing. The different groups A-F represent different patterns found. All photographs by James Indsto.

Group A – Flowers with a dramatic bullseye pattern.

1. *Epigeneium nakahari*. a = HVR, b = near-UV (M) (p. 175 and Plate 102, p. 173).
2. *Dendrobium lindleyi* (Sect. *Callista*). a = HVR, b = near-UV (B) (p. 175 and Plate 103, p. 173).
3. *Dendrobium chrysotoxum* (Sect. *Callista*). a = HVR, b = near-UV (B) (p. 175 and Plate 104, p. 173).
4. *Dendrobium chryseum* (Sect. *Dendrobium*). a = HVR, b = near-UV (M) (p. 175).
5. *Dendrobium bellatulum* (Sect. *Formosae*). a = HVR, b = near-UV (M-B) (p. 175 and Plate 105, p. 173).
6. *Dendrobium truncatum* (Sect. *Rhopalanthe*). a = HVR, b = near-UV (B) (p. 175).
7. *Dendrobium primulinum* (Sect. *Dendrobium*). a = HVR, b = near-UV (VB). Compare the colour plate of this species: the labellum absorbs near-UV strongly, whereas the colour is relatively monochromous in the HVR (p. 175 and Plate 106, p. 173).

Group B – The whole flower absorbs near-UV strongly.

8. *Dendrobium heterocarpum* (Sect. *Dendrobiuma*). a = HVR, b = near-UV (L). Compare the colour plate. The outer flower parts are very glossy in the near-UV, but matt in the HVR. A passing insect would probably see the high near-UV reflectance of the outer flower parts, but the labellum may be perceived as matt and dark in both the HVR and the near-UV (p. 175 and Plate 108, p. 173).

9a

9b

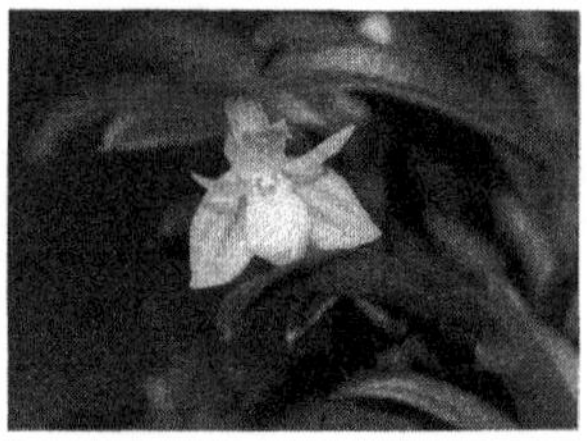

10a

10b

11a

11b

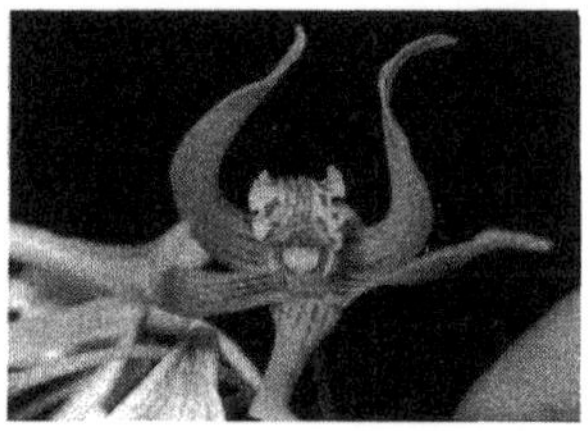

12a

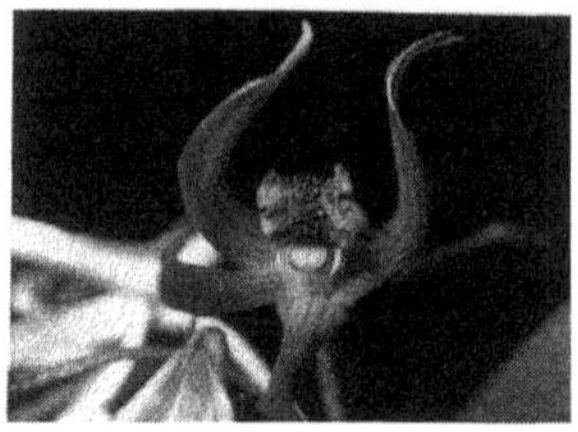

12b

13a

13b

14a

14b

Group C – Less dramatic bullseye pattern.

9. *Dendrobium rhodostictum* (Sect. *Latouria*). a = HVR, b = near-UV (B). The callus ridge of the labellum absorbs strongly in the near-UV (A) (p. 175 and Plate 107, p. 173).

Group D – Outer flower parts show moderate near-UV reflectance, the whole of the labellum absorbs somewhat in the near-UV.

10. *Cadetia chionantha*. a = HVR, b = near-UV (p. 175).
11. *Diplocaulobium bicolor*. a = HVR, b = near-UV (p. 175).
12. *Dendrobium cucumerinum* (Sect. *Rhizobium*). a = HVR, b = near-UV (p. 175 and Plate 109, p. 173).

Group E – Contrast between flower and stem + leaves.

13. *Dendrobium pentapterum* (Sect. *Oxyglossum*). a = HVR, b = near-UV. Example of one of the many New Guinean species with flowers which absorb strongly in the near-UV, in contrast to the moderatively reflective stems and leaves (p. 175 and Plate 110, p. 173).

Group F – Resemblance between HVR and near-UV.

14. *Dendrobium obtusisepalum* (Sect. *Calyptrochilus*). a = HVR, b = near-UV. Example of species which have very close tonal resemblance between HVR and near-UV black and white images where near-UV reflectance is moderate to high (p. 175 and Plate 111, p. 173).

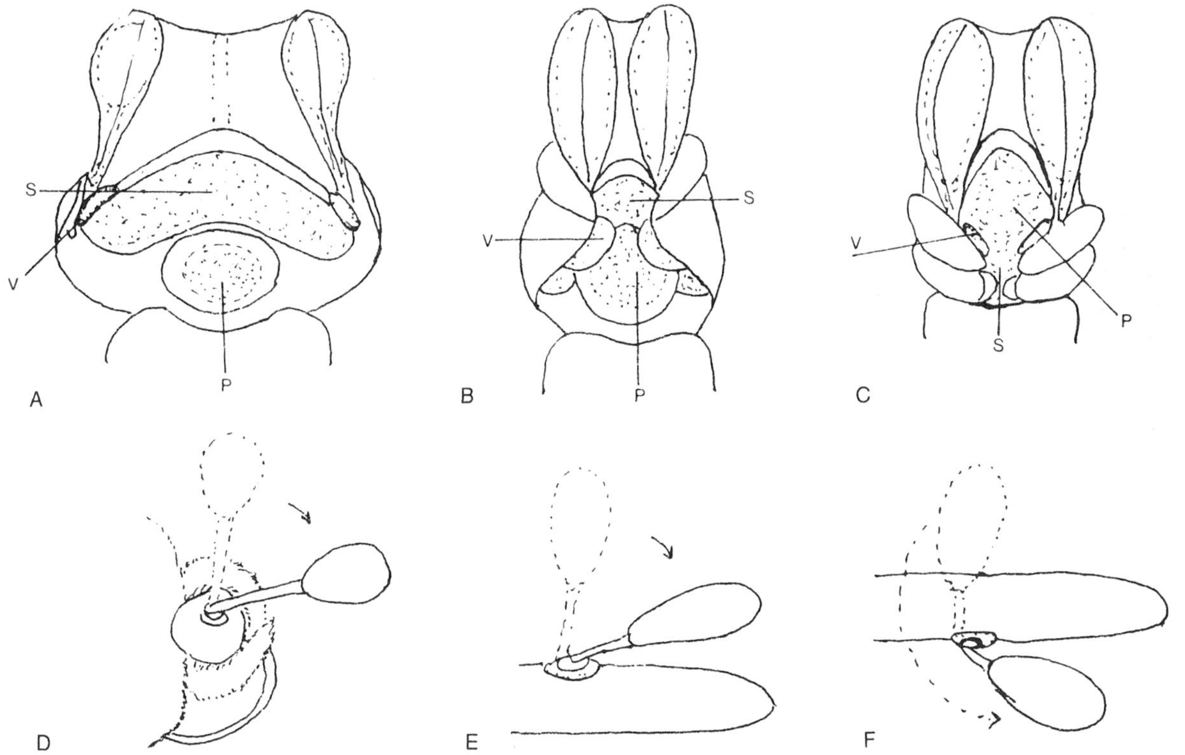

Figure 39. Diagrams of types of pollen attachment to insects in *Platanthera* species (after Inoue 1983). A and D) Type e: viscidia remotedly located; pollinaria attaching to the compound eyes of insects. B and E) Type p-1: viscidia situated on upper side. C and F) Type p-1:viscidia on lower side. Spur orifice without obstruction at center. Pollinaria attaching to proboscis. Arrows indicate movements of pollinaria after deposition. P = area of injection of probosces, S = stigma, V = viscidium.

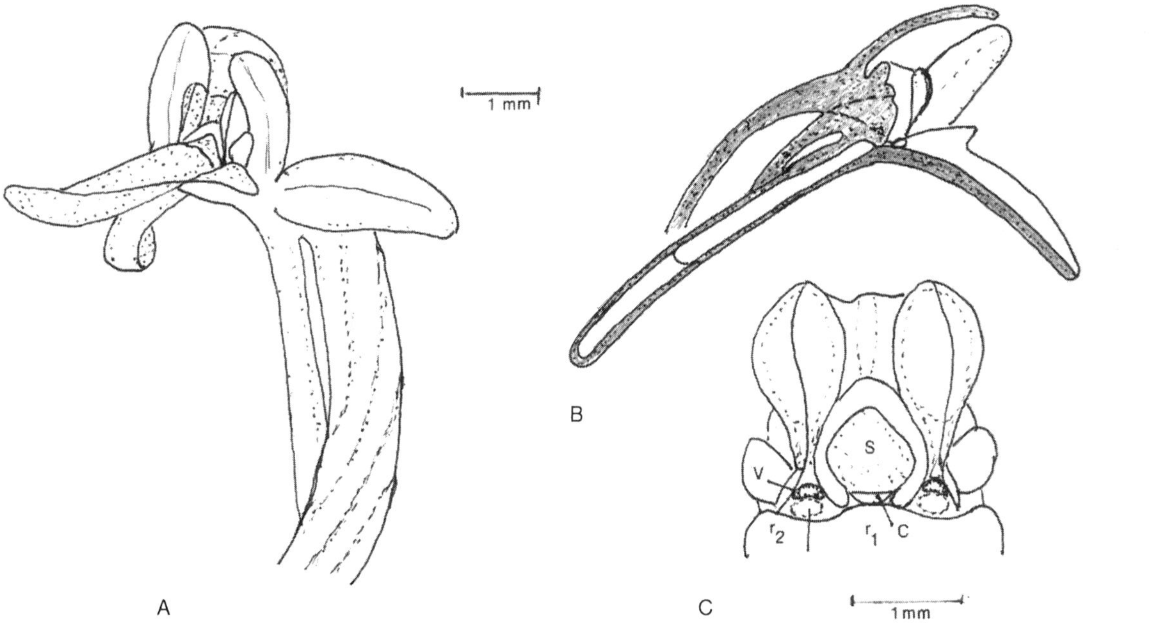

Figure 40. Diagrams of types of pollen attachment to insects in *Platanthera* species (after Inoue 1983). Type p-2: insertion of proboscis restricted to lateral sides of lip. A-C) *P.ussuriensis*. A) *P.ussuriensis*. Front view of the flower. B) *P.ussuriensis*. Median section of flower. C) Front view of column. c = central projection below lateral stigma lobes, p = area of injection of proboscis, r_1 = hump at center of lip base, r_2 = marginal projection of lip base, S = stigma, V = viscidium.

pollinating insects may lead to continuous transformation of the flowers and in this way rule intraspecific and interspecific variation.

Other species of **Platanthera**

From the three described pollination modes found in *Platanthera*, the pollinator class of unknown species may be predicted. Short-spurred species with spurs shorter than 3 mm, like *P.iinumae* and *P.takedae*, will probably be pollinated by small insects other than lepidopterans.

There are some reports of pollination of other species of *Platanthera* by moths previous to the study of Inoue of 1983. They refer to *P.florentii*, which is pollinated by Pyralidae, *P.minor* (by Geometridae, Noctuidae and Pyralidae) and ditto *P.mandarinorum* (var. *brachycentron*, *neglecta*, or *hachijoensis*).

For several species the pollinators are as yet unknown, but may be guessed at on the basis of their flower morphology.

Classification of Japanese **Platanthera**

Inoue based his classification on the results of the pollination studies combined with the analysis of morphological characters. He recognises 21 species, seven subspecies and six varieties.

The short-spurred species are considered as more primitive and adapted to other insects than lepidopterans early in the evolution of the genus. Ancestral forms adapted to lepidopterans may have had the Type p-1 mode of pollination, whereas Type-e and Type p-2 flowers may be seen as specialisations and further lines of evolution, the latter being especially successful because of the reduction in loss of pollinaria that are deposited on the eyes of the moths.

EPIDENDROIDEAE: CYMBIDIOID PHYLAD

MALAXIDEAE

Genus **Malaxis**

Malaxis monophyllos

This is a circumboreal species. For pollination see Chapter 4.

Genus **Hammarbya**

Hammarbya paludosa

Hammarbya (=*Malaxis*) *paludosa* (see also Chapter 4).

CALYPSOEAE

Genus **Calypso**

Calypso bulbosa

Calypso bulbosa, a circumboreal species is represented with a var. *japonica* (=*speciosa*), as well as with the type var. *bulbosa*, the latter from Tsugaru Peninsula, north of Honshu (Maekawa 1971). Bee-pollinated (for details see Chapter 4).

Genus **Cremastra**

Cremastra appendiculata

Sugiura (1996) observed three species of bees and two species of syrphid flies visiting the pendulous, tube-shaped flowers of *C.appendiculata* var. *variabilis* (Plates 128-130, p. 196). The flowers have a strong unique perfume and secrete nectar in the saccate base of the labellum. The flowers bloom under cool and dark conditions, i.e. understorey of forests. Only queens of *Bombus diversus diversus* received pollinaria on one of their fore-femora. Visiting frequency was extremely low, however. A carpenter bee, *Xylocopa appendiculata circumvolans*, consistently robbed nectar by making a perforation in the flower base. Fruit set was quite low.

Genus **Oreorchis**

Oreorchis patens

Of the 10 species of this Asian genus one species, *O.patens*, is known from Japan, growing in deciduous forest. The flowers are scented with a faint, pleasant scent and secrete nectar into a concavity at the basal column. The nectar is advertised with an orange-yellow colour. They have a hinged labellum. These last two characteristics are shared with the syrphid-pollinated flowers of *Epipactis thunbergii*.

Sugiura et al. (1997) observed visitation by eight species of insects on Mt. Misaka, Shimane. Pollinators are short-tongued insects, especially small syrphids (Plates 131 and 132, p. 196). The main pollinator was *Melanostoma scalare* (Syrphidae), which received the pollinaria on the thorax. A small bee, *Nomada flavoguttata japonensis*, was also observed as pollinator. Fruit set per rachis was 32.2 % on average. However, this figure should decrease finally because of heavy parasitism by an agromyzid fly *Japanagromyza tokunagai*. They lay eggs into the ovary of the orchid.

Genus Tipularia

For *T.japonica* see Chapter 4

CYMBIDIEAE: CYRTOPODIINAE

Genus Cymbidium

Cymbidium pumilum
Sasaki et al. (1991, 1993) report that drones of the Japanese honeybee *Apis cerana japonica* are attracted to the aroma of the flowers of *C.pumilum* during their mating flights in spring. Some drones cluster on the flower racemes and others insert their heads deep into the flowers. Drones with pollinaria on their scutellum were observed to visit other orchid flowers, effecting pollination. Individual workers and swarming colonies are also attracted by the flower aroma. The allopatric western honeybee *Apis mellifera*, however, is not attracted.

The orchid flowers produce no nectar. Only extrafloral nectaries secrete nectar, but the bees rarely collect it.

EPIDENDROIDEAE: EPIDENDROID PHYLAD

ARETHUSEAE: BLETIINAE

Genus Bletilla

Bletilla striata
This species is recorded from China, Tibet, Taiwan and Japan, where it is commonly cultivated in gardens and parks. It is rarely seen naturally.

Ishiwaka & Tanaka (in Sugiura 1995) have recorded honey bees, *Apis mellifera*, to visit and take away pollinia.

Sugiura (1995) observed 26 species of hymenopteran, dipteran and lepidopteran visitors on the flowers in Kobe (Plates 133-135, p. 196). The flower is nectarless and the species seems to deceive by food deception of a generalised character. Seven species of aculeate Hymnoptera were able to pollinate because of their body seize fitting the space between column and labellum. From the bees with appropriate size (i.e. intermediate) male longhorn bees, *Tetralonia nipponensis*, were the most effective ones, by abundance of visitation and legitimate intrafloral behaviour. The females of this species and the other six bee visitors contributed to a lesser extent, because of lower visiting frequency, as well as by grooming and thus removing pollinia. Mean fruit set was near 25% for individual scapes, with 20% bearing no fruits at all.

Genus Calanthe

Calanthe triplicata
Sugiura & Miyanaga (1996) observed a male *Radena similis similis* (Lepidoptera: Danaidae) with a pollinarium on its proboscis taken from this species.

Genus Phaius

Phaius tankervilleae
This widespread species is also found in the south of Japan.

Bee pollination by *Xylocopa* bees has been reported as well as auto-pollination and cleistogamy.

CHAPTER 8

Orchids of Austral(as)ia and New Zealand

If nature ever showed her playfulness in the formation of plants, this is visible in the most striking way among the orchids. The manifold shape of these flowers arouses our highest admiration. They take on the form of little birds, of lizards, of insects. They look like a man, like a woman, sometimes like an austere, sinister fighter, sometimes like a clown who excites our laughter. They represent the image of a lazy tortoise, a melancholy toad, an agile, ever chattering monkey. Nature has formed orchid flowers in such a way that, unless they make us laugh, they surely excite our greatest admiration

Jakob Breyne (1678), translated by Oakes Ames

Australia – including New Caledonia and Vanuatu

The magnificent book *Native Orchids of Australia* by David L. Jones (1988) was my guide to help structure the information taxonomically, especially within the genera. Once again this structure was, however, subordinated to Dressler's classification (1993). For more recent taxonomic information I used the catalogue of Australian Orchidaceae (Clements 1989).

The Australian continent is regarded as remarkably poor in orchid taxa in proportion to its size (Dafni & Bernhardt 1990).

However, recently a lot of new species have been described and according to Clements (pers. comm.) more are to come. Expectations are that in the end about 2000 species will be recognised, instead of about 800, as is usually stated as the number at present. At the moment 1200 species are already known, although opinions on this differ. A debate on orchid taxonomy is going on in Australia.

The epiphytic species are confined, primarily, to rain forest relics on the north coast and to the moist-temperate jungle gullies as far as Southeast Gippsland (eastern Victoria) and are not regarded as endemic, but, together with certain terrestrial tropical genera (e.g. *Habenaria*) as disjuncts of the palaeotropical flora isolated after the break-up of the Indo-Malaysian land-bridge.

The remaining terrestrial genera show their greatest diversity along the southern coast. About 29 genera are considered autochtonous elements, endemic to temperate Australasia.

A dominant trend in Australian terrestrial orchids concerning pollination is floral mimesis.

In southern Australia geophytic orchids flower throughout the year and seasonal and intraseasonal variation in taxonomic diversity are marked by general shifts in pollination syndromes that occur in three overlapping pulses. These pulses are dominated respectively by fungus gnats, bees and wasps and insects with short mouthparts from different orders. The first flowering group of orchid taxa exploits feeding, mating and ovipositional habits of the gnats, the second (and largest group) consists of species that are nectarless and are either food mimics or sexual mimics, the third are nectariferous with small flowers in usually tight racemes. Nectar collects at the base of the labellum instead of in spurs or tubes (except in *Spiranthes sinensis*).

Autogamy is observed in at least one species of all major Australian genera except one. The majority of genera show a trend toward self-compatibility and many orchids with friable pollinia appear to be pre-adapted for mechanical autogamy (fail-safe mechanism of delayed self-pollination). This may even be considered as a partial mechanism promoting sympatric speciation in genera with high intraspecific variation (Dafni & Bernhardt 1990).

Allopatric speciation via successful long distance dispersal, followed by genetic drift (maintained by self-pollination) may have contributed to the diversity of many orchid genera. The role of introgressive hybridisation in the speciation is controversial.

Primitive orchids with a single fertile anther

GASTRODIEAE

Formerly, Dressler (1981) classified them as a primitive tribus within the subfamily Epidendroideae, but he (1993) has now placed this tribus in a separate group of primitive orchids with a single fertile anther. All Gastrodieae are saprophytic, which means that reduced characters make it a taxonomically difficult group.

GASTRODIEAE: GASTRODIINAE

The subtribe is pantropical, primarily in tropical Asia and Australasia. There are six genera, on which the information about pollination is lacking, with the exception of *Gastrodia*.

Genus Gastrodia

Gastrodia sesamoides is a saprophytic orchid of the temperate regions of Australia (and New Zealand, but here grows a slightly different species *G.* aff.. *sesamoides*), especially abundant in years of good rainfall. Its flowering is stimulated by fire. The flowers possess a pleasant spicy perfume. It is pollinated by a small xylocopid bee, *Exoneura* sp. (*Ceratinini*, *Xylocopinae*). The bee gathers pseudopollen from the callus of the lip (Jones 1981, 1985) and carries it in between the hairs of the hind legs. Entry to the flowers is upside down from the top and the bee initially feeds on the first part of the callus and then moves deeper into the flower, to the second part of the callus, where it spends most of the time. They scratch the callus with their legs. On leaving, the insect turns around and emerges head first and upright. In doing so it contacts first the rostellum where the thorax is smeared with glue and then contacts the anther, receiving the bright orange pollinia. About two-third of the flowers were pollinated in the population investigated. Bee pollination is probably true for most species (Christensen 1994).

No data are available for the second Australian species: *G.queenslandica.*

Hatch (1954) notes that the columns of *G.cunninghamii* and *G.minor* of NZ, as well as *G.orobanchoides* of the New Hebrides have the column compressed, illustrating the requirements of self-fertilisation.

GASTRODIEAE: EPIPOGIINAE

This is a small subtribe with representatives in Eurasia, tropical Africa, tropical Asia and Australasia. About five species in three genera.

Genus Epipogium

Epipogium roseum
Jones studied this orchid in north-eastern Queensland. It is strictly self-pollinating and has a simple mechanism: the stigma increase in size until the two lobes contact the pollinia. These may swell after contact is made and lose all coherence. The pollen becomes firmly attached to the stigma.

Pollinaria cannot be withdrawn by insects. The flowers open only for a couple of days. See also page 163.

NERVILIEAE

The tribe contains only one genus of 65 species of tropical Asia, Australasia and Africa.

Some forms are self-pollinating and in Africa small wasps and bees have been observed.

Some species produce nectar, others lack this and apparently achieve pollination by false advertisement.

Nothing is known about the pollination of the five Australian species.

N.aragoana would appear to be self-pollinating and the others are dependent on insects (Jones 1988).

VANILLEAE

VANILLEAE: GALEOLINAE

Saprophytes. Four genera with together 27 species.

Genus Erythrorchis

This genus was determined in 1986 by Garay and contains only three species, one of which occurs as an endemic in Australia. Members of the genus are saprophytic climbers.

Erythrorchis cassythoides is pollinated by small bees which are attracted by the perfume (Jones 1994).

More advanched orchids with a single fertile anther

SPIRANTHOIDEAE

TROPIDIEAE

This pantropical tribe consists of two small genera: *Corymborkis* and *Tropidia*, each of which is represented by one species in Australia. About their pollination nothing is known, except that the species are allogamous. *Corymborkis* is supposedly sphingophilous. *Corymborkis veratrifolia* has green and partly white flowers, with a tubular lip and a spicy, agreeable perfume.

CRANICHIDEAE: GOODYERINAE

This large subtribe is pantropical and northern, but occurs mainly in tropical Asia. There are 35 genera in two alliances.

Genus **Goodyera**

Goodyera is primarily pantropical and northern and species occur in America, Europe, Asia.

In Australia two species occur, of which nothing is known regarding pollination.

In Europe and North America bumblebee pollination is reported.

Genus **Anoectochilus**

Only one species (out of ca 25) is endemic to Australia: *Anoectochilus yatesiae*, which appears to be self-pollinating.

Genus **Cheirostylis**

One species, out of ca 22, occurs in Australia. This endemic *Cheirostylis ovata* is a deciduous lithophytic or terrestrial orchid with unknown pollination mode.

Genus **Hetaeria**

One out of 20 species extends into Australia. Pollination unknown.

Genus **Vrydagzynea**

One out of 40 species occurs in Australia. This species *Vrydagzynea paludosa* is probably autogamous.

Genus **Zeuxine**

Three out of 76 species occur in Australia. Pollination of the species is not known.

CRANICHIDEAE: SPIRANTHINAE

This primarily tropical American subtribe comprises 28 genera. In Australia only one species occurs belonging to this subtribe.

Spiranthes sinensis (syn. **S.australis**)
The species is pollinated by small native bees that probe deeply in the nectariferous flowers. Forms with short-lived flowers, which do not open widely and that are also known as *Spiranthes australis*, are self-pollinating (Jones 1988).

Coleman (1931), in discussing the pollination, mentions Fitzgerald who held the opinion that the species self-pollinates, because he could not find a trace of parts essential for the intervention of insects, but Darwin already commented that this should be examined more carefully because nectar glands are present. The pollinary mechanism of *S.australis* is almost exactly like that of *S.autumnalis* described by Darwin. Coleman found that in spite of the unusual high percentage of seed set the species is entirely dependent for pollination upon the intervention of insects, and that self-pollination is not possible, at least not in young flowers, from the position of the pollinia with regard to the stigma (which rises in older flowers and recedes from the labellum, thus enlarging the entrance). It may however be that Fitzgerald and Coleman studied different forms.

Coleman saw small Halictine bees as well as larger bees visiting the flowers. Some flowers, however, are pollinated by night-flying moths. The flowers are protandrous by means of the *androecium* maturing earlier than the gynoecium, so that the older flowers are pollinated by pollen from the younger flowers. In *S.sinensis/australis* the pollinating bees observed by Coleman did not work as systematically upwards in the spirally arranged inflorescence as in *S.autumnalis*, in which the protandry thus promotes cross-pollination. In older flowers the insect cannot release the pollinia.

Adams & Lawson (1993) mention *Apis mellifera* en *Coelioxys albolincata* as probable pollinators.

Hatch (1954) states that *S.lancea* (=*S.sinensis*) from Eurasia has evolved from *S.aestivalis* and that in the tropics the flowers open fully and are cross-pollinated, whereas in south-east Australia the flowers tend to become progressively self-fertile (eventually cleistogamic) and open only at the tip. By flowering the plants under glass it is possible to make the flowers open fully.

CRANICHIDEAE: PACHYPLECTRONINAE

A genus of two spurred species, *Pachyplectron*, from New Caledonia, about the pollination of which nothing is known.

ORCHIDOIDEAE

DIURIDEAE: CHLORAEINAE

About 80 species within six genera, mostly from South America, but with representation in New Caledonia in one genus only: *Megastylis*. Pollination not known.

DIURIDEAE: CALADENIINAE

About 120 or more species of terrestrial orchids in 10 genera from Australasia and parts of tropical Asia. Part of the genera have definite column wings, whereas other genera have not. Most data on pollination are about *Caladenia*, indicating sexual attraction of, or pseudocopulation by, male thynnid wasps. Each thynnine-pollinated orchid (see also Drakaeinae) can, according to Bower (1992) be expected to have one primary pollinator to which it is adapted in terms of *kairomone* composition and physical structure. However, it is likely that in genera which are still developing, species may change fast in either direction by regular adaptation to different pollinators available.

Pollination by winged male ants is observed in *Leporella fimbriata*. *Eriochilus* and some species of *Caladenia* are pollinated by bees, wasps or syrphid flies that are apparently attracted by false advertisement.

According to Dafni & Bernhardt (1990) the pseudocopulatory systems are derived from food mimic systems of ancestral species. The shift from floral mimesis to sexual mimesis reflects a canalisation of a floral syndrome emphasising lower diversity of potential pollinators.

Genus **Adenochilus**

The genus *Adenochilus* consists of two species, one endemic to Australia and one to New Zealand. The flowers of the Australian species *Adenochilus nortonii* are allogamous, but details concerning its pollination are unknown. David Banks (Internet 1998), however, observed small black flies in numbers, hovering around flowers. Two out of 15 flowers set fruit.

Genus **Burnettia**

The genus contains only one species, which is endemic to south-eastern Australia. *Burnettia cuneata* is a leafless saprophyte and is not autogamous. Details unknown.

Genus **Caladenia**

There are probably about 120 species, of which only 70 were described in 1994 (Jones 1994). The genus is endemic to Australia with the exception of 4 (-10, in 1998!) species from New Zealand, one of which extends in New Caledonia, Indonesia and Malaysia.

Morphology

All species are terrestrial with a single basal leaf, usually hairy. All have a labellum that is hinged to the base of the column by a small claw. The labellum is often three-lobed and the upper surface of the lamina is decorated with conspicuous calli, either sessile or stalked and of various shapes, colours and arrangements.

There is no viscidium and no caudicle, the rostellum is poorly developed and the pollinia lie free in the anther cells.

The anther flaps hang like curtains in front of the pollinia and an insect pushing into the flower holds them shut. On backing out again it draws the curtain flaps apart and permits the pollinia to drop down. They attach to the back that is made sticky by the insect passing the stigma (Erickson 1965).

Pollinators

Brightly coloured species are pollinated by a range of nectar-seeking insects, principally native bees, although some species are favoured by wasps, and beetles, flies and ants visit the flowers transferring pollinia with a lesser degree of success (Bates 1985a). The majority of large-flowered species (spider orchids) is pollinated by sexually attracted thynnid wasps. Although this is mostly called pseudocopulation, the term is in general not appropriate because the male wasps often try to take the wingless females first to another place, a nectar source, before copulating. The picking up of the female is normally only the first step in the mating process, although some females are transported *in copula*. Mating pairs feed on local flowers, usually nectar-producing species of *Leptospermum*, *Eucalyptus* and *Vertecordia*. The pairs can be collected, but it is difficult to assign a collection to a known orchid pollinator (Stoutamire 1981).

The females are usually smaller than the males. Sometimes the males carry food to the female before copulating. In other species both sexes feed on the nectar source where copulation takes place. Females prepare for mating by climbing on low vegetation and assuming a vertical presenting position ('calling'). The females may rub the rear pair of legs over the abdomen, perhaps spreading sexually attracting secretions from the abdominal glands. At the same time their open jaws may be releasing mandibular gland products. Male wasps fly a zigzag flight path upwind toward a presenting female. When the males approach the females, they grasp her, begin to copulate or carry her away to a feeding site.

The male wasps are attracted to *Caladenia* flowers by kairomones (imitation pheromones) that usually are situated in the glandular areas on perianth tips. Often clublike structures of these tips are covered with ovoid unicellular trichomes and they function as osmophores. They are odourless to humans. The labellum often has a recurved tip with rows of dark calli that give the general impression of an insect resting on the petal surface. These structures are believed to imitate the wingless female thynnid wasps. However, Stoutamire (1981) observed the female of the pollinator of *C.dilatata* and found little resemblance to the flower structures involved. Erickson (1965) supposes that the 'palisade' of calli may be a precaution that keeps out visitors too small to accomplish the flower's fertilisation.

Some small-flowered food-deceptive species are pollinated by small bees and flies. Even beetles have been reported. Some species are self-pollinating or have self-pollinating forms. Catling (1990) mentions as such: *C.bicalliata*, *C.catenata* var. *minor*, *C.latifolia* and *C.pusilla*.

Food-advertising species bear relatively small non-mobile labella lacking nectaries. Flowers of such species are usually brightly coloured, in contrast to the dull ones that rely on sexual attraction. They have fragrances that vary from sweet to musky, with some species being odourless to humans. The sepals are often the source of the odours and usually bear long multicellular trichomes. Calli of contrasting colour on the labella may produce secretions for short-tongued visitors, or they may be chewed, although there is little evidence for mechanical damage after visits. They primarily appear to function as nectar guides (Stoutamire 1983).

In some species the labellum calli are yellow-topped and probably represent pseudopollen (Bates 1985a). Several species seem to be floral mimics of the small colourful bush flowers.

Bates (1984b) states that all of the bees he collected on the colourful bee-pollinated flowers of the section *Eucaladenia* were male, and in some cases (e.g. *C.congesta*) were seen to clasp the labellum and either shake it vigorously or exhibit probing movements of the abdomen.

Pollination groups

According to Bates (1994-1995) some groups of species produce no kairomones and are either bee-pollinated or selfed. Other groups produce not very specific kairomones (*C.patersonii* complex) and attract both wasps and bees, and a third group or set of groups, always with red tipped labella, attracts wasps only with highly specific kairomones. The latter groups may attract more than one species of wasp, but in general the attraction is weak for all but one species, which is the only one to effect pollination. In 90% of *baiting experiments* wasps are full responders only for the actual *Caladenia* species of the locality involved, so presumably the distributions of the orchids and pollinators parallel each other. Multiple attraction of wasp species seems very low and specificity high. So far, sympatric major responders have been identified for eight *Caladenia* species, but only three species have shown multiple attraction.

Experiments

In baiting experiments only a small shift of less than 100 m is enough to change the interest of a local wasp population. A new location for the pheromone (or kairomone) stimulates the males into believing that a female has just emerged and is therefore available. And since obviously the fastest wasp will be rewarded they will appear within 1-3 minutes, if present. This behaviour is the basis for the success of *choice experiments.*

Hybridisation
Natural hybrids have been found between several species, including those between food-advertising and sexually attractive species, although most often in sections sharing the same pollination syndrome. There is a suggestion that increased hybridisation occurs at disturbed sites, where factors that would normally limit cross-pollination are altered or removed (Backhouse & Jeanes 1995).

Small-flowered species

Caladenia alba
Pollination by a syrphid fly (*Syrphus* sp.) is suggested by Uhlherr (1967) and probably already observed by Fitzgerald in 1882, but unsatisfactorily described by him (Rogers 1931).

Caladenia carnea
This widespread and variable species (Plate 136, p. 197) is pollinated by a small bee, a *Trigona* species (Adams et al. 1992). Food mimicry seems to be the basis of the system by means of advertisement with bright colours and fragrance (Adams & Lawson 1993). Many forms are described as different varieties (Jones 1988, Clements 1989) or as different species (Clements 1989).

Bartareau (pers. comm.) observed pollination by *Homalictus* sp.

Caladenia cucullata
Pollination by a wasp species (Bates 1985a).

Caladenia minor
C.minor (*C.catenata* var. *minor*, *C.carnea* var. *minor*) from New Zealand is either autogamous or cleistogamous. The Australian plants of this variety are also self-pollinated, but the type variety is bee-pollinated.

Caladenia congesta
Pollinated by a *Hylaeus* bee species, according to Bates (1982, 1985a). The dark labellum calli resemble the black body of the bee. All bees captured were males, but sexual attraction is unlikely to be involved. The syndrome involved is probably *(general) food source mimicry* (Adams & Lawson 1993).

Caladenia deformis
Pollinated by a bee, *Lasioglossum* (*Halictus*) *subinclinans.* Food source mimicry by means of fragrance and bright colour, falsely suggesting the presence of nectar (Rogers 1931, Adams & Lawson 1993). Matthews (in Bates 1984a) noted that the bee used the pubescent calli on the labellum as a foothold to help it push far enough into the flower to collect the pollinia.

Caladenia dimorpha
Fitzgerald observed visitation by a house-fly on flowers placed in a room. The fly, after alighting on the lip, 'was carried by its spring against the column and, becoming entangled in the gluten of the stigma, and struggling to escape, removed the pollen in its masses from the anther and smeared them on the stigma'. He incorrectly believed such large insects to be the principal agents of fertilisation in the genus (Rogers 1931). He was also wrong in supposing some degree of 'spring' (Erickson 1965). The curved labellum is merely balanced and may be swung away from the column by the slightest weight on its tip, or swung over against the column if that weight be shifted further towards the base of the labellum. She describes it as a one-man see-saw. This device with the shifting point of balance ensures that the insect enters without contacting the stigma and anthers, and yet cannot leave without touching both in the right order.

Caladenia flava
Pollination by a bee, *Paracolletus nicholsonii*, is reported by Erickson (1965). She watched two sizes of bees at work. The smaller male bees alighted on the front sepals and circled the tube formed by the bases of the column and labellum. They mounted the projecting tip of the labellum but did not attempt to climb over the palisade of the calli that barred the way into the tube. 'These smaller bees were persistent, frustrated visitors who received nothing for their pains'. The larger female bees crawled directly on to the labellum and surmounted the palisade, widened the opening forcibly and 'supped' from the bottom of the tube (but since no nectar is provided (Bates 1984a) they received no reward either). Erickson (1960) noted that the large sentinel calli of this species prevent insects which are too small from entering. The taller calli

nearest the column also serve to lift the insect up to the correct height to contact stigma and pollinia. In areas where the local race of pollen vector has large males these calli may be short. Where the wasps are small the calli may be large (Bates 1984).

Other species with a row of large calli and a labellum-tube are *C.paniculata*, *C.unita*, *C.latifolia* (see below) and *C.reptans*.

Caladenia latifolia

Cady and Rotherham as well as Wallace have reported beetle pollination (in Armstrong 1979). It appears to be a derived condition, i.e. represented in taxa, the ancestors of which were pollinated by more highly developed insects. Bates (1985a) mentions self-pollination.

Large-flowered species

Caladenia barbarossa

Dragon orchid: Sargent (1907, 1918) reported a large black wasp to be the pollinator. 'The lip is so arranged that when the wasp alights on the lamina it swings forward towards the centre of the flower and the movement brings the wasp's back forcibly into contact with the stigma. The stigmatic fluid makes the back sticky and the sticky patch contacts the disc-like pollinia. The disc is withdrawn from the anther as the wasp backs out. The pollinia then fall on the patch and become fixed'.

Sargent adds to this description the following remarks: 'The central labellar appendage, which the wasp seizes with its mandibles, seems a true gland, secreting nectar so slowly that crystallisation takes place practically simultaneously. These crystals appear to be sucrose'. 'I have never yet found them (= the peg-like calli on the labellum) gnawed by insects (excepting cases where a large portion of the labellum has been damaged – evidently not by a pollinator): I believe they are simply 'licked'.'

It is clear that Sargent did not understand the true nature of the wasp's behaviour, but he is aware that he does not know everything: 'My investigations are at present far from complete'. Stoutamire leaves open the question of whether Sargent understood the wasp's behaviour, writing that he 'either did not recognise what he saw happening or did not wish to admit it in print'.

Stoutamire (1974, 1975, 1979, 1983), who is mentioned as the discoverer of sexual attraction in *Caladenia* (but see Curtis 1964 and Coleman 1930), observed the same pollinator, which turned out to be *Thynnoides bidens* (Plate 137, p. 197), and recognised its behaviour as sexually motivated. 'The visiting wasp alights on the labellum, curves the abdomen under the tip in probing movements, and then grasps the mobile labellum, attempting to carry it away. The wasp is catapulted into the column opposite the labellum and the pollen is attached to the thorax by stigmatic secretion'.

The labellum imitates a female thynnid wasp and the male attempts both to lift the labellum off the flower and to copulate. It is the struggle that leads to the wasp falling forward under the column and thus effecting uptake of pollinia and/or pollination.

It should be mentioned that Edith Coleman (1930) saw 'a small bee, or fly' visit the flowers of some plants sent to her, and stated that their actions obviously were 'actuated by similar motives to those of the insects concerned in the pollination of the various species of *Cryptostylis*'. She writes: '...from certain movements it was clear to me that the insect was not interested in edible tissue, nor

Plates

128. Flower raceme of *Cremastera appendiculata* var. *variabilis*. Photo N.Sugiura (p. 188).
129. A queen of *Bombus diversus diversus* with pollinaria from *C.appendiculata* var. *variabilis*. Photo N.Sugiura (p. 188).
130. A queen of *Bombus diversus diversus* visiting the flowers of *C.appendiculata* var. *variabilis*. Photo N.Sugiura (p. 188).
131. A syrphid fly pollinator *Melanostoma scalare* visiting the flower of *Oreorchis patens*. Photo N.Sugiura (p. 188).
132. *Melanostoma scalare* with pollinaria of *Oreorchis patens* on their thoraces. Photo N.Sugiura (p. 188).
133. Flowers of *Bletilla striata*, showing the flower structure. Photo N.Sugiura (p. 188).
134. Copulation of *Tetralonia nipponensis* on the labellum of *Bletilla striata*. Note the pollinia on the thorax. Photo N.Sugiura (p. 189).
135. A female *Tetralonia nipponensis* with attached pollinia of *Bletilla stiata* on the body surface. Pollinia partly removed by grooming with the legs. Photo N.Sugiura (p. 189).

128	130
	132
129	133
131	134
	135

was the close grip of the body on the labellum necessary for the purpose of merely resting. There could, to my mind, be only one possible reason for the visit'. From this and following remarks about the 'need of reproduction', as well as from her earlier work on *Cryptostylis* to which she refers, it is perfectly clear that she should be recognised as the discoverer of sexual attraction and mating behaviour in *Caladenia.*

Stoutamire reports (in Bower 1992) that *C.barbarossa* attracts males of the same wasp as in *C.falcata* in baiting trials. Hybrids between the two, however, are unknown.

Caladenia bicalliata
Some forms of this species are cleistogamous and the flowers never open (Bates 1984).

Caladenia cairnsiana
This 'zebra orchid' is pollinated in much the same way as *C.barbarossa* by an unidentified thynnid wasp (Stoutamire 1979, 1983). They try to mate with and remove the pseudofemale on the labellum. The wasps fly low above the ground and have difficulty in locating flowers above the flying level.

A lip may be partially torn from the claw by competing males.

Caladenia caudata
In Tasmania the species is reportedly pollinated by *Lophocheilus villosus* (Bates 1994-1995).

Caladenia corynephora
A large black-and-yellow banded wasp of undetermined genus pollinates this species through sexual attraction (Stoutamire 1983). Bates (1994-1995), however, names *Lestricothynnus modestus* as the pollinator in Western Australia.

Caladenia dilatata
Coleman (1930) reports the pollination of *C.dilatata* var. *rhomboidiformis*, to be similar to that of *C.barbarossa.* 'It is visited by a large ichneumon fly, the body of which is banded with yellow'. This appears to be *Lissopimpla excelsa*, the same ichneumonid wasp that pollinates several *Cryptostylis* species by pseudocopulation (Curtis 1964).

'Spider orchids', as these and other orchids of the similar type are called, may trap blowflies hopelessly (e.g. when placed in vases), a habit which has caused the wrong belief that they feed on flies like spiders do.

Erickson (1965) believes the long and slender calli probably exert a slight catapulting effect on insects that grip the basal calli.

Stoutamire (1975, see also Bower 1992 and Bates 1994-1995) reports that *C.dilatata* attracts different wasps in different geographical areas. The species complex is in his view constituted of several different geographical races like *C.dilata*, *C.falcata*, *C.stricta* and *C.gracilis.* Each of the geographical races appears to have a local, specific pollinator.

According to Jones (1991, also Carr 1991), however, there are in *Caladenia* several complexes of previous unrecognised species (biospecies in the sense of Mayr) each with a specific pollinator, attracted by different kairomones. *C.dilatata* hybridises with *C.patersonii* (Stoutamire 1983). For more details see Bates (1994-1995) who presents information on 10 species within the *C.dilatata* complex and their pollinators. They attract a broad spectrum of wasp genera including *Aeolothynnus*, *Eirone*, *Guerinius*, *Lestricothynnus*, *Lophocheilus*, *Neozeloboria*, *Thynnoides*, *Zaspilothynnus* and *Tachonymia.* In general these responders and pollinators show specificity in their preference for certain species of orchids.

C.dilatata s.s. in Tasmania are visited by different wasps and investigations there cast doubts on the occurrence of true *C.dilatata* in South Australia.

C.tentaculata, *C.verrucosa*, *C.phaeoclavia*, *C.clavula* and *C.dilatata* have been clearly shown to

136	137	139
138	140	141

Plates
136. *Caladenia carnea*, a bee-pollinated species. Photo N.A.van der Cingel, August 1998, Tavinghy Mountains, Queensland, Australia (p. 195).
137. *Thynnoides bidens* wasp (male) on *Caladenia barbarossa.* Photo R.J.Bates, Jerramungup, West Australia (p. 196).
138. *Caladenia* aff.*tentaculata.* Photo N.A.van der Cingel (p. 199).
139. *Myrmecia* (male) ant pollinating *Leporella fimbriata.* Photo R.J.Bates, Mylor, South Australia (p. 200).
140. *Arthrochilus huntianus*, showing insectiform labellum. Photo R.J.Bates, Victoria (p. 203).
141. *Caleana major.* Photographer unknown (p. 203).

be valid species by means of choice tests (Bates 1994-1995).

A newly described species *C.rileyi* attracts a new species of *Thynnoides*.

Caladenia doutchae
Stoutamire (1975, 1983) reports probable sexual attraction by an unidentified thynnid wasp as a mechanism for its pollination. Bait experiments show attraction of male wasps.

Caladenia filamentosa
The variety *tentaculata* of this species is reported to be pollinated by a small native bee *Euryglossa rejecta* (Erickson 1965). Bower (1996) suggests that only thynnines (related to *Ariphron = Chilothynnus*) are actively attracted. Although there is no evidence of sexual behaviour of wasps towards flowers of this species, this makes sexual deception more likely than floral mimicry.

Caladenia fitzgeraldii and **C.montana**
Choice experiments (Bower 1996a) demonstrated that these morphologically similar species are representing different species.

Caladenia huegelii
The species complex utilises a sexual attraction mechanisms (Stoutamire 1983). The flowers attract males of a black thynnid wasp, *Zaspilothynnus* sp. A male wasp flies upwind toward a flower, crossing and recrossing until it is less than a meter downwind of it. The search flight, obviously guided by fragrance, then changes to a rapid and irregular circling. Visual cues come into play on approach. The wasp crawls up the labellum, but does not appear to be attempting copulation. The dark tip of the flower probably serves as a visual female equivalent and the wasp is trying to pick it up. Female thynnid wasps are wingless and the dark-coloured insects in calling position on low vegetation contrast with their background in much the same way as does the dark *C.huegelii* labellum spot.

Stoutamire reports hybridisation with *C.patersonii*.

Caladenia integra
Stoutamire (1983) reports a large, swift-flying black and yellow *Guerinius* species (Thynninae) to be attracted and repeatedly approaching the flowers, but this wasp leaves quickly without alighting and is of inappropriate size and weight to be considered the probable pollinator.

Caladenia leptochila
Bates (1994-1995) points to the curious fact that despite 15 years of observation and testing no pollinators have been found, although the species is not self-pollinating.

Caladenia lobata
In this 'Butterfly spider orchid', males of *Thynnoides preisii* and *T.bidens* fly the usual zigzag, upwind path, alight on the labellum tip and quickly crawl to the maroon calli lining the labellum base (Stoutamire 1974, 1983). The initial landing pulls the labellum down, exposing the long calli. As the insect crawls upward, the balance changes and labellum plus insect are flipped under the column. Pollen is attached to the thorax. Flowers are visited repeatedly by the male thynnids, which often cruise in pairs, one insect following another. The wasps appear to develop avoidance reactions after the initial encounter with the flower, subsequent visits being more hesitant and less likely to result in pollination. In his 1983 article Stoutamire no longer refers to *T.preisii*.

T.bidens is a shared pollinator with *C.barbarossa*. A rare hybrid between the two species is known. Stoutamire reports also natural hybrids with *C.patersonii*.

Caladenia macrostylis
The species is similar to *C.multiclavia* and replaces this geographically westward, but it is vertically oriented. It is pollinated by a *Tachonymia* species.

Caladenia magnifica
In Victoria the species is pollinated by *Tachonymia ? moorans* (Bates 1994-1995).

Caladenia multiclavia
The labellum in this species has a horizontal surface over the vertical claw while the column base is horizontal, perpendicular to the ovary axis. The labellum rocks and trembles with air movements. The insectlike appearance is enhanced by this movement. Broad lateral column wings form a cup into which the searching wasps are dropped after alighting on the labellum. An undetermined species of *Tachonymia* is the pollinator (Stoutamire 1981, 1983). Pollen is attached to the wasp thorax.

Caladenia patersonii
This species, according to Stoutamire (1975) attracts a variety of syrphid flies and (unidentified) bees to the flowers, which have an odour suggest-

ing fermentation. He found a population of *C.lobata* in the Stirling Mountain range with fresh flowers, including also late flowering *C.patersonii* and a supposed hybrid swarm of these two species. *C.lobata* is a species pollinated by pseudocopulating wasps, whereas *C.patersonii* is not. Different species using different attracting strategies are thought to be able to produce hybrids if the pollinator responds to both strategies: male thynnid wasps are nectar feeders and pseudocopulate as well. Normally, however, hybridisation is restricted by flower phenology and geographical isolation.

A parallel situation is observed between *C.patersonii* and *C.dilatata* in which a cross has been reported.

Dafni & Bernhardt (1990) describe the pollination as an intermediate stage in pseudocopulatory systems towards pseudocopulation, intergrading sexual deceit with *nonmimicry deception* (or better guild mimesis). The flowers have a mode of presentation, with lip calli and filamentous sepals and lateral petals, that is typically associated with pseudocopulatory syndromes within the genus. The flowers are, however, pollinated by tiphiid males and other insects of both sexes, including bees and syrphid flies searching for food.

Caladenia phae(o)clavia

Bower (1992) discovered the existence of a second, different taxon within populations of *C.tentaculata.* Choice experiments revealed that this taxon, named *C.phaeoclavia*, attracted the male wasp *Lophocheilus anilitatus*, trying to mate with the flowers and collecting pollinia, whereas *C.tentaculata* attracted three species of *Thynnoides*. In 1996 Bower discovered that both *C.phaeoclavia* and *C.parva* attract *Lophocheilus analitatus*, which fact suggests synonymy.

Caladenia reticulata (complex)

This species complex (*C.calcicola*, *C.cardiochila*, *C.reticulata*, *C.rigida*, *C.richardsiorum*, *C.stellata*, *C.valida*) has been investigated by Bates (1994-1995). Five species of *Phymatothynnus* were attracted to species of the complex. *P.victor* is the major responder to and pollinator of *C.reticulata. P.nitidus* is pollinator of *C.stellata. C.valida* attracts *P.pygidialis*. Other species also attract *Phymatothynnus* species and it appears that many cryptic species will be detected in further research.

Phymatothynnus is a genus with many cryptic species as well (for more details see Bates 1994-1997).

Bickerton (1997) studied the pollination of *C.rigida.* The flower has two glands at the base of the column which are thought to be pseudopollen or the source of kairomones. Bickerton caught five species of insects, a probably fruit-predating beetle, two native bees (among which a *Homalictus* sp.), an introduced honey bee, and a native wasp. Native bees were observed pollinating. The wasp, a male thynnid, was exhibiting mating behaviour. Early season wasps may be naïve. The species is endangered by loss of habitat and by herbivory, probably by kangaroos.

Caladenia radiata

Stoutamire observed a large black wasp approaching this orchid, but the insect was not captured. Sexual attraction is therefore indicated and the well developed sepal clubs appear to function as osmophores (Stoutamire 1983).

Caladenia roei

This species, concurrently flowering with *C.doutchae,* is thought to be pollinated by unidentified pseudocopulating thynnid wasps of a different, slightly larger species than in the latter sympatric orchid (Stoutamire 1975).

Caladenia stricta

Caladenia stricta (syn. *C.dilatata* var. *stricta*) is pollinated by a thynnid wasp (Bates 1987).

Caladenia tentaculata

The species is probably a complex of previous unrecognised species (Plate 138, p. 197), each of which has a specific pollinator. Such morphologically very similar biospecies differ in their kairomones. Bower (1992) split off one taxon as *C.phaeoclavia* (see above). In populations of *C.tentaculata* he found the flowers to attract one or more of three species of *Thynnoides*, i.e. *T.pugionatus*, *T.rufithorax* and *T.fumipennis*.

Initially five names of Thynnoides were assigned to the pollinating specimens, but these have been reduced to three by synonymy.

See also above under *C.dilatata*.

Light (1998) mentions that recent studies of this species have revealed that pollinators within a local group of orchids tend to avoid visiting a second flower by 'once deceived-twice shy'-behaviour, but less so between distant groups. Such counter-deception behaviour would lead to wider dispersion of pollen. The source of this observation is however not mentioned.

Caladenia tesselata
Fitzgerald, around 1882, observed pollen on stigmas together with scales from a cocoon of a dipteran insect and believed this to have been from the pollinator, emerged from the shelter of the dorsal sepal. He believed this method of fertilisation perhaps to occur frequently, but adds that such a method 'would hardly be conjectured if not observed' (Rogers 1931).

Caladenia uliginosa
Bates (1989) reports to have caught a pergid sawfly on a flower of *C.uliginosa*, but considers it unlikely that it was the normal pollinator.

Genus Elythranthera

This genus is limited to two species from south-western Australia. The flowers of both species are pollinated by small native bees (Jones 1988). This is probably a case of food mimicry (Buchman 1983). Overlap in flower characters with small blue-flowered *Caladenia* species leads to *intergeneric* hybrids (see also *Glossodia*).

Genus Eriochilus

The genus contains six endemic Australian species. According to Jones (1988) the flowers of *Eriochilus* are pollinated by small bees while they probe with extended mouth parts between the tightly oppressed column and labellum. Erickson (1965) mentions *Hylaeus dorothae* as pollinator of *Eriochilus dilatatus*. She describes the effectiveness of the small native bees. She found the stigmas of 40 out of 42 flowers 'filled to the brim' with transferred pollinia. 'The unbroken clusters had been placed inside the narrow openings as neatly as sailors of a past generation placed tiny models of ships inside bottles'.

Cross-pollination is promoted by a mechanism in which the anther flaps shield the viscid discs from contact as the bee enters, but when it withdraws they ensure contact.

Experiments by Rowe & Calder (1988) prove *E.cucullatus* to be dependent on insects for pollination, although it is selfcompatible. Natural pollination varies from zero to around 70%. The presence of a strong sweet odour suggests false advertising and the white colour suggests pollination by an evening or night flying insect searching for nectar.

Genus Glossodia

Two species are known, *G.major* and *G.minor*, both endemic to eastern Australia. Flowers of both species are pollinated by small native bees (Jones 1988).

Stoutamire (1976) includes them in the group of food-offering species which, however, do not produce liquid nectar and are characterised by unusual (blue) colours. In my opinion this group could better be called food advertising instead of food-offering, since mimicry is not excluded.

The flowers of *G.major* are fragrant by day.

According to Dafni & Bernhardt (1990) some *Caladenia* species overlap in floral presentation, column morphology, and labellum position, leading to disjunctive, intergeneric hybrids.

Genus Leporella

The genus contains only one species, which is endemic to southern Australia: *Leporella fimbriata* (Fig. 41). The unique orchid has rather inconspicuous flowers.

Pollinator
The flowers are pollinated by male flying ants of the species *Myrmecia urens* (Plate 139, p. 197) while they are attempting to copulate with the labellum of the flowers.

The species has been widely investigated.

Bates (1979) was the first to record observations of pollination by winged male ants and suggested sexual attraction.

Woolcock (1980) reported a hymenopteran, very close to the Ichneumoneideae, to have been caught, lying torpid on the labellum of a flower after a sudden fall in temperature, with pollinia on its back. This also was probably a male ant.

Paekall et al. (1987) reported pollination by pseudocopulation by winged males of the ant species mentioned. They studied the interaction between orchid and pollinator to test two hypotheses about the apparent rarity of ant-pollination systems.

Ant-pollination hypotheses
The first hypothesis states that pollen can hardly adhere to ants because they are small, smooth and flightless, and that pollen transfer between plants does not occur because distances between plants are relatively large. Only plants adapted to overcome such shortcomings could be pollinated effec-

Figure 41. 1. *Leporella fimbriata*: a = petal, b = lip (after Van der Cingel 1988). 2. *Prasophyllum fimbria*: a = flower from aside, b = flower in front view (after Van der Cingel 1991c).

tively. Such plants therefore should have traits like the formation of a mat or cushion, with interlacing of branches from neighbouring individuals and sessile, small flowers with low nectar volumes and few concurrently open flowers. Such traits would promote outcrossing by forcing ants to move between flowers and plants to gain sufficient reward.

The second hypothesis is based on studies demonstrating reduced viability and germination of pollen, due to secretions from the ants' metapleural glands with antibiotic and inhibitory effects on pollen and/or on fungi spores and hyphes on which the orchid depends. Since metapleural glands are absent in *M.urens* males and since *L.fimbriata* does not possess the above adaptations to ant pollination, the second hypothesis seemed to be supported by the evidence. However (Peakall et al. 1990), the absence of metapleural glands makes no difference since the ant integument has a strong disruptive effect on artificially applied pollen. During pollination the orchid, however, secures the pollen mass to the ant surface with stigmatic secretions that insulate the pollen from harmful substances. The conclusion of the second hypothesis therefore, still stands. Note: see also ant pollination in *Microtis parviflora*.

Pseudocopulation
Adaptations of the orchid to ant pollination include flowering coinciding with the emergence of winged ants, production of attracting pheromones, and modified petals as pseudocopulatory organs.

Typical pollination behaviour sequences begin with a circling flight, followed by a zigzag flight to the flower (Bates 1987, Peakall 1989). The ant alights on the inflorescence stem and quickly crawls to the flower where it adopts a copulation position sideways along the wide labellum, followed by (pseudo)copulatory probing. In this position pollen from the ant's thorax may be deposited on the stigma. On leaving the labellum pollen is usually removed. From the flowers a variable percentage up to a maximum of 70% is pollinated. Pollinator limitation is indicated. *L.fimbriata* is self-compatible, but not autogamous. Seed set observed was comparable between selfs and crossings, mean percentage of viable embryos being 50% and 40% respectively.

The glandular petal tips of the flowers, as in *Caladenia*, are the source of the long distant chemical attractant. The labellum acts only at short range as a stimulus for copulatory behaviour. Visual cues are apparently short range in sexually motivated ants. The mating system in *M.urens* obviously fits the orchid, females probably attracting males to mate on low vegetation. Bates (1987) observed that *Leporella* flowers placed near *Myrmecia* nests were grasped by worker ants after climbing to the flowers and were carried to the nest. Other flowers did not elicit this response.

Reproduction
L.fimbriata propagates vegetatively by underground tubers from dense but discrete patches of plants. Clones may extend up to 16 m^2. Most pollinator movements are within clones. This behaviour, combined with self-compatibility of the orchid, suggests high selfing rates. However, outcrossing by ant pollinators, extensive dispersal by wind of the small seeds, and elimination of selfed progeny reduce the effect of selfing and keep genetic variation at a high level. Peakall & James (1989) investigated the genetic structure of four populations and found genetic variability (heterozygosity) consistent with *panmixis*.

Genus **Lyperanthus**

The genus consists of about five species, four of which are endemic to Australia (*L.antarcticus*, of New Zealand, however, was lately renamed as *Waireia stenopetala*). Australian species are deciduous, terrestrial orchids with large obovoid tubers.

The flowers of *Lyperanthus nigricans* are pollinated by a native bee which forces a passage in between the labellum and column. About the pollination of the other species nothing is known. Dafni & Bernhardt (1990), referring to Stoutamire (1975, 1983) who does not however mention *Lyperanthus*, suggest the possibility of pseudocopulation. The labella appear intermediate between motile and nonmotile systems, and according to them pseudocopulation in this genus (and two other genera) may combine wasps from different families or several genera of tiphiids with varying modes of courtship.

Genus **Rimacola**

The genus contains a solitary species endemic to south eastern Australia and is closely related to the foregoing.

Rimacola elliptica is pollinated by a small black wasp.

DIURIDEAE: DRAKAEINAE

This subtribe, consisting of over 30 species in five or six genera (depending on inclusion of *Paracalaeana* in *Calaeana* or not) was formerly included in the Caladeniinae (Dressler 1981), but because of the motile lip of which a part mimics a female thynnid wasp, with, as a consequence, pollination by pseudocopulation by male thynnid wasps, the group has been separated (Dressler 1993). It is now thought that pseudocopulation could have evolved independently in the two groups.

Genus **Arthrochilus**

The genus contains six Australian species, two of which are undescribed. Two of the known species extend into New Guinea.

Pollination

A. huntianus (formerly *Spiculaea huntiana*; Plate 140, p. 197), the elbow orchid, is pollinated through pseudocopulation by male thynnid wasps of the genus *Rhagigaster*, as was first observed and photographed by Rotherham in 1949 (Rotherham 1967, 1968 and Northen, 1971b). The wasp grasps the insectiform labellum with its head near the glands at the top of the lip. The momentum of the insect landing on the hinged labellum causes it to be swung up towards the column and the thorax is directed between the claw-like appendages of the column where it contacts the pollinia, which then adhere on the back of the thorax.

Probably all species have the same kind of mechanism (Jones 1988).

Genus **Caleana**

A genus restricted to Australia. The orchids are known as 'duck orchids' or 'flying ducks' because of the appearance of the flowers. Five species are included in the genus, two undescribed.

The species with a glandular labellum were for some time separated into *Paracaleana*, but recently brought back into *Caleana* (Clements 1989).

Parasitic thynnine wasps (Typhiidae: Thynninae) pollinate (Stoutamire 1983). No details are reported about the former *Paracaleana*. Catling (1990) mentions *Paracaleana* (=*Caleana*) *minor* as self-pollinating. Cady (1965) mentions *P.*(=*C.*)*sullivani* and *P.*(=*C.*)*nublingii* to contain no pollinia and to produce abortive anthers. Jones (1977, 1988) considers both taxa to be freak apomictic developments of *P.minor*.

Caleana major

Fitzgerald noted that the lip was movable (Plate 141, p. 197) and experimented with lady beetles forced to climb the plants, in order to find out about the pollination. He saw that the labellum flipped back into the bowl formed by the column wings, thereby imprisoning the insect for some time until the irritable labellum had retaken its original position. The beetles escaped without taking away the pollinia. A friend of Fitzgerald observed a large 'fly' in the flower.

Cady (1965b) observed a male sawfly, probably wrongly identified as *Lophyrotoma leachii*, landing on the labellum of *C.major* with the head facing the narrow end of the labellum lamina and then, apparently due to its own forward momentum, the sawfly was forcibly taken by the labellum into the broad column wings in a position with its thorax pressed against the pollinia masses. It was held thus for approximately 90 seconds before it could free itself.

During the struggling some of the pollinia masses were scattered over the stigma, resulting in self-pollination. The sawfly then took the rest of the pollinia masses with it on leaving. The reset time of the labellum depends on the time of day (Cady 1965a).

Stoutamire (1983) suggested sexual attraction on the basis of the colour scheme of the flower (maroon and green).

Bates (1989), over a period of 20 years of observations of populations in south Australia, located only one seed capsule. In eastern Australia he experimented with transferred flower spikes near populations of *C.major* and after much patience observed pollination by a pergid sawfly, a *Pterygophorus* sp. (Plate 142, p. 212). He observed the same kind of struggling behaviour as Cady during the 60 seconds before the insect freed itself. A sawfly was captured at a second flower. A third individual about 100 m further along the track had traces of pollinia on its thorax and was seen to pollinate. There is no obvious source of pheromones on the flower, but from the downwind approach it seems that some volatile chemical is released by the flower.

Flowers of *C.major* demonstrate extreme longevity, a fact that in view of the rarity of pollination is understandable enough.

Genus Chiloglottis

Twenty-nine species (Bower 1995), confined to Australia and New Zealand.

Forms of *C.cornuta* are known to be self-pollinating. Catling (1990) mentions also *C.diphylla* as autogamous. All other species are pollinated by male thynnine wasps species which try to copulate with calli on the labellum. These calli seem to mimic female wasps in some respects. Due to taxonomical problems within this wasp genus, which has many still undescribed species, it is not clear yet whether there is always a 1:1 relationship between orchid and pollinator, but this does not seem unlikely in most cases.

There is also a possibility of geographical replacement of pollinators in some widespread *Chiloglottis* species (Stoutamire). However, so far this has not been found in any species tested, and usually the same wasp was found to be the pollinator throughout the entire range of the orchid. According to Dafni (1984) *Chiloglottis* resembles the European genus *Ophrys*, in that the pollinator reacts only to a specific geographical race, while other races have different pollinators. The complex of different biological races is supposedly undergoing speciation. As in *Ophrys*, recent investigations (Bower 1993-1997) lead to the conviction that more biospecies can be distinguished on the basis of attracting different pollinators (Bower 1995). It is found that in baiting (choice) experiments, in most cases more than one wasp species is attracted (Bower 1995), but this does not imply more than one pollinator species. One species, *C.pluricallata*, attracted even 11 species of wasps. A wasp species must occur within the distribution and habitat of the orchid.

The behavioural responses of the visitors to flowers vary enormously within and between wasp species. Most visiting wasp species respond only weakly and are not likely to effect pollination (*minor responders*). Among *confirmed pollinators*, the proportion of visitors which complete all the behavioural steps leading to removal of pollinia is quite low (around 5%). Some *major responders* were found to occur outside the range and habitats of the orchid to which they respond in tests and cannot be pollinators (*allopatric major responders*). It is thought that there is overlap in the sex pheromones of different wasp species. Probably there are a limited number of distinctive pheromone systems among *Neozeloboria* wasps, with several wasp species sharing some systems. Several *Chiloglottis* species may belong to some pheromone systems. Bower identifies 10 such systems.

Thus far, *sympatric major responders* have been determined for 19 *Chiloglottis* species. Sixteen of these have only one major responder, the other three two each (Bower 1996).

Baiting experiments (Plates 15-20, p. 28) can also reveal the presence of *cryptic species* which are difficult to distinguish morphologically and might be dismissed as variants within other species. Most of these races superficially (i.e. to the human eye) resemble an insect and do not produce nectar. The olfactory stimulus is the dominant one. It is the sole factor in long-range attraction and works on innate pollinator behaviour. Optical cues are secondary and act primarily at close range, while the tactile stimulus is important after the pollinator lands on the labellum.

Investigations are complicated by the fact that not only orchid species, but also their thynnine wasp pollinators may be cryptic or sibling species, representing hidden biodiversity (Bower 1997). This can be detected by exposing flowers in field experiments (e.g. in sympatric, allopatric and sequential choice tests). This phenomenon appears to be more widespread and may be typical for pollinators which rely mainly on chemical signals for mate recognition.

Note: Some of the results described below are preliminary. They are taken from reports that are not for publication. It is therefore the sole responsibility of the author to use these data.

Chiloglottis diphylla

Small and large forms of this species are attracted mainly attracted by the same two wasp species of the genus *Neozeloboria*. On the basis of experiments this species should be considered distinct from the other species of the *C.reflexa* group (see below).

Chiloglottis formicifera

Dockrill (1956) was the first to report pollination by pseudocopulation in the genus. He observed, during a sunny break in a rainy day, several ichneumons struggling to get on top of each other on labella of *C.formicifera*. In the first observation three wasps arrived shortly after each other and the head of no. 3 was pushed up under the anther and the pollinia adhered to the insect's head. Almost instantly it flew to another flower about 25 yards away, at which another wasp was already in position. In the struggling that followed the head of no. 3 was momentarily pushed against the stigma and the pollinia were transferred to it. In a manner comparable to several misnomers in the European genus *Ophrys*, the name *formicifera* was wrongly chosen long before the pollination mechanism was discovered.

Stoutamire (1975) mentions a *Neozeloboria* species (Thynninae) to be the probable pollinator. It was observed to respond to the pheromones of the orchid (see also Bower 1992).

Bower (1994) considers *C.formicifera* s.l. (Plate 20, p. 28) as a group of five related species on the basis of baiting experiments: *C.formicifera*, *C.trullata*, *C.truncata*, *C.platyptera* and *C.trapeziformis* (see below). *C.formicifera* is a widespread, variable species, occupying a diverse range of habitats, and may contain unrecognised taxa. Indications for two cryptic species were found in 1996. *C.formicifera* and *C.platyptera* are pollinated by different *Neozeloboria* species. *C.* aff. *formicifera*, possibly of hybrid origin, seemingly shares pollinators of both species.

C.truncata attracts a *Neozeloboria* nr (= near) *cryptoides*. See also Plates 17-20 (p. 28) taken by the author in 1998 during a visit to Appin, Australia.

Chiloglottis gunnii

This species is geographically widespread and ranges from New South Wales to Tasmania. Geographic races differ in several characteristics like peduncle length, flower colour and position of calli. Stoutamire (1975, 1981) observed *Eirone* wasps in Tasmania. When flowers of several races were used as a bait Stoutamire found the *Eirone* to visit only one Tasmanian race, whereas three morphologically different races in Tasmania and New South Wales each attracted different wasps. The species *C.cornuta*, flowering concurrently and sympatrically with two Tasmanian *C.gunnii* races, did not attract any wasp and appears to be self-pollinating.

In a 1996 paper Bower includes *C.valida* and *C.pluricallata* in the *C.gunnii* complex, but see below for details about these species (*C.valida* complex).

Chiloglottis palachila

This is a group of populations from the northern and central tablelands of NSW. Baiting experiments indicate the existence of two different species: *C.palachila* and *C.* aff. *palachila*. They are probably both pollinated by *Ariphron* species, but this should still be confirmed for *C.* aff. *palachila*. For this species a wasp of the genus *Chilothynnus* is an allopatric major responder and this fact provides evidence for the reproductive isolation of both taxa (Bower 1995).

Chiloglottis platyptera

This species, closely related to *C.formicifera*, is pollinated by a new species of *Neozeloboria* (Bower 1992, 1994) (see above).

Chiloglottis pluricallata

Pollinated by *Neozeloboria impatiens* (Bower 1992, 1994, see also below). Two taxa, *Chiloglottis* aff. *pluricallata* (nrs 1 and 2), attract a completely different array of wasps in baiting experiments and are reproductively isolated from *C.pluricallata* s.s. One of these species (nr 1) shows close kairomonal relationship with *C.valida*. Both species, however, have a different confirmed pollinator and occur allopatrically.

Chiloglottis reflexa and related species

The flowers of *C.reflexa* mimic the appearance of female thynnine wasps. *C.reflexa* mimics *Neozoloboria* nr *proxima* and its sex pheromone. Male wasps attempt mating with the flower.

Occasionally *Neozeloboria proxima* s.s. may respond weakly to *C.reflexa* (and *C.seminuda* as well), probably caused by a slight similarity in pseudopheromones (kairomones) among orchid and/or wasp species (Bower 1997).

Bower (1993-1997) has described baiting experiments with several related species of the *C.reflexa* group, consisting of eight species and differing from each other in number and shape of the labellum calli, the shape of the labellum lamina, and the length and curvature of the lateral sepals and their osmophores: *C.reflexa*, *C.trilabra*, *C.diphylla*, *C.sylvestris*, *C.seminuda*, *C.anaticeps*, *C.sphyrnoides* (all in south eastern Australia) and *C.longiclavata* (north Queensland).

Six wasp species of the genus *Neozeloboria* were attracted to taxa in this group. The results indicate the existence of several distinct species. Some of these are consequently treated separately. In 1996 the probable pollinators of *C.sylvestris* and *C.sphyrnoides* were discovered. It is possible that variation in *C.sylvestris* includes unidentified cryptic species.

C.anaticeps and *C.diphylla* mutually attract two wasp species (*Neozeloboria* ? n sp. 33 and *Arthrothynnus latus*), of which the latter is more strongly attracted to *C.diphylla* and has been observed to remove its pollinia.

Male flight behaviour

Handel & Peakall (1993) used the flowers of *C.reflexa* to investigate mating behaviour. In field choice experiments males prefer to visit flowers that are very low in the habitat, 15 cm, vs flowers that are placed at 55 cm or 105 cm.

Females spend much of their time underground and are not easy to find. Studying of mating behaviour therefore may be greatly helped by this kind of experiment. If the flowers are presented within populations of the wasp the males are easily attracted to the flowers, often in a few seconds.

Throughout the study in the field, males were observed perching on the understorey vegetation up to 1.5 m high. Some copulas were captured on the foliage of trees approximately 2-3 m high. In searching for mates, however, the males focus on a narrow flight zone close to the ground. Mate search behaviour seems constrained by an innate stereotypic response, and any dissociation of the cues prevents a complete response. Males responded to the long-range stimulus even when only high flowers were available, but few landed on the source, suggesting that short-range cues are the most tightly constrained.

A narrow flight zone close to the ground may be typical for thynnine species. Subsequent additional fieldwork with two other orchid species, *C.trilabra* and *C.trapeziformis* (Peakall & Handel 1993) has shown that their thynnine pollinators also prefer low flowers. Stoutamire (1974) too noted preference for low flowers in thynnine wasps pollinating *C.gunnii*. These wasps were unable to locate flowers held above their normal ground level.

Species like *C.gunnii* (see above) and *C.sylvestris* rarely exceed 5 cm in height. These taxa probably mimic thynnine females that call from the ground. The male's behaviour may have selected for floral height in *Chiloglottis*.

Males of *Neozeloboria* nr *proxima* were found to avoid the orchid flowers following an initial visit, despite the continuing presence of the flowers. The same behaviour occurs when females are presented but are prevented from being taken. The data suggest a learning response probably associated with male choice. This behaviour will influence the amount of fruit set and pollinator limitation may be expected, although no data are reported.

Chiloglottis seminuda

According to Bower (1992, 1997) this species is pollinated by a new undescribed *Neozeloboria* species. During baiting trials also other wasps were attracted. At one of the places where Bower tested, the thynnid wasp *Rhagigaster unicolor* visited *C.seminuda*. It displayed vigorous copulatory behaviour on the labellum, but was too large to be an effective pollinator. At another locality a new species of *Eirone* (Thynninae) was attracted, probably an incidental visitor. Similarly, *Asthenothynnus westwoodi* was strongly attracted in a population outside the normal range of the orchid.

Chiloglottis trapeziformis

The species is again pollinated by a *Neozeloboria* species. Stoutamire (1975) found a wasp to respond to the pheromones of the orchid. It was a species of wasp that is different from the one attracted to *C.formicifera*. Bower identified the pollinators of both species. Experiments indicate that *Neozeloboria cryptoides* can pollinate this species as well as *C.valida*. Where both species occur *C.* x *pescottiana* may be formed. This hybrid is the only one so far found in *Chiloglottis*. Peakall et al. (1997) confirmed the hybrid origin by allozyme analysis. Despite this confirmation little evidence for backcrossing is found. They concluded from the available evidence that hybridisation has not been a major evolutionary factor in the diversification of sexual deception world-wide.

Chiloglottis trilabra

This species is pollinated by *Neozeloboria proxima*, both on the mainland and in Tasmania. It is also attractive to another new species of *Neozeloboria* and occasionally other wasp species were attracted in choice experiments done by Bower (1992, 1994, 1997). It is distinct from *C.reflexa*. *Neozeloboria proxima* is capable of pollinating both *C.trilabra* and *C.reflexa*, although in NSW where they occur sympatrically they are consistently pollinated by different wasp species (Bower 1995). There may well be geographical variation in kairomones within one species

There is a possibility that the taxon currently known as *C.trilabra* is in fact the true *C.reflexa*, since it is very common in Tasmania and was formerly not recognised as a species separate from *C.reflexa*.

As mentioned above, pollinators, when given a choice, prefer to visit flowers at a certain height, in this case between 15 cm and 22 cm high, rather than flowers above or below this height range. Pollinators therefore have the potential to exert strong stabilising selection to floral height. Floral height within the genus, however, appears relatively invariant. Pollinator preferences can drive stabilising selection only when height is both heritable (which probably is the case) and pollen removal or deposition is limited. Pollinator-mediated selection does not seem to occur and different factors may be responsible for an evolutionary compromise (Peakall & Handel 1993).

Chiloglottis valida complex

According to Bower this is a group of at least four species, probably more (five or six): *C.valida*, *C.pluricallata*, *C.chlorantha* and *C.turfosa*. *C.palachila* (see above) is a distinctive species group.

In baiting trials Bower (1992) tested flowers of *C.valida* from a different range of localities. Wholly black wasps, identified as *Neozeloboria monticola* or *N.nitidula* were attracted. He observed 29 wasps attracted, part of which had pollinia removed from the thorax by the flowers. It is possible that the two wasp species are synonymous and in that case it is not an example of multiple pollinators, but simply imperfect taxonomy.

When two pairs of flowers of both orchids *C.valida* and *C.pluricallata* were exposed in separate choice experiments *C.valida* attracted 18 *N.nitidula* or *N.monticola* male wasps, whereas *C.pluricallata* attracted 10 *N.impatiens* male wasps. The latter wasp is reddish with yellow spots. In 1995 a new taxon was discovered within the *C.valida* group, *C.* aff. *valida* and a probable pollinator collected (Bower 1996).

In baiting experiments *C.chlorantha* and *C.turfosa* were completely isolated, without cross-attraction of their pollinators being observed to other species of the group (Bower 1995). *C.chlorantha* was discovered in 1995 to consist of two taxa, of which the newly discovered form *C.* aff. *chlorantha* attracts a different wasp species (Bower 1995-1997).

C.pluricallata attracts a diversity of wasps, but only one of these is sympatric and is almost certainly the pollinator, whereas two out of three allopatric major responders are confirmed pollinators of other *Chiloglottis* species.

Genus **Drakaea**

There are six species of 'hammer' orchids, including two undescribed ones (Jones 1988), all restricted to Western Australia. All are pollinated by male thynnid wasps in attempting to fly off with the female-mimicking labellum for in-flight copulation.

Sargent (1918), in speculating about the pollination mechanism, misinterpreted his findings. He thought the basal calli bore crystalline nectar and concluded that when an insect settled on the labellum it was 'allowed a meal in peace' and, when seeking to leave, its legs being somewhat entangled by the shaggy hairs', it would be temporarily attached to the labellum, and 'so is constrained to fly in an arc which quickly brings it into forcible contact with the stigma, where it soon manages to free itself'.

Stoutamire (1975) observed the pollination of two species.

The labellum base is a claw to which the expanded terminal portion is attached by a hinge. The labellum projects more or less horizontally, the position varying in different species, but the conspicuous terminal part is always maroon and decorated with 'tufts of calli, glandular protuberances or warty excrescences'. The flexible joint makes it possible for the labellum to be carried in an arc in front of the column. Movement is passive and the lip returns to its original position when released.

Drakaea elastica

Male wasps, *Zaspilothynnus nigripes*, when crossing the pheromonal scent stream trailing downwind of the flower, begin a zigzag flight pattern and thus fly upwind in the direction of increasing concentration. In the vicinity of the flower the flight becomes more direct. The male then, on seeing the pseudofemale, alights on the decoy, often curving the abdomen and probing the sides and the end of the structure. After several seconds it grasps the decoy and tries to carry it away. The wasp is then catapulted into the column, where pollen is fixed to the thorax by means of stigmatic secretion.

Stoutamire (1981) once observed a female wasp with four males competing for her. After release the female crawled up a grass stem and rubbed her rear legs, presumably releasing or spreading pheromone ('calling') and Stoutamire noted striking similarity in colour, position and general resemblance of this wasp to the flowers of the orchid.

Drakaea glyptodon

This species is pollinated by *Zaspilothynnus trilobatus* in a similar manner as described above for *D.elastica.*

Stoutamire placed flowers on the floor of his rented car and was followed down the road by wasps. They entered open car windows readily and located the flowers on the floor.

Mating behaviour

Peakall (1990) conducted experiments on the mating behaviour of male *Z.trilobatus* during encounters with females and orchid flowers. The flightless female called for mates by releasing a sex pheromone. Winged males respond rapidly to calling females and carry away 60% in the first minute and 100% within five minutes. Males locate flowers mostly in the first minute of presentation. The males grasp the flower but do not copulate. Although 93% of the males landed only 21% attempted to grasp the female decoy (only 16% completed the full sequence of behaviour that culminates in an attempt to carry away the female decoy, the step essential to pollination) and only 6% attempted copulation. Ejaculation was never observed.

The males grasp the female and transport them, in copula, to a nectar source (stands of the shrub *Hakea prostrata* e.g.), where the females are fed by regurgitation. Females are then returned to the ground where they burrow in search of host scarabeid beetle larvae which are paralysed before oviposition.

In one of the tests, the females, olfactory and visual stimuli were spatially separated by placing calling females under an opaque vial cap with a hole to allow the odour to be emitted and presenting dead (non pheromone emitting) females glued to a stick.

Refractory period

Males leave the area immediately after visiting and do not pay attention to nearby flowers within a refractory period of several hours at least. A learned response is suggested. This may be a site-specific response leading to the avoidance of a stimulus at a particular location and in the normal wasp-mating system such a refractory period may prevent males from returning to inappropriate females. Only occasionally wasps moved to a second or third flower and none of the wasps observed completed the sequence of behaviour essential for pollination of the second visited flower. Geitonogamy or pollination within clones therefore seems less likely. The males move up to 132 m (mean distance 32 m) in search of females. In larger populations of the orchid, distances of pollen flow can be an order of magnitude greater than in food-based pollination systems (which are in the order of 1 m).

Pollination may even occur when flower densities are low, for even single flowers attract pollinators. This feature may be critical for the survival of very small or scattered populations.

Genus **Spiculaea**

One species, endemic to south-western Australia, *Spiculaea ciliata*, is pollinated by male thynnid wasps by means of pseudocopulation.

The 'elbow-orchid', as it is named locally because its hinged insectiform labellum lamina has appendages of which the rear one bends 'at the elbow', is well adapted to survive in a seasonally dry environment.

The bulbous maroon posterior of the labellum is a decoy female wasp and produces a sex pheromone which attracts only males of one species of thynnid wasp. The wasp is an undescribed species of *Thynnoturneria*, according to G. Brown (Stoutamire 1985a, b). The male wasp picks up the decoy and as he rises his wings slip into the column hooks and takes up pollen with his thorax. Any pollen present on the thorax adheres to the stigma in a next visit.

Northen (1971b) described its growth habit as burning 'its candle at both ends', since, as the dry season advances, the plant cuts itself off from the underground tubers by a completely drying and withering stem and consequently has to rely on the water and food stored in the succulent stem.

She mentions two other species in the genus: *Spiculaea huntiana* and *S. irritabilis*, as do Nicholls (1969) and Schlechter (1992), but Jones places the latter two species in the genus *Arthrochilus*, to which *Spiculaea* is closely related.

DIURIDEAE: PTEROSTYLIDINAE

Genus **Pterostylis**

The 'greenhoods' (Plate 143, p. 212) form a subtribe with one genus and some 120-140 species or even more. A number of species still await description. Australia has 100 species (Clements 1989).

They are terrestrial orchids with tubers. Within Australia the genus is best developed in the southeast, with a few species in the south-west. Some eastern species extend into tropical regions. About 29 species (some undescribed) are largely endemic to New Zealand. Some species are found in New Caledonia and New Guinea.

The nectarless greenhoods form temporary fly traps. The lips are sensitive and they throw flies against the columns. The time for reset of the irritability after closure of the flowers varies in the different species (Sargent 1909 mentions for *P.turfosa* 25 minutes, *P.reflexa* 35 minutes, *P.constricta*

40 minutes, *P.vittata* two hours, *P.recurva* three hours). Details of the mechanism at work are given below, with the descriptions of the pollination of the different species. Pollination visits last 30-90 seconds, irrespective of reset time. Gnats scramble about until they find the exit tunnel (Jones 1981).

According to Jones the genus is basically pollinated by small gnats and only some of the larger species by mosquitoes.

Light thrown through the clear crystalline panels in the hoods concentrates on the inside and may attract the insects (window flowers).

According to Stoutamire (1982) the genus is dominated by two clearly defined species groups, plus several species which do not fit well into either.

One group includes plants which produce single-flowered stems from basal leaf rosettes with tubers produced at the ends of elongate roots. These plants form colonies. Flowers are erect, with an overarching hood. Lateral sepals are erect and often thread-like. The lips are largely concealed within the tubular flowers.

The second group, called the 'rufa' group from the maroon colouring in the flowers, is better represented in drier areas and the stems are multiple-flowered. The flowers have reflexed lateral sepals below the overarching hoods. The lips are completely exposed outside the hood. Some flower after the rosette has withered. Bentham (18th century) experienced serious difficulties in separating the different species during his preparation of the Flora Australensis and lumped them all under the one species *P.rufa*, hence the name of this exclusively Australian group of drier habitats.

All members of this group may be pollinated by pseudocopulation with the highly irritable labellum (Beardsell & Bernhardt 1983, ref. in Dressler 1993). Mosquitoes and mosquito-like fungus gnats are the most frequently reported visitors. They are attracted by a pheromone-like perfume, combined probably with the insectiform labellum.

The flowers of *P.boormannii* attract only male fungus gnats (Bates 1977, ref. in Dressler 1993).

Dafni & Bernhardt (1990) have hypothesised about the derivation of the pseudocopulatory syndrome in *Pterostylis*. They suppose a female fungus-gnat mimic with a female wasp mimic ancestor (ancestral Caledeniinae) and a latent selection potential for pollination by male fungus gnats, with small labella attracting the physically smaller genera of tiphiid wasps.

P.barbata and *P.recurva* are mentioned as species different from both dominating groups.

A few species are self-pollinating. Catling (1990) mentions *P.aphylla* en *P.foliata* as autogamous.

Pterostylis acuminata

Pollination of the species by a female mosquito, *Culex* sp., was reported by Coleman (1934), but see below (*P.falcata*).

Pterostylis curta

Bernhardt (1995) suggests that *P.curta* is pollinated by fungus gnats of the genus *Mycomya* (Mycetophilidae).

The labellum is irritable and self-motile. Reset and regain of irritability occurs 2-3 hours after triggering the first time and the flowers of cultivated plants fail to self-pollinate. The viscidia will be attached dorsally to the thorax.

Pterostylis falcata

Pollination of the genus *Culex* (Culicidae) in the same way by a female mosquito was reported in the same article (Coleman 1934), but according to Vogel (1976) the insect depicted belongs to the Mycetophilidae.

Pterostylis grandiflora

Coleman (1928) reports the lip to be irritable at certain times and that a slight touch of almost any part of it causes it to spring up against the column.

In many instances the visiting insect comes in contact with the side of the labellum without being imprisoned. The species belongs to the 'tubular' group.

Pterostylis longifolia

Probably pollinated by a mycetophylid fly (Adams & Lawson 1993).

Darwin (1888) described the pollination mechanism of the species after Fitzgerald and Cheeseman. It belongs to the group with exposed lip (Plate 144, p. 212).

Two of the petals and one of the sepals form a hood which encloses the column. The distal portion of the labellum affords a landing place for insects. It readily springs up when touched and carries away the insect, which is imprisoned in the otherwise completely closed flower. The labellum remains shut from 0.5-1.5 hours, and on reopening is again sensitive. The insect can only crawl through the narrow passage formed by two projecting membranes on each side of the upper part of the column. In doing so it will remove the pollinia. In a second flower it will leave pollen on the adhesive stigma.

Pterostylis mutica
Gumprecht made a photograph of a mycetophylid insect with pollinia of this species. It belongs to the section Antennaea (Vogel 1976). According to Bates a gnat, after landing on hood or labellum, probes the surface with its proboscis. Then it grasps the dark-green appendage on the lip, which then suddenly flicks upwards and traps the gnat inside the transparent hood. After a few seconds of fluttering about inside the flower, the gnat tries to escape through the gap between the column projections. It folds the wings tightly and extends its legs backwards to become streamlined. With kicking movements it works its way through and emerges with pollinia on its thorax. A perfume, reported by Van der Pijl & Dodson (1966), is, according to Jones (1981), difficult to discern. The same is true for *P.trullifolia.*

Pterostylis nutans
Stoutamire (1982) mentions a faint musty odour in the flowers of this species, that belongs to the tubular-flowered group.

In these species a brush-like projection extends outward from the labellum base and is probably the trigger set off by small dipterans. Traps reset in a short time.

P.curta and *P.pedunculata* also fall in the group of tubular-flowered species. According to Vogel (1976) *P.nutans*, *P.curta* and *P.vittata* (sect. Antennaea), which have only a non- or slightly irritable labellum are adapted to fungivore insects. The labellum has a stalked, and at the end ramified appendage at the base which presumably is the source of attraction to the insect. Vogel suggests that in species of this section it should be tested if the appendix functions as a fungose trap that triggers certain postures in the visitors.

Hyett (1960) found a mosquito-like insect in the hood of a flower, buzzing furiously. Eventually the buzzing ceased and the labellum of the flower was seen to move, not with a snap, as in *P.longifolia*, but slowly.

Pterostylis parviflora
Although most *Pterostylis* species are scentless to human noses, *P.parviflora* is reported to have a smell resembling semen.

The odours of *Pterostylis* species are likely to be species specific, for species of sexually attracted gnats seem to pollinate specific species only. Among sympatric species interspecific hybridisation may, however, occur.

Pterostylis procera
Bartareau & Jackes (1994) observed flowering and pollination during three years in north-east Queensland. Most plants produced new shoots each year, but many did not flower every time. The flowers were pollinated by flies and mosquitoes. Unpollinated flowers remain open for 15 to 20 days after anthesis. Insects collected were Coleoptera, Diptera, Hemiptera, Lepidoptera and Thysanoptera, but only two unknown fly species and the mosquitoes *Aedes notoscriptus* and *Anopheles annulipes* were confirmed as pollinators. Also, three unknown *Culex* species and *Aedes* (*Verallina*) species were collected with pollinia and are likely pollinators. All these insects were females. The pollinators are caught by the closing of the sensitive labellum and the only exit from the flower is up along the labellum and against the column. No obvious reward is collected. Between sequential visits to the flowers resulting in successful pollination, the pollinators take a relatively long time (0.5-4.5 hours). The irritable labellum resets in 4-8 hours.

On average two-third of the flowers were visited and one-fifth produced capsules. Experimental pollination resulted in 100% capsule set. Individuals are both self-fertile and outbreeding. Capsule production is in part the result of pollination limitation.

Pterostylis reflexa and **P.turfosa**
According to Sargent these species are pollinated by gnats.

Pterostylis rufa
Sargent (1909) observed tiny Diptera alight on the labella of this and other species (*P.sargenti*, *P.vittata*) without having set the labella in motion. To describe the irritability of the labellum he mentions insects which flew off again, because they did not touch the appendage on the labellum on which they had landed. 'Experiments immediately after its departure showed the labellum to be in a highly irritable condition... I weighed one like that which I saw hurled into a *P.sargenti* blossom: its weight was just one milligram. So far as I could see it touched the sensitive spot with one foreleg only, so that the pressure upon the spot must have been considerably less than one milligram. From this, some idea may be gleaned of how ticklish the labellum is'.

Pterostylis sargenti
To his observations above Sargent (1909) added that the large appendages of this species are not sensi-

tive to a gentle touch, but that the sensitive spot lies between them at the base of the ridge, which he regards as a third appendage.

Several times he had seen an insect struggle through the tunnel of the flower, but more often these insects showed no desire to leave their prison. Once he saw a dipteran hurled into a flower. It immediately set about sucking juice. He watched the flower for 25 minutes, at the end of which time the insect managed to escape by squeezing past the labellum, the flower having been damaged at one spot.

All Diptera seen in this species belonged to one species and did not pay attention to other *Pterostylis* species. Other observations of different dipterans visiting different greenhoods led him believe that each species is pollinated by its own species of dipteran, except for some closely related species that hybridise.

Pterostylis vittata
This species was the first of the genus of which the pollination was actually observed and described (Sargent 1909). He noticed a small gnat which after a second landing on the labellum was caught and, after fluttering for a few seconds in the lower chamber, commenced to climb through the tunnel formed by the column wings, with its back towards the column. It had to struggle violently to get through the narrow top of the tunnel, but within three minutes after its capture it was free again. It bore two pollen-masses on its back. Sargent added notes on the pollination of *P.reflexa* and *P.turfosa* in 1934. The pollinators were gnats and a large brown dipteran, but these were destroyed before identification (Hyett 1960).

DIURIDEAE: ACIANTHINAE

This subtribe of five genera from Australasia and tropical Asia consist of about 150 species.

Although the flowers of *Acianthus* and *Corybas* are very different, both seem to attract fungus gnats and are reported to be pollinated by these insects. The flowers of *Corybas* may mimic fungi. Jones suggested that the Mycetophylidae have co-evolved with these (and other) galeate orchids.

Some taxonomists treat this subtribe as part of the Triphoreae (Burns-Balogh and Hesse). It may even be an artificial instead of a natural subtribe. Jones (1988) places *Acianthus* significantly amongst 'Other miscellaneous terrestrials'.

Genus Acianthus

These 'mosquito' or 'mayfly' orchids, as they are called, secrete nectar and attract small gnats that pollinate them. *A.amplexicaulus*, *A.sublestus* and *A.viridis* are self-pollinating. Cheeseman witnessed small flies moving from flower to flower in a group of *A.sinclairii*, resulting in high fruit set.

Acianthus caudatus
This small orchid forms loose colonies and may flower in mass following bushfires. It attracts microdipterous insects by its scent. These insects follow the scent trail flying into the wind and land on the inflorescences in search for suitable flowers to feed on. Only some species are large enough to effect pollination and of these only one has been actually found consistently bearing pollinaria. The other species, females of *Sylvicola dubius* (Anisopodidae) and female *Sciara* (Sciaridae) pick up pollinaria only occasionally.

The pollinator is a female fly, a *Mycom(y?)a* species (Jones 1974). The flowers smell of 'something dead' at night and especially in humid or hot weather. Pollinia are picked up during feeding on nectar which is secreted from the small lip. In the young flower the relative positions of column and lip prevent withdrawal of the pollinia. About two days after opening the column arches in such a way that the rostellum can be brushed by the visiting insect. The lip can move downward to let the pollinator in and changes in shape to form a small basin at the base, in which the nectar accumulates. The nectar is produced by two small erect glands on the upper margin of the labellum. During feeding the insect makes upward movements and touches the rostellum with the thorax. Intoxication by the nectar lasts long enough to allow the glue of the rostellum to cement the pollinarium to the animal.

Genus Corybas

This relatively large genus is widely distributed from northern India to New Zealand, including southern China, the Malay Archipelago, Polynesia, New Guinea, Australia (15 species) and Macquarie Island.

The flowers are pollinated by small gnats. Some believe that the flowers mimic the fungi that are part of the life cycle of the pollinating fungus gnats.

Vogel (1978) e.g. notes that the flowers of most species are pollinated by Mycetophilidae and that some mimic mushrooms. The flowers are nectarless and lateral sepals and petals of several species are reduced and rudimentary, so that they resemble mushrooms more closely.

The way in which the flowers attract the gnats is not known, although Jones (1971) believes that the *auricles* at the base of the labellum of some species like *C.diemenicus* not only allow a small amount of light to penetrate into the interior of the flower, making it pretty gloomy, but also allow a small air current to circulate and diffuse the attracting scent from the small glands near the opening of the auricles. Several species in New Zealand have been found to be pollinated by Mycetophilidae, but these are species that do not clearly show mimesis of fungi.

Some species (Dransfield et al. 1986) have two spurs at the base of the lip (section *Corybas*), others have two open auricles or slits instead of spurs (section *Steleocorys*).

C.despectans from SW Australia and a species from New Zealand *C.saprophyticus* (see also Hatch 1952) are autogamous (Catling 1990).

Corybas aconitiflorus, C.abellianus and **C.diemenicus**

These species are mentioned by Vogel as examples of mushroom-mimicking species. They show two characteristics of the syndrome: clear shine and humid cavities.

Coleman (1931) described the pollination mechanism of *C.aconitiflorus* (as *Corysanthes bicalcarata*). She observed that in 50% of the flowers pollen was removed (male function), but in less than 1% pollen was found on the stigma (female function). She found intact pollinia in almost all fertilised flowers, so self-fertilisation, as claimed by Fitzgerald, is unlikely.

Moreover, cross-pollination is suggested by the large number of intermediate forms between the different species, especially those in which a great similarity is found in the reproductive parts, i.e. *C.fimbriatus*, *C.diemenicus*, *C.dilatatus* and *C.pruinosus*.

Jones (1970) observed a fungus gnat (Mycetophilidae) caught by a tiny (spider?) web on the surface of the labellum of *C.diemenicus*. It was entrapped by a rear tarsal segment and had apparently struggled vigorously, but to no avail, and its wings were quite tattered. The pollinarium of the flower was attached to its back and was in good order. The insect was about 4 mm and seemed perfectly adapted to pollination by fitting neatly into the confined space. The gnats seem to be attracted to decaying smells (sapromyophily), but no scent perceptible to the human nose was observed. There appear to be two small glands situated on each side of the base of the column which may be nectaries or scent-glands. They are moist and supply some fluid.

C.aconitoflorus is known to supply some nectar from a projection below the stigma after puncture.

In *C.diemenicus* there is a similar projection and liquid oozes from any surface of the column after puncture but it does not appear to be nectar. The mechanism adopted by the orchid to ensure cross-pollination seems simple and relatively ineffective considering the small number of swollen ovaries observed.

It possibly may be interpreted as an example of brood site deception (Dafni & Bernhardt 1990), although Jones did not mention this.

Genus **Cyrtostylis**

The five species of this genus, known from Australia and New Zealand only, are pollinated by insects, probably small gnats, which are attracted by nectar from two glands near the basis of the labellum.

Cheeseman noted minute species of Diptera as frequent visitors (1874).

Plates

142. Male sawfly about to set off the mobile labellum of *Caleana major*. Photo R.J.Bates, Mylor, South Australia (p. 203).
143. *Pterostylis concinna*. Photo N.A.van der Cingel, Trim Greenhood. Appin, NSW, Australia (p. 208).
144. *P.longifolia*. Note the exposed position of the labellum. Photo N.A.van der Cingel, Tall Greenhood. Appin, NSW, Australia (p. 209).
145. *Cryptostylis leptochila* with copulating wasp (*Lissopimpla*). Photo R.J.Bates, Avarat, Victoria (p. 213).
146. *Diuris maculata*, one of the species mimicking wild peas (Fabaceae). Photo N.A.van der Cingel (p. 216).

142	143
	145
144	146

DIURIDEAE: CRYPTOSTYLIDINAE

Two genera with 16 species from Australasia, Pacific Islands and Tropical Asia. Formerly placed in the Cranichidinae by Dressler (1981), but recently (Dressler 1993) moved them to the Diurideae on the basis of anatomical evidence.

Genus **Cryptostylis**

In Australia five species are found, one extending into New Zealand.

All species are pollinated by males of an ichneumonid wasp *Lissopimpla excelsa* (*L.semipunctata*), trying to copulate with the flowers.

According to Stoutamire (1974, 1975, 1979) a promiscuous system such as that in *Cryptostylis* can function only if there are strong barriers to hybridisation; otherwise species would cross extensively in areas of overlapping distribution. Hybrids are unknown and it appears that the species are isolated by genetic barriers that prevent cross-pollination.

Stoutamire points to the possibility of unrecognised physiological races of the wasp species, which however seems unlikely from the experiments by Coleman and (later) himself, in which different species of *Cryptostylis* were offered to wasps.

'Bait flowers of three species were presented in local bushland and *L.excelsa* wasps visited the bait. Wasps often indicated a preference for one flower over another. This may be due, perhaps, to differing pheromone production, which may be influenced by flower age, time of day, or some other factor, but individual wasps were observed several times to visit a preferred flower of one species and then approach flowers of another species'.

Still another hypothesis is mentioned by Stoutamire: 'A mechanical barrier may also function in which the first pollinarium attached to the insect reduces the chances of a second attachment. If this interference occurs in mixed orchid populations where one species predominates, the pollinaria of the dominant orchid species would have the greatest chance of transport by the available wasp population. The successful fertilisation of the less common species would be reduced through competition for pollinators and mixed orchid populations would be selected against. Such a mechanical control would be significantly only if the orchid species sharing the common pollinator flowered concurrently – the situation in *Cryptostylis*.' Again, this seems unlikely, as Stoutamire himself in fact points out, because of selection of different flowering times to be expected, although this would be counteracted by the dependence of the flowers on the timing of *Lissopimpla* males and females emergence. Moreover wasps are observed with several pollinia attached. Genetical barriers seem involved.

'It appears that the species cannot hybridise, even if wasps visit two or more species in succession. Pollen massulae will be lost during illegitimate visits, but sufficient pollen remains on the pollinarium for legitimate pollination of later-visited flowers' (Stoutamire 1975).

Cryptostylis leptochila

This is the first species in the genus for which the mode of pollination (Plate 145, p. 212) was observed and for that reason it is treated here first. It was discovered by Edith Coleman.

This endemic from south-eastern Queensland, New South Wales, east Victoria and Tasmania is widespread and sometimes locally common in open forests. It flowers from the end of December till the mid of March.

147	148	151
149	150	
152	153	

Plates

147. Eggs and Bacon, *Dyllwynia retorta*. One of the Australian wild peas mimicked by *Diuris* species. Photo N.A.van der Cingel (p. 216).
148. *Dendrobium canaliculatum*. Photo N.A.van der Cingel, September 1998, Queensland (p. 234).
149. *D.kingianum*. Photo N.A.van der Cingel, August 1998, Lamington National Park, Queensland (p. 234).
150. *D.kingianum*. Photo N.A.van der Cingel, August 1998, Habitat, Tavinghy Mountains, Queensland (p. 234).
151. 'Big bang' flowering in *Dendrobium speciosum*. Photo N.A.van der Cingel, August 1998, Tavinghy Mountains. Queensland (pp 233 and 237).
152. Gregarious flowering *D.speciosum*. Photo N.A.van der Cingel, August 1998, Maryborough, Queensland, Australia (pp 233 and 237).
153. Gregarious flowering *D.aemulum*. Photo N.A.van der Cingel, Tavinghy Mountains, Queensland (p. 233).

History of discovery

Coleman (Coleman 1927, 1928, 1929b, Godfery 1929a and Watson 1961), whose daughter observed the phenomenon first and drew her attention to it, witnessed the ichneumon *Lissopimpla excelsa*, than still named *L.semipunctata*, entering backward into the flower of *C.leptochila*. The insects came from the trees and not from the ground and landed on the flowers. The end of the wasp's body appeared curved and pressed against the rostellum, the labellum gripped by the claspers.

The insect quivered and then became motionless. At the time Coleman did not understand the behaviour, but asked the entomologically better informed of her readers for help to interpret the behaviour observed. She was not aware of having observed pseudocopulation and thought of oviposition instead. She looked in vain for eggs and larvae. But it is in an attempt to copulate with the flower that the wasp pollinates (Froggatt 1927).

Pollinia were attached to the abdomen with the viscidium and deposited on the stigma of the same or a next visited flower. The whole act on the flower lasted about one second. Although aware that the wasps observed were males she looked for eggs laid by female wasps and for moth larvae, which are sought for as prey, but neither were found. Coleman hypothesised that perfume, although not very pronounced, had lured the insect to the flowers. She noted that the orchid labellum somewhat resembled the visiting insect's body. The fact that the insects were not deterred from subsequent visits by any disturbing finger, suggested some important reward.

In a second paper she refers to pollination in the genus *Ophrys* from Europe and comes to the conclusion that the male wasps 'are answering to an irresistible sex-instinct' and that mimicry and scent lure the insect into a copulatory attempt. 'Even to our eyes the likeness is apparent. To the inferior eyesight of the insect the resemblance may be still more convincing', she writes. We now know that it is probably less a matter of insufficient eyesight (wasps have good eyes) than of 'Gestalt' perception and inborn reaction to a stimulus the wasps may never have met before. Males emerge in advance of females and during that period find most *Cryptostylis* flowers.

Experiments with transfer of flowering spikes to other places demonstrated the role of scent in attracting the male wasps.

In a more detailed description of the behaviour of the wasp she tells: 'The wasp (attracted by scent), flies directly to the flower selected. (On seeing it), grasping the labellum with its two front pairs of legs, it then draws up the long hind legs so that they clasp the upright posterior margins of the labellum. This brings the abdomen of the insect into the falcate curve, which enables the tip to penetrate the stigma of the orchid flower, near the base of the labellum. The upper surface of its body, at the last segment, rests on the viscid disc of the labellum (...). With the effort needed to free itself, the wasp exerts pressure on the rostellum. This pressure opens the anther cells and releases the pollinia (whose caudicles are attached to the disc), and these are withdrawn by the insect, adhering to the last segment of the abdomen. (...). The pollinia adhere always symmetrically, at right angles to the body of the insect. On exposure to the air, the viscid disc contracts and the caudicles become depressed – though this movement is less apparent than in many other orchids. With the straightening of the insect's abdomen, the pollinia are brought parallel to it, and slightly projecting beyond the tip – in the exact position for striking the stigma of the next flower visited'. The depression act lasts about 30-40 seconds.

The anther cells of the flower lie below the stigma and the prominent rostellum, projecting over the anther, prevents the reception of its own pollen by the stigma.

Coleman and Rogers discussed these pollination phenomena after the latter doubted their nature and came to full agreement on the facts. Concerning the presence of spermatozoa of the wasp on the flower, Coleman (1928b) had little doubt after examining smears of droplets from the tip of the abdomen of the wasp and from the stigma and the fleshy basis of the labellum under the microscope.

The slides were compared with sketches of spermatozoa from dissected males. There is however a note about difficulties in discriminating them from vegetable matter, pollen grains etc. It is reasonable still to doubt the presence of spermatozoa on the flowers (see below).

In 1938 photographs from Coulson (in Coleman 1938) were published in which it can be seen that in pseudocopulation the male claspers are extruded as in normal copulation.

Fruiting

Seed set is usually low, which may be explained by the emergence of female wasps in the period after the first flowers have opened.

C.leptochila flowers relatively late compared with

other species and this may also play a role in the low percentage of fruits.

Cryptostylis erecta

Species from central- and south-eastern Queensland, Victoria (east), NSW. Locally common in a variety of habitats.

Coleman (1930a, 1930b) describes the pollination of this species. In January 1930 she witnessed the removal of pollinia by *Lissopimpla excelsa*, the wasp in this species sitting nearly upright and removing pollen on the tip of the abdomen. She experimented with fresh flowers and observed several removals as well as interchange of visits to three other species of the genus.

Morphology

'There is a large upright labellum and the narrow petals and sepals in the flower are insignificant. The column is enclosed and partly hidden by the folded lower part of the lip. The projecting rostellum prevents self-fertilisation. The pale green lip resembles a shell, with longitudinal stipes in reddish brown and purple and is semitransparent. On each side of a central ridge there are small dark red spots, which clearly stand out. There is some suggestion of mimetic resemblance with the pollinating insect, but the scent, not perceived by humans is the main attraction'

Isolation

Coleman observed withdrawal of one pollen mass, whereas in other species she only saw removal of whole pollinaria. In West and South Australia species are geographically isolated, but in Victoria and New South Wales species grow intermingled or near to each other. Anther and stigma mature at different periods in different species, so isolation is in the first place phenological. Moreover the species are genetically isolated. Artificial cross-pollinations have led to offspring resembling only the mother plants and the hybrids are not recognisable as such.

Leonard (1970) observed pollination in pot plants in an area where the species does not grow and Woolcock (1980) reports that an ichneumoid wasp was trying frantically to get through the wirescreen of his sunroom to a pot of *C.erecta* plants inside.

Cryptostylis hunteriana

This is a leafless saprophytic species from moist soils, endemic to NSW and Victoria (east).

Nicholls (in Wallace 1978 and in Watson 1961) observed pollination in this species in 1938. Again *L.excelsa* is involved as a vector.

Cryptostylis ovata

The species is endemic to south-western West Australia, growing in a range of habitats.

Coleman (1929a, 1930b) exposed a raceme with one flower that had not entirely faded, together with racemes of *C.leptochila* and *C.subulata*. It appeared very attractive to ichneumons (*L.excelsa*) which were aroused and which made attempts to enter. Goadby (in Coleman 1930a) informed her that he observed pollination of pot plants in his garden by wasps that reverse themselves on the flower and then back. This was 2900 km away from the natural growing place and about 160 km from the supposed hatching place of the wasps.

Cryptostylis subulata

This very widespread Australian endemic has large flowers. Coleman (1929) observed pollination. The species flowers before *C.leptochila.* The wasp (*L.excelsa*) is positioned upside down under the labellum and has to back up a considerable distance to reach the pollinia. The viscidium adheres slightly higher than in *C.leptochila*, but experiments show that after the usual drying and contraction of the disc the pollinia project over the lip in a position well adapted to strike the stigma in either species.

Whitehead (1964) observed pollination in his bush house and concluded that the wasp was clearly attracted by scent. This was also in an area where the genus is not represented within 130 km.

Wallace (1978) also observed pollination in his glasshouse 100 km from the nearest growth place in the wild. He points to the chain character of attraction by successively scent (pheromone), eyesight and touch and refers to Van der Pijl & Dodson (1966) who report on the observation of ejection of seminal fluid by the pseudocopulating wasps, which even preferred the flowers to the real females (supernormal stimulus). Reports of such observations, however, all seem to be based on one instance, seen by Coleman. If ejaculation appears to be a usual phenomenon in *Cryptostylis* pollination, this would create a powerful counter-evolutionary stimulus, for it would directly influence the wasp species fitness (see also above).

DIURIDEAE: DIURIDINAE

Three genera with more than 40 species.

Genus Epiblema

Epiblema grandifolia, the only species, endemic to south-western Australia, is not self-pollinating, but its pollinators are unknown.

Burns-Balogh & Bernhardt (1988) classify it with the Thelymitreae.

Dressler puts it in the Diuridinae on the basis of its column structure.

According to Burns-Balogh and Bernhardt this species may be a generalist food mimic with a mode of presentation comparable to the pollination systems as described for *Calopogon*, *Arethusa* and *Pogonia* (Chapter 4), but the floral form suggests mimicry of co-blooming papilionid legumes (cf. *Diuris*).

Genus Diuris

This is a large genus of about 40 species with usually attractive colourful flowers. One species is endemic to Timor (Indonesia), the rest is native to Australia, especially in the south-east and south-west.

There is taxonomic confusion concerning several taxa within the genus. Pollination is primarily by small bees, in certain species lured by the (nectarless) flowers mimicking those of flowers of the pea family (Fabaceae).

According to Nilsson (1992) *Diuris* species mimic guilds of spring-flowering legumes, which means that several models are mimicked by one fraud. Associative learning by pollinators of a model controls the selection of the mimic, whose variation is thus constrained by the model(s).

Dafni & Bernhardt (1990), following Beardsell et al. (1986) and Armstrong (1979), have pointed out that the mimicry of legumes by *Diuris* leads to an interesting dichotomy in the spectrum of pollinia vectors. They are pollinated by naïve, generalist wasps and halictid bees, but also by specialised female colletids, which feed their young primarily on legume pollen.

Diuris abbreviata

Hamilton (Erickson 1965) noted that this species is often fertilised by auto-pollination by folding back the stigma to the pollen masses.

Diuris alba

Coleman (1930) found pollinia removed in 13 out of 40 flowers, proving that they were visited by insects, probably halictine bees.

Diuris aurea

Coleman (1933b) observed halictine bees and believed that pseudocopulation was involved. This was deduced from the behaviour and male sex of the bees observed, but it has never been confirmed.

Diuris longifolia

Erickson (1965) supposes this species to be pollinated in a similar fashion as in *Diuris pedunculata*. She observed a leaf cutter bee *Megachile remeata* to visit the flowers shortly during a swift flight, depressing the labellum. Pollinia were already removed from the flowers and she therefore did not observe pollinia to be taken away.

Diuris luteola

Pollinated by *Allodapulla* sp., *Lasioglossum* spp., *Nomia* sp. (Bartareau, pers.comm.)

Diuris maculata

Beardsell et al. (1986) studied the pollination of *D.maculata* in south-eastern Australia, where it is one of the commonest orchids (Plate 146, p. 212). The flowers are vary variable, usually yellow with reddish brown markings, and they are markedly similar to those of several genera of Fabaceae ('wild peas') in both colour and shape (Plate 147, p. 213).

Several bees, Colletidae, were collected with (1-3) pollinia of the orchid deposited on the head while visiting *Daviesia virgata*. These were:

- Three ♂♂ and one ♀ *Leioproctus* sp. (Colletidae) (one ♂ on *D.maculata*);
- Six ♀♀ *Trichocolletes* sp. (Colletidae).

Some other bee species were found on *D.mimosoides*:

- One ♀ *Apis mellifera* (Apidae) and one ♂ *Gasteruption* sp. (Gasteruptiidae).

The small shrub *Pultenaea scabra*, with similar flowers to those of the *Daviesia* species, was also visited by the same kinds of bees.

In total about one-third of the bees observed carried pollinia.

The bees were seldom observed to land on the orchid. The pollen they carried was mainly from the wild peas.

It is therefore strongly suggested that *D.maculata* flowers are pollinated by insects attracted to them by their mimicry of the nectar-producing wild peas.

The variability (polymorphy) of *D.maculata* flowers may play a role in the associative learning of the foraging bees, leading to more pollinations.

The orchid is not autogamous, but geitonogamy may occur. Especially after bushfires mass-flowering is observed, which poses a problem, for legume flowers are uncommon then, due to the death of mature plants. It is supposed that the lack of models in such cases leads to the attraction of naive bees. Pollination in burnt areas appears to be very low, which is to be expected when the model is scarce in relation to the mimic.

Since several interspecific hybrids are known in *Diuris*, the pollination of other species may be similar. Beardsell (1975) has described a hybrid between *D.maculata* and *D.pedunculata* (*D.palachila*) and related hybrid swarm as a threatened population.

Diuris oporina

Pollinated by *Allodapulla* sp., *Lasioglossum* spp., *Nomia* sp. (Bartareau, pers.comm.)

Diuris palachila

Coleman (1932) found eight pollinia removed in 10 open flowers examined.

Diuris pedunculata

Coleman (1930, 1932, 1933) frequently found halictine bees, *Lasioglossum* (*Halictus*) *languinosus*, in the yellow flowers of the variety *gigantea* of this species. Those of the smaller nominal form are visited less frequently, they open less widely. In one flower of *gigantea* she sometimes found as many as three or four of these small bees, which were not held fast by viscid matter but appeared to be stupefied, remaining so for several hours after being taken from the flowers. All seemed to be of one sex. However, pseudocopulation, although suggested, has never been observed.

The flowers contain no free nectar. Around the top of the receptacle is, according to Coleman, a fleshy glandular ring ('honey-disc') which secretes a fluid apparently palatable to the bees. However, to reach the nectar, Coleman believes, tissues must be pierced. Counterpressure of the labellum may be too much for the bee, so that it remains with the mouth-parts firmly gripped, to perish in the flower. The pollinia are fixed to the head. On sunny days about 50% or more of the pollinia may be removed and 20% of the stigma receive pollen.

In general the bees can easily escape but sometimes they are accidentally imprisoned, due to being wedged, gripped, between the base of the sessile labellum and the column. Only by force could the mouth-parts have entered the narrow opening. She found 14 male bees among a total of 15 bees collected on or near *D.pedunculata* flowers in 1930. The only female captured was taken in a dandelion flower and belonged to another species and this female did not carry pollen. It is thus suggested that in one way or another sex was involved, although the word pseudocopulation was not used by Coleman in her papers on *D.pedunculata*. In a second article of 1933 she mentions a total of 22 male bees out of 23, including the bees found in 1930. Coleman counted 12 pollinia removed in 20 open flowers.

It is doubtful whether the bees find any reward in the flowers. Bees are probably lured by mimicry of rewarding flowers.

Bates collected a total of 45 male *Lasioglossum* bees on the blossoms of this species and concluded that sex pheromones were involved in attracting these insects. Nash, however, could not find support for this conclusion (Armstrong 1979).

Rayment (1932) mentions an earth-digging *Paracolletes* spec. as responsible for the pollination of *D.pedunculata* (and *D.sulphurea*) near Melbourne.

Hamilton (1887) mentioned beetle pollination (in Armstrong 1979), but as secondary trait, i.e. as a derived condition: represented in taxa, the ancestors of which were pollinated by more highly developed insects (Faegri & Van der Pijl 1966). Highly developed here refers to the relation with flowers.

Diuris punctata

Coleman mentions removal of two pollinia in 16 open flowers examined by her (1932), indicating pollinator limitation.

Diuris setacea

Coleman mentions observed removal of two pollinia from 18 flowers.

Diuris sulphurea

Coleman observed bees different from the species visiting *D.pedunculata* on *D.sulphurea* removing pollinia. In 18 open flowers, examined in 1930, she found pollinia removed in five. She observed the bees curving their abdomen over the apex of the labellum and was convinced that they were males and that the motive for their visits was a sexual attraction (Erickson 1965). But see above.

Rayment (1932) mentions a *Paracolletes* sp. as pollinator.

Diuris tricolor (=D.sheaffiana)

Hamilton found this species to be fertilised by

small pollen-eating beetles, *Hiparetrus globulus*, which resort to the flowers to eat the pollen and rostellum etc. (Fitzgerald 1893, Erickson 1965).

Genus Orthoceras

This used to be considered as a monotypic genus, distributed in Australia, New Zealand and New Caledonia, but nowadays the species from New Zealand is considered different and renamed as *O.novae-zeelandiae*. *Orthoceras strictum*, the 'horned orchid' is self-pollinating.

Rogers (1913) studied the development of the flower. Dehiscence of the anther is completed as soon as the flower has expanded. The rostellum bends forward and the pollen-masses drop from their loculi, one on each side of the small *mesial* ridge on the back of the stigma, and adhere to it. Numerous pollen tubes penetrate the back in the neighbourhood of the ridge and cause the pollinia to become firmly adherent. The yellow gland from the rising lip presses the stigma back against the pollen-masses. The flower closes again and shuts off the sexual organs from external influences.

DIURIDEAE: THELYMITRINAE

About 60 terrestrial species in two genera distributed in Australasia and into Malaysia. Burns-Balogh & Bernhardt (1988) present an interesting model of the possible floral evolution in this tribe, based on the breeding and pollination systems.

Genus Calochilus

There are 11 species of 'beard orchids', all terrestrial, of which ten occur in Australia, one of these extends to New Guinea, three extend to New Zealand. The eleventh species occurs in New Caledonia. In Australia they are most abundant in the temperate zone, with three species in the northern tropical forests. Pollination occurs in two ways, either by pseudocopulation through scoliid wasps or by self-pollination. Self-pollinating are *C.campestris*, *C.holtzei* (=*C.caeruleus*, ?=*C.imberbes*), *C.paludosus* and *C.robertsonii* (Catling 1990), but see below.

The 'sham eyes' of the staminodal collar, the glossy labellum calluses, and the dark, broad, thick 'beard' on the lip are adaptations to pseudocopulation. The species are highly self-compatible and self-pollination may occur by mechanical autogamy, in the absence of scoliid wasps or by wasp-mediated autogamy. (Burns-Balogh & Bernhardt 1988).

Dafni & Bernhardt (1990) hypothesise that the pseudocopulatory system may be derived from the food mimic system in a *Thelymitra*-type ancestor.

Calochilus caeruleus

Pollination by pseudocopulation of this species (described then as *C.holtzei*) was observed near Darwin in northern Australia by Jones & Gray (1974). As with other members of this genus, the species self-pollinates if insect pollination has not occurred within three days of opening, thus assuring 100% chance of success. Its pollination was effected by a scoliid wasp of the genus *Campsomeris*. The insect could not be identified because of insufficient knowledge of northern members of the genus. The insect is large and hairy and both sexes are winged. The parasitic wasps, which lay eggs in beetle larvae, are known to feed on nectar and pollen. The pollination process was observed on several occasions. The wasps circle three or four times around the flower before they land on the labellum with the head near the column. They make copulatory movements with the abdomen and usually stay for about 10-15 seconds. The orchid releases a very potent perfume. The attraction to inflorescences that are placed in sunlight is so strong that the wasps appear within minutes and on one occasion about 30 wasps were captured in a single day on one solitary inflorescence. Sometimes three or four wasps fought for possession. This behaviour strongly reminds one of the behaviour of scoliid wasps which copulate with the European Mediterranean *Ophrys speculum* (=*O.ciliata*), the species in which the phenomenon of pseudocopulation was first observed. The breaking up of the friable pollinia and the immediate pushing of the pollen onto the stigma cause self-pollination to occur, but a considerable amount of pollen adheres to the wasp's head and can be transferred to the stigmas of other flowers. This probably will cause prevention of inbreeding depression in the orchid species. Self-pollination will occur eventually if wasps do not visit.

Calochilus campestris

Fitzgerald (1878) supposed this species as well as *C.paludosis* to be solely self-fertilised and found the rostellum to be not removable and not attached to the pollen, the pollen-masses themselves being not removed by moderate force when connected with the stigma, which occurs early.

Fordham was first to describe the pollination of this species by *Campsomeris tasmaniensis*, a scoliid wasp. The females of these wasps lay eggs in the larvae of other insects and such a wasp was observed and thought to try to lay eggs in the labellum of the orchid. This was referred to as pseudoparasitism by Van der Pijl & Dodson (1966). The sex of the wasp however was not clear (Gumprecht 1977), although Rupp, in the foreword of Fordham's paper, mentioned the identification as male. Fordham observed nine visits by wasps, of which two were captured, one with pollinia on its forehead, and these were clearly identified as male wasps. Fordham observed stabbing movements with what he supposed to be stings, but must have been part of the genitalia. Fordham also observed deposition of pollinia on a virgin stigma.

Bower (1993) observed and photographed the pollination of this species by male wasps of the species mentioned by Fordham. He did so in 1988 in New South Wales. The photographs clearly show pseudocopulation. The wasp curls the abdomen around the labellum. It entered the flower and pushed its head below the angled column. This dislodged the pollinia, smearing fragments on the face of the wasp between the bases of the antennae. The wings are then held still and the abdomen remains curled below the labellum. The wasp then backed out with the wings waving vigorously. The wasp remounted the labellum for a second time, its abdomen again curled in the same way. The head was not seen now below the column and after backing out the wasp rested on the outside for a moment and flew away. Pollen fragments were found smeared over the stigma. It is however not known if there was any pollen on the stigma prior to the wasp's visit, but there is no doubt that it removed the pollinia. At least one pollinium was intact in the anther before the visit. It can be safely deduced that Fordham also observed pseudocopulation.

Cady (1972) observed 100% fertilised ovaries and concluded self-pollination for the species in the area studied. The pollinia of a dozen of plants investigated were extremely difficult to remove by outside means. He tried this at varying intervals during the life of the flowers without success. He describes the mechanism for self-pollination in this species.

Calochilus holtzei

The species is reportedly autogamous, but according to Jones (1981) this autogamy is a fail-safe mechanism in case insect pollination does not occur. It is pollinated by scoliid wasps of the genus *Campsomeris*. The labellum is immobile and covered with coarse reddish hairs. Male wasps are attracted by a pheromone and try to copulate, contacting the anther. The friable pollinia drop pollen on the stigma, but a significant amount adheres to the insect and can be transferred to the stigmas of other flowers.

Genus Thelymitra

The 'sun-orchids' (about 70 or more species) are mostly endemic Australian orchids, but several species (23 are recognised in 1998, according to a list of I. St George) occur in New Zealand. One species extends into New Guinea, New Caledonia, and the Philippines. Dafni and Calder mention Indonesia (Java and Timor). Their pollination occurs in different ways.

Self-pollination

Self-pollination is observed in e.g. *T.canaliculata*, *T.circumsepta*, *T.flexuosa*, *T.fuscolutea*, *T.longifolia*, *T.javanica*, *T.media*, *T.mucida*, *T.nuda*, *T.rubra* and *T.venosa* (the latter probably wind-assisted). These species are partially autogamous and autogamy functions as a back-up in case no insects pollinate. *T.luteocilium* and *T.pauciflora* are obligatorily autogamous.

Catling (1990) mentions *T.retecta* and a new unpublished species as autogamous without further details. *T.matthewsii*, *T.carnea* and *T.luteocilum* are cleistogamous (Cropper & Calder 1990). The friable pollen of self-pollinating species may fall on the stigma before the flowers opens, if it opens at all. The flowers of several species open only on sunny days. Such species usually depend on both self- and cross-pollination. The transition between these two extremes is so gradual that it has previously not been noticed as occurring in the same species (Fitzgerald 1875-1893). Darwin (1877) described modifications of the column that promote autogamy.

Actinomorphy

The labella of *Thelymitra* species are not very different from other perianth parts. This is often interpreted as a lack of adaptation to pollinating zygomorphous insects, but may as well be an adaptation towards mimesis of flowers with actinomorphic perianths. The labellum has no special significance as an attractant, as an attention-holder, insect positioner or landing platform (Bates 1984c). The column is mostly quite conspicuous and may differ in colour to the rest of the flower. It bears prominent wings. In most species two prominent arms (sterile stamens) flank the anther, frequently adorned by glands or calli and sometimes ending in hair tufts, providing mimicry and probably disseminating pheromones, according to Jones (1981), but this expectation relates to a false (outdated) interpretation of the behaviour of the pollinators (see below).

The trend toward actinomorphy together with interspecific variation in morphology and ornamentation of the mitra will orient bee pollinators and consequently pollinia will be deposited either on the underside of head, thorax, or abdomen, or dorsally on the abdomen. Sympatric species may share pollinators, but opportunities for interspecific hybridisation may be lowered by this differential placement of pollinia (Dafni & Bernhardt 1990). However, Bates (1985) observed a single bee visiting three different species within a single foraging bout and he suggests that interspecific hybridisation may occur when syrphid flies drink stigmatic fluid from previously self-pollinated flowers and pass pollinia fragments to the stigmas of congeners.

Self-pollinating species have flowers that rarely open and possess granular loose pollinia, whereas the cross-pollinated species have flowers that open widely and the tightly-packed pollinia are attached to a viscid disc and can be removed as a single unit by the pollinators, usually native bees.

Mimicry

Other species mimic foodplants of visiting bees and some species are thought to be pollinated by pseudocopulating insects.

Van der Pijl & Dodson (1966) have described the trichomes and crest of the column as a type of pseudoanther, advertising pollen.

Bates (1984) supposes imitation of a bee by the column and this might probably lead to either pseudocopulation or pseudoaggression (territorial behaviour).

But according to Burns-Balogh & Bernhardt (1988) these observations need reinterpretation in light of their observations of attempts of bees trying to manipulate the mitra (staminodal complex) as a kind of pollen harvesting by thoracic vibration. This is a normal foraging behaviour of females of *Lasioglossum* (Halictidae) and *Leioproctus* (Colletidae) called *buzz-pollination*. Such *Thelymitra* sp. mimic co-blooming Liliaceae or Asparagales with 'pseudopollen'.

Fly pollination

According to Christensen (1994) 50 species of *Thelymitra* may be pollinated by flies, but this assertion must be erroneous. Syrphid flies may indeed be co-pollinators of *Thelymitra*, but only few observations are reported. E.g. pollen monads from *Thelymitra* species were found on four species of syrphid flies during observations of pollinators on *Prasophyllum odoratum* (Bernhardt & Burns-Balogh 1986b).

Stigmatic fluid

Mature flowers may produce droplets of stigmatic fluid and insects have been observed consuming such secretions but it is debatable whether this serves as a nectar substitute (Dafni & Bernhardt 1990). Its consumption appears to be a form of robbery.

Several hybrids between allogamous species have been described and some are very colourful (Bates 1985).

Thelymitra antennifera

This species was studied by Dafni & Calder (1987). It is widespread and favours heath and heathy forests.

The column of the yellow flowers bears dark brown erect arms which give it a special appearance ('rabbit ears'). There is no reward in the form of nectar, pollen or pseudopollen, but the flowers emit a pleasant fragrance. However, stigmatic fluid may play a role as an exception to what is stated above (Dafni & Bernhardt 1990). They open on sunny days, when the conditions favour pollinator activity.

Dafni and Calder observed and collected pollinators with insect nets on other yellow-flowered species in the neighbourhood of the orchids. The pollinators carried pollen of several foodplants (*Hibbertia stricta* and other *Hibbertia* spp., *Goodenia geniculata*, *Platylobium obtusangulum*, *Pimelea glauca*, *Stackhousia monogyna* and *Helichrysum* species, etc.).

Pollinaria of *T.antennifera* are deposited ventrally on the head and/or thorax.

Among the pollinating insects were: solitary bees (*Lasioglossum* (*Chilalictus*) species, solitary wasps (*Eurys* spec.) and hover flies (*Syrphus damaster*).

Pollination is low, on average 3.2% of the flowers are pollinated and the rate varies between 1.7-11.0.

Mimicry

It was concluded that *T.antennifera* is a general or unspecialised mimic of the search-image of co-blooming, yellow flowered foodplant species. General similarity to rewarding species seems to offer opportunities to exploit different models depending on differences in the community of potential foodplants over the entire geographic range of a mimicking species. This syndrome probably relies to a greater extent on similarity in colour than on form, although colour variability could greatly contribute to the effectiveness of pollination. A role of naivety of the pollinators and of low density of the mimic in relation to the model(s) in ensuring pollination success by reducing learning to avoid the nonrewarding flowers is supposed, and supported by the results.

The mechanism by which self-pollination is prevented is described in its different stages by Rogers (1913).

Thelymitra apiculata, **T.spiralis** and **T.variegata**

According to Mark Clements (pers. comm.) the flowers of *Th.apiculata* mimic lilylike flowers that are visited by buzz-pollinators (*Calathea* species?). The same is true for *Th.spiralis* and *Th.variegata.*

Thelymitra aristata

According to Bernhardt & Burns-Balogh (1986), Dafni and Calder observed female *Lasioglossum* spp. trying to 'harvest pollen' from the staminodal hoods of this species.

Jones and Bates have suggested mimicking of bees by the staminodal hood and pseudocopulation or pseudoaggression as a pollination syndrome.

Adams & Lawson (1993) suppose food mimicry, which is the more probable theory.

Thelymitra circumsepta and **T.ixioides**

Sydes & Calder (1993) studied the strongly differing reproductive strategies of these similar blue flowering species. *T.circumsepta* is a vulnerable and localised species of sedgeland swamps and comparable wet sites. *T.ixioides* is a widespread common species of open woodland-heath communities. There is a strong contrast in the floral display of both species of which the first is autogamous and the second insect-pollinated.

In *T.circumsepta*, with a mean of about seven flowers per spike, rarely more than one flower is open on any one day and flowers open for one day only. *T.ixioides*, with fewer flowers per spike (about five), has a greater number of flowers open on a spike at one time, the flowers staying open for a period up to one week. *T.circumsepta* flowers mostly with only one concurrent flowering species, whereas *T.ixioides* flowers together with many other woodland species in the same habitat.

In *T.ixioides* about 50% or more of the flowers had their pollinaria removed in the different sites studied and 20-28% had pollen on the stigma. Unpollinated flowers did not set fruit. Fruit set was 28%. In *T.circumsepta* the pollen tubes grow from within the anther onto the stigma. Pollinia were never seen to be removed from the anther, nor was pollen observed on the stigma, but nevertheless 96% of the flowers did set fruit, the remaining 4% being mostly damaged or grazed.

In one of the study-sites *T.ixioides* flowers were visited by hoverflies (*Syrphus* sp.), which did not pollinate and two native bees (*Exoneura* and *Lasioglossum* spp.) which effectively did. Insects of the same species, captured within *T.ixioides* populations, collected pollen from the whole floral community, showing general feeding behaviour. These *polylectic* insects also collected pollen tetrads of the orchid. Stigma scrapes of senescent orchid flowers had orchid pollen on the stigmas, usually over 1000 pollen tetrads per flower. The orchid flowers offer no reward. *T.ixioides* does not resemble other flowers within the communities examined, neither in colour, nor in shape, except flowers of *Dichopogon strictus* at one of the sites. Mimicry of this plant however seems unlikely. It is more likely a matter of deception than of mimicry, the low density of *T.ixioides* ensuring reduction of the rate at which the insects discover the deception and learn to avoid. The reduced attractiveness of floral display in *T.circumspecta* flowers is understandable in an autogamous species. The development of autopollination in this species is probably favoured by the lack of co-flowering species in an area depauperate in insects where it lives.

Catling (1990) mentions self-pollination of *T.ixoides* with reference to Bates (1978). Whether this refers to one of the many varieties or not is not clear.

Thelymitra epipactoides

Cropper & Calder (1990) studied the floral biology of this rare and endangered species of coastal heaths of south-eastern Australia.

Polymorphism

This orchid with highly variable visual display through polychromatic flowers (six morphs: brown, blue, green, and copper for ±70% or mixed green-brown and brown-green) and variable inflorescence size (7-31 flowers) attracts foraging polylectic bees of the genus *Nomia*. The bees were observed to carry pollen and to pollinate successfully. They are deceived by the non-rewarding orchid. These bees normally gather pollen and store it in both the crop and the scopa of the hind legs. On the flower, tactile, visual and probably olfactory stimuli direct bees to the false anther (column wings). Other visiting insect species observed on the flowers did not carry pollinaria or their remains. Pollen loads of insects visiting the orchid consisted primarily of pollen from yellow flowered species and from *T.epipactoides*. A small number of pollen grains from purple, blue, white and pink flowers was found as well. Individuals carried up to eight different pollen types.

Apart from the orchid flowers (primarily brown-green and brown), yellow and blue were the most frequent flower colours represented in the community.

The flowers of the orchid do not resemble, in shape or colour, any other plant in the community. The flowers of the orchid are greater in density and frequency than any other colour types or species in the area sampled (600 m^2). The column may resemble a poridical anther and the bees were attempting to 'milk' this false anther.

It therefore seems that the pollination of *T.epipactoides* is a case of deception (sometimes called confusingly 'non-model mimicry', Dafni 1986) instead of mimicry. The bees are supposed to visit unfamiliar species during foraging. Cropper and Calder correctly state that these are not necessarily naive bees, for most species of bees will profit from such visits and will therefore continually monitor food availability during normal foraging activities. Especially in this case, where different colour morphs are present, the bee will not learn as fast to avoid the flowers compared to deceptive plants with constant flower colour or morphology.

As Kay (1976) has pointed out, if the pollinators may alternate between different morphs (disassortative pollination), a stable polymorphism with an excess of heterozygotes would be maintained.

The flowers produce a sweet smell after they have fully expanded. There are no observations of bees collecting any of the potentially edible tissues like stigmatic mucus or trichomes.

Seed set

T.epipactoides produces viable seeds from xenogamous, geitonogamous and autogamous pollinations, but depends on insects to achieve pollen transfer. Seed set is relatively high (35% of flowers which opened). There is no indication of mechanical autogamy.

Ecology

Cropper et al. (1989) and Calder et al. (1989) also studied the ecology and biology of *T.epipactoides* and formulated guidelines for the management of this endangered species. It does not seem, however, that the pollination syndrome itself is a major factor in survival, but the absence of biotic vectors, apparently due to alteration of the fire regime at coastal sites, probably plays a major role in threatening future survival.

Thelymitra flexuosa

According to Jones (1988) the flowers open tardily on hot days and in cool cloudy weather they self-pollinate without opening.

Thelymitra fuscolutea

The mechanism by which self-pollination is attained is described by Rogers (1913). Although the dark-brown spotted yellow flowers are conspicuous enough to suggest cross-pollination by insects, it appears to be self-pollinating. The flowers open in hot weather and nearly all become fertilised (99 out of 104 had fruits). The hot weather dries and pulverises the pollen. This falls upon the rostellum and stigma, facilitated by the forward bending of the rostellum. The rostellum may undergo *necrosis* (degeneration by blackening and dying cells). Contraction of the rostellum resulting from this process as well as bending forward of the anther over the stigma increase the possibilities for falling down of pollen. Minute crawling insects assist and hasten the breaking up of the pollen.

Thelymitra javanica

Docters van Leeuwen (1933) observed pollination of this Indonesian species by the syrphid *Chamaesyrphus nigripes*.This was the first observation of pollination by an insect within the genus.

Thelymitra luteocilium

The flowers open rarely, only on hot days. They are fertilised in the bud before the full size of the mature flower has been attained. The mechanism of dehiscence of the anther and the falling down of the granular pollen by gravity and by active and passive movements onto the stigma is described by Rogers (1913).

Thelymitra x macmillani

This is a rare natural hybrid, resulting from crossings between *T.antennifera* and *T.nuda*. The mechanism of its pollination appears the same as in *T.antennifera* (Rogers 1913).

Thelymitra media

Jones and Bates (in Bernhardt & Burns-Balogh 1986) have suggested that the trichome brushes of the staminodal hood are insectile in form and that the hood mimics a bee. This orchid as well as some other species possibly would be pollinated when fast-flying male bees attempt to copulate with the crest of the hood or fight with a 'pseudorival'.

Adams & Lawson (1993) however, suppose food mimicry as the pollination syndrome, which is more likely.

Thelymitra nuda

The species forms sometimes dense clumps which produce a lot of flowers, attractive to native bees (Philips 1996).

Bernhardt & Burns-Balogh (1986) have investigated the pollination of this species. On sunny days, when the ambient temperature exceeds 20°C, the flowers open. In such circumstances they are visited by female bees of the genus *Lasioglossum* (three species of the subgenus *Chilalictus*: Halictidae) and by syrphid flies. The flowers are brightly coloured, blue/purple. The scented column is contrasting in colour. Staminodes and the filament of the fertile stamen are fused and form an inflated hood over the anther (mitra). The four lobes of this hood bear many white trichomes, forming a kind of false anthers. The bees land directly on the staminodal hood and try to forage on the hairy lobes on the tip as if collecting pollen. They apply thoracic vibration to the two central yellowish lobes (the crest) and actively scrape the hairs with the forelegs. The pollinia are attached dorsally to the abdomens. Co-blooming blue-purple flowered Liliaceae (*Dichopogon fimbriatus*, *Thysanotus pattersonii*) are visited by the same bees, which gather pollen from poricidal anthers by buzzing and it is supposed that the flowers of *T.nuda* mimic these nectarless, but pollen-rewarding flowers. The bees also collect pollen from nectariferous Compositae and Leguminosae. It is supposed that *T.nuda* is probably pollinated by polylectic naive female bees that have not yet learned to discriminate their flowers and that this exploitation acts as a second strategy.

According to Dafni & Bernhardt (1990) bees that apply thoracic vibration to a series of nectarless, poridical flowers will continue to carry over this behaviour to the first few flowers they visit that do not bear a true buzz-pollination form of presentation. Buchmann (1983) has referred to this as the '*carryover phenomenon*'.

Pollinaria removal is relatively high and exceeds 30%.

The syrphid flies land on the labellum or lateral sepals. They crawl to the stigma and probe this, trying to drink the mucilage. They also probe the trichome brushes or central lobes of the hood sometimes. They do not, however, take away pollinia.

In an examination of 70 open flowers Bernhardt and Burns-Balogh found pollinaria removed in 22. On the stigmas of these flowers fragments of pollinia were found. Wasted pollinaria were found on perianth segments in 12 cases.

What distinguishes *T.nuda* from other orchid taxa with false anthers is the location of the false reward on the column instead of on or within the labellum. Selection seems to have favoured modification of the column and this may explain the tendency towards actinomorphy.

Bartareau (pers. comm.) observed pollination by *Allodapulla* sp. and *Lasioglossum* sp.

Thelymitra pauciflora

Thelymityra pauciflora var. *holmesii* (=*Th.holmesii*) includes two different kinds of plant (Catling 1990). One opens briefly under optimal conditions of heat, sunshine and humidity. This permits cross-pollination. If this does not occur auto-pollination proceeds. The other genotype is cleistogamous. According to Mark Clements (pers.comm.) the species mimics *Wahlenbergia stricta*. Pollen of both plant species were found on the same bee.

Thelymitra venosa

This species has been confused with *T.cyanea*, a much more widespread species. Until recently the true identity of the species was not recognised (Jones 1988). It is therefore not clear whether the following remarks refer to *T.venosa* or to *T.cyanea* or to both. There is confusion also concerning the self-pollination or cross-pollination.

Jones (1971) refers to the reports of Rogers and Nicholls about the observed mechanism of self-pollination in the buds. Rogers (1913) described the mechanism as entirely adapted to self-pollination, analogous to that of *T.luteocilium*.

This fact is difficult to tally with involvement of the species in hybridisation. Jones made a comprehensive study in 1970. Open flowers had for 35-40% no pollen on the stigma. Close examination showed that in about 20% the anther sacs were completely empty; in 10% the anther itself was abortive while the stigma was still functional and the remaining 19% had anthers which had not yet shed their pollen. He concluded from the absence of pollen in 20% of the open flowers that these were male sterile and could be cross-pollinated in principle by other species. In some places the species is growing together with *T.ixioides* and although the chances are low, a pollinating agent bearing pollinia of *T.ixioides* could visit *T.venosa* if they grow intermingled.

Burns-Balogh & Bernhardt (1988), however, think that the claim of male sterility in 20% of the population must be reassessed in light of a photograph taken by Bates (1984). It is probably another case of trying to empty anthers by vibration.

According to Adams & Lawson (1993) Bates observed pollination by an unidentified bee.

DIURIDEAE: RHIZANTHELLINAE

An Australian subtribe, consisting of one genus with two species. Both species grow below ground, are *achlorophyllous* and leafless and have a fleshy stem.

Burns-Balogh et al. (1986) think that probably the underground flowering is an adaptation to specific ecological conditions such as dry periods during flowering time.

Genus **Rhizanthella**

Rhizanthella gardneri

The orchid was discovered in 1928 by the farmer John Trott while clearing land. The discovery aroused great excitement among both the general public and scientist. Several interesting essays on its mysterious life-cycle have been published, e.g. by Bernhardt (1989) and Dixon et al. (1990). It was rediscovered after 20 years and more systematically studied since 1959 by a deliberate search programme, funded by the World Wildlife Fund.

The orchid is associated with the broom honeymirtle *Melaleuca uncinata* (Myrtaceae), and probably is a parasite of this shrub. It is dependent on a mycorrhizal *Rhizoctonia* fungus that forms a mycorrhiza with the shrub. The inflorescence forms a head (capitulum) of many (40-120) spirally (in 8-12 rows) arranged small flowers, the pollen is soft and mealy. The flowers may be pollinated by fungus gnats that enter via cracks in the soil.

George (1980) and George & Cooke (1981) experimented with a muslin covering over the orifice above a bloom and showed removal of pollinia by a small sapromyophilous fly, a *Megaselia* spec. (Phoridae). Dixon found a fungus gnat (Cecidomyiidae) and a wasp (Braconidae: Alysinae) capable of removing pollinia. Dixon also found termites (*Drepanotermes* spec.) bearing pollinaria. Termites are probably regular pollinators, since they can transport pollen massulae from flower to flower. They systematically visit flowers. Nevertheless less than 5% of the flowers are pollinated and setting fruit. This is the only known example of probable pollination by termites in the Orchidaceae.

The flowers offer no nectar and seem to deceive pollinators (Burns-Balogh et al. 1987).

The cut surface of the flower emits a formalin-like scent. Seed development takes 6-7 months, which is an unusually long time for orchids.

Termites have been observed to consume dried parts of the capitulum, but not the fruits. Intact and viable seeds are voided in faeces of small marsupials (*Notomys* mice spec.) after voluntary feeding on ripe capitulae. They as well as termites probably contribute to the dispersal of the seeds.

Rhizanthella slateri

This species was discovered by Slater in 1931 and described by Rupp as *Cryptanthemis slateri*. Five specimens were found in soil rich in organic matter and in association with the rhizomes of the hyacinth orchid *Dipodium punctatum* which is also leafless and mycotrophic.

It is unknown whether there is any myrtaceous shrub or tree involved in its life-cycle. There are however plenty in the habitat. *Rhizanthella slateri* is a waxy white plant, hidden under accumulations of dead leaves and other debris. It was found only a few times. Rupp supposed the ripe fruit to break the crust of the soil and disperse its seed by a breeze of wind, but it seems not unlikely that termites or other insects will do the job.

According to Mark Clements the species is pollinated by a fly, as observed by one of his assistants, who caught the pollinator (pers. comm.)

DIURIDEAE: PRASOPHYLLINAE

About a 100 terrestrial species in three genera in Australasia

Note: Analyses have shown that the genera *Prasophyllum* and *Genoplesium*, formerly united as sections within *Prasophyllum*, should probably even be placed in different subfamilies (Burns-Balogh & Funk 1986). I will follow Jones (1988), who puts both genera within the Prasophyllinae. However, Jones & Clements (1989) have reinterpreted *Genoplesium* and more carefully circumscribed the genus since. Jones (1991) described 15 new species of *Genoplesium* and 11 of *Prasophyllum.* Most of the *Genoplesium* species mentioned below are now once again included in *Prasophyllum.*

Genus Genoplesium

The socalled 'midge orchids' (formerly *Prasophyllum* sect. Micranthae) are very small and inconspicuous terrestrials with a single leaf. Most of the (about 35) species are endemics of Australia, a couple are found in New Zealand and one in New Caledonia.

Dafni & Bernhardt (1990) state that the flowers possibly contain some nectar.

They are pollinated by small (vinegar) flies (drosophilids and chironomids) that receive the pollinaria on the thorax. They are attracted by fruity perfumes, the dark-coloured flowers or the trembling labellae (hinged on a small claw, so that they are mobile). The flowers are non-resupinate. Some species are either self-pollinating or apomictic.

G.baueri is merely saprophytic.

Genoplesium (=Prasophyllum) acuminatum
Bates observed several tiny chloropid flies visiting this species. The flies position themselves upside down on the labellum and then push into the flowers as far as they can reach. Then, as if trying to get further into the flowers, they make a series of jerky movements with the legs, causing the labellum to tremble. These flies may also visit flowers of the next species.

Genoplesium (=Prasophyllum) archeri
Garnet mentions as probable pollinators the flies *Claviceps flavipes* and *Oscinosoma subpilosa* and Cady and Rotherham a *Conioscinella* fly sp. (Adams & Lawson 1993)

Hybrids between *G.archeri* and *G.despectans* do exist and may be due to the drosophilous flies (Chloropidae) that visit both species (Bates & Weber 1980) or by the pollinators mentioned (Christensen 1994).

Genoplesium aureoviride (=Prasophyllum pumilum)
This species is probably self-pollinating (Jones 1972).

Genoplesium baugleholei (=Prasophyllum nudum)
This species with granular pollen is self-pollinating (Jones 1972). The pollen is weakly attached to the caudicle. The mechanical changes that promote self-pollination when the flower is open are: opening of the anther case, movement of the rostellum bringing the pollinia in position above the stigma and 'disproportional' outgrowth of the stigma.

Genoplesium (=Prasophyllum) despectans, G.(=P.) morrisii and G.(=P.)nigricans
These species are insect-pollinated (Jones 1972). Adams and Lawson refer to Garnet, mentioning the same flies as for *P.archeri* (see above).

Genoplesium (=Prasophyllum) fimbriatum
The flowers of this species have a strong smell of lemon (probably due to aldehydes like citral and citronellal), which may indicate a specific pollinator. They are not attractive to the pollinators of the previously mentioned species (Christensen 1994, Jones 1971).

Prasophyllum anomalum and *P.bowdeniae* are both self-perpetuating apomictic forms (teratological forms) of *G.(=P.)fimbriatum* (Jones 1977).

Genus Microtis

The genus comprises 14 or more species (see also New Zealand), of which at least one is endemic to New Zealand, 13 occur in Australia, mostly endemics, but two species extend to Japan, China and Malaysia.

The so-called onion orchids have dull coloured, usually greenish flowers. They bear functional nectaries.

Some species are self-pollinating, two species are even apomictic. Other species are pollinated by beetles, small ichneumons or ants.

Microtis parviflora
Jones (1975) described the pollination of this species by (wingless) ants. Ant pollination is rare and this is the first case described for orchids. The rarity is believed to be connected with the antibiotic excretions of the metapleural glands of ants. These

substances kill potentially pathogenic micro-organisms, but are also detrimental to pollen.

The wingless worker ants of many ant species are attracted to the flowers by a sweet perfume. They feed on the abundant nectar secreted by the labellum. A sunken area between the two lobes of the lip callus and the fleshy column fills with nectar. The many spirally arranged flowers remain open for 20-30 days, but are only visited in the first 3-4 days of anthesis. Ants are known to be frequent visitors of nectar secreting flowers. Flightless worker ants of nine species were captured on the inflorescences. Three species of ants contributed to the pollination of *Microtis parviflora*: *Iridomyrmex gracilis*, *Meranoplus* sp. and *Rhytidoponera tasmaniensis* (in order of visiting frequency). About 35-45% of the individuals bore pollinia. Microdipterans (Mycetophylidae) were also captured, but without pollinia. The viscidium of the pollinia of the first two mentioned small ant species is cemented to the frons, in the last mentioned larger species to the jaws. In the latter insect the pollinium was sometimes deliberately wiped off to various parts of the plants. The ants worked the young flowers systematically, but ignored the older ones. One pollinium can pollinate several flowers. Some ants were recorded to spend up to 35 minutes on a single inflorescence.

On the two sites studied 94% of the flowers had the pollinia removed in the first site, 66.6% in the second, 84% and 64% respectively of the flowers were insect-pollinated.

Pollinaria could not be withdrawn from flowers more than three days old due to the onset of autogamy. Over the first days of anthesis the upper margins of the stigma grows slowly around the lower edge of the rostellum and engulfs the viscidium. The pollinia begin to lose coherence and the granular packets become separated. About four days after anthesis pollinia and stigma contact and autogamy occurs. The self-compatible orchid is not parthenocarpic and pollen is needed for fertilisation. Geitonogamy is considerable, but cross-pollination is not easily measured by the observational method used, although it definitely was seen to occur.

Peakall & Beattie (1991) used a pollen labelling technique to find out more about the genetic consequences of the ant pollination. A high proportion of the ant movements was interclonal, but some transfer between inflorescences of differing genotypes contributed to a low incidence (max. 8%) of outcrossing. Although ant foraging was restricted, the amount of outcrossing is comparable to a mixed mating system as reported for a wide array of insect pollinators. The lack of genetic diversity (or polymorphy) in the populations studied was obviously related to clonality.

In an earlier study (1989) the same authors demonstrated that the pollen of *Microtis parviflora* is unaffected by antibiotic secretion from the metapleural glands of the pollinator *Iridomyrmex gracilis*. The pollinia are separated from the ant integument by a short stalk and are always carried on the frons, far from the metapleural glands.

Microtis unifolia

The pollinator spectrum of this species includes wasps and beetles (Beardsell & Bernhardt 1982 in Dafni & Bernhardt 1990). See also below.

Pollination of a hybrid swarm

Bates (1981) studied pollination in a putative hybrid swarm of three *Microtis* species: *M.parviflora* var. *densiflora*, *M. rara* and *M. unifolia*, growing sympatrically. The first one grew in grassy flats along a creek, the second one in a tea-tree thicket nearby and the third one on rocks 100 m above the creek. Intermediate forms were found in between the localities. They were visited by flies, beetles, gnats and predatory spiders. Black male ichneumonid wasps and orange male and female Brachonidae wasps were the only pollinators. On hot afternoons 20% of the flower spikes of all three species and the hybrids were visited by wasps and up to 50% of these carried pollinia. The wasps had a slight preference for *M.rara* and for sweet smelling intermediates. Nectar was absent. The wasps, after alighting and moving to the labellum, went upward until they were positioned in the groove between the calli of the lip. Here they touched the viscidium disk and readily removed the pollinia which were attached to the head. In next visits parts of the pollinia come off on the stigma. Some wasps collected as many as five sets of pollinia. On flowers examined, usually all but the top few flowers had part of their pollinia removed.

Genus **Prasophyllum**, *section* **Prasophyllum**

This genus of so-called leek orchids, with non-resupinate flowers and a single erect leaf, has about 45 species with the majority being endemic to Australia. The genus occurs also in New Zealand. They are restricted to temperate and subtropical regions.

Pollination is mainly by wasps and bees attracted

by the scenting and nectariferous flowers. Other insects are also attracted (hoverflies, crane flies, beetles). Honeybees of the introduced species *Apis mellifera* may be detrimental by taking away pollinia and rubbing them off. They odour-mark the flowers visited and thus prevent native bees from pollination (D.L. Jones, pers. comm.).

Flowering is stimulated by summer fires. Some species mimic the inflorescences of grasstrees (*Xantorrhoea*), which attract similar insects.

The flowers are non-resupinate (upside down on the spike) and the lip does not function as a landing platform. The labellum is solidly fixed at the base so that it is immovable and the central callus is well developed into a fleshy plate which sometimes occupies most of the surface. The reversal of the flower, either a retrograde movement or a progressive adaptive change, has led to the dorsal (and lateral) sepals forming a hooded protection over the column. The sides of the labellum may form, together with the labellum, a kind of channel which holds the insect in a direct line with the rostellum and thus guarantees effective pollination. The insect will find a series of nectar beads leading it to a large swollen gland full of nectar and positioning it to remove the pollinia, which after mechanical curling will be transferred to the stigma of another flower in a next visit. This rather speedy movement is caused by the contraction of the rostellar gland which adheres the pollinium to the insect's head. When the pollinia are not removed the apex of the anther becomes deflexed and the caudicle moves in such a way that the pollinia are exerted without external aid. They remain exposed above the anther or poised vertically. In *P.rostratum* (*=P.gracile*) they are frequently deposited on the stigma (Rogers), but in *P.elatum* they remain poised at about 45° and in *P.drummondii* (*=P.muelleri*; *=P.paludosum*) they are held vertically over the apex of the rostellum without causing self-pollination.

Coleman (1933) believed that due to an increase in the number of pollinating insects and consequent adaptation of flower-parts, once self-pollinating species may become dependent on insects again.

Catling (1990) mentions *P.colensoi* from New Zealand as autogamous.

Prasophyllum alpinum
Pollinated by the ichneumoid wasp *Pterocormus promissorius*, probably a case of sexual mimicry (Adams & Lawson 1993).

Jones (1972) observed the male wasps with pollinia attached to their mouthparts, seeking nectar and pollinating. The wasps seemed to follow a perfume trail and landed on the lowest flowers and worked up the spike in a spiral manner. Not every flower was examined, but the wasps seemed to probe only the attractive ones. Of 120 flowers 15% had a whole pollinarium deposited on the stigma and the rest had only a few grains. The flowers are self-fertile and form mature seeds if pollinated with their own pollen. Jones further observed a small chrysomelid beetle (*Edusella tridens*) feeding on the copious nectar from the callus plate on the lip, as well as two dipterans and a honey bee (the latter taking up a pollinarium on the mouthparts). Furthermore ants (*Myrmecia pilosula* and a second unidentified species) were seen on the inflorescences, and although pollinia were observed on the jaws of these ants, their habit of cleaning the mandibles on the vegetative parts of the plant, before proceeding to other blossoms, prevented pollen exchange (Jones 1972, Armstrong 1979).

Prasophyllum australe
Bates (1984) observed ichneumoid and thynnid wasps pollinating *P.australe*. Experiments showed that no seeds are formed if the flowers are protected from insects. The glandular plate is raised and facilitates connection between the insect's body and the rostellum (Coleman 1933).

Coleman (in Erickson 1965) observed rove beetles as (accidental?) pollinators.

Prasophyllum brainei (=P.lindleyanum)
A small flowered species with nevertheless efficient insect pollination.

Coleman (1933a) mentions for this and comparable small flowering species (*P.fuscum*, *P.frenchii*) pollen removal of up to 70-80% and up to 40% observed pollen on stigma's.

Prasophyllum constrictum
Synonymous with *P.gracile* (illegitimate name).

Rogers (1913) observed honey-loving beetles, *Trogoderma adelaide* carrying pollinia of *P.gracile* (Coleman 1933).

Prasophyllum drummondii
This species is synonymous with *P.muelleri* and *P.paludosum*.

Walton Rowe (Coleman 1933a) observed chrysomelid beetles of the species *Ametalla spinolae* on this orchid as pollinating agent. There is no well developed glandular plate, but instead the rostellum bends down to the visiting insect.

Prasophyllum elatum

Bates (1984) studied this spectacular, common and widespread nectariferous species which is faintly and sweetly scented. It is usually about 1 m in height (up to 1.80 m!) and most profusely flowers after bushfires (with up to 80 flowers).

Fire stimulated plants often have dark leaves in the season after fire. This is supposed to be crypsis and camouflages them from herbivores (cf. *Corycium* of South-Africa with black cryptic flowers under the same conditions).

Bates observed wasps to be the most effective pollen vectors on many spring-flowering species of *Prasophyllum*. Stoutamire (Bates 1984) noted that both native and introduced wasps are pollinators of the genus. The flowers of *P.elatum* attract a wide spectrum of insects, but only wasps (male Thynninae) and more rarely bees transferred pollinia. Beetles and flies, although feeding on the flowers, seldom removed pollinia and if they did were ineffective in transferring them.

The flower colours (green, cream, brown, maroon) are known to be particularly attractive to wasps. It is supposed that the copious nectar producing *Xanthorroea* in the vicinity of which the orchids tend to grow are mimicked by the orchids. This is a case of Müllerian mimicry.

Occasionally much smaller ichneumoid wasps were seen on the flowers and more frequently small native bees, but both seldom were able to remove pollinia.

The thynnid wasps (sometimes carrying wingless females) fly directly to the flowers without circling about. They probe the callus plate with its free nectar and puncture the surface of the lipbase which contains hidden nectar in the epidermal cells. Pollinia are in this position attached to the frons. They fall forward by their own weight, thereby taking the right position for pollinating. The pollen is stripped off layer by layer in following visits, so that pollen from one flower can be transferred to a dozen or more plants.

Prasophyllum fimbria

Peakall studied its pollination by nectar-feeding wasps and bees by means of labelling the pollinia with histochemical stains in six different colours. The pollinia are spatially separated from the viscidium by a stipe so that labelling can be done without interfering with pollinarium removal. Pollen flow thus can be monitored by following the movement of the coloured pollen. This was done in several populations of the orchid in western Australia. Pollinaria removal (of labelled and normal pollen) was observed in 10% of the flowers visited at only 30% of the visited spikes. Vectors did not visit every flower of an inflorescence. Wasps visited about 45%, bees only one-third of the available flowers. Wasps occasionally flew long distances, up to 21 m compared to a maximum of 11 m for bees, but the mean wasp flight distance was not significantly different from that of bees.

An amount of 85 labelled pollinaria were removed by pollen vectors from 16 test spikes, with a total of 125 flowers on 47 spikes receiving labelled pollen. An average of two flowers received pollen for every pollinium removed, but up to six flowers received pollen from a single pollinium.

Multiple pollinations increased the heterogeneity of pollen transfer to stigmas. A flower thus may receive both self- and cross-pollen (from several sources).

On average geitonogamous transfers accounted for 22% of all the pollinations observed. It was calculated that the percentage of flowers receiving some outcrossed pollen approached 85%.

Prasophyllum fitzgeraldii

A small form of this species is apomictic (Bates 1984). The larger, nominal form is probably insect-pollinated.

Prasophyllum frenchii

This species, which is synonymous with *P.hartii*, is pollinated by wasps.

Bates (1984) mentions both Ichneumonidae and Thynninae. No seed set without insects.

Prasophyllum fuscum

This species is pollinated by wasps, both Ichneumonidae and Thynninae (Bates 1984), including female Ichneumonidae!

According to Bates it is obligate insect-pollinated, but Coleman mentions a form (*P.* aff. *fuscum*) which is apomictic.

Prasophyllum goldsackii

A species with closed or semiclosed flowers. At least partly apomictic (Bates 1984).

Prasophyllum macrostachyum

A very variable species with different forms. Partly apomictic (Bates 1984).

Prasophyllum occidentale

Wasp-pollinated (Bates 1984). Ichneumonid and

thynnid wasps have been observed. Experiments prove the obligate character of insect pollination.

Prasophyllum odoratum
Pollinated by syrphid flies. Male colletid beetles are of secondary importance (Bernhardt & Burns-Balogh 1986b).

Prasophyllum pallidum
Self-pollination has been demonstrated (Bates 1984).

Prasophyllum patens
Mentioned by Bates (1984) as insect-pollinated (Ichneumonidae and Thynninae).

Prasophyllum validum
Pollinated like the previous species.

ORCHIDEAE: HABENARIINAE

This large subtribe with more than 900 species in 23 genera in two tentative alliances is African and pantropical, with extensions into Eurasia and North America. In Australasia species are found of two genera, one each of the two alliances.

Genus Perystylis

The genus comprises about 65 terrestrial deciduous species distributed in China, Taiwan, India, Malaysia, Polynesia and New Guinea. Three species are found in tropical Queensland, one of which is endemic. Nothing is known about pollination.

Genus Habenaria

This very large genus of about 600 terrestrial and deciduous species of open grassy habitats has 11 representatives in Australia in the northern tropical part, with one species extending south. The majority are found in Queensland. Hardly anything is known about the pollination of native species of Australia. Usually the species of this genus are pollinated by moths. Flower colour and other syndrome characteristics of Australian species point in the same direction.

Dixon and Buirchell mention an unidentified butterfly and a ditto moth for *Habenaria* aff. *ferdinandi* (Adams & Lawson 1993).

Armstrong (1979) considers fly pollination as probable, referring to several American and English authors. But all authors cited refer to *Platanthera* (syn. *Habenaria*) *obtusata* and this statement needs proof.

It is one of the six genera in Australia with functional nectaries, but unlike the other genera they bear long spurs.

DISEAE: DISINAE

Only one accidentally introduced self-pollinated species from this southern African subtribe, *Monadenia bracteata* is found in western Australia.

EPIDENDROIDEAE: CYMBIDIOID PHYLAD

MALAXIDEAE

World-wide about 960 species in six genera, of which *Liparis*, *Malaxis* and *Oberonia* are represented in Austral(asi)a.

Genus Liparis

Of this cosmopolitan genus of about 250 species 10 occur in Australia, of which nine are endemic. Most Australian species occur in the tropical region (seven), or subtropical (five), only two extend into temperate regions. There are terrestrial as well as epiphytic species. Most epiphytic species grow on trees and/or rocks (=lithophytic) in rainforests. Two are entirely lithophytic. Pollination seems to be mainly by small flies and mosquitoes. Hemipteran pollinators are also known. The insects are attracted by the usually strong, unpleasant floral perfumes and they feed on nectar from the labellae.

Liparis reflexa
Wallace (1974) studied the pollination of this lithophyt. Its flowers open for 18-30 days, the lowest ones first, but the majority open at once. The flowers are green to yellow, yellowing with age, which may be an adaptation to pollinating insects, colour attraction increasing while scent diminishes. They have a strong unpleasant odour. Nectar is secreted from the labellum ridges.

Visitors: Sarcophagidae (Diptera) were the most common visitors and were observed to remove pollinia. Pollination success measured was 22%.

Drosophilidae are less active and too small to pollinate. They seemed uninterested in the labellum juices. Chironomidae are also too small, but eat from the nectar. Sugar-ants (*Iridiomyrex*) visit as well, but do not pollinate.

The pollinator works down from the labellum apex to the base. On backing out the dorsal thorax brushes the rostellum, picks up sticky fluid, lifts the anther cap and adheres the falling pollinia.

Plants (with 53 flowers) made accessible to ants only did not set seed. Of plants made accessible to flies only, nine out of 47 flowers set seed. Of plants made inaccessible to all insects none of the 37 flowers set seed.

Akam (1990) observed a *Dixa* mosquito-like Diptera species as pollinator, whereas Adams found Mycetophilidae (Adams & Lawson 1993).

CYMBIDIEAE: BROMHEADIINAE

Genus Bromheadia

Of this tropical Asian genus of 12 terrestrial species one extends to Australia: *B.pulchra*. Pollination unknown, but probably by bees.

CYMBIDIEAE: EULOPHIINAE

The pantropical subtribe Eulophiinae is mainly an Old World group. There are about 264 species in six genera. Four genera have Australian representatives.

Genus Dipodium

Of the about 25 species, distributed in South East Asia and Australasia seven are found in eastern Australia, of which six are endemic. The majority of these (five) are leafless terrestrial saprophytes. One is a climber and another one is leafy. Pollination by small bees.

Dipodium punctatum

Bernhardt & Burns-Balogh (1983) observed a female megachilid bee, *Chalicodoma* (*Hackeriapis*) *derelicta*, on *Dipodium punctatum*. The bee visited the flowers on three stalks and was then captured: it carried a viscidium just above its antennae, below the middle ocellus. Also pollen of other non-orchid foodplants were found on the bee. This orchid species has extremely variable flowers, i.e. in background colouring (white to rose and red) and spotting (Adams 1986).

According to Bates (1985) *Dipodium* species advertise falsely with shiny yellow-green areas on the column or base of the labellum just below the stigma. The lack of nectar forces the bees to push further under the column and thus pick up pollinia on their heads. Nectaries are found on the scape at the base of floral bracts, protecting the flowers from ants, which are kept busy by these extra-floral nectaries.

Dipodium stenochilum

Jones (1988) observed wasps removing pollinia from *D.stenochilum*.

Genus Eulophia

Over 200 species form this large genus of mostly African and Asian orchids. Three occur in Australia of which two are endemic. Little is known about the pollination of Australian species. In Africa and Asia (?) carpenter bees have been observed as pollinators.

Genus Geodorum

The genus contains about 10 species, distributed from India to New Guinea. One extends to Australia. They are deciduous terrestrials.

Geodorum densiflorum

The species is pollinated by small native bees (Jones 1988, as *G.neocaledonicum*).

Genus Oeceoclades

About 30 species, distributed from Africa to Asia. One extends into Australia: *Oeceoclades pulchra*, formerly incorporated in *Eulophia* and recently placed in *Oeceoclades*. The species is self-pollinating.

CYMBIDIEAE: CYRTOPODIINAE

About 140 species in 12 genera are distributed pantropical, mainly in the Old World. *Cymbidium*, with about 50 species, is represented in Australia by three endemics.

Genus Cymbidium

All three species of Australia, *C.canaliculatum*, *C.madidum* and *C.suave*, are epiphytic. They are pollinated by small native bees of the genus *Trigona*. These bees use pre-existing cavities for nesting and forage on nectar and pollen resources from a wide range of plant species (Rayment 1932, Bartareau 1994c). In orchids, however, fragrance is the main attractant. All three species are self-fertile and outbreeding (Adams 1990b).

Cymbidium canaliculatum

This species is reported to be resource-limited in the production of seeds and to be pollinated by a *Trigona* bee species (Adams et al. 1992, Adams & Lawson 1993). Bartareau (1993b) names *T.hockingsi.*

Cymbidium madidum and C.suave

Macpherson observed the pollination of *C.madidum*, then called *C.iridifolium*, by small unidentified native honey bees in 1934.

Reward

The bees were hovering and alighted on the expanded outer end of the labellum, obviously attracted by the sweet perfume. The labellum due to its jerky movements entrapped the bee, bringing its back into contact with the stigma upon which pollen grains were deposited and the bee then struggled backward, receiving a new supply of pollen. The bees were seen gnawing at the viscid secretion along the median line of the lip. The exudate is transported from the front legs to the second pair and fastened in the pollen carriers of the rear legs. Macpherson and Rupp believe the substance to be used as 'bee glue' to close up small spaces and cracks in the nest. The bee species was later identified as *Trigona hockingsi.*

According to Bartareau (1993b) both *T.hockingsi* and *T.carbonaria* are pollinators of both orchid species.

Bartareau (1995) studied the floral biology of these species. They are selffertile and outcrossing. According to Bartareau no rewards are produced and floral reward deception is supposed (Bartareau does not mention the observations and hypothesis of Macpherson and Rupp concerning the nestlining material as a supposed reward, although he knew their publications and observed the chewing and scratching on the labellum ridge). The species display flowers during a long period.

Pollinators and visitors

The fragrant flowers are visited by butterflies, honey bees, wasps, native bees and a single hoverfly. They are pollinated by nectar- and pollen-feeding native bees, *Trigona carbonaria* and *T.hockingsi.* Both remove pollinaria and pollinate flowers during warm, sunny days. The bees have an *opportunistic group foraging strategy*, i.e. they search independently for food but if an exceptional source is found they rapidly recruit their nestmates by communicating. An opportunistic strategy helps explain a generalist flower-visiting habit with a tendency to explore potential new food resources. From the foraging behaviour observed and from observed resource limitation and abortion of excess capsules it is concluded that a fragrance chemical reward system is not probable.

The species have overlapping flowering periods, but differ in habitat. There are no models in the habitat and therefore mimicry is not supposed (as in *C.insigne*). It is concluded that pollination is based on deception, probably through occasional bee visits during foraging on other food plants. *Apomixis* and parthenocarpy are absent. In unvisited flowers no capsules are produced.

Pollination

The bees approach, hover and land on open flowers. They spend 5-10 minutes chewing and scratching on the ridge of the labellum below anther and stigma. The labellum moves up and down with the activity of the bee on entering. The head is brought close to the column. Bees have to struggle out of the tunnel and then receive the viscidium with pollinia adhered to the scutellum surface between the wings. They usually preen after a visit and then crawl or fly to a second flower. In *C.madidum* about 60% of 97 bees and in *C.suave* about 80% of 172 bees had two pollinia attached. Of the flowers 65-80% had pollinia removed and 5-19% produced a capsule. Under conditions of pollen deposition being less certain than pollen removal, producing more flowers than capsules may increase pollinator attraction and reproductive fitness through male function. Flowers pollinated early in the flowering season have the greatest probability of producing capsules. Pollination removal correlated significantly with inflorescence size.

In experimental hand-pollinated self- and cross-pollinated flowers capsule set increased by 22-50% over natural pollination. The higher capsule set influences future growth, flowering and fruiting negatively. Under limited resource conditions natural selection may favour deceptive pollination systems.

CYMBIDIEAE: ACRIOPSIDINAE

Genus **Acriopsis**

The 6-12 species are distributed from SE Asia to New Guinea. One species extends to Australia: *A.javanica*. It is epiphytic. Pollination unknown. Not self-pollinating.

EPIDENDROIDEAE: EPIDENDROID PHYLAD

ARETHUSEAE: BLETIINAE

About 400 species in 21 genera. Four genera are represented in Australia.

Genus **Calanthe**

A large genus of 150 species, widely distributed, mainly in Asia. In Australia only one species is found: *C.triplicata*. *C.veratrifolia* is a synonym. Pollination not known. Lepidopterans may be expected as pollinators.

Genus **Pachystoma**

This genus comprises only three species, found from India to New Guinea. One extends into Australia: *P.pubescens* (syn. *P.holtzei*). Pollination not known.

Genus **Phaius**

A genus of about 50 species from Africa and Asia (India to New Guinea). Five occur in Australia, of which three are endemic: *P.australis*, *P.bernaysii*, *P.pictus*). All are self-pollinating with usually hardly opening flowers.

The mechanism of self-pollinating is described by Rogers (1921).

Jones (1976) described in detail the self-pollination of a form of *P.pictus*. The upper margins of the stigma expand and engulf the pollen after early movement of the anther. No movements of the pollinia or caudicles are involved. The stigma is never gluey. Jones supposes the species is not always all self-pollinating.

P.tankervilleae is a widespread species, known from several Asian countries (India, China, Malaysia, Indonesia, New Guinea). There is a tendency in this genus to produce supplementary anthers or staminodia (Rogers 1921).

Genus **Spathoglottis**

About 45 species from India to the pacific region. Two extending into Australia.

S.pauliniae is self-pollinating. The pollination of *S.plicata* is unknown, but the species is not self-pollinating.

COELOGYNEAE: COELOGYNINAE

A large genus with about 285 species in 20 genera, of which only one is extended with a solitary species into Australia.

Genus **Pholidota**

The only Australian species *P.imbricata* is not self-pollinating, but the pollinators are not known.

EPIDENDROIDEAE – DENDROBIOID SUBCLADE

PODOCHILEAE: ERIINAE

Note: Dressler (1993) distinguishes a socalled dendrobioid subclade within the Epidendroideae with three tribes: Podochileae (very similar to the Epidendreae), Dendrobieae and Vandeae.

The Eriinae comprise about 700 species in 10 genera. Only one genus is represented in Australia.

Genus **Eria**

A large genus of about 375 species from tropical Asia. In Australia six species occur in north-eastern Queensland, of which four are endemic. One more species occurs on Christmas Island.

Little is known about pollination, three species appear to be self-pollinating: *E.eriaeoides*, *E.inornata* (=*E.kingii*), *E. queenslandica*.

PODOCHILEAE: PODOCHILINAE

Of about 140 species in six genera from tropical Asia and Australasia only one species is known in Australia: *Podochilus australiensis*. The epiphytic species is not self-pollinating.

PODOCHILEAE: THELASIINAE

Tropical Asian and Australasian subtribe with about 235 species in six genera. Three genera are represented in Australia.

Genus **Phraetia**

Two endemic epiphytic species occur in mainland Australia. Three more species on Christmas (one) and Norfolk Island (two). Pollination unknown.

Genus **Rhynchophraetia**

One species in Australia. An epiphytic and entomogamous species. Details of pollination unknown.

Genus **Thelasis**

One epiphytic and self-pollinating species, *T.carinata*, occurs in Australia. *T.capitata* is known from Christmas Island (and Indonesia).

DENDROBIEAE: DENDROBIINAE

A large subtribe with over 1150 species in six genera, distributed in tropical Asia and Australasia.

Genus **Cadetia**

Four species are known from Australia, of which two are endemic. *C.collinsii* is self-pollinating (Lavarack 1981). The pollinia in this species fuse with the stigmatic surface. Pollination otherwise unknown.

Genus **Dendrobium**

This large genus with over 1000 (Jones states: over 1400) species has its centre of distribution in Asia, but about 60-70 species occur in Australia, of which the majority are endemic.

According to Clements (pers. comm.) this genus is polyphyletic and at least five major evolutionary lines can be recognised. Dendrobiums with terete leaves have been separated recently as *Dockrillia*.

Pollinators

As far as is known small native bees are the pollinators. Some species have self-pollinating forms. Most species, however, are self-incompatible (and even to a high degree incompatible in interspecific pollinations) and this is expressed by flower abscission after self-pollination. The reaction is induced by auxin and can even be caused by dead pollen. (Johansen 1990, Adams 1990b). Natural hybrids, of which there have been found many, appear to be even more strongly self-incompatible.

Syrphid flies, thynnid wasps, and even a weevil have also been observed as pollinators. There is also a record of bird pollination.

Big bang

'*Big bang*' or mass flowering seems to play a major role in attracting pollinators in many species (Plates 151-153, p. 213).

Mass flowering occurs under consistently warm conditions with intense fragrance and short longevity of the flowers. This seems favourable to attract insects, but is probably a case of false advertising (Slater & Calder 1988). It would, however, be worthwhile to calculate the investment of the orchid in fragrance, compared to the usual investments in rewards.

Adams (1991) states that recent analyses of floral fragrance have revealed that flower probing, labellar grazing and manipulative activities of bees on floral parts are very likely to represent gathering of fragrance chemicals for use in their social and reproductive interactions. Some of the major fragrance components of *D.speciosum* and *D.kingianum* are known to be produced, collected, stored, and used as trail pheromones in the communication of social and semi-social bees. This is probably an analogue of the perfume gathering of euglossine bees in tropical America. The major components identified for the fragrances are esters of short chain fatty acids, aromatic (benzene) derivatives and terpenes (essential oils). Some compounds are restricted to a single species, but there are several major ones shared by several species and their natural hybrids (Adams 1990b).

Pollination mechanism
Insects enter the flower in upright position. If the insect is of the correct size it receives some glue onto the thorax on entering. On backing out, hairs or a protruding part of the thorax contact the rostrum of the anther cap. The anther pivots and ejects the pollinia onto the thorax of the insect. Deposition of the pollinia in the sunken stigma is achieved by catching them on the rim and releasing them from the struggling insect. The labellum pivots on its hinge as the insect moves (Jones 1983).

Dendrobium adae
Pollination by a *Trigona* bee species was observed by Adams et al. in 1990 (Adams et al. 1992, Adams & Lawson 1993). The fragrance is described as pleasant (Jones 1983).

Dendrobium bigibbum
Jones (1983) observed pollination by a fast flying fairly large bee of the family Colletidae or Halictidae.

Dendrobium canaliculatum
Bartareau studied the pollination of two varieties of this orchid species: var. *canaliculatum* (Plate 148, p. 213) and var. *nigrescens*, both in the wild and in private collections. The varieties are both pollinated by bees of two species: *Hylaeus chromaticus* and *H.ruficeps*. (Pollination by *Hylaeus* was already observed by Gray; Jones 1983, mentions the capture of two species by Gray: *H.ruficeps* and *H.theodori*.) The bees remove pollinia and pollinate these co-occurring varieties which overlap in flowering time. Experimental crossings with self- and cross-pollination within and between the varieties demonstrate self-fertility and outbreeding. Capsule set in these experiments was 50-60%, with significant abortion rates in self-pollinated plants, indicating some self-incompatibility.

The fragrant but nectarless plants were visited by butterflies, syrphid flies, honey bees and native bees. Only the *Hylaeus* bees were pollinators. They visited 2-3 flowers per plant and did not discriminate between the varieties.

The descriptions of pollinator behaviour and pollination mechanism are the same as for *D.kingianum* and *D.speciosum*, as mentioned below. Pollinia are attached to the scut(ell)um.

It is supposed that pollination by multiple *Hylaeus* species and outbreeding has been a factor in the evolution of floral variation in *D.canaliculatum*. It seems that both varieties are expressions of flower colour variability as an adaptation to reduce chances of pollinators learning to avoid the nectarless flowers.

A third variety *foelschei* is prevented from interbreeding with the other ones by geographical isolation, but this does not warrant species status to this taxon in Bartareau's view. However, Clements (1989) describes them as different species.

Adams & Lawson (1993) mention *Hylaeus theodorii* (and *H.ruficeps*) as suggested pollinators (observations of Jones).

Indsto & Weston (1999) observed a weak near-UV bullseye pattern in *D.canaliculatum* and suggest this to be an adaptation to bee pollination.

Dendrobium discolor
Jones (1983) observed regular visiting by hornets, wasps and large bees, however, without removing pollinia.

Dendrobium gracilicaule
Jones (1983) observed visits by *Trigona bees.*

Dendrobium kingianum
Jones (1983) observed *Trigona* bees visiting freely.

Adams & Lawson (1987) observed a full pollination sequence in a private collection. The pollinator was a small variant of the honey bee *Apis mellifera*. Intense fragrance and a mass display of colour ('big bang') are the major attractants. A raised green, pink or yellow linear callus on the middle lobe of the labellum may serve as a guide. Lateral lobes on the labellum form a tunnel (Plate 149, p. 213). The labellum is hinged and moves up and down with the activity of the bee, allowing it to reach with its head near the column foot. Pollinia are deposited on the dorsal thorax when the bee backs out. In natural clones, which have smaller flowers, the small variants of honey bees are too large to enter the flowers. The orchid is self-incompatible (at least to a high degree in most parts of its distribution range, according to Adams 1990b) or infertile for its own pollen. In natural populations fruit set is very low (2-10%).

Adams & Lawson (1988) observed also *Trigona* bees as effective pollinators in the natural range of distribution (Plate 150, p. 213). The bees rarely exchange pollen between flowers that are more than 1.5 m apart, and preferentially exchange between flowers separated by less than 0.5 m. Pollen removal amounted to over 60%, but pollinia deposition was less than 10%. Seed capsules developed in only 1% at the time of investigation. Honeybees are considered to be only moderately effctive as pollinator, because of many displaced pollinia and pollen caps found as a result of difficult entry in the flowers, which lowers the frequency of seed capsules.

Smedley (1989) observed pollination of three clones of *D.kingianum* in his backyard by the honeybee *Apis mellifera*. The bee took pollinia from one clone and deposited the pollinarium in the flowers of another one and carried pollinia of this second clone to the first, after which it flew to the third clone, carrying fresh pollinia of the first. This pollinarium got lost by knocking against the anther cap and fell off. New pollinia were withdrawn and the bee disappeared from the scene. In this last clone pollination was unsuccessful because of the proportionally smaller aperture available to the bee for entrance. This is consistent with the observations of Adams and Lawson.

Of the flowers 43% (18) had been visited within 40 minutes. There were five pollinations.

Rentoul (1976) observed birds visiting epiphytic growing *D.kingianum* in his private collection. He believes the honey-eaters to be pollinators because the flowers visited set so much fruit that the flower stems were incapable of supporting the pods. He speculates on the evolution of the originally lithophytic species and supposes that destruction of the original habitat by almost total clearance of rainforests may have led to the elimination of pollinators, i.e. enormous numbers of insects, moths, butterflies and birds and adaptation of the plant species by aerial growths. This vegetative reproduction could have taken over the functions of flower and seed in part. It is however doubtful if birds are able to withdraw pollinia and will do so on a regular basis in the absence of nectar.

Dendrobium lichenastrum

Pollinated by native bees *Trigona* sp. (Jones & Gray 1976). The fragrance is musky (Jones 1983).

The same bee species pollinates *D.toressae* and *D.monophyllum*. This very variable orchid species grows epiphytically or lithophytically on large boulderings bordering streams. The variety *lichenastrum* was studied in its habitat. Insects may spend 5-6 minutes or more in each flower suggesting that they become stupefied or intoxicated by some substance within.

Smythe (1970) noticed an insect fall to the ground during his observations. The insect continually moves backwards and forwards as if feeding on the base of the labellum in the tunnelshaped flower. Plants exposed to sunlight situations set more pods than those in dark shade, indicating the significance of the fragrance on attraction.

The syndrome and mechanism are the same as for other species of *Dendrobium*.

The species has a weak near-UV bullseye pattern (Indsto & Weston 1999).

Dendrobium (=Dockrillia) linguiforme

Epiphytic or lithophytic. Cream, strong musk-like scented flowers, nectarless.

Lister (1987) observed tiny pollen-feeding weevils (*Cytallia sydneyensis*) on the flowers of *D.linguiforme* var. *linguiforme*. The mass-flowering highly reflective flowers attract insects by scent and sight. One weevil carried a pollinarium on its back. The weevils seem attracted to the labellum and manipulate the calli ridges with their mouthparts, making their way into the flower. The weevil that carried pollinia was captured and kept. It pollinated a flower and later took pollinia from a fresh flower.

Gilbert (1958) observed a hoverfly *Syrphus viridiceps* as pollinator and ichneumonid and thynnid wasps as well as small beetles as visitors.

Jones (1974) mentions the bee species *Euryglossina hypochroma* and *E.lynettae* (Colletidae, subf. Euroglossinae), all females, as pollinators. These usually feed on Myrtaceae. They seem to follow a perfume trail towards the orchids. They force themselves with extended mouthparts right into the base of the orchid flowers. After 1-15 seconds visiting the bees clean themselves.

It seems that the orchid utilises different insects and that there is no pollinator specificity.

Bartareau (pers. comm.) observed *Hylaeus albonitens* on var. *nugentii* (*=Dockrillia nugentii*). The pollinia attach to scutellum and abdomen of native bee visitors as they exit a flower, the pollinia are deposited on the stigma of a subsequent flower visited. Plants are self-fertile and outbreeding.

Dendrobium luteocilium

Smythe (1970) observed a native bee (not identified) pollinating *D.luteocilium*.

Dendrobium mirbellianum

According to Jones (1983) the species is self-pollinating, but this may not apply to all plants.

Dendrobium mohlianum

Slade (1980) describes pollination by a honey-eater in the New Hebrides (Vanuatu) and illustrates this by photographs. He observed these birds fly in and hop on the plant, run through the flowers with their active beaks, and depart, all in a few moments. They leave hardly a flower untouched. The same bird effectively pollinates flowers of *Ascocentrum* 'Sagarik Gold' in his garden.

Dendrobium monophyllum

The species grows on boulders and trees in humid open forest and rainforest margins.

Pollination

Bartareau observed pollination by a *Trigona* bee species (Bartareau 1993, Adams & Lawson 1993). The sweet fragrance of the flowers is apparent around 8:00 hours. and reaches a peak at 14:00-15:00 hours. A range of flies and bees were observed visiting (*Apis mellifera*, *Amegilla* sp., *Homalictus* sp., *Hylaeus* sp., *Lasioglossum* sp.), but only worker caste *Trigona* bees were confirmed to pollinate. The nectaries attract ants which transport scale insects. Aphids, assassin bugs, butterflies, dendrobium beetles, grasshoppers and spiders were also collected while visiting the plants. Spiders and assassin bugs captured *Trigona* bees, the other ones were feeding on perianths and capsules. The *Trigona* bees approach by typical hovering behaviour and land on the flowers. They crawl towards the tip of the labellum, chew and scratch near the basal part and slowly move into the open flower, apparently attracted to something at the base of the column, where they seem temporarily trapped, spending 20-30 minutes there. The labellum holds the insect firmly against the column as its scutum protrudes into the sticky stigmatic cavity, while it struggles to leave the flower. The anthers are deposited on the curved edge of the scutum where stigmatic fluid is smeared. In a second visit the pollinia come into close contact with the column and parts are placed into the sunken stigmatic cavity. Pollination is mostly geitonogamous, with no seed set as a result. Cross-pollinations are scarce, about eight *Trigona* visitations being observed for one successful pollination.

The author observed visitation by *Trigona* bees in *D.monophyllum* during a visit to Tad Bartareau's study site in the Herberton area of Atherton Tablelands, North Queensland, in early September 1998.

Reproduction

In a three-year study Bartareau (1995) studied the reproduction of the species. The self-incompatible rewardless flowers have a capsule-to-flower rate of about 1:14. Pollinators visited about 80% of the flowers, and capsule production was significantly related to inflorescence size and pollinaria removal. About nine pollinator visits occurred per capsule. The inflorescence size classes most successful in capsule production were also the most frequent in natural populations. Experimental outcrossing increased capsule set by 45% (but was limited to about 53%). Such experimental increase led to a significant decrease in subsequent growth and flowering. It is thus suggested that an increase in capsule set above 10% will not necessarily correlate with greater reproductive fitness because of the increased cost of capsule production. The production of surplus flowers appears to function in pollinator attraction and increases fitness through male function.

Dendrobium pugioniforme (=Dockrillia pugioniformis)

Jones (1983) observed a small unidentified bee visit the flowers.

Dendrobium ruppianum (=D.jonesii sps. **jonesii)**

Jones & Gray (1977) studied the pollination of this highly variable common orchid of open forests (formerly also *D.speciosum* var. *fusiforme*). The flowers emit a heavy spicy perfume, released only in the morning on warm days. It is pollinated by Halictidae (*Homalictus* spec.). The dark, small bees, attracted by the perfume, probably seek the dark glandular spot at the base of the column foot, of which the exudate contains unknown and probably delectable substances. It may be a reward system as is claimed for *D.kingianum* and *D.speciosum* by Adams (1991).

The lateral lobes of the labellum form a tube through which the insect forces its way to the foot, passing the middle lobe of the raised callus plate which narrows the gap between labellum and column. The insect brings its thorax near the anther and smears glue onto the thorax on entering. The anther pivots and ejects the pollinia as the insect backs out. The pollinarium becomes attached to the thorax and will be guided against the upper lip of the stigma on a subsequent visit. The bees only need about 10 seconds and then usually move to another flower of a separate raceme.

The author observed visitation of this species by Eastern Spinebills (Meliphagidae) in early September 1998 on a fieldtrip in a primary wood, when visiting Paluma N.P. (Qld.) and environment in the company of B. Lavarack and several other people. The birds were probing into the flowers several times. It is doubtful, however, if these nectar-feeding birds can pollinate this species.

Dendrobium smilliae

The flowers of this species are hard and waxy and somewhat cylindrical, less zygomorphic than most orchid flowers. They lack nectar guides, are red coloured and the inflorescence is more or less cylindrical with a robust stem suitable for alighting birds. The flowers are readily accessible on leafless pseudobulbs.

Smythe (1970) formulated a hypothesis about the same geographical range for the orchid and a bottle bush tree, the latter being limited to that range area by the soil type. Smythe then asked members of his orchid society to help finding the probable bird pollinator in the Townsville area, where a large honey flora is cultivated. Mrs Scott than observed the yellow honey-eater *Meliphaga flava* (=*Lichenostonum flavum*) on *D.smilliae*. The husband, Mr Scott, observed the pollinia being removed.

Dendrobium speciosum

Macpherson & Rupp (1934, 1936) observed visits by *Trigona hockingsi* and *T.symei*, but the bees after escape from the orchid cleaned themselves and got rid of the pollen-masses. They then nevertheless visited other flowers. They also observed larger *Euryglossa subsericea* entering and picking up pollinia and depositing pollen on the stigmas of flowers visited next.

Smythe (1970) observed a native bee, *Trigona carbonaria* (initially twice wrongly named), emerging from a flower of *D.speciosum* with two large pollinia on its thorax, through bending touching the insect's head. The captured and kept insect appeared annoyed by the load, but was unable to remove it. Numerous other visitors were attracted by the heavily scenting flowers, obviously too small to pollinate.

Jones (1983) also observed *Trigona* bees pollinating.

According to Adams & Lawson (1987) there are in this species small variants with narrow tubes, e.g. var. *hillii* (=*D.tarberi*). The flowers of this variety cannot be entered by honey bees and these often knock pollinia to the ground in the struggle. This is detrimental to the orchid, because it removes from these flowers a supply of pollinia for effective pollinators.

Slater & Calder (1988) observed nectar-seeking insects, guided to the central reproductive area by high UV-reflection near the centre and by a bright yellow labellum ridge.

D.speciosum provides a 'big bang' effect or gregarious flowering through visual display of over 1000 flowers per plant and by fragrance (Plates 151 and 152, p. 213). Osmophores are found over the lateral petals and sepals. Bees remain on flowers for about five seconds. The flowers are thought by Slater and Calder to be non-rewarding, but Adams (1991) suggests use of fragrance for social interactions. The species relies on infrequent outcrossing for successful pollination because of self-infertility. Fruit set is low, as Darwin already noted (Slater 1988).

Polllinators and visitors

Pollination is effected by bees of appropriate size. The orchid is visited by a diversity of insects. The non-pollinating dendrobium beetle *Stethopactus formosa* is resident on the plant. Mosquitoes, flies and wasps visit from dawn to mid morning. They are casual visitors.

There is a marked correlation between the intensity of the floral scent and the activity of hoverflies and bees (*Apis mellifera*, *Trigona carbonaria*, *Homalictus brisbanensis*, *Lasioglossum*, *Hylaeus*). Only native bees were observed to effect pollination. Size of the bees in relation to the size of the floral tube is an important characteristic and different bee species seem to pollinate the different varieties. The native bees collected from flowers of *D.speciosum* varied from 4-10 mm long and the bees had a variety of social types and general food collecting behaviours. The visiting bees land on the flower and enter in search of food. They then reverse from the flower immediately and proceed to another flower. If the bee was of appropriate size for the floral tube, it would come into contact with the stigma, built as a sunken cup on the ventral surface of the column, which contains a copious amount of viscid material, and obtain a smear of this stigmatic liquid onto which the pollinia would drop as the anther cap was hinged back (Slater 1995). The stigmatic mucilage is also essential in the dehydration and germination of the pollen. Detached cells of the stigma may provide food for the growing pollen tubes. There is no obvious feeding behaviour (but see Adams 1991, cited above and below). *Homalictis brisbanensis* was observed to pollinate var. *hillii*. With similar behaviour observed for all six species of bees, it seems likely that bee size in relation to flower size is indeed the critical factor in determining pollination success.

Fruit set
Slater and Calder found only 1-2 seed capsules per plant. Adams & Calder (1982) found 2-14% seed set. Fitzgerald noted less than 1% (1891). There is a set limit to the number of seed capsules which develop to maturity following the pollination of large numbers of flowers. Adams (1989) proposes that restriction of plant resources (resource-limitation) in reproductive activity is part of a survival mechanism which balances investment in seed production with needs for vegetative growth, flowering and seeding in future generations.

Variability
In a review Adams (1991) comes to the conclusion that *D.speciosum* forms a complex of regional and highly variable races. Six varieties are recommended for discerning different taxa.

Clements (1989) reviewed the species and has several new species. One of these *D.curvicaule* (*D.speciosum* var. *capricornicum*) is pollinated by a *Trigona* sp. (Adams 1991). Two other ones *D. pedunculatum* (*D.spec.* var. *pedunculatum*) and *D.rex* (*D.spec.* var. *grandiflorum*) probably also. Some varieties appear to be hybrids. One described variety has moved to *D.jonesii* as a subspecies.

Multiple bee pollinators, mainly *Trigona* species, have been observed in four varieties studied so far. The fragrance chemistry of three varieties, *speciosum*, *grandiflorum* and *hillii*, is almost identical in composition and appears to represent a reward system for bees who use the chemicals in their social interactions.

The species is able to regenerate remarkably from basal shoots and from seed after severe bushfires (Adams & Lawson 1984).

Dendrobium stuartii
This species with dull white or greenish flowers without perfume is self-pollinating (Jones 1983)

Dendrobium teretifolium (=Dockrillia teretifolia)
Jones suggests pollination by *Trigona* bees. He observed *T.carbonaria* in a white flowered form. (Jones 1983, Adams et al. 1992)

Dendrobium tetragonum
The floral structure allows for pollination by *Apis mellifera*.

The highly variable species has recently been split into three species (Clements 1989): *D.tetragonum*, *D.cacatua* and *D.capitis-york*.

Dendrobium toressae
Pollinated by the common *Trigona* bees that also pollinate *D.lichenastrum* (Bartareau 1993). Both *Trigona carbonaria* and *T.hockingsi* are involved. Pollinator behaviour, pollination mechanism and pollination syndrome are the same as in *D.kingianum* and *D.speciosum.*

Bartareau (1994) studied the reproductive biology of *D.toressae.* In 166 flowers 72% of the pollinaria were removed, 43% initiated capsules, but only 19% produced capsules. Herbivores and insect predators (spiders, bugs) reduce capsule set. Abortion rates were high (51%). Per capsule produced, about four pollinator visitations with removal of pollinia were recorded.

Experimental self-pollinations give reduced capsule production compared to cross-pollination. As in other *Dendrobium* species the flowers are adapted to outcrossing (xenogamy). Fires appear to affect the epiphytic orchids more than any other environmental factor by damaging plants and by destroying nests of *Trigona* bees in hollow trees.

Dendrobium wassellii
Jones (1983) observed small bees in the flowers of cultivated plants.

Genus **Diplocaulobium**

Two non-endemic species are found in Australia. They are epiphytic. Pollination unknown, but not autogamous.

Genus **Flickingeria**

Two epiphytic species occur in NE Queensland. One on Christmas Island. Not autogamous.

DENDROBIEAE: BULBOPHYLLINAE

This pantropical subtribe of especially the Old World comprises more than 1100 species in 10-15 genera. There is only one genus represented in Australia.

Genus **Bulbophyllum**

Of the more than 900 species Australia has 28 (NSW and Queensland), mostly endemics.

Most species seem to conform to one or another of the fly-pollination syndromes (myophily, sapromyophily). The flies are attracted by scents, colour or by breeze-generated movements of mobile parts of the lip. Many species are nectariferous.

Bulbophyllum baileyi
Smythe observed fruit flies of the species *Strumeta* visit the flowers each morning, four or five to one flower. Pollen was attached to their backs. A fly that pollinated a flower was left suspended in space attached to the stigmatic surface via the pollinia and remained there until it died. The ovary swells considerably in a couple of days and may even push the pollinia off. Smythe states that the fruit flies appear to be the exclusive agent in pollinating the orchid, at least in the area of Townsville and speaks about observations needed to determine the frequency of mortality. It is, however, hard to imagine a pollination system based on the death of a pollinator as a regular phenomenon.

Bulbophyllum longiflorum
Jones & Gray (1976) describe the pollination of the large flowers of *B.longiflorum* (=*B.clavigerum*, =*B.umbellatum*) by a fly *Silba excisa* (Lonchaeidae). The black fly is about 7 mm and has an extremely hunched thorax.

The pollination mechanism is similar to that of other Bulbophyllums of the Cirrhopetalum section.

The insect is attracted by odour and lands somewhere on a flower of the inflorescence, walking to the labellum and starting feeding on the nectar, usually moving towards the base, until the point of balance of the hinged lip is passed and the insect's own weight causes the labellum to tip in towards the column. The fly is guided by two long stelidia of the column and then held firmly between column and labellum. In forcing itself outward it contacts the viscidium and receives the pollinarium on its thorax. In the next flower the pollinarium is forced into the sunken stigma when the fly struggles backward. The process was observed on several occasions at various times, always under very humid atmospheric conditions just before, during or after a shower of rain. The pollinator may visit several flowers on the same inflorescence, slowly walking from flower to flower.

Bulbophyllum macphersonii
An epiphytic and lithophytic species of Queensland, also found in mangroves in Papua New Guinea.

Bartareau (1994) describes pollination by midge flies of the species *Forcipomyia* (*Euprojoanissia*) *santesi*. The non-resupinate flowers with a hinged and tremulous labellum open for a length of ten days. They are xenogamous, but without rewards. They have no obvious fragrance. Pollination occurs in warm weather ($t > 20°C$).

The fly often hovers around open flowers. Up to three insects were observed to visit a single flower. After landing the fly probes the upper and lower surfaces of the lip near the apex and slowly moves inward. The scutum protrudes into the stigma. The fly may search about 10 minutes for rewards and than struggles out. After numerous trials it reverses and wriggles itself free, picking up pollinia with the scutum.

Bulbophyllum weinthalii
This species grows epiphytically on the Hoop Pine *Araucaria cunninghamii*.

The flowers are pollinated by a common blowfly *Calliphora tibialis* (food mimicry syndrome), of which the larvae are reported as parasitising earth worms (Jones 1985).

The flower has a smell similar to fish meal, which is only released on warm days ($t > 15°C$).

Nectar is provided in large quantities from the tongue-shaped labellum with a deeply grooved vee down the centre, in which the droplets are collected. Some nectar may flow down to the column foot and collect in two small pouches formed by the lateral sepals.

The blowflies are attracted to the flowers by the smell they release and actively seek them out. They land somewhere and then wander erratically about, touching various surfaces with their proboscis, Reaching the flower they actively suck the nectar, crawl onto the tip of the labellum causing this to tip outwards under their weight. The fly crawls upwards probing in the groove. Passing the point of balance the labellum tips and the fly is slammed against the column where it becomes temporarily wedged. The struggle leads to the picking up of the pollinarium on the thorax or in a second visit to pollen deposition. Larger blowflies are also attracted but do not pollinate, smaller ones may become entrapped and perish.

VANDEAE: AERIDINAE

Over 1250 species in about 103 genera, mainly from tropical Asia and Australasia.

Genus **Chiloschista**

The genus contains only three species, one is Australian. They are small leafless epiphytes with flat green photosynthetic roots. Pollination unknown. Not self-pollinating.

Genus **Drymoanthes**

Four species in the genus: one in Australia, one in New Caledonia and two in New Zealand. Small evergreen epiphytes. *D.minutus* of Australia is probably self-pollinating.

Genus **Papillilabium**

A monotypic genus endemic to Australia. A small epiphyte (*P.beckleri*) that is not self-pollinating, but the pollinators are not known.

Genus **Parasarcochilus (=Sarcochilus)**

Syn: *Pteroceras*. Small epiphytes. Two species, both endemic to Australia: *P.hirticalcar* and *P.spathulatus*. Both are not autogamous.

Clements (1989) has included them in *Sarcochilus*.

Genus **Peristeranthus**

A monotypic genus, endemic to Australia. A small epiphyte that is not autogamous: *P.hillii*. It has flowers with functional nectaries.

Peristeranthes hillii
Wallace (1980) reported visits by two species of beetles to cultivated plants, 150 km outside the orchid's natural range. A male and female *Agonischius australis* (Elateridae) were trying to copulate on the inflorescence and were not observed to take away pollinia. An unidentified *Methiorrhynchus* sp. carried pollinia. Many flowers had pollinia removed. The anther cap remains intact for some time before it falls off and may be deposited on the stigma of a next visited flower. The delay of several minutes before the pollinia are in the position to pollinate promotes cross-pollination. Wallace formulates a subsyndrome for small beetle-orchid pollination on the small basis of these observations.

Forster (1987) observed several *Methiorrhynchus rufipennis* beetles on inflorescences in the natural habitat. The beetles were feeding on nectar in individual flowers. They had several pollinia attached to the antennae. Fitzgerald mentioned the same pollinator 105 years earlier (Adams & Lawson 1993).

Genus **Phalaenopsis**

The genus of about 35 species found from India to New Guinea has one species extended to Australia: *P.rosenstromii* (*P.amabilis* var. *papuanum*). Pollination unknown.

Genus **Plectorrhiza**

Of the three species of this genus two are endemic to mainland Australia (*P.brevilabris*, *P.tridentata*) and the third (*P.erecta*) to Lord Howe Island. They are small epiphytes that are not autogamous.

Genus **Pomatocalpa**

About 60 species from India to Polynesia, with two species in Australia, one endemic. Epiphytic. *P.macphersonii* is pollinated by small native *Trigona* (*Carbonaria*) bees.

Jones (1981) observed a bee with the dark pollinarium of this species attached to the head entering a flower of *Rhinerrhiza moorei*. Hybridisation is prevented by different placement of the pollinarium.

Genus **Rhinerrhiza**

Two species, both in Australia. One extends overseas. *R.divitiflora* is entomogamous, but pollinators are unknown. *R.moorei* is pollinated by a small native bee (Jones 1988).

Rhinerrhiza moorei
This gregarious flowering species of which the flowers last only one day emit a very spicy perfume. *Trigona* (*Carbonaria*) bees are attracted by the perfume and land on leaves, racemes or perianths before entering the flowers. They move to the labellum and force their way to the base. They probe the yellow glandular mass on the labellum and further down in the sac. They back out after 30-60 seconds and usually take a pollinarium away, adhering to

the thorax. They then spend several minutes cleaning antennae, mouthparts, head and front legs.

Genus **Saccolabiopsis**

A genus with eight species, from India to New Guinea and Australia. Two endemics in Queensland: *S.armitii*, *S.rectifolia*. Small epiphytes, not self-pollinating.

Genus **Sarcochilus**

Australian genus of 16 small epiphytic species, including *Parasarcochilus*. Very little is known of pollination.

Jones (1981) mentions a bee of the genus *Carbonaria* (=*Trigona*) for *S.moorei*, but has now put the species in the genus *Rhinerrhiza* (Jones 1988). Most species appear highly fertile, both in selfing and outcrossing to the same and other species and easily hybridise, although many outcrossings produce no seeds (Adams 1990b).

Genus **Schistostylis**

A monotypic genus. The endemic Australian *S.purpuratus* is not self-pollinating, but no details are known about its pollination.

Genus **Schoenorchis**

Twenty species from India to New Guinea and two species extending into Australia. *S.sarcophylla* appears to be self-pollinating. Pollination of *S.micrantha* (syn. *densiflora*) unknown.

Genus **Taeniophyllum**

A large genus, ranging from India to New Guinea, Japan and Australia with 170 species, of which five are found in Australia.

The species are poorly known. They are leafless epiphytes with green roots. *T.muelleri* is self-pollinating.

Genus **Thrixspermum**

A genus of about 100 epiphytic species, ranging from India to Polynesia. Three species extend into Australia, one is found on Christmas Island. No self-pollination, but otherwise unknown.

Genus **Trichoglottis**

Sixty species from India to New Guinea. One in Australia, *T.australiensis*. It is not self-pollinating.

Genus **Tuberolabium (=Trachoma)**

Seven species from India to New Guinea and Australia. Three species in Australia of which the pollination is unknown, except that they are not self-pollinating.

Genus **Vanda**

About 60 species from India and China to New Guinea.

One species extends into Australia: *V.hindsii*. An epiphyte from Queensland that is not self-pollinating.

New Zealand

Although already many genera and species from New Zealand have been mentioned in the section on Australia, there are good reasons to present a separate section on this country, as will be seen below.

The New Zealand orchids show a great deal of conformity to those of Australia. We find about 125 species in New Zealand of which probably one-third are shared with Australia. In general we find the same genera with in part the same and in majority different species, mostly endemics. Only some five New Zealand genera are not represented in Australia.

It has become evident that the prevailing mechanism of pollination in the native orchids of New Zealand is self-pollination.

Molloy (in St George & Mc Crae 1990) estimates that about 60% or more of all species are predominantly self-pollinating and self-fertilised, which is considerably higher than the 5-20% Catling supposes for the whole family of Orchidaceae. This places the orchid flora of New Zealand in a unique position. Many of these self-pollinating forms are non-resupinate, which is understandable from the adaptive nature of resupination in orchids, towards insect pollination. Self-pollination is generally considered to be a derived condition. It is a means whereby plants survive in unfavourable environments, in particular those in which potential insect pollinators are scarce.

Especially bees and wasps (400 species of Hymenoptera) are poorly represented in New Zealand, compared to Australia (11,500 species). Native bumblebees are absent. The Hymenoptera of New Zealand represent 0.4% of the world total of species. Probably flies, moths, beetles and weevils, which are the larger insect groups (on average 1.3% of the world population of species) are relatively overrepresented as pollinators of New Zealand orchids in comparison to Australia or the rest of the world.

Insect pollination prevails in the epiphytes and in two of the largest genera *Pterostylis* and *Corybas*. Most nectar-producing species are dependent on insects. Most of the records of fragrance involve self-pollinating species, which may reflect derivation from insect-pollinated ancestors.

Godley (1979) noted that of the orchids shared with Australia some are self-pollinating in both countries. Other species are insect-dependent in Australia, but self-pollinating in New Zealand, and still others are insect-dependent in both countries and pollinated by the same or a different insect. In the last case the orchid is either dependent on occasional pollinators (and vegetative growth) or pollinated by different insects. A few orchids demonstrate both conditions, insect-dependence and self-pollination in different parts of New Zealand. E.g. *Thelymitra pulchella* is insect-dependent in Northland, autogamous south of that region.

Knowledge of the pollination mechanisms is far from complete. Even the orchid flora itself is insufficiently known.

Molloy (1990) gives an overview of the pollination systems of New Zealand orchids, which I shall present here in adapted order, for reasons of conformity with the classification system used in this Atlas. When no references are given, the source of the data is Molloy himself, who based his data on observations of wild and cultivated plants, simple experiments and herbaria specimens.

Primitive orchids with a single anther

GASTRODIEAE: GASTRODIINAE

Genus **Gastrodia**

Gastrodia cunninghamii and *G.minor* (Hatch 1954) and a third species not yet described and provisionally named *G.*'long column' are self-pollinating. Moore and Edgar have reported that *G.cunninghamii* has a disagreeable odour, which could indicate pollination by flies as well. The pollination system of *G.* aff. *sesamoides* is undetermined as yet. It is a local species. In Australia *G.sesamoides* is pollinated by bees, but the species from New Zealand is now regarded as different (I. St George).

More advanched orchids with a single fertile anther

SPIRANTHOIDEAE

CRANICHIDEAE: SPIRANTHINAE

Genus **Spiranthes**

Spiranthes sinensis (Hatch 1954, as *S.lancea*) and an as yet undescribed *Spiranthes* 'Motutangi' are autogamous.

ORCHIDOIDEAE

DIURIDEAE: CALADENIINAE

Genus **Adenochilus**

Adenochilus gracilis is autogamous.

Genus **Aporostylis**

Aporostylis bifolia is, according to Godley (1979), cross-fertilised (Thomson 1878), Molloy refers to this as geitonogamy.

Genus **Caladenia**

Some 10 species are recognised, several as yet undescribed.

Caladenia bartletii (was *Caladenia carnea*) and *C.lyallii* are insect-pollinated.

Scanlen (1996) observed a minuscule purple eyed black fly with pollinia in a flower of *C.lyallii*.

C.alata, C. aff. *carnea* (was *C.catenata*), *C.atradenia* (was *C.iridescens*) and *C.minor* are self-pollinating. The same holds for *C.chlorostyla*.

Genus Waireia

Waireia stenopetala, formerly called *Lyperanthus antarcticus*, is self-pollinating and frequently even cleistogamous (Thomson 1878).

DIURIDEAE: DRAKAEINAE

Genus Caleana

Caleana minor reproduces spontaneously by apomixis, but can also be cross-fertilised (Godley 1979). Cheeseman has speculated that gnats would become trapped, but no observations support this.

Thomson (1927) suggested such gnats to be Culicidae.

Genus Chiloglottis

Three species occur in New Zealand. *Chiloglottis cornuta* is autogamous (Thomson 1878, 1880). The pollination of *C.formicifera* and *C.valida* is unobserved. There are no recent records of *C.formicifera* in New Zealand. St George points to the fact that *C.valida* in Australia is pollinated by a thynnid wasp, whereas this species (as *C.gunnii*) is nevertheless insect-pollinated in NZ where no thynnids are found (Molloy & Johns 1983, found no fruit set after exclusion of insects).

DIURIDEAE: PTEROSTYLIDINAE

Genus Pterostylis

Charles Darwin (1862, 1888) used the observations of Cheeseman (1872, 1874, 1880) to describe the pollination mechanism in *Pterostylis* in his famous book on orchid pollination. Many species of this genus are known to be insect-pollinated: *Pterostylis alobula, P.areolata, P.australis* (Thomson 1927), *P.banksii* (inefficiently: Cheeseman 1872, Thomson 1878, Hatch 1945, Godley 1979), *P.brumalis, P.micromega* (formerly abusively considered as *P.furcata*), *P.graminea* (Cheeseman 1872, Scanlen 1996), *P.irsoniana, P.oliveri, P.patens, P.trullifolia* (Thomson 1927), *P.porrecta* (was *P.* aff. *graminea*), *P.* aff. *montana* and *P.agathicola* (was *P.'rubricaulis'*, or *P.graminea* var. *rubricaulis*).

Cheeseman (1872) artificially enclosed small insects into the flowers of *P.alobula* (as *P.trullifolia*, St George 1996) and observed the mechanism of pollinia removal. He also observed dead insects in many flowers, part of these carrying pollen. All of these were Diptera, probably of one species. He supposed less than one in four flowers to produce seeds and about three in four to have 1-4 pollinia removed.

The following species are predominantly self-pollinating: *P.cardiostigma* (Cooper 1983), *P.foliata, P.humilis* (Hatch 1945), *P.montana, P.puberula* (was confused with *P.nana*), *P.tasmanica* (confused with *P.barbata* and *P.plumosa* (Hatch 1987) *P.tristis, P.venosa, P.tanypoda* (=*P.* aff. *cycnocephala*), *P.paludosa* (=*P.'linearis'* and *P.agathicola*.

The pollination system of *P.nutans* is undetermined (occasional Australian vagrant, recently rediscovered).

According to Primack (1983) a wide variety of insects visit the flowers of New Zealand montane plants. *Pterostylis* species however exhibit specialised pollinator relationships with an insect order. Craneflies are observed in the montane *Pterostylis* species.

Pterostylis humilis

Is excellently adapted for self-fertilisation (Hatch 1945). While it is possible for insects to effect cross-pollination, there seems little doubt that at least 75% of the capsules are impregnated by pollen from their own anthers. The anther is directly above the stigmatic plate and the pollinia are extremely sensitive; the slightest touch to the flower is sufficient to cause them to fall.

DIURIDEAE: ACIANTHINAE

Genus Acianthus

Acianthus sinclairii is insect-pollinated (Cheeseman 1874). Cheeseman observed numerous Diptera flitting from flower to flower and taking out pollinia on their backs. Seed set was 85%. *A.viridis* is autogamous.

Genus **Corybas**

Corybas acuminatus, *C.macranthus* (Thomson 1878), *C.rivularis*, *C.trilobus* (Cheeseman 1874), *C.iridescens*, *C.orbiculatus* (was *C.*'short petals'), *C.*'Trotters' (St George 1988, fungus gnat) and *C.* aff. *trilobus* are insect-pollinated. *C.*'whiskers' is attractive for crane flies (Tipulidae) (Scanlen 1996).

C.cheesemanii, *C.cryptanthus* (Hatch 1952, as *C.saprophyticus*), *C.oblongus* and *C.rotundifolius* (*C.* aff. *unguiculatus*) are autogamous. Scanlen (1996) observed a fungus gnat on a flower of *C.cheesemanii*, this suggesting possible entomogamy.

Of the rare species *C.carsei* (=*C.fordhamii*?) the pollination system is still undetermined.

Corybas oblongus, C.rivularis and **C.macranthus**

Vogel (1978) (see also Thomson 1927 and the original paper of Miller in reprint in NZNOG Journal 1996) mentions observations of Miller (1918) of a Mycetophilid *Exechia thomsoni* with pollinia in a flower of *C.oblongus* and of unidentified Culicidae (obviously used as a name for fungus gnats) with pollinia in flowers of *C.rivularis* and *C.macranthus*. There is no nectar, but the labellum, if punctured, exudes a considerable amount of sweetish purple juice, which is probably delicious to insects (Thomson 1927).

In *C.macranthus* Thomson (1878, 1927) found dead and living flies glued to the stigma and wings and legs of flies in flowers, which were probably remnants of spider meals. He supposed these insects to be caught in endeavours to escape from spiders, for withdrawing flies all removed pollinia with their heads. Male function was 63% in 143 flowers.

Corybas iridescens and **C.papa**

Fuller (1994) observed pollination of *C.iridescens* (as *Corybas* '*A*') in 1979 in a fernery by (female) *Mycetophila diffusa*, a fungus gnat. The flies showed no interest in *C.papa* (*C.* 'Mt Messenger'). Other female *Mycetophila* species (*M.colorata*, *M.subspinigera*) also were caught on or near flowers, but without pollen. Similar observations were made in 1980. Entry to the flower is demanding for the gnats, involving considerable folding of the long legs, so the enticement must be exceptionally strong. The gnats bear pollinia on their thorax. Their excited behaviour may point in the direction of attraction by pheromones. St George photographed fungus gnats pollinating a devious late-flowering form of *C.trilobus*, which may be a hybrid of *C.trilobus* and *C.iridescens*.

Genus **Cyrtostylis**

Cyrtostylis oblonga is insect-pollinated, chiefly by minute species of Diptera, which are attracted by the abundance of nectar (Cheeseman 1874).

C.reniformis also. Both species have been misnamed formerly as *Acianthus reniformis* varieties.

DIURIDEAE: CRYPTOSTYLIDINAE

Genus **Cryptostylis**

Cryptostylis subulata is insect-pollinated. Graham (1983), after field observations that failed to reveal any pollinators, experimented with plants transported in pots about 150 km from the natural growth place. He observed pollinations by male *Lissopimpla excelsa*, the same pollinator as known from Australia. The wasps usually alight directly on the labellum and face towards the protruding horn near the apex. They then back rapidly, inserting the tip of the abdomen into the centre of the flower and grasp the outer edges of the labellum with the hind legs. The pseudocopulatory activity lasts up to 90 seconds, but on average only takes eight seconds.

Of 73 flowers, 33 had the pollinia removed and six capsules developed. Only flowers near the base of the racemes were fertilised. Experiments demonstrated attraction of the ichneumoids by scent from the flowers.

DIURIDEAE: DIURIDINAE

Genus **Orthoceras**

Orthoceras novae-zeelandiae is self-pollinating (Johns & Molloy 1983).

DIURIDEAE: THELYMITRINAE

Genus **Calochilus**

Calochilus herbaceus, *C.paludosus* and *C.robert-*

sonii are self-pollinating and probably occasionally insect-pollinated (St George 1994). All three species are also known from Australia.

Wilson (1993) observed and photographed a female *Dilophus nigrostigma*, a 'pollen fly' with long mouth parts on a flower of *C.robertsonii*. It perched head down on the underside of the labellum. This insect is considered a possible pollinator but proof is lacking.

Genus Thelymitra

Thelymitra malvina, *Th.* aff. *longifolia* and *Th.* aff. *pulchella* are insect-pollinated, but *Th.malvina* may fall back on self-pollination. It has a strong fragrance of honeysuckle. *Th.* aff. *pulchella* has crumbly pollen and may also self-pollinate (St George 1998).

Godley names also *Th.venosa*, but there has been confusion with *Th.cyanea.*

Th.aemula, *Th.carnea*, *Th.cyanea* (Thomson 1881), *Th.nervosa* (=*Th.decora*), *Th.circumsepta* (=*Th.formosa*), *Th.hatchii*, *Th.longifolia* (Cheeseman 1880), *Th.matthewsii*, *Th.pauciflora*, *Th.pulchella*, *Th.* aff. *ixioides*, *Th.*'Ahipara', *Th.*'darkie', *Th.'intermedia'*, *Th.*'rough leaf' and *Th.'sanscilia'* are all autogamous.

Of *Th.* x *dentata* the pollination system is undetermined. The name is based on hybrid material (*Th.pauciflora* × *Th.pulchella*) and all specimens examined have sterile pollen.

Hatch (1952) notes that in the NZ *Thelymitra forms* moisture reduces the pollen to a sterile pulp, and since the simple perianth when open provides no cover for the anther, the flower is frequently forced to remain closed. Cleistogamy results from such conditions and probably accounts for the relatively small amount of differentiation to be observed in NZ, where also few hybrids are reported.

Thelymitra longifolia

Cheeseman (1880) studied its pollination and wrote extensively on the variety in ecology and phenology, the occurrence of both (rare) insect-pollination and (common) self-pollination and its consequences for species evolution.

To me his article appears a good antidote against the present habit of splitting species in an extreme way, which risks losing sight of evolution.

Thelymitra aff. longifolia

This species may contain more than one taxon. Thomson (1927), e.g. found lilac- and pink-flowered, very conspicuous forms, which had pollinia removed.

St George (1995, 1996) observed pollination by small black insects and mentions Bruce Irwin who observed solitary bees. The photographs St George took show such a native bee. It is probably a *Leioproctus fulvescens*. It moved vigorously against the column. St George suggests mimicry of manuka, the usual source of pollen for the bee. Removal of pollinia on the abdomen is suggested.

DIURIDEAE: PRASOPHYLLINAE

Genus Microtis

At the moment seven different species are recognised in NZ. But the status of some is unclear.

Traditionally only the three following species were mentioned: *Microtis oligantha* (Johns & Molloy 1983), *M.unifolia* (Thomson 1878) and *M.parviflora*. They are self-pollinating species.

Thomson (1927) observed pollinia removed in just opened flowers in *M.unifolia*, leaving room for some insect pollination as well.

An interesting observation was made by St George (1996). He saw visitation and pollinium removal of *M.unifolia* by an introduced braconid wasp *Cotesia ruficrus* (three insects), brought into NZ to control noctuid pests and also a visit by an *Euryischya* sp., an unusual parasitic chalcidoid male wasp. The flowers scent sweetly and secrete nectar.

Genus Prasophyllum

Prasophyllum colensoi (Thomson 1878), *P.nudum* and *P.pumilum* are autogamous species. The last two are not on the recent list of NZ orchid species of St George (1988).

Thomson (1927) observed pollinia removed from some flowers of *P.colensoi* and states that the flowers are occasionally visited by insects.

Prasophyllum aff. *patens*, a local species, is insect-pollinated.

St George (1996) observed visitation by several species of insects, attracted by the strong almond fragrance and suggests pollination by a beetle species (*Dasytes*).

Out of 40 flowers on two spikes 24 had pollinia removed.

EPIDENDROIDEAE: CYMBIDIOID PHYLAD

CALYPSOEAE

Genus **Danhatchia**

Danhatchia (=*Yoania*) *australis* is considered self-pollinating.

The column is adapted for this purpose, the wings almost enclosing the stigma. The concave labellum, however, in conjunction with the dorsal sepal, forms a narrow tube which could be insect-attractive, so just what happens in nature is open to observation (Hatch 1963).

EPIDENDROIDEAE: EPIDENDROID PHYLAD

EPIDENDREAE: GLOMERINAE

Genus **Earina**

Godley (1979) mentions *Earina autumnalis* and *E.mucronata* as cross-fertilised species, referring to Thomson (1927).

Molloy refers to geitonogamy for the first species. But Thomson was unable to demonstrate this in an experiment in 1878 from which he concluded self-incompability.

Ducker and Marshall (Scanlen 1996) observed a crane fly (Tipulidae) supping nectar from *E.mucronata*. Scanlen himself took a photo of this insect on the flower. The same flower was inhabited by very small straw coloured insects, probably thrips (?).

Another crane fly was observed by Scanlen on a flower of *E.autumnalis*, without sipping nectar however.

DENDROBIEAE: DENDROBIINAE

Genus **Dendrobium (=Winika)**

Dendrobium cunninghamii is insect-pollinated (Thomson 1878).

Recently the species has been renamed as *Winika cunninghamii* in a new monotypic genus. The flowers secrete a considerable amount of nectar.

Thomson observed e.g. pollinia removed in (5) 10 out of (22) 80 flowers in (1878) 1880.

DENDROBIEAE: BULBOPHYLLINAE

Genus **Bulbophyllum**

Bulbophyllum pygmaeum (Cooper 1981) and *B.tuberculatum* (Johns and Molloy 1983) are predominantly self-pollinating species.

VANDEAE: AERIDINAE

Genus **Drymoanthus**

Drymoanthus adversus and *D.flavus* (was *D.*'spotted leaf', described by Thomson (1880) as *Sarcochilus adversus*) are insect-pollinated. The inconspicuous flowers are slightly fragrant and produce nectar and appear self-incompatible (Thomson 1927).

The language of orchid pollination

All words are pegs to hang ideas on
Henry Ward Beecher (1887)

Orchid literature is full of specific biological concepts and also numerous specific orchid terms. Not every term can be defined without the use of long sentences and those who want to acquaint themselves with the topic of this book need to come to terms with this specific vocabulary. An extended glossary can be of help in achieving this.

Achlorophyte: A plant lacking chlorophyl and therefore fully dependent on fungi for its resources, because of the deficiency in photosynthetic capacity
Achlorophyllous: Without chlorophyl
Acropetal: from the base in the direction of the apex
Actinomorphy: Radial symmetry
Adaptation: The process by which an organism becomes fitted to its environment; a character (structure, habit) fitted for some special circumstance
Adaptive radiation: Evolution towards different structures (e.g. pollination systems) within a related group (e.g. genus) as an adaptation to different selection pressures (e.g. pollinators)
Adaxial: Dorsal or backside position, turned to the main axis
Adventitious embryony: Agamospermy (see next) by vegetative propagation/multiplication of cells of the nucellus, the surrounding cells of the egg cell or ovulum, usually leading to polyembryony
Agamospermy: Development of an embryo or seed from a normal (diploid) cell without fertilisation; apomixis in which embryos and seeds are formed asexually
Allele: Gene at a certain locus on a chromosome
Allogamy: Cross-fertilisation
Allopatry: The occurrence of taxa in different geographic areas
Allopolyploidy: Hybridisation in which the bastard has two genomes from each parent. The parents have different sets of chromosomes (genomes). The bastard is tetraploid (four sets)
Allotropous: Used to describe attraction of insects of different orders; allo = other, tropy = movement released by an external stimulus
Allozyme: Enzymes genetically coded as alleles as inferred from electrophoretic analysis of enzyme phenotypes
Androclinium: Anther bed; portion of the column surrounding the anther
Androecium: The collective stamens, i.e. the male parts of the flower
Androeuglossinophily: Attracting male euglossine bees only
Anthecology: Name given to pollination biology by Robertson in 1904
Anthecotype: Flower-variant of a species, adapted to a certain environment
Anther: Pollen-bearing part of the stamen (male organ)
Antherophore: Anther-bearing structure. From anthera (= anther) and phorum (= carrier)
Anthesis: Flowering or opening of the flower-buds
Anthophilous: Attracted by and/or feeding on flowers
Anthropomorphism: From anthropos(=man) and morph (=form); the attribution of a human character to animal behaviour
Apogamy: Reproduction without fertilisation by male cells
Apomixis: Asexual seed production from unfertilised (egg) cells
Apomorph(y): Derived or secondary character (apo = away from). See Cladistic analysis
Appetitive behaviour: Behaviour directed at the search of a goal. First phase of motivated behavioural sequence
Asexual propagation: See Vegetative reproduction
Assortative pollination: Pollination in which the pollinator prefers one (or some) of the morphs in a

polymorphic species during each series of visits
Auricle: Small lateral outgrowth of the column, possibly a sterile anther (in Habenaria e.g.)
Autapomorph: Advanced, evolved character which is derived or secondary, and limited to a single monophyletic group
Autogamy: Self-fertilisation (with pollen)
Autotrophy: The capacity of plants to photosynthesize – literally 'self feeding'
Baiting experiments: Field tests in which flowers are artificially presented to pollinators in their habitats
Batesian mimicry: Mimicry of Bates. See Mimicry
Bichromatic vision: Colour vision based on two types of colour-sensitive cells
'Big bang': Gregarious or mass flowering; occurs usually under consistently warm conditions with intense fragrance and short longevity of the flowers and attracts many insects
Biogeography: Study of the distribution patterns of living organisms on earth in present and/or past
Biological species concept. See Biospecies
Biometry: Method of analysis to determine population variation by measuring characters
Biospecies: Biospecies are groups of interbreeding natural populations that are reproductively isolated from other such groups (Mayr 1970)
Biotope: Part of the biosphere, in which the conditions for life are more or less equal, characterised by certain organisms that form a biocoenosis (= community)
Bract: A small leaflike or scale-like structure at the base of the flower
Bursicle (or bursicula): Sack-like covering over the viscidium (in Habenaria e.g.)
Buzz-pollination: Method of pollen-collecting by bees in which they make shaking movements by thoracic vibration, resulting from contraction of the flight muscles, the vibration being transmitted by the legs to the visited flower with the result that the ripe anthers release the pollen by splitting open or via pores. Buzzing can be heard from a distance
Callus: Fleshy outgrowth e.g. of the labellum
Calyculus: Cuplike structure
Cantharophily: Pollination by beetles
Carryover phenomenon: Conservative behaviour of buzz-pollinators, carrying over buzz-pollination to the first few flowers visited that do not bear a true buzz-pollination form of presentation
Caudicle: Tail-like part of the pollinium, usually composed of viscin and some pollen, not to be confused with the stipe
Chasmogamy: Pollination occurring after opening of the flower
Chemotaxonomy: Study of chemical substances in plants in relation to taxonomy in order to identify relations between organisms of different taxa
Choice experiments: Baiting experiments in which different flowers are presented simultaneously or sequentially
Clade: A group of taxa of monophyletic origin, sharing the same ancestry (clade = branch)
Cladistic analysis: Classification on the basis of derived (= apomorph) characters only
Cladogenesis: Speciation by splitting up of a taxon into two separate lineages during evolution
Cladogram: Graphic picture of evolutionary relationships between organisms of the same clade
Classification: The distribution of plants and animals into classes or divisions of the taxonomical system
Cleistogamy: Self-pollination in an unopened flower
Clinandrium: Androclinium
Clone: Aggregation of genetically identical individuals, derived by vegetative propagation
Co-evolution: Combined evolution involving more than one species by interaction through mutual relations (e.g. competition) which generate selection pressure on the species involved
Column: Central portion of the orchid flower, partial or complete fusion of stamen and style; gynandrium; gymnostemium
Commensalism: Kind of symbiosis in which one species profits, but not at the expense of the other species
Competition: Striving for the same object between species, affecting their survival
Conditioning: Learning process in which a certain stimulus is associated with a certain response by reinforcement through reward or punishment to form a conditioned reflex
Connate: United or jointed parts of sepals, petals, leaves
Conspecific: Of the same species
Constancy: Tendency of the pollinator towards constantly visiting a particular kind of flowers during a certain period as a consequence of learning by classical conditioning
Consummatory (act): Response to a stimulus directed at a goal after this has been located, often stereotyped (fixed action pattern)
Convergence: Development of similarity in nonrelated species, as a consequence of adaptation towards the same environment(al factors). Also parallel evolution

Co-pollinator: Pollinator which is not the one to which a plant is adapted but takes a minor role in pollinating it
Corymb: A raceme in which the stalks of the lower flowers are longer than those of the upper ones
Costs of reproduction: Amount of energy invested in progeny
Cretaceous: Last division of the Mesozoic era, 70-135 million years ago when orchids are thought to have appeared
Cross-fertilisation: Fertilisation by pollination with pollen from a different plant
Crypsis: Adaptation of organisms to make them unconspicuous to enemies
Cryptic species: Species that is so similar to a different species that it has been overlooked or not yet recognised
Cupula: Cuplike structure
Cytotaxonomy: Study of chromosomes and their behaviour in processes of speciation
Deception: Attraction of pollinators in spite of offering non-rewarding flowers
Demographic data: Statistical data of the dynamics of populations
Density dependence hypothesis: Hypothesis stating that at low population densities deception pollination systems have no fitness cost (low-density hypothesis)
Differential reproduction: One of the postulates of Darwin's theory of evolution is the differing number of progeny of individuals as a basis for selection
Disassortative pollination: Pollinators alternate between the different morphs of a polymorphic species
Dufour glands: Abdominal glands in bees
Elaiophores: Glands producing a fatty oil
Electrophoresis: Separation of proteins by their migration movements in an electric field
Embryology: The study of development of embryos
Endosperm: Nutritive tissue of seeds
Entomogamy: Fertilisation by insects
Entomophily: Dependence or reliance on insects for pollination (or transport of seeds)
Epichile: Terminal (distal) portion of the labellum (e.g. *Epipactis*)
Epiphytic: Living on trees
Euglossinophily: Attracting euglossine bees
Facilitation: Promotion of performance of a certain behaviour
False prey syndrome: Mimesis (mimicry) by a flower of the prey of a predatory insect to provoke pollination: pseudocarnivory
Fecundity selection: Selection on reproductive success (fecundity) of individuals, either through reduction of male or female function.
Female fitness: (Differential) reproduction success of the investment of the mother plant in the offspring
Female function: Seed set
Fidelity: Constancy for one single type of flower
Filament: Part of the stamen; sterile thread bearing the anther
Fitness: Survival value measured as differential reproduction success (between organisms of the same species
Fixed action patterns: Stereotyped species-specific movements, hardly susceptible for modification by learning
Flower fidelity: See Fidelity
Flower induction: A number of factors may induce flowering like chemical factors, temperature and light
Founder effect: The result of the fact that only a small fraction of the genetic variation (variance) of a species population is present in the few individuals colonising a new area
Gamokinetic: From gamo (= unite) and kineoo (= to move); movements to unite the sexes
Geitonogamy: Fertilisation by pollen from the same plant or clone, but not from the same flower
Gene-pool: The total of genes of a breeding population
General food source mimic: Flower that mimics a group of similar flowers of different species, either by Müllerian or Batesian mimicry
Genetic drift: Random fluctuations in gene frequencies in effectively small populations (e.g. due to isolation from a larger population)
Genetic erosion: The disappearance of genetic variance by loss of parts of the population
Genetics: The study of heredity
Genetic variance: Variation of (allelic) genes within a population
Genotype: All the individuals sharing the same genetic constitution; also used in reference to a particular gene or allele
Geographic isolation: Formation of a separate population (reproductively isolated) by splitting up in space
Geophily: Positive growth response towards the gravitational field of the earth
Gestalt: A structure or form perceived as a whole on the basis of a search image.
Glenion: More or less circular structure on the front surface of the lip, a 'little eye', in some Pleuro-

thallidinae, that is supposed to play a role in attracting the pollinator
Guild mimesis: See Nonmimic deception
Gynaecium: The female part of a flower, i.e. the pistil
Gynandrium: Column formed by the combination of androecium and gynaecium
Gynandro-euglossinophily: Attraction of female euglossine bees
Gynoecium: See Gynaecium; (gyn = female; oecium = house)
Gynostemium: Column; sometimes also Gynostegium
Habituation: Conditioned (un-)learning in reaction to a stimulus without reward or punishment
Hand-pollination: Artificial pollination
Hemipollinarium: A half of a pollinarium that can be taken away separately (as in Platanthera's) because it has a distinct viscidium
Herkogamy: Condition of spatial separation of sexual parts (anther and stigma) to prevent self-fertilisation
Hermaphroditic: Having both sexes on the same flower
Heterosis: Condition in which the heterozygous form of two alleles gives the organisms a better chance of surviving than both homozygous forms
Heterotrophy: The dependence of plants on external sources for their organic substances
Heterozygous: Combination of two different alleles in one organism
Hexose-dominant: Nectar rich in hexoses (= monosaccharides: viz. fructose and glucose), referring to the sucrose/hexose ratio as being < 0.1, i.e. containing less than 10% sucrose
Homolog(ous)(y): Trait or character derived from the same embryonic structure; related by common ancestry
Homoplasy: Trait that has evolved more than once in a clade
Homozygous: Combination of two identical alleles in one organism
Hybridisation: producing of progeny from parents of different species
Hygroscopy: Movement in reaction to air humidity
Hypochile: Basal part of the labellum (e.g. *Epipactis*)
Imprinting: Fast learning-process during a sensitive period, which is difficult to reverse. May have dramatic consequences for later behaviour. A reward is not necessary
Inbreeding: Opposite of outcrossing; breeding with close relatives (e.g. within clones or by geitonogamy) or by self-pollination
Inbreeding depression: the observable effects of inbreeding, due to homozygosity of damaging recessive genes
Infrageneric: See Intrageneric (infra = within)
Insight learning: Learning by reasoning, i.e. combining spontaneously two or more isolated experiences to form a new experience, which is effective for obtaining a desired end
Instinctive behaviour: Species-specific stereotyped innate behaviour
Integument: Around the embryo two layers, called integuments develop, an inner and an outer, the first usually disorganises, the second becomes the seed coat. As a result there is some space between seed coat and embryo
Intergeneric: Between genera
Intrageneric: Within genera
Introgression: Transfer of genes from one species to another by hybridisation and backcrossing. Fertile progeny crosses back with one of the parents. Genes spread in one direction. Reciprocal introgression goes in both directions
Kairomones: Pheromone imitations, produced by flowers to attract insects
Karyology: Study of substances and processes in the plant (or animal) cell's nucleus, like DNA-content, numbers of chromosomes etc.
Klinotaxis: Orientating movement of an organism directed by information gathered by continuously turning to either side
Labellum: Median petal, the usually largest of the tepals, a morphological structure specially adapted to pollinators
Latent learning: Learning by association of indifferent stimuli or situations without a visible reward. What is learnt remains latent until a later moment
Learning behaviour: Ability to modify behaviour on the basis of experience
Lek behaviour: A lek is a place where certain spectacular male birds (e.g. hummingbirds) gather to display in order to attract females to mate. The same behaviour is known from euglossine bees. It appears a means of assuring mating between animals where one of the sexes has a vagabond life style
Lineage: A line of common descent
Long-range attraction: Attraction of insects by flowers from a relatively long distance by odour or visual cues
Long-term memory: Memory of events in the past of long duration

Low density hypothesis: See Density dependence hypothesis
Lumper: Taxonomist who tends to limit the number of species
Magnet species hypothesis: Hypothesis concerning a deceptive species being attractive to a pollinator when growing near similar rewarding species
Major responders: Insects that show behaviour or other evidence in baiting experiments that pollinia removal occurs or could occur. Allopatric major responders occur outside the range of the orchid involved in the test and are not pollinators. Sympatric major responders are either confirmed, putative or potential pollinators
Male efficiency factor: Ratio between the percentage of pollinated flowers and the percentage of flowers functioning as donors of pollen
Male fitness: (Differential) reproduction success of the investment of the father plant in the offspring
Male function: Deposition of pollen
Massulae: Compound grains of pollen tetrads, held together by elastic material
Mate mimicry: Mimicry of flowers imitating females of pseudocopulating insect species
Megasporocyte: Early phase of the seed which develops into the embryo
Melittomy(i)ophily: Pollination syndrome whose features suggest pollination by syrphids and bee flies
Melittophily: Bee-pollination
Mesial: Pertaining to, situated or directed towards the middle, esp. the middle line of a body
Mesochile: Middle part of the labellum
Mesonotum: Middle part of the dorsal surface of an insect
Micropyle: Small opening at the rim of the integuments of the embryonic seed
Mimicry: Adaptation leading to resemblance, for imitation or simulation to serve for protection or deception. Two forms exist: Batesian mimicry, in which the function is to simulate, in order to profit one-sidedly and Müllerian mimicry, imitation to profit mutually
Minor responders: Insects attracted to flowers in baiting experiments which show no evidence of behaviour indicating that pollen removal could occur
Miocene: Geological period from about 26 million to 7 million years ago
Modern synthesis: Evolution theory based on the Darwinistic theory of selection and mutation incorporating insights from population genetics, ecology, embryology, palaeontology and other relevant sciences
Monophyletic: Derived from one common ancestor. Refers to a group with one or more synapomorphic traits from one common ancestor and all its descendants. If not all descendants are included the group is called paraphyletic
Monophilous: A flower attracting one kind of pollinator
Morphology: Study of organisms by comparison of forms and building plans
Morphometry: Measuring of morphological characters for taxonomical purposes; cf. numerical taxonomy
Müllerian mimicry: Mimicry of Müller. See Mimicry
Multipaternal pollination: Pollination by pollen from more than one parent plant
Mutualism: Symbiosis in which both organisms profit mutually
Mycetomy(i)ophily: Pollination syndrome suggesting pollination by fungus gnats and fungus-loving flies
Mycetophily: See Mycophily
Mycorrhiza: Symbiotic combination of roots and infecting fungus
Mycophily: Attracted by fungi (as food or place for oviposition)
Mycotrophy: Fungus-assisted nutrition
My(i)omelittophily: See Melittomy(i)ophily
My(i)ophily: Fly-pollination (syndrome)
Myrmecophytic: Living with ants
Naiveté exploration: Exploration of flowers by inexperienced pollinators with polytrophic and/or polylectic foraging habits
Natural selection: Selective action caused by population pressure in the struggle for survival
Near-UV absorbance: Absorption of near-UV light (370-390 nm). In *Dendrobium* this may lead to a dramatic bullseye pattern in the central part of the flower
Near-UV reflectance: Reflectance of near-UV light (370-390 nm) as observed in many Dendrobiinae, leading to patterns in the flowers that probably function as visual cues for pollinators
Necrosis: Degeneration of cells by dying and becoming black
Nectar guide: Markings or structures on the labellum or petals, leading the pollinator to the nectar (in orchids often simulating)
Nectary: Nectar-container, usually but not necessarily a spur or concave structure

Nominalistic species concept: An abstract concept in which a species as a non-existing entity is invented to describe a large number of individuals collectively

Nonmimic(ry) deception: Deception based on naiveté; early blooming before the naïve pollinator establishes a rewarding search-image. The 'mimic' species lacks any specific model. Also called nonmodel mimicry/deception and guild mimesis

Nonmodel mimicry/deception: See Nonmimic deception. The pollinator has a generalised search image for flowers based on experience with a guild of species and which includes the (deceiving) species concerned

Numerical taxonomy: Taxonomy based on comparing measures of morphological characters

Oligotrophy: Plants adapted to nutrient-deficient habitats

Ommatidia: Single subunits of the insect eye

Ontogeny: Development of the individual from embryo till death

Opportunistic group foraging strategy: Strategy in which the foraging animals search independently, but after finding a foodsource communicate this and recruit help

Ornithophily: Bird-pollination

Orthogenesis: Supposed internally directed evolutionary trend

Osmophores: Scent producing organs or cells

Outcrossing: Cross with a different genotype

Ovary: Part of the flower that develops into the fruit

Oviposition: Egg-laying; egg-deposition

Ovule: Undeveloped seed

Panmixis: Thorough mixing of genotypes in a population

Parallel evolution: See Convergence

Paraphyletic: See Monophyletic

Parasitism: Symbiosis of organisms in which one species (a parasite) profits at the cost of the other (the host)

Parsimony: Methodological criterion in cladistic taxonomy; principle by which a choice is made in favour of the simplest of solutions taking the smallest amount of steps in a cladogram to determine a phylogenetic relationship

Parthenocarpy: Development of fruits without fertilisation. The expression is or should be limited to fruits with infertile seeds

Parthenogenesis: Development from egg cell without fertilisation

Perception: Process of receiving information by stimulation of the sense organs

Perianth: Collective term for sepals and petals (together tepals)

Pericarp: Wall of the fruit, cover of the seed. From peri (= around) and carpus (= fruit)

Petal: Coloured inner leaf of the flower

pH: Measure of acidity (negative logarithm of H^+ (ion) concentration)

Phalaenophily: Moth-pollination (by moths other than sphingids)

Phenetic method: Method of analysis to determine population variation

Phenotype: The observable functional and structural properties of an organism, resulting from environmental influences on the genotype

Pheromones: Sex attractants, fragrant substances excreted by female insects to attract males

Photoperiodism: The relative length of light and dark periods, which may determine flowering. There are short-day, long-day and neutral-day plants

Phylogeny: Development of the species (or group of living organisms) in evolution

Pistil: Female part of the flower (ovary, style and stigma)

Plant-geography: Study of patterns of spatial distribution of plants in time on earth

Pleistocene: Ice Age, 2 million to 5000 years ago

Plesiomorph: Original (primary) evolved character

Pleurotriby: pollinia attached laterally to the head

Pollen: One-celled spores borne in the anthers, containing the male gametes

Pollen limitation: Limitation of fruit set by the amount of pollen. See Pollination limitation

Pollinarium: Complete set of pollinia, including viscidium and stipe (if present)

Pollination biology: The subject of this Atlas; the study of all aspects regarding the pollination of flowers

Pollination limitation: Condition in which the production of seeds or fruits is limited by the number of pollinator visits

Pollination syndrome: A complex of features that suggests adaptation to a particular class of pollinator

Pollinator limitation: See Pollination limitation

Pollinia: Packages of pollen masses (2-12) in the anther

Polycarpy: From poly (= many) and carpus (= fruit)

Polylectic: (Insects with) a wide choice of foodplant species

Polymorphism: Variability in plants, expressed in the existence of different morphs in one species, e.g. yellow and red flowers
Polyphyletic: Derived from more than one ancestral species
Polyploidy: More than two sets of chromosomes (*n* = three or more)
Polytrophic: (Insects) with a wide array of foods
Population dynamics: The study of changes in populations
Potential pollinators: Pollinators showing pollination behaviour, e.g. pseudocopulation with the labellum (in *Chiloglottis*), but removal or deposition of pollinia not yet observed
Praespecies: Subspecies- morphological or ecological race (cf. also biospecies)
Prementum: Distal part of the labium in insects
Prey-imitation: Attraction of pollinators by mimicking a prey in which the insect lays eggs
Pronotum: Dorsal side of the first thorax segment
Protandry: Condition in which the stamen(s) mature(s) before the pistil reaches maturity
Protogyny: Also Proterogyny; condition in which the pistil matures before the stamen(s)
Pseudoantagonism: Aggressive action taken by insects (bees) against flowers which they perceive as intruders into their territories or as enemies
Pseudocarnivory: Aggressive action taken by insects against flowers which they perceive as a prey
Pseudocopulation: Attempts of insects to copulate with an insect-mimicking flower
Pseudoparasitism: Attempts of parasitic flies or wasps to lay eggs in structures of flowers simulating their normal prey
Pseudopollen: Pollen-imitating structures in a flower to deceive pollen-feeding insects
Psychophily: Butterfly-pollination
Punctuated equilibrium: A model of evolution in which species are relatively stable and long-lived and in which new species appear during concentrated outbursts of rapid speciation
Putative pollinators: Insects carrying pollinia that likely have come from the local population of the orchid involved
Remote species hypothesis: Hypothesis concerning a deceptive species attracting pollinators when growing in remote habitats where attractive flowering species are rare
Rendez(-)vous flowers: Flowers that are visited by male insects to meet females
Resource-limitation: Condition in which the number of seeds or fruits is limited by the amount of received or uptaken nutrition
Resource parasitism: Interaction for pollinator visitation between two plant species in which one no-reward mimic draws away pollinators from a rewarding model (= Batesian mimicry)
Resupination: The twisting (mostly 180°) during the ontogeny (=individual embryonic development) of the flower that usually brings the labellum into the downward position
Rhinomy(i)ophily: Pollination syndrome suggesting pollination by long-tongued flies
Rostellum: Portion of column, separating anther and functional stigma, often beaklike; sometimes it produces viscin, sometimes viscin and retinacles (retinaculae), sometimes one or two bursicles (bursiculae) surrounding viscin and retinaculae and sometimes retinaculae with stipes (appendices)
Sapromyophily: Pollination by carrion flies
Scanning electron microscopy: Technique for visualising an object in three dimensions by reflected electrons sent through it
Search image: Generalised image stored in the memory used to compare observations with, formed by conditioning
Selection pressure: (Total of) limiting environmental factor(s) relating to the survival of a species
Self(in)compatibility: (In)capacity to produce seeds by self-pollination
Self-sterility: Selfincompability
Sepal: Outer leaf-element of the flower, usually green(ish)
Sex-attractants: Substances from an animal species (male or female) attracting the other sexe
'Sexploitation': Exploitation of (different forms of) the sexual behaviour of pollinators for pollination purposes
Sexual attraction: Attraction between the sexes of a species
Sexual selection: Selection of female or male characters
Short-range attraction: Attraction of insects by flowers from nearby
Short-term memory: Memory of recent events, easily disturbed by various influences
Sibling species: Species with the same morphology, but different scents and pollinators (cryptic species)
Speciation: The formation of species through evolution. Can be either allopatric or sympatric if the splitting up of one species into two happens

either by geographical isolation or by phenological isolation, e.g. different pollinators for early and late flowers

Sphingophily: Pollination by Hawk moths (Sphingidae)

Splitter: Taxonomist who tends to split up a taxon considered as a single species by others into as much species as possible

Spur: Slender tubular or sack-like projection of a flower, containing nectar (often not so in orchids), usually situated at the base of the labellum

Stabilizing selection: Selection against individuals with extreme characters

Stamen: Male element of the flower, usually consisting of anther and filament

Staminodia: Part homologous with a stamen, but infertile

Stelidia: Appendages of plumed structures (hairy callus) on the column of some orchids, considered as rudimentary stamens that may function as food for the pollinator

Stigma: Receptive part of the pistil

Stigmatophore: From stigmatis (= stigma) and phorum (= carrier). In *Habenaria* the fertile stigma lobes are borne on long stalks

Stipe: Releasing part of rostellum, connecting viscin and top of pollinium

Subspecies: Geographical race

Sucrose-dominant: Containing less than 0.1% sucrose

Summation: Capacity of the central nervous system to add up stimuli that come in at different times (temporal summation) or from different places (spatial summation)

Supernormal: Stimulus that elicits a stronger reaction than the adequate stimulus

Sympatry: The occurrence of taxa in the same geographical area

Symplesiomorphy. A trait derived from an earlier common ancestor and that is therefore shared with other taxa than the most recent common ancestor

Synapomorphy: An evolutionary new shared trait of organisms derived from the most recent common ancestor

Syndrome: Group of concomitant symptoms, e.g. pollination syndrome = group of characters attracting a particular class of pollinators

Syntope: Living in the same place (cf. Sympatry)

Systematics: Science of classification in which the organisation, history and evolution of living organisms are studied in order to create a natural system

Taxon: Unit of classification (of all levels)

Taxonomy: Science of classification

Terrestrial: Living on or partly in the ground

Tetrads: Unit of four pollen grains, sometimes sticking together

Traplining: Regular flying over long distances of euglossine bees, hawkmoths or hummingbirds (etc.) along certain foraging routes from flower to flower

Trichomes: Plant hairs (or scales)

Trichromatic vision: Colour vision based on three types of colour-sensitive cells, each sensitive to different wavelengths; bees for example have receptors for UV, blue and green

Tropotaxis: Orientating movement of an organism directed by simultaneously compared information from paired senses

Turgor: Pressure on the wall of cells, due to osmotic concentration of the cell-contents

Typological species concept: a concept in which a species is based on a generalised ideal type

Unipaternal pollination: pollination by pollen from a single parent plant

Vegetative propagation (reproduction): Reproduction by vegetative means, e.g. tubers, bulbils or pseudobulbs

Vibrissae: Vibrating structures functioning as visual attractants

Viscidium: Viscid part of the rostellum, removed with the pollinia and attaching these to the pollinator

Viscin: Binding, viscid material of pollen grains in pollinia

Window-theory: Theory about transparent spots in trap-flowers, attracting trapped insects and directing them to follow a route through the flower which leads them along the column

Xenogamy: Pollination by pollen of an other individual (clone)

Zygomorphy: Bilateral symmetry

References

But science, which from the very beginning has been content with knowledge and ascends to heaven when it appears in print, leaves nothing out

Erhart Kästner, 1956

Ackerman, J.D. 1975. Reproductive biology of *Goodyera oblongifolia*. *Madroño* 23: 191-198.

Ackerman, J.D. 1977. Biosystematics of the genus *Piperia* Rydb. (Orchidaceae). *Bot. J. Linn. Soc.* 75: 245-270.

Ackerman, J.D. 1981. Pollination biology of *Calypso bulbosa* var. *occidentalis* (Orchidaceae): a food deception system. *Madroño* 28: 101-110.

Ackerman, J.D. 1983a. Euglossine bee pollination of the orchid *Cochleanthes lipscombiae*: A food source mimic. *Am. J. Bot.* 70(6): 830-834.

Ackerman, J.D. 1983b. Diversity and seasonality of male euglossine bees (Hymenoptera: Apidae) in central Panama. *Ecology* 64: 274-283.

Ackerman, J.D. 1983c. On the evidence for a primitively epiphytic habit in orchids. *Syst. Bot.* 8: 474-477.

Ackerman, J.D. 1983d. Specifity and mutual dependency of the orchid-euglossine interaction. *Biol. J. Linn. Soc.* 20: 301-314.

Ackerman, J.D. 1984. Mechanism and evolution of food-deceptive pollination systems in orchids. *Am. J. Bot.* 71(5): 2 (Suppl.).

Ackerman, J.D. 1985. Pollination mechanisms of tropical and temperate orchids. *Proceedings 11th WOC*: 98-101. *Miami, Florida.*

Ackerman, J.D. 1986a. Mechanisms and evolution of food-deceptive pollination systems in orchids. *Lindleyana* 1(2): 108-113.

Ackerman, J.D. 1986b. Coping with the epiphytic existence: pollination strategies. *Selbyana* 9: 52-60.

Ackerman, J.D. 1989. Limitations to sexual reproduction in *Encyclia krugii* (Orchidaceae). *Syst. Bot.* 14(1): 101-109.

Ackerman, J.D. 1990. Short- and long-term limitations to fruit production in a tropical orchid. *Ecology* 71: 263-272.

Ackerman, J.D. & M. Galarza-Pérez 1991. Patterns and maintenance of extraordinary variation in the Caribbean orchid *Tolumnia (Onc.) variegatum. Syst. Bot.* 16: 182-194.

Ackerman, J.D. & M.R. Mesler 1979. Pollination biology of *Listera cordata* (Orchidacae). *Am. J. Bot.* 66(7): 820-824.

Ackerman, J.D. & A.M. Montalvo 1983. Limitations to natural fruit production in *Epidendrum ciliare. Am. J. Bot.* 70: 43.

Ackerman, J.D. & A.M. Montalvo 1990. Short- and long-term limitations to fruit production in a tropical orchid. *Ecology* 71: 263-272.

Ackerman, J.D. & J.C. Montero-Oliver 1985. Reproductive biology of *Oncidium variegatum*: moon phases, pollination, and fruit set. *AOS-Bull.* 54: 326-329.

Ackerman, J.D., J.A. Rodríguez-Robles & E.J. Meléndez 1994. A meager nectar offering by an epiphytic orchid is better than nothing. *Biotropica* 26(1): 44-49.

Ackerman, J.D., A. Sabat & J.K. Zimmerman 1996. Seedling establishment in an epiphytic orchid: an experimental study of seed limitation. *Oecologia* 106: 192-198.

Ackerman, J.D., M.R. Mesler, K.L. Lu & A.M. Montalvo 1982. Food-foraging behaviour of male Euglossini (Hymenoptera: Apidae): Vagabonds or trapliners. *Biotropica* 14: 241-248.

Adams, H. & E. Anderson 1958. A conspectus of hybridisation in the Orchidaceae. *Evolution* 12: 512-518.

Adams, P.B. 1986. *Dipodium punctatum* (Smith) R.Br.-Alba form in East Gippsland. *The Orchadian* 8: 169-172.

Adams, P.B. 1989. The capacity of *Dendrobium speciosum* Smith to carry seed capsules to maturity. *Orchadian* 9: 201-204.

Adams, P.B. 1990. Evolution, pollination biology and future of Australian Orchidaceae. In *Proc. 1st Aust. Native Orchid Conf.*: 64-86.

Adams, P.B. 1991. Variation, multiple pollinators and breeding system in *Dendrobium speciosum* Smith: A biological review. *The Orchadian* 10: 124-140.

Adams, P.B. & S.D. Lawson 1984. The effects of bushfire on Victorian epiphytic and lithophytic orchids. *Orchadian* 7: 282-286.

Adams, P.B. & S.D. Lawson 1987. Pollination of *Dendrobium kingianum* Bidw. by the Honey Bee (*Apis mellifera*). *The Orchadian* 8: 250-251.

Adams, P.B. & S.D. Lawson 1988. Multiple bee pollinators of *Dendrobium kingianum* Bidw. in the natural habitat. *The Orchadian* 9: 103-107.

Adams, P.B. & S.D. Lawson 1993. Pollination in Australian Orchids: A critical assessment of the literature 1882-1992. *Flora (Jena)* 188(4): 383-386.

Adams, P.B., T. Bartareau & K.L. Walker 1992. Pollination of Australian orchids by *Trigona* (*Tetragona*) Jurine bees (Hymenoptera: Apidae). *Aust. Ent. Mag.* 19: 97-101.

Adams, R.M. 1975. The reproductive biology of the epiphytic orchids of Florida 1. A statement of the problem. *AOS-Bull.* 44: 699-707.

Adams, R.M. & G.J. Goss 1975. The reproductive biology of the epiphytic orchids of Florida 2. A method for observing night pollination. *AOS-Bull.* 44: 1091-1093.

Adams, R.M. & G.J. Goss 1976. The reproductive biology of the epiphytic orchids of Florida 3. *Epidendrum anceps* Jaquin. *AOS-Bull.* 45: 488-492.

Akam, M. 1990. A possible pollinator of *Liparis reflexa. J. Orchid Species Soc. Vict.* 9(4): 3.

Alexandersson, R. & J. Ågren 1996. Population size, pollinator visitation and fruit production in the deceptive orchid *Calypso bulbosa. Oecologia* 107: 533-540.

Allen, P.A. 1951. Pollination in *Coryanthes speciosa. AOS-Bull.* 19: 528-536.

Antlfinger, A.E. 1991. Temporal patterns of reproduction and above ground growth in the terrestrial orchid *Spiranthes cernua. Am. J. Bot. Suppl.* 78: 74.

Arditti, J. 1966. Orchids. *Scient. Amer.* 204: 70-78.

Arditti, J. 1992. *Fundamentals of orchid biology.* New York: John Wiley & Sons.

Arditti, J., B.H. Flick & D.C. Jeffrey 1971. Post-pollination phenomena in orchid flowers. *New Phytol.* 70: 1125-1141.

Argue, C.P. 1995. Notes on the pollination of grasspink, *Calopogon tuberosus* (L.) BSP., rose Pogonia, *Pogonia ophioglossoides* (L.) Juss., and dragon's mouth, *Arethusa bulbosa* L. *North Amer. Native Orchid J.* 1: 274-285.

Argue, C.P. 1996. Pollination notes on Minnesota orchids: hart-leaved twayblade, *Listera cordata. North Amer. Native Orchid J.* 2: 20-27.

Armstrong J.A. 1979. Biotic pollination mechanisms in the Australian *flora. NZJ Bot.*: 17: 467-508.

Atwood, J.T. 1982. How is Paphiopedilum pollinated? *AOS-Bull.* 51: 1057-1058.

Atwood, J.T. 1983. *Dendrobium rupicola.* AOS-Bull. 52: 22-23.

Atwood, J.T. 1985a. The range of *Cypripedium kentuckiense. AOS-Bull.* 54: 1197-1199.

Atwood, J.T. 1985b. *Paphiopedilum rothschildianum.* Broodsite deception. *Nat. Geogr. Res.* 1: 274-254.

Auclair, A.N. 1972. Comparitive ecology of the orchids *Aplectrum hyemale* and *Orchis spectabilis. Bull. Torrey Bot. Club* 99: 1-10.

Backhouse, G. & J. Jeanes 1995. *The orchids of Victoria.* Melbourne Univ. Press.

Baker, H.G. 1963. Evolutionary mechanisms in pollination biology. *Science* 139: 877-883.

Baker, H.G. 1983. An outline of the history of anthecology, or pollination biology. In L. Real (ed.), *Pollination Biology*: 7-28. Academic Press.

Baker, H.G. & I. Baker 1985. Floral nectar sugar constituents in relation to pollinator type. In C.E. Jones & R.J. Little (eds), *Handbook of Experimental Pollination Biology*: 117-141. New York: Nostrand Reinhold Cy.

Banks, D.P. & D.I. Smedley 1997. *Dendrobium lawesii* F. Meull. Its history and notes on cultivation. *The Orchadian* 12: 152-153.

Bänziger, H. 1994. Studies on the natural pollination of three species of wild Lady-slipper orchids (*Paphiopedilum*) in Southeast Asia. In *Proc. 14th WOC, Glasgow 1993*, pp. 101-102. HMSO Edinburgh.

Barclay-Estrup, P., T.E. Duralia & A.G. Harris 1991. Flowering sequence of the orchid genus *Goodyera* in Thunder Bay District, Ontario. *Rhodora* 93: 141-147.

Bartareau, T. 1993a. Some observations on the pollination of *Dendrobium monophyllum* F. Muell. in north-east Queensland. *The Orchadian* 10: 446-450.

Bartareau, T. 1993b. Pollination of Australasian epiphytic orchids by *Trigona* Jurine bees in North-East Queensland. In W.K. Harris & N.J. Grundon (eds), *Proc. Second Austr. Native Orchid Conf. and Show*: 85-89. Nat. Orchid Soc. of Toowoomba Inc. Queensland.

Bartareau, T. 1994a. The reproductive ecology of *Dendrobium toressae* (Bailey) Dockr., a geographically restricted species in north-east Queensland. *The Orchadian* 11: 106-112.

Bartareau, T. 1994b. Pollination of *Bulbophyllum macphersonii* Rupp by a midge fly (*Forcipomyia sauteri*). *The Orchadian* 11: 255-258.

Bartareau, T. 1994c. Foraging behaviour and habitat requirements of the Australian orchid bee (*Trigona carbonaria*). Progress Report to Australian Orchid Foundation.

Bartareau, T. 1995a. Pollination and breeding systems in varieties of *Dendrobium canaliculatum* and their implications on the taxonomic status of the group. *The Orchadian* 11: 381-387.

Bartareau, T. 1995b. The pollination of *Cymbidium madidum* Lindley and *C.suave* R.Br. by native bees – a case of floral reward deception? *The Orchadian* 11: 487-495.

Bartareau, T. 1995c. Pollination limitation, costs of capsule production and the capsule-to-flower ratio in *Dendrobium monophyllum* F.Muell. (Orchidaceae). *Austr. J. Ecol.* 20: 257-265.

Bartareau, T. & B. Jackes 1994. Some observations on the pollination of *Pterostylis procera* M. Clements and D. Jones in North-east Queensland. *The Orchadian* 11: 198-199.

Barth, F.G. 1991. Insects and flowers. The biology of a partnership. Princeton University Press.

Bates, R.J. 1979. *Leporella fimbriata* and its ant pollinators. *J. of the Native Orch. Soc. of South Austr.* 3: 9-10.

Bates, R.J. 1981. Observations of pollen vectors on a putative swarm of *Microtis*. R.Br. *The Orchadian* 7: 14.

Bates, R.J. 1984a. Pollination of *Prasophyllum elatum. The Orchadian* 8: 14-17.

Bates, R.J. 1984b. Pollination of *Caladenia*: an overview. *The Orchadian* 7: 269-270.

Bates, R.J. 1984c. Australia's colourful sun-orchids. *Thelymitra. Austr. Orchid Rev.* 49: 109-112.

Bates, R.J. 1985a. Colourful Caladenias. *Aust. Orchid Rev.* 50: 6-11.

Bates, R.J. 1985b. Colourful *Thelymitra* hybrids. *The Orchadian* 8: 119-121.

Bates, R.J. 1985c. Dishonesty and deceit in Australian orchids. *The Orchadian* 8: 112-113.

Bates, R.J. 1987a. *Caladenia stricta*: new status for an Australian terrestrial orchid. *The Orchadian* 8: 247-249.

Bates, R.J. 1987b. Observations on the pollination biology and ecology of *Leporella fimbriata* (Lindley) A.S. George in South Australia. *South Austr. Naturalist* 61: 40-45.

Bates, R.J. 1989. Observations on the pollination of *Caleana major* R.Br. by male sawflies (*Pterygophorus* sp.). *The Orchadian* 9: 208-210.

Bates, R.J. 1995. Report on the results of field investigations of pollinators of the genus *Caladenia* (Orchidaceae: Caladeniinae) in 1994-1995. Australian Orchid Foundation.

Bates, R.J. & J.Z. Weber 1980. A putative hybrid between *Prasophyllum archeri* and *P.despectans. Orchadian* 6: 188-189.

Beardsell, D. 1975. Remnants of a *Diuris* series of Warrandyte. *Vict. Nat.* 92: 244-246.
Beardsell, D.V., M.A. Clements, J.F. Hutchison & E.G. Williams 1986. Pollination of *Diuris maculata* R.Br. (Orchidaceae) by floral mimicry of the native legumes *Daviesia* spp. and *Pultenaea scabra* R.Br. *Austr. J. Bot.* 34: 165-173.
Bender, J.D. 1985. The reproductive biology of *Cypripedium candidum* Muhl ex Willd. (Orch.). *Ohio J. of Science* 85(2): 12.
Benzing, D.H. 1984. To what extent has vegetative divergence kept pace with orchid cladogenesis? *Am. J. Bot.* 71(5): 2 (Suppl.).
Benzing, D.H. 1987. Major patterns and processes in orchid evolution: A critical synthesis. In J. Arditti (ed.), *Orchid Biology: Reviews and Perspectives IV*: 35-77. London: Cornell University Press.
Bernhardt, P. 1989. Wily violets and underground orchids. Revelations of a botanist. William Morrow and Cy, Inc. New York.
Bernhardt, P. 1995. Notes on the anthecology of *Pterostylis curta* (Orchidaceae). *Cunninghamia* 4: 1-8.
Bernhardt, P. & P. Burns-Balogh 1983. Pollination and pollinarium of *Dipodium punctatum* (Sm.) R.Br. *Victorian Naturalist* 100: 197-199.
Bernhardt, P. & P. Burns-Balogh 1986a. Floral mimesis in *Thelymitra nuda* (Orchidaceae). *Pl. Syst. Evol.* 151: 187-202.
Bernhardt, P. & P. Burns-Balogh 1986b. Observations on the floral biology of *Prasophyllum odoratum* (Orchidacae, Spiranthoideae). *Pl. Syst. Evol.* 153: 65-76.
Berry, P.E. & R.N. Calvo 1991. Pollinator limitation and position dependent fruit set in the high Andean orchid *Myrosmodes cochleare* (Orchidaceae). *Pl. Syst. Evol.* 174: 93-101.
Bickerton, D. 1997. A pollination study of the endangered orchid *Caladenia rigida. Journ. Nat. Orch. Soc. South Austr.* 21: 74-76.
Bierzychudek, P. 1980. Pollinator limitation of plant reproduction effort. *The American Naturalist* 117: 838-840.
Bierzychudek, P. 1981. Asclepias, Lantana and Epidendrum: A floral Mimicry Complex? *Biotropica* 13(2): 54-58.
Borba, E.L. & J. Semir 1998. Wind-assisted fly pollination in three *Bulbophyllum* (Orchidaceae) species occurring in the Brazilian campos rupestris. *Lindleyana* 13: 203-218.
Bower, C. 1992. The use of pollinators in the taxonomy of sexually deceptive orchids in the subtribe Caladeniinae (Orchidaceae). *The Orchadian* 10: 331-338.
Bower, C.C. 1994. Report on the results of field investigations of pollinators of the genus *Chiloglottis* (Orchidaceae: Caladeniinae) in 1993-1994. Austr. Orchid Foundation.
Bower, C.C. 1995a. Determination of the pollinators of sexually deceptive orchids in the subtribe Caladeniinae. Progress Report 1994-1995. Australian Orchid Foundation.
Bower, C.C. 1995b. Experimental determination of specific pollinators in the sexually deceptive orchid genus *Chiloglottis*, and their use in taxonomy. *Lecture notes Austr. Nat. Orchid Soc. Wide Bay Group Inc. Maryborough, Qld.* 48-59.
Bower, C.C. 1996a. Determination of the pollinators of sexually deceptive orchids in the subtribe Caladeninae. Progress Report, 1995-1996. Australian Orchid Foundation.
Bower, C.C. 1996b. Demonstration of pollinator-mediated reproductive isolation in sexually deceptive species of *Chiloglottis* (Orchidaceae: Caledeniinae) *Aust. J. Bot.* 44: 15-33.
Bower, C.C. 1997. Determination of pollinators of sexually deceptive orchids in the subtribe Caladeniinae. Report 1996-1997. Australian Orchid Foundation.
Bower, C.C. & P. Branwhite 1993. Observations on the pollination of *Calochilus campestris* R.Br. *The Orchadian* 11: 68-71.
Bower, C.C. & G.R. Brown 1997. Hidden biodiversity. Detection of cryptic thynnine wasp species using sexually deceptive female mimicking orchids. Australian Orchid Foundation.
Boyden, Th.C. 1980. Floral mimicry by *Epidendrum ibaguense* (Orchidaceae) in Panama. *Evolution* 34(1): 135-136.
Boyden, Th.C. 1982. The pollination biology of *Calypso bulbosa* var. *americana* (Orchidaceae): initial deception of bumblebee visitors. *Oecologia (Berl)* 55: 178-184.
Braga, P. 1976. Estudos da flora orquidológica do Estado do Amazonas.1 Descriçao e observaçao da biologia floral de *Stanhopea candida* Barb. Rodr. *Acta Amazonica* 6: 433-438.
Braga, P. 1977. Aspectos biologicos das Orchidaceae da Amazônica Central. *Acta Amazonica, Manaus* 7: 1-89.
Brantjes, N.B.M. 1978. Sensory responses to flowers in night-flying moths. In A.J. Richard (ed.), *The pollination of flowers by insects. Linn. Soc. Symp. Series* 6: 13-19.
Brown, J.H. & A. Kodric-Brown 1979. Convergence, competition, and mimicry in a temperate community of hummingbird-pollinated flowers. *Ecology* 60: 1022-1035.
Buchmann, S.L. 1983. Buzz pollination in angiosperms. In C.E. Jones & R.J. Little (eds), *Handbook of Experimental Pollination Biology*: 73-114. New York: Van Nostrand Reinhold.
Buchmann, S.L. 1987. The ecology of oil flowers and their bees. *Ann. Rev. Ecol. Syst.* 18: 343-396.
Burkill, I.H. 1919. Some notes on the pollination of flowers in the Botanical Garden, Singapore, and other parts of the Malay Peninsula. *Str. Sett. Gard. Bull. Singapore* 2: 165-176.
Burns-Balogh, P. & P. Bernhardt 1985. Evolutionary trends in the androecium of the Orchidaceae. *Pl. Syst. Evol.* 149: 119-134.
Burns-Balogh, P. & P. Bernhardt 1988. Floral evolution and phylogeny in the tribe *Thelymitreae* (*Orchidaceae: Neottioidae*). *Pl. Syst. Evol.* 159: 19-47.
Burns-Balogh, P. & V.A. Funk 1986. The phylogenetic analysis of the Orchidaceae: a summary. *Lindleyana* 1: 131-139.
Burns-Balogh, P, D.L. Szlachetko & A. Dafni 1987. Evolution, pollination, and systematics of the tribe *Neottieae* (Orchidaceae). *Pl. Syst. Evol.* 156: 91-115.
Bush, S.P. & J.H. Beach 1995. Breeding systems of epiphytes in a tropical montane wet forest. *Selbyana* 16: 155-158.
Cady, L. 1965a. The flying duck orchids. *Austr. Plants* 3: 174-177.
Cady, L. 1965b. Notes on the pollination of *Caleana major*. *The Orchadian* 2: 34.

Cady, L. 1972. Notes on the pollination of *Calochilus campestris* R.Br. *The Orchadian* 4(6): 2-5.

Calder, D.M., S.C. Cropper & D. Tonkinson 1989. The ecology of *Thelymitra epipactioides* F. Muell. (Orchidaceae) in Victoria, Australia. *Austr. J. Bot.* 37: 19-32.

Calvo, R.N. 1990a. Four-year growth and reproduction of *Cyclopogon cranichoides* (Orchidaceae) in South Florida. *Am. J. Bot.* 77(6): 736-741.

Calvo, R.N. 1990b. Inflorescence size and fruit distribution among individuals in three orchid species. *Am. J. Bot.* 77(10): 1378-1381.

Calvo, R.N. 1990c. Matrix models of terrestrial orchid population dynamics: reproduction and transitions to subterranean states. *Am. J. Bot.* 77: 44 (Suppl.).

Calvo, R.N. 1993. Evolutionary demography of orchids: Intensity and frequency of pollination and the cost of fruiting. *Ecology* 74(4): 1033-1042.

Calvo, R.N. & C.C. Horvitz 1990. Pollinator limitation, cost of reproduction, and fitness in plants: A transition-matrix demographic approach. *Am. Nat.* 136: 499-516.

Campbell, E.O. 1980. New Zealand Orchids. *Proc. NZ International Orchid Conference* 1980: 5-15.

Carlson, M. 1943. The morphology and anatomy of *Calopogon pulchellus. Bull. Torrey Bot. Club* 70: 349-368.

Carrasco, S.A. & J. Ackerman 1997. Deception pollination: Density dependent hypothesis. *Paper IV Taller International de Orquídeas. Pinar del Río. Cuba.*

Case, M.A. 1994. Extensive variation in the levels of genetic diversity and degree of relatedness among five species of *Cypripedium* (Orchidaceae). *Am. J. Bot.* 81(2): 175-184.

Catling, P.M. 1980. Rain-assisted autogamy in *Liparis loeselii* (L.) L.C. Rich. (Orchidaceae) *Bull. Torr. Bot. Club* 107: 525-529.

Catling, P.M. 1982. Breeding systems of northeastern North American *Spiranthes* (Orchidaceae). *Can. J. Bot.* 60: 3017-3039.

Catling, P.M. 1983a. Pollination of northeastern North American *Spiranthes* (Orchidaceae). *Can. J. Bot.* 61: 1080-1093.

Catling, P.M. 1983b. *Spiranthes ovalis* var. *erostellata*, a new autogamous variety from eastern USA. *Brittonia* 35: 120-125.

Catling, P.M. 1987. Notes on the breeding system of *Sacoila lanceolata* (Aublet) Garay (Orchidaceae). *Ann. Miss. Bot. Gard.* 74: 58-68.

Catling, P.M. 1990. Auto-pollination in the Orchidaceae. In J. Arditti (ed.), *Orchid Biology: Reviews and Perspectives V*: 123-158. Portland, Oregon: Timber Press.

Catling, P.M. & J.R. Brown 1983. Morphometrics and ecological isolation in sympatric *Spiranthes* (Orch.) in South Western Ontario. *Can. J. Bot.* 61: 2747-2759.

Catling, P.M. & V.R. Catling 1989. Observations of the pollination of *Platanthera huronensis* in southwest Colorado. *Lindleyana* 4(2): 78-84.

Catling, P.M. & V.R. Catling 1990. A review of breeding systems in North American orchids. *Am. J. Bot.* 77: 44 (Suppl.).

Catling, P.M. & V.R. Catling 1991a. A synopsis of breeding systems and pollination in North American orchids. *Lindleyana* 6: 187-210.

Catling, P.M. & V.R. Catling 1991b. Anther-cap retention in *Tipularia discolor. Lindleyana* 6: 113-116.

Chan, C.L., A. Lamb, P.S. Shim & J.J. Wood 1994. *Orchids of Borneo.* Vol 1. Royal Gardens Kew.

Chase, M.W. 1985. Pollination of *Pleurothallis endotrachys. AOS-Bull.* 54(4): 431-434.

Chase, M.W. 1986. Pollination ecology of two sympatric synchronously flowering species of *Leochilus* in Costa Rica. *Lindleyana* 1: 141-147.

Cheeseman, T.F. 1872. On the fertilization of the New Zealand forms of *Pterostylis. Trans. NZ Institute* 5: 352-357.

Cheeseman, T.F. 1874. On the fertilization of *Acianthus* and *Cyrtostylis. Trans. NZ Institute* 7: 349-352.

Cheeseman, T.F. 1880. On the fertilization of *Thelymitra. Trans. NZ Institute* 13: 291-296.

Chen, S.C. & T. Tang 1982. A general review of the orchid flora of China. In J. Arditti (ed.), *Orchid Biology: Reviews and Perspectives II*: 39-81. Cornell University Press, London.

Cheng, C.S. & C.Z. Tang 1988. A revision of the genus *Vanda* (Orchidaceae) of China. *Orchid Digest* 52: 38-40.

Chinery, M. 1983. *Elseviers insektengids.* Amsterdam: Elsevier (Translation of: *Insects of Britain and Northern Europe*. London: Collins)

Christensen, D.E. 1992. Notes on the reproduction biology of *Stelis argantata* Lindl. (Orchidaceae: Pleurothallidinae) in eastern Ecuador. *Lindleyana* 7: 28-33.

Christensen, D.E. 1994. Fly pollination in the Orchidaceae. In J. Arditti (ed.), *Orchid biology: Reviews and Perspectives VI*: 415-454. New York: John Wiley and Sons.

Cingel, N.A. van der 1983. Door de ogen van het insekt. *Orchideeën* 45(4): 123-129.

Cingel, N.A. van der 1994. Wandelen in Malawi. *Orchideeën* 94(4): 16-17.

Cingel, N.A. van der 1995. *An Atlas of orchid pollination: European orchids.* Rotterdam: A.A. Balkema.

Cingel, N.A. van der 1997. Zweefvliegen en *Epipactis helleborine. Eurorchis* 9: 10-11.

Claessens, J. & J. Kleynen 1998. Column structure and pollination of *Corallorhiza trifida* Châtelain. *Journal Europäischer Orchideen* 30: 629-637.

Clements, M.A. 1989. Catalogue of Australian Orchidaceae. *Australian Orchid Research* 1: 1-160.

Cole, F.R. & D.H. Firmage 1984. The floral ecology of *Platanthera blephariglottis. Am. J. Bot.* 71(5): 700-710.

Coleman, E. 1927. Pollination of the orchid *Cryptostylis leptochila. Victorian Naturalist* 44: 20-22.

Coleman, E. 1928a. Pollination of *Cryptostylis leptochila. Victorian Naturalist* 44: 333-340.

Coleman, E. 1928b. Pollination of an Australian orchid by *Lissopimpla semipunctata* Kirby; communicated with introductory note by prof. E.B. Poulton and appendix by A.M. Lea. *Trans. Royal Entom. Soc. London* 76: 533-539.

Coleman, E. 1928. *Pterostylis grandiflora* R.Br. *Vict. Nat.* 45: 111-114.

Coleman, E. 1929a. Pollination of *Cryptostylis subulata. Victorian Naturalist* 46: 62-66.

Coleman, E. 1929b. Pollination of an Australian orchid *Cryptostylis leptochila* F. Muell. *Journal of Botany* 67: 97-99.

Coleman, E. 1930a. Pollination of some West Australian orchids: *Victorian Naturalist* 46: 203-206.

Coleman, E. 1930b. Pollination of *Cryptostylis erecta.* R.Br. *Victorian Naturalist* 46: 236-238.

Coleman, E. 1930c. Further observations on the pollination of Australian orchids by a male Ichneumoid. *Proc. Royal Entom. Soc. London* 5(1): 114.

Coleman, E. 1931a. The pollination of *Corysanthes bicalcarata* (R.Br.). *Vict. Nat.* 48: 59.

Coleman, E. 1931b. Pollination of *Spiranthes australis* Lindl. *Victorian Naturalist* 47: 204-209.

Coleman, E. 1932. Pollination of *Diuris pedunculata. Victorian Naturalist* 49: 179-186.

Coleman, E. 1933a. Pollination of orchids: genus *Prasophyllum. Victorian Naturalist* 49: 214-221.

Coleman, E. 1933b. Further notes on the pollination of *Diuris pedunculata* R.Br. *Victorian Naturalist* 49: 243-245.

Coleman, E. 1938. Further observations on the pseudocopulation of the male *Lissopimpla semipunctata* Kirby (Hymenoptera, Parasitica) with the Australian orchid *Cryptostylis leptochila* F.v.M. *Proc. Royal Entom. Soc. London* 13: 82-83.

Comber, J.C. 1990. *Orchids of Java.* Royal Botanical Gardens Kew.

Cooper, D. 1983. *Pterostylis cardiostigma* – a new species of Orchidaceae from Wellington, New Zealand. *NZJ. Bot.* 21: 97-100.

Corbet, S.A., I. Cuthill, M. Fallows, T. Harrison & G. Hartley 1981. Why do nectar-foraging bees and wasps work upwards in inflorescences? *Oecologia* (Berlin) 51: 79-83.

Corbet, S.A., H. Chapman & N. Saville 1988. Vibratory pollen collection and flower form: bumble-bees on *Actinidia*, *Symphytum*, *Borago* and *Polygonatum. Funct. Ecol.* 2: 147-155.

Cribb, Ph. 1999. Lindsley's life-long love affair with orchids. In W.T. Stearn 1999, *John Lindley 1799-1865. Gardener-Botanist and Pioneer Orchidologist.* Antique Collector's Club Ltd. Woodbridge, Suffolk.

Cribb, P.J. 1987. *The genus Paphiopedilum.* Royal Bot. Gdns Kew. Collingridge.

Cribb, P.J. 1997. *The genus Cypripedium.* Royal Bot. Gdns Kew. Timber Press.

Cribb, P.J. & P. Gasson 1981. Unusual asexual reproduction in the East African orchid, *Cynorkis uncata. Kew Bull.* 36(4): 661-663.

Cropper, S.C. & D.M. Calder 1990. The floral biology of *Thelymitra epipactoides* (Orchidaceae) and the implication of pollination by deceit on the survival of this rare orchid. *Plant Syst. Evol.* 170: 11-27.

Cropper, S.C., D.M. Calder & D. Tonkinson 1989. *Thelymitra epipactoides*: the morphology, biology and conservation of endangered species. *Proceedings of the Royal Society of Victoria* 101: 89-101.

Curry, K.J. 1987. Initiation of terpenoid synthesis in osmophores of *Stanhopea anfracta* (Orchidaceae): A cytochemical study. *Am. J. Bot.* 74(9): 1332-1338.

Curtis, R.J. 1964. The pollination of orchids. *The Orchadian* 1: 72-75.

Dafni, A. 1984. Mimicry and deception in pollination. *Ann. Rev. Ecol. Syst.* 15: 259-278.

Dafni, A. 1987. Pollination in *Orchis* and related genera: evolution from reward to deception. In J. Arditti (ed.), *Orchid biology. Reviews and Perspectives IV*: 81-104. London: Cornell University Press.

Dafni, A. 1992. *Pollination ecology: A practical approach.* The practical Approach Series. New York: Oxford University Press.

Dafni, A. & P. Bernhardt 1990. Pollination of terrestrial orchids of southern Australia and the Mediterranean region. Systematic, ecological, and evolutionary implications. *Evol. Biol.* 24: 193-252.

Dafni, A. & D.M. Calder 1987. Pollination by deceit and floral mimesis in *Thelymitra antennifera* (Orchidaceae). *Pl. Syst. Evol.* 158: 11-22.

Darwin, C. 1871. *The descent of Man and selection in relation to sex.* London: John Murray.

Darwin, C. 1888. *Fertilisation of orchids. The various contrivances by which orchids are fertilised by insects.* 2nd. edition. London: John Murray.

Daumann, E. 1968. Zur Bestäubungsbiologie von *Cypripedium calceolus* L. *Öst. Bot. Zeitung* 115: 434-446.

Davenport, D. & J.M. Foley 1979. Fringe benefits of cataract surgery. *Science* 204: 454-457.

Davenport, D. & Y. Kohanzadeh 1982. Orchids, bilateral symmetry and insect pollination. *J. Theor. Biol.* 94: 241-252.

Davis, R.W. 1986. The pollination ecology of *Cypripedium acaule* (Orchidaceae). *Rhodora* 88: 445-450.

Dawkins, R. 1986. *The blind watchmaker.* London: Penguin Books.

Day, C. 1972. Observations of self-pollination in *Paphiopedilum* species. *AOS-Bull.* 41: 592.

Delabie, J.H.C. 1995. Ants attraction to the extrafloral nectaries of *Epidendrum cinnabarinum* in an area of resting vegetation. *Anais de Soc. Ent. do Brasil* 24: 479-487.

Dieringer, G. 1982. The pollination ecology of *Orchis spectabilis* L. (Orchidaceae) *Ohio J. Sci.* 82: 218-225.

Dixon, K.W., J.S. Pate & J. Kuo 1990. The western Australian fully subterranean orchid *Rhizantella gardneri.* In J. Arditti (ed.), *Orchid Biology, Reviews and Perspectives V.* Portland, Oregon: Timber Press, Inc.

Dobson, H.E.M. 1987. Role of flower and pollen aromas in hostplant recognition by solitary bees. *Oecologia (Berlin)* 1987(72): 618-623.

Dockrill, A.W. 1956. The cross-pollination of *Chiloglottis formicifera* Fitzg. *Austr. Orch. Rev.*: 26-27.

Docters van Leeuwen, W.M. 1933. The biology of *Epipogium roseum* (D. Don) Lindl. *Blumea* suppl.1: 57-62.

Dod, D.D. 1976. *Oncidium henekenii*-bee orchid pollinated by bee. *AOS-Bull.* 45: 792-794.

Dod, D.D. 1986a. Incredible adaptations of orchids to their pollinators. *Moscosoa* 4: 50-60.

Dod, D.D. 1986b. Afidos y tripidos polinazan orquideas en las Pleurothallidinae (Orch.) *Moscosoa* 4: 200-205.

Dodson, C.H. 1962a. Pollination and variation in the subtribe Catasetinae (Orchidaceae) *Ann. Missouri Bot. Gard.* 49: 35-56.

Dodson, C.H. 1962b. The importance of pollination of the orchids of tropical America. *AOS-Bull.* 31: 525-534, 641-647 and 731-734.

Dodson, C.H. 1965. Studies in orchid pollination. The Genus *Coryanthes. AOS-Bull.* 34: 680-687.

Dodson, C.H. 1966a. Studies in orchid pollination. *Cypripedium*, *Phragmopedium* and allied genera. *AOS-Bull.* 35: 125-128.

Dodson, C.H. 1966b. Studies in orchid pollination. The genus *Anguloa. AOS-Bull.* 35: 624-626.

Dodson, C.H. 1967. Studies in orchid pollination. The Genus *Notylia. AOS-Bull.* 36: 209-214.

Dodson, C.H. 1975. Coevolution of orchids and bees. In

L.E. Gilbert & P.H. Raven (eds), *Coevolution of Animals and Plants.* Univ. of Texas Press. Austin. Texas.

Dodson, C.H. 1978. The Catasetums of Tapakuma, Guyana. *Selbyana* 2: 159-168.

Dodson, C.H. & R. Escobar 1987. The Telipogons of Costa Rica. *Orquideologia* 17: 3-69.

Dodson, C.H & G.P. Frymire 1961. Natural pollination of orchids. *Bull. Missouri Bot. Gdn* 49: 133.

Dodson, C.H., R.L. Dressler, G.H. Hills, R.M. Adams & N.H. Williams 1969. Biologically active compounds in orchid fragrances. *Science* 164: 1243-1249.

Douglas, R. 1980. Orchid pollination. *S. Afr. Orchid J.* 11: 68-74.

Dransfield, J., J.B. Comber & G. Smith 1986. Corybas West of Wallace's Line. Reprint of Kew Bulletin 41(3): 575-613.

Dressler, R.L. 1966. Some observations on *Gongora. Orchid Digest* 30: 220-223.

Dressler, R.L. 1968a. Observations on orchids and euglossine bees in Panama and Costa Rica. *Rev. Biol. Trop.* 15: 143-183.

Dressler, R.L. 1968b. Pollination by euglossine bees. *Evolution* 22: 202-210.

Dressler, R.L. 1971. Dark pollinaria in hummingbird-pollinated orchids or do hummingbirds suffer from strabismus? *Am. Naturalist* 105: 80-83.

Dressler, R.L. 1976a. Una *Sievekingia* nueva de Colombia. *Orquideología* 11: 215-225.

Dressler, R.L. 1976b. How to study orchid pollination without any orchids. In K. Senghas (ed.), *Proceedings of the 8th World Orchid Conference*: 534-537.

Dressler, R.L. 1977. El género *Polycycnis* en Panamá y Costa Rica. *Orquideología* 12: 117-133.

Dressler, R.L. 1979. *Eulaema bombiformis*, *E.meriana* and Müllerian mimicry in related species (Hymenoptera: Apidae). *Biotropica* 11: 144-151.

Dressler, R.L. 1981. *The Orchids. Natural history and classification.* Cambridge Mass. Harvard University Press.

Dressler, R.L. 1982. Biology of the orchid bees (Euglossini). *Ann. Rev. Ecol. Syst.* 13: 373-394.

Dressler, R.L. 1983. *Palmorchis* in Panama mit einer neuen Art, *Palmorchis nitida*, an einem unerwarteten Standort. *Die Orchidee* 34: 25-31.

Dressler, R.L. 1993. *Phylogeny and Classification of the Orchid Family.* Melbourne: Cambridge University Press.

Du Puy, D. & P. Cribb 1988. *The genus Cymbidium.* Timber Press.

Duque, O.H. 1993. Polinizacion en Pleurothallis. *Orquideologia* 19(1): 55-69.

Echevarría, L., E. Arrieta & M.A. Díaz 1997. Limitación en la produción de frutos de *Ionopsis utricilarioides* (Sw.) Lindl. en una comunidad de epífitas en àreas del Jardin Botànico Nacional. *Resúmenes IV Taller Internacional de Orquídeas. Pinar del Rio. Cuba.*

Erickson, R. 1965a. *Orchids of the West* 2nd ed. Paterson Brokensha Pty. Ltd Perth.

Erickson, R. 1965b. Insects and Australian orchids. *The Austr. Orchid Rev.* 30: 172-174.

Evoy, W.H. & B.P. Jones 1971. Motor patterns of male euglossine bees evoked by floral fragrances. *Anim. Behav.* 19: 579-584.

Faegri, K. & L. van der Pijl 1966. *Principles of pollination ecology.* Pergamon Press.

Falb, D.L. & D.J. Leopold 1993. Population dynamics of *Cypripedium candidum* Muhl. ex Willd., Small White Ladyslipper, in a Western New York Fen. *Natural Areas Journal* 13(2): 86.

Firmage, D.H. 1990. The influence of ramet biomass and microhabitat on the reproductive patterns of *Platanthera blephariglottis* (Orchidaceae). *Am. J. Bot. (Suppl.)* 77: 45.

Firmage, D.H. & F.R. Cole 1984. The survivorship and reproduction of *Platanthera blephariglottis* (Orchidaceae). *Am. J. Bot.* 71(5): 2.

Firmage, D.H. & F.R. Cole 1985. Pollination, fruit set and infloresence size of *Calopogon pulchellus* (Orchidaceae) *Am. J. Bot.* 72(6): 852.

Firmage, D.H. & F.R. Cole 1988. Reproductive success and inflorescence size of *Calopogon tuberosus* (Orchidaceae) *Am. J. Bot.* 75(9): 1371-1377.

Fisher, B.L. & J.K. Zimmerman 1988. Ant/orchid associations in the Barro Colorado National Monument, Panama. *Lindleyana* 3(1): 12-16.

Fisher, B.L., L. da Silveira Lobo Sternberg & D. Price 1990. Variation in the use of orchid extrafloral nectar by ants. *Oecologia* 83: 263-266.

Fitzgerald, R.D. 1878. *Calochilus paludosus* and *Calochilus campestris.* In *Australian orchids* 1(4). Sydney: Government Printer,

Fitzgerald, R.D. 1893. *Diuris tricolor.* In *Australian orchids* 2(5). Sydney: Government Printer.

Folsom, J.P. 1984. Una reinterpretacion del estatus y relaciones de las taxa del complejo de *Platanthera ciliaris. Orquidea (Mex)* 9: 321-345.

Folsom, J.P. 1985a. Pollination floral strategy and pollen flow in *Dichaea* sp. (Orchidaceae) *Am. J. Bot.* 72: 953-954.

Folsom, J.P. 1985b. Pollination biologies and relationships of plants in the Yellow-Fringed-orchid complex. *Am. J. Bot.* 72(6): 53.

Folsom, J.P. 1994. Pollination of a fragrant orchid. *Orchid Digest* 1994: 83-99.

Forster, P.I. 1988. Pollination of *Peristeranthes hillii* (Orchidaceae) by *Metriorrhynchus rufipennis* (Coleoptera: Lycidae). *The Coleopterists Bull.* 42(2): 166.

Frogatt, R.W. 1927. The attraction of the flowers of the orchid *Cryptostylis leptochila* for ichneumon wasps (*Lisssopimpla semipunctata*). *The Austr. Nat.* 7: 6-7.

Füller, F. 1966. *Malaxis, Hammarbya, Liparis. Die Orchideen Deutschlands* 6. Die Neue Brehmbücherei 360. Ziemsen Verlag.

Füller, F. 1978. *Platanthera, Gymnadenia, Leucorchis, Neotthianthe. Orchideen Mitteleuropas* 8 (2nd ed.). Die Neue Brehmbücherei 411. Ziemsen Verlag.

Füller, F. 1980. *Montan-alpine und Nordisch-alpine Orchideen. Orchideen Mitteleuropas* 10 (2nd ed.). Die Neue Brehmbücherei 445. Ziemsen Verlag.

Fuller, G. 1994. Observations on the pollination of *Corybas* 'A'. *NZNOG Journal* 52: 18-22.

Gandawidjaja, D. & J. Arditti 1982. Post-pollination phenomena in orchid flowers. XI. Autogamy in *Phaius tankervilliae* (Aiton) Bl., Orchidaceae. *Am. J. Bot.* 69: 335-338.

Garay, L.A. 1970. El genero *Trevoria. Orquideología* 5: 3-13.

Garside, S. 1922. The pollination of *Satyrium bicallosum* Thunb. *Ann. of the Bolus Herb.* 3: 147-154.

Gassner, A. 1982. Observations on the pollination of *Disa stolzii* in Southern Malawi. *The Orchid Review*. July 1982: 230.

George, A.S. 1980. *Rhizanthella gardneri* R.S. Rogers – the underground orchid of western Australia. *AOS-Bull.* 49: 631-646.

Gerlach, G. & R. Schill 1991. Composition of orchid scents attracting euglossine bees. *Bot. Acta* 104(5): 379-391.

Gilbert, P.A. 1958. Australian Dendrobiums. *Dendrobium linguiforme* Sw. *A.O.S. Bulletin* 27: 472-475.

Gilbert, L. 1992. *The Orchid Man. The Life, Work and Memoirs of the Rev. H.M.R. Rupp 1872-1956*. Kangaroo Press.

Gill, D.E. 1989. Fruiting failure, pollinator inefficiency, and speciation in orchids. In D. Otte & J. Endler (eds), *Speciation and its consequences*: 458-481. Sinauer, Sunderland MA.

Gill, D.E. & W.P. Stoutamire 1990. Are the breeding systems of deceptive orchids evolutionary stable? Experimental tests of inbreeding depression in *Cypripedium acaule*. *Abstracts Am. J. of Bot.* 77 (Suppl.): 45.

Godfery, M.J. 1929a. Pollination of *Cryptostylis leptochila*. *Orchid Rev. (London)* 37: 163-166.

Godfery, M.J. 1929b. Note. In E. Coleman's 'Pollination of an Australian orchid *Cryptostylis leptochila*. *Journ. of Bot.* 67: 97-100.

Godley, E.J. 1979. Flower biology in New Zealand. *NZJ. Bot.* 17: 441-466.

Gonzalez Diáz, N. & J.D. Ackerman 1985. Pollination biology of *Oeceoclades maculata* (Orchidaceae): Rain assisted autogamy. *Am. J. Bot.* 72: 954-955.

Gonzalez Diáz, N. & J.D. Ackerman 1988. Pollination, fruit set and seed production in the orchid *Oeceoclades maculata*. *Lindleyana* 3: 150-155.

Gorham, J. 1976.Orchid pollination by *Aedes* mosquitos in Alaska USA. *Am. Midl. Nat.* 95: 208-210.

Goss, G.J. 1977. The reproductive biology of the epiphytic orchids of Florida 5: *Epidendrum difforme* Jaquin. *AOS-Bull.* 46: 631-636.

Goss, G.J. 1977. The reproductive biology of the epiphytic orchids of Florida 6: *Polystachya flavescens* (Lindley) J.J. Smith. *AOS-Bull.* 46: 990-994.

Goss, G.J. & R.M. Adams 1976. The reproductive biology of the epiphytic orchids of Florida 4: Sexually selective attraction of moths to the floral fragrance of *Epidendrum anceps* Jaquin. *AOS-Bull.* 45(11): 997-1001.

Gould, J.L & C. Grant Gould 1992. *De honingbij, een samenleving van kleine giganten.* Natuur en Techniek, Maastricht/Brussel. Dutch version of: *The Honey Bee.* 1988. Scientific American Library, New York.

Graham, D.K.F. 1983. Interaction between an ichneumonid wasp and the Australian tongue orchid in New Zealand. *Records Auck. Inst. Mus.* 20: 217-222.

Gregg, K.B. 1982. Reproductive biology of the orchid *Cleistes divaricata* (L.) Ames var. *bifaria* Fernald growing in a West Virginia meadow. *Castanea* 54(2): 57-78.

Gregg, K.B. 1984. Reproductive biology of the orchid *Cleistes divaricata* (L.) Ames growing in a West Virginia meadow. *Am. J. Bot.* 71: 79.

Gregg, K.B. 1990. How to defraud a deceitful orchid. *Am. J. Bot.* 77: 46.

Gregg, K.B. 1991a. Reproductive strategy of *Cleistes divaricata* (Orchidaceae). *Am. J. Bot.* 78: 350-360.

Gregg, K.B. 1991b. Defrauding the deceitful orchid: Pollen collection by pollinators of *Cleistes divaricata* and *C.bifaria*. *Lindleyana* 6: 214-220.

Gumprecht, R. 1977. Seltsame Bestäubungsvorgänge bei Orchideen. Sonderdruck *Die Orchidee* 28. Beilage zu Heft 3.

Hagerup, O. 1952a. Bud autogamy in some northern orchids. *Phytomorphology* 2: 51-60.

Hagerup, O. 1952b. The morphology and biology of some primitive orchid flowers. *Phytomorphology* 2: 134-138.

Hall, A.V. 1959. Observations on the distribution and ecology of Orchidaceae in the Muizenberg Mountains, Cape Peninsula. *J. of S. Afr. Bot.* 25: 265-278.

Handel, S.N. 1983. Pollination ecology, plant population structure, and gene flow. In L. Real (ed.), *Pollination Biology*: 163-212. Academic Press.

Handel, S.N. & R. Peakall 1993. Thynnine wasps discriminate among heights when seeking mates: test with a sexually deceptive orchid. *Oecologia* 95: 241-245.

Hapeman, J.R. 1997. Pollination and floral biology of *Platanthera peramoena* (Orchidaceae) *Lindleyana* 12: 19-25.

Hatch, E.D. 1945. Notes on New Zealand orchids. *Trans. Roy. Soc. NZ* 75: 367-370.

Hatch, E.D. 1952. A new species of *Corybas* Salisbury. *Trans. Roy. Soc. NZ* 79: 366-368.

Hatch, E.D. 1952. The New Zealand forms of *Thelymitra* J.R. and G. Foster. *Trans. Roy. Soc. NZ* 79: 386-402.

Hatch, E.D. 1954. Further notes on *Gastrodia* R Brown and *Spiranthes* L.C. Richard. *Trans Roy. Soc. NZ* 82: 613-615.

Hatch, E.D. 1963. Notes on New Zealand orchids – II. *Trans. Roy. Soc. NZ* Bot. 2: 185-189.

Heinrich, B. 1975a. Bee flowers: A hypothesis on flower variety and blooming times. *Evolution* 29: 325-334.

Heinrich, B. 1975b. Energetics of pollination. *Ann. Rev. Ecol. Syst.* 6: 139-170.

Heinrich, B. 1979. *Bumblebee economics*. Cambridge, Mass.: Harvard University Press.

Hennesy, E.F. & T.A. Edge 1989. *The slipper orchids*. Acorn Books.

Herrera, C.M. 1989. Pollinator abundance, morphology, and flower visitation rate: analysis of the 'quantity' component in a plant-pollinator system. *Oecologia* 80: 241-248.

Hills, H.G., N.H. Williams & C.H. Dodson 1968. Identification of some orchid fragrance components. *AOS-Bull.* 37: 967-971.

Hills, H.G., N.H. Williams & C.H. Dodson 1972. Floral fragrance and isolating mechanisms in the genus *Catasetum*. *Biotropica* 4: 61-76.

Hocking, B. 1968. Insect-flower associations in the high Arctic with special reference to nectar. *Oikos* 19: 359-388.

Hogan, K.P. 1983. The pollination biology and breeding system of *Aplectrum hyemale* (Orchidaceae) *Can. J. Bot.* 61: 1906-1910.

Hogen, R. 1970. A pollinator of *Dendrobium speciosum*. The Orchadian 3: 119.

Holman, R.T., W.P. Cunningham & E.S. Swanson 1981. A closer look at the glandular hairs on the ovaries of Cypripediums. *AOS-Bull.* 50(6): 683-687.

Holttum, R.E. 1949. Gregarious flowering of the terrestrial orchid *Bromhaedia finlaysoniana*. *Gdns. Bull. Singapore* 12: 295-302.

Holttum, R.E. 1953. *Flora of Malaysia I. Orchids*. Governm. Printing Off. Singapore.

Howcroft, N.H.S. 1976. A New Guinea *Bulbophyllum*. *Orchadian* 5: 70-72.

Hyett, J. 1960. Pollination of the nodding greenhood. *Victorian Naturalist* 76: 240-241.

Indsto, J.O. & P.H. Weston 1999. Near-ultraviolet Reflectance in Dendrobium (Orchidaceae). In Wilson, K.L. & D.A. Morrison (eds), Systematics and Evolution of Monocots – *Vol. 1 of Proceedings of the Second International Conference on the Comparitive Biology of the Monocotyledons*, Sydney, Sept. 1998, pp. 326-334.

Inoue, K. 1981. Beetle pollination of *Platanthera chorisiana* (Orchidaceae). *J. Jap. Bot.* 56: 213-218.

Inoue, K. 1983. Systematics of the genus *Platanthera* (Orch.) in Japan and adjacent regions, with special reference to pollination. *J. Fac. Sci. Univ. Tokyo Sect. III. Bot.* 13: 285-374.

Inoue, K. 1985. Reproductive biology of two Platantherans (Orchidaceae) in the island of Hachijo (Japan). *Jpn. J. Ecol.* 35: 77-84.

Inoue, K. 1986a. Experimental studies on male and female reproductive success: Effects of variation in spur length and pollinator activity on *Platanthera mandarinorum* ssp. *hachijoensis* (Orchidaceae). *Pl. Sp. Biol.* 1: 207-215.

Inoue, K. 1986b. Different effects of sphingid and noctuid moths on the fecundity of *Platanthera metabifolia* (Orchidaceae) in Hokkaido. *Ecol. Res.* 1: 25-36.

Inoue, K. & T.P. Lin 1980. Orchid genus *Platanthera* of Taiwan. *J. Phytogeogr. Tax.* 28: 1-13.

Ivri, Y & A. Dafni 1977. The pollination ecology of *Epipactis consimilis* Don (Orchidaceae) in Israel. *New Phytologist* 79: 173-177.

Jaffe, K., C. Pavis, G. Vansuyt & A. Kermarrec 1989. Ants visit extrafloral nectaries of the orchid *Spathoglottis plicata* Blume. *Biotropica* 21(3): 278-279.

Janzen, D.H. 1971. Euglossine bees as long-distance pollinators of tropical plants. *Science* 171: 203-205.

Janzen, D.H. 1980. When is it coevolution? *Evolution* 34: 611-612.

Janzen, D.H. 1981. Bee arrival at two Costa Rican female *Catasetum* orchid inflorescences, and a hypothesis on euglossine population structure. *Oikos* 36: 177-183.

Janzen, D.H., P.J. de Vries, M.L. Higgins & L.S. Kimsey 1982. Seasonal and site variation in Costa Rican euglossine bees at chemical baits in lowland deciduous and evergreen forests. *Ecology* 63: 66-74.

Janzen, D.H., P.J. de Vries, D.E. Gladstone, M.L. Higgins & T.M. Lewinsohn 1980. Self- and cross-pollination of *Encyclia cordigera* (Orchidaceae) in Santa Rosa National Park, Costa Rica. *Biotropica* 12(1): 72-74.

Jeffrey, D.C., J. Arditti & H. Koopowitz 1970. Sugar content in floral and extrafloral exudates of orchids: pollination, myrmecology and chemotaxonomy implication. *New Phytol.* 69: 187-195.

Johansen, B. 1990. Incompability in *Dendrobium* (Orchidaceae). *Bot. J. Linn. Soc.* 103: 165-196.

Johnson, S.D. 1992a. The pollination of *Disa filicornis*. *SAOJ* 25: 22-23.

Johnson, S.D. 1992b. Pollination of *Disa filicornis* (Orchidaceae) through deception of mason bees. *S. Afr. J. of Bot.* 58: 541-542.

Johnson, S.D. 1992c. Plant-animal relationships. In R.M. Cowling (ed.), *The ecology of fijnbos*. pp. 175-204. Cape Town: Oxford University Press.

Johnson, S.D. 1992d. A butterfly with a passion for red. *Afr. Wildlife* 46: 176-179.

Johnson, S.D. 1993a. Climatic and phylogentic determinants of flowering seasonality in the Cape flora. *J. Ecol.* 81: 567-572.

Johnson, S.D. 1993b. Red Disas and brown butterflies. *South Afr. Orchid J.* 24(1): 9-12.

Johnson, S.D. 1993c. A belated discovery: The pollinator of *Herschelianthe graminifolia*. *South Afr. Orchid J.* 24(3): 83-84.

Johnson, S.D. 1993d. Preliminary observations on the pollination of *Disperis capensis* (Orchidaceae). *South Afr. Orchid J.* 25(1): 22-23.

Johnson, S.D. 1993e. Carpenter bee pollination of *Herschelianthe graminifolia* (Orchidaceae) on the Cape Peninsula. *Flora* 188: 383-386.

Johnson, S.D. 1994a. Evidence for Batesian mimicry in a butterfly-pollinated orchid. *Biol. J. Linnean Soc.* 53: 91-104.

Johnson, S.D. 1994b. The pollination of *Disa versicolor* by anthophorid bees in South Africa. *Lindleyana* 9: 209-212.

Johnson, S.D. 1995a. Moth pollination of the cryptic Cape orchid *Monadenia ophrydea*. *Flora* 190: 105-108.

Johnson, S.D. 1995b. Observations of hawkmoth pollination in the South African orchid *Disa cooperi*. *Nordic J. Bot.* 15: 121-125.

Johnson, S.D. 1996a. Pollination, adaptation and speciation models in the Cape flora. *Taxon.* 45: 59-66.

Johnson, S.D. 1996b. Pollination of *Satyrium* species: Latest findings. *South Afr. Orchid J.* 27(3): 70-73.

Johnson, S.D. 1996c. Bird pollination in South African species of *Satyrium* (Orchidacaea). *Plant Syst. and Ev.* 203: 91-98.

Johnson, S.D. 1997a. Self-pollination in *Disa woodii*. *South Afr. Orchid J.* 28: 94.

Johnson, S.D. 1997b. Insect pollination and floral mechanisms in South African species of *Satyrium* (Orchidaceae). *Pl. Syst. Evol.* 204: 195-206.

Johnson, S.D. 1997c. Pollination ecotypes of *Satyrium hallackii* (Orchidaceae) *Bot. J. Linn. Soc.* 123: 225-235.

Johnson, S.D. & W.J. Bond 1992. Habitat dependent pollination success in a Cape orchid. *Oecologia* 91: 455-456.

Johnson, S.D.& G. van Ede 1993. The Pollination of *Satyrium hallackii* at Verloren Vallei. *South Afr. Orchid J.* 24(4).

Johnson, S.D. & M. Fraser 1996. Sunbirds pollinate orchids. *Africa-Birds and birding* 1: 14-15.

Johnson, S.D. & K. Johnson 1993. Beauty and the beast: a Cape orchid pollinated by horseflies. *Veld & Flora* 79: 38-39.

Johnson, S.D. & W.R. Liltved 1997. Hawkmoth pollination of *Bonatea speciosa* (Orchidaceae) in a South African coastal forest. *Nordic J. of Bot.* 17: 5-10.

Johnson, S.D. & H.P. Linder 1995. Systematics and evolution of the *Disa draconis* complex (Orchidaceae). *Bot. J. Linn. Soc.* 118: 289-307.

Johnson, S.D. & K. Steiner 1994a. Efficient pollination in the mass flowering Cape Orchid *Disa obtusa* Lindl. by *Bibio turneri* Edwards (Diptera: Bibionidae). *Afr. Entomology* 2: 64-66.

Johnson, S.D. & K. Steiner 1994b. Pollination by megachilid bees and determinants of fruit-set in the Cape orchid *Disa tenuifolia*. *Nordic J. Bot.* 14: 481-485.

Johnson, S.D. & K. Steiner 1995. Long-proboscid fly pollination of two orchids in the Cape Drakensberg mountains, South Africa. *Pl. Syst. Evol.* 195: 169-175.

Johnson, S.D. & K. Steiner 1997. Long-tongued fly-pollination and evolution of floral spur length in the *Disa draconis* complex (Orchidaceae). *Evolution* 51: 45-53.

Johnson, S.D., K. Steiner & H. Kurzweil 1994. Auto-pollination in *Disa* species (Orchidaceae) of the Cape mountain cloud-zone. *Lindleyana* 9(1): 3-6.

Johnson, S.D., K. Steiner, V.B. Whitehead & L. Vogelpoel 1998. Pollination ecology and maintenance of species integrity in co-occuring *Disa racemosa* L.f. and *Disa venosa* Sw. (orchidaceae) in South Africa. *Ann. Missouri Bot. Gdn* 85: 231-241.

Jones, C.E. & R.J. Little 1985. *Handbook of experimental pollination biology*. New York: Nostrand Reinhold.

Jones D.L. 1965. Orchid pollination. *Austr. Plants* 3(22): 73-74.

Jones D.L. 1970a. Unique perfume in *Prasophyllum fimbriatum* R.Br. *Vict. Nat.* 87: 328.

Jones D.L. 1970b. The pollination of *Corybas diemenicus* (H.M.N.) Rupp and W.H. Nicholls ex H.M.R.Rupp. *Victorian Naturalist* 87: 372-374.

Jones D.L. 1971a. The pollination of *Corybas diemenicus* (Lindl.) Rupp et Nicholls ex Rupp. *The Orchadian* 3: 156-159.

Jones, D.L. 1971b. A study of the self-pollination of *Thelymitra venosa* R.Br. and some notes on its implications. *Victorian Naturalist* 88: 217-219.

Jones, D.L. 1972a. The self-pollination of *Prasophyllum alpinum*. *Victorian Nat.* 89: 260-263.

Jones, D.L. 1972b. The self-pollination of *Prasophyllum beaugleholei*. *Victorian Nat.* 89: 144-146.

Jones, D.L. 1974a. The pollination of *Acianthus caudatus* R.Br. *Vict. Nat.* 91: 272-274.

Jones, D.L. 1974b. The pollination of *Dendrobium linguiforme*. *Austr. Orchid Rev.* 39: 70-72.

Jones, D.L. 1975. The pollination of *Microtis parviflora*. *Ann. Bot. (London)* 39: 585-589.

Jones, D.L. 1976. The self-pollination of *Phaius pictus* T.E. Hunt. *Austr. Orchid Rev.* Dec.1976: 210-213.

Jones, D.L. 1977. Miscellaneous notes on Australian Orchidaceae II. Reduction of six teratological forms of synonymy. *Orchadian* 5: 126-128.

Jones D.L. 1981. The pollination of selected Australian orchids. *Proc. Orchid Symposium, 13th Int. Bot. Congress Sydney*: 40-43.

Jones D.L. 1983. The pollination and occurence of natural hybrids in Australian species of the genus *Dendrobium*. *Proceedings 8th Aust. Orchid Conf.*: 127-131.

Jones D.L. 1985a. Pollination of *Bulbophyllum weinthalii* Rogers. *Vict. Nat.* 103: 99-101.

Jones D.L. 1985b. The pollination of *Gastrodia sesamoides* R.Br. in southern Victoria. *Vict. Nat.* 102: 52-54.

Jones, D.L. 1985c. The self-pollination of *Epipogium roseum* (D. Don) Lindley. *Orchadian* 8: 91-95.

Jones, D.L. 1988. *Native Orchids of Australia*. Reed Books Pty Ltd. NSW.

Jones, D.L. 1991. New taxa of Australian Orchidaceae. *Australian Orchid Research*, Vol. 2: 1-203.

Jones, D.L. & B. Gray 1974. The pollination of *Calochilus holtzei* F. Muell. *AOS-Bull.* July 1974: 604-606.

Jones, D.L. & B. Gray 1976a. The pollination of *Bulbophyllum longiflorum* Thouars. *AOS-Bull.* 45(1): 15-17

Jones, D.L. & B. Gray 1976b. The pollination of *Dendrobium lichenastrum* (F. Muell.) Krzl. *AOS-Bull.* 45: 981-983.

Jones, D.L. & B. Gray 1977. The pollination of *Dendrobium ruppianum* A.D. Hawkes. *AOS-Bull.* 46: 54-57.

Jongejan, P. 1994. Specializations in ways of attracting insects for pollination in the genus *Bulbophyllum*. In *Proc. 14th WOC, Glasgow*: 383-388, HMSO, Edinburgh.

Judd, W.W. 1971. Wasps (Vespoidea) pollinating Helleborine *Epipactis helleborine* (L.) Crantz, at Owen Sound, Ontario. *Proc. Entomol. Soc. Ont.* 102: 115-118.

Kaiser, R. 1993. *The scent of orchids. Olfactory and chemical investigations*. Elsevier.

Kay, Q.O.N. 1978. The role of preferential and assortative pollination in the maintenance of flower colour polymorphisms. In A.J. Richards (ed.), *The pollination of flowers by insects*: 175-190. Ac. Press. London.

Kallunki, J.A. 1976. Population studies in *Goodyera* (Orchidaceae) with emphasis on the hybrid origin of *Goodyera tesselata*. *Brittonia* 28: 53-75.

Kallunki, J.A. 1981. Reproductive biology of mixed-species populations of *Goodyera* (Orchidaceae) in Northern Michigan. *Brittonia* 33: 137-155.

Keenan, P.E. 1989. Butterflies of the orchid World. *AOS-Bull.* 58(8): 767-771.

Kimsey, L.S. 1980. The behaviour of male orchid bees (Apidae, Hymenoptera, Insecta) and the question of leks. *Animal Behav.* 28: 996-1004.

Kjellson G., F.N. Rasmussen & D. Dupuy 1985. Pollination of *Dendrobium infundibulum*, *Cymbidium insigne* (Orchidaceae) and *Rhodondrendon lyi* (Ericaceae) by *Bombus eximus* (Apidae) in Thailand: a possible case of floral mimicry. *J. Trop. Ecol.* 1: 289-302.

Klier, K., M.J. Leoschke & J.F. Wendel 1991. Hybridization and Introgression in White and Yellow Ladyslipper Orchids (*Cypripedium candidum* and *C.pubescens*). *J. of Heredity* 82: 305-318.

Klots, B. & E.B. Klots 1960. *De wereld der insekten*. W. Gaade. Den Haag.

Knerr, J.N. 1981. The genus *Bulbophyllum*: A living phantasy. *AOS-Bull.* 50: 1051-1056.

Knoll, F. 1922. Fettes Öl auf den Blütenepidermen der Cypripedilinae. *Öst. Bot.Zeitschrift* 1922: Heft 4(6): 121-129.

Koehler, D.J. & D. Davenport 1983. Ultraviolet mimicry by *Bulbophyllum lepidum*? *AOS-Bull.* 52: 359-363.

Kramer, R.D. 1990. The unique pollination mechanisms of *Paphiopedilum sanderianum* (Rchb.f.) Stein. *Orchid Digest* 1990: 115-117.

Kroodsma, D.E. 1975. Flight distances of male euglossine bees in orchid pollination. *Biotropica* 7: 71-72.

Kugler, H. 1951. Blütenökologische Untersuchungen mit Goldfliegen (Lucilien). *Ber. Dtsch. Bot. Ges.* 64: 327-341.

Kugler, H. 1956. Über die Optische Wirkung von Fliegenblumen auf Fliegen. *Ber. Dtsch. Bot. Ges.* 101: 387-398.

Kullenberg, B. 1961. Studies in *Ophrys* pollination. *Zool. Bidr. Upps.* 34(1): 1-340.

Kurzweil, H. & S.D. Johnson 1993. Auto-pollination in *Monadenia bracteata*. *South Afr. Orchid J.* 24(1): 21-22.

Lammi, A. & M. Kuitunen 1995. Deceptive pollination of *Dactylorhiza incarnata*: an experimental test of the magnet species hypothesis. *Oecologia* 101: 500-503.

Lang, D.C. 1980. *Orchids of Britain*. Oxford Un. press.

Larson, K.S. & R.J. Larson 1990. Lure of the locks; showiest ladies-tresses orchids, *Spiranthes romanzoffiana*, affect bumble bee, *Bombus* spp., foraging behavior. *Can. Field Nat.* 104: 519-523.

Larson, R.J. 1992. Pollination of *Platanthera dilatata* var. *dilatata* in Oregon by the noctuid moth *Discestra oregonica*. *Madroño* 39: 236-237.

Larson, R.J. & K.S. Larson 1987. Observations on the pollination biology of *Spiranthes romanzoffiana*. *Lindleyana* 2(4): 176-179.

Leonard, D.R. 1970. A recent observation of the pollination of *Cryptostylis erecta* by a wasp. *The Orchadian* 3: 111.

Lertzman, K.P. & C.L. Gass 1985. Alternative models of pollen transfer. In C.E. Jones & R.J. Little (eds), *Handbook of experimental pollination biology*: 474-490. New York: Nostrand Reinhold.

Light, M.H.S. 1998. Promise of reward-scent, nectar and pollen. *Orchid Rev.* 106: 330-336.

Linder, H.P. 1985. X *Herschelianthe vogelpoelii*. A new generic hybrid in the Disinae. *South Afr. Orchid J.* 16: 102-103.

Lister, P. 1987. Observations on the pollination of *Dendrobium linguiforme* Sw. *The Orchadian* 1987: 258-261.

Lock, J.M. & J.C. Profita 1975. Pollination of *Eulophia cristata* (Sw.) Steud. (Orchidaceae) in southern Ghana. *Acta Bot. Neerl.* 24: 135-138.

Löjtnant, B. & N. Jacobsen 1976. The biology and taxonomy of the orchids of Greenland. *Jahresberichte des Naturwissenschaftlichen Vereins in Wuppertal: Die Orchideen der Randgebiete des europäischen Florenbereiches*, Heft 29: 17-41.

Luer, C.A. 1986a. *Icones Pleurothallidinarum I. Systematics of the Pleurothallidinae*. Missouri Bot. Garden.

Luer, C.A. 1986b. *Icones Pleurothallidinarum II. Systematics of Masdevallia*. Missouri Bot. Garden.

Luer, C.A. 1991. *Icones Pleurothallidinarum VIII. Systematics of Lepanthosis, Octomeria* subgenus *Pleurothallopsis, Restrepiella, Restrepiopsis, Salpistele* and *Teagueia*. Missouri Botanical Garden.

Luer, C.A. 1993. *Icones Pleurothallidinarum X. Systematics of Dracula*. Missouri Bot. Garden.

Macior, L.W. 1971. Co-evolution of plants and animals-systematic insights from plant-insect interactions. *Taxon* 20: 17-28.

Macpherson, K. & H.M.R. Rupp 1934. The pollination of *Cymbidium iridifolium* Cunn. *The North Queensland Naturalist* 3(7): 26.

Macpherson, K. & H.M.R. Rupp 1936. Further notes on orchid pollination. *The North Queensland Naturalist* 4(43): 25-26.

Maekawa, F. 1971. *The wild orchids of Japan in color*. Seibundo-shinkosha, Tokyo.

Manning, A. 1973. *An Introduction to animal behaviour* (2nd ed.). London: Edward Arnold Publishers.

Manning, J.C. & D.J. Brothers 1986. Flora relations of four species of *Rediviva* in Natal (South Africa). *J. Ent. Soc. South Afr.* 49: 107-114.

Manning, J.C. & P.H. Linder 1992. Pollination and evolution in *Disperis* (Orchidaceae) or why are there so many species? *S. Afr. J. Sc.* 88: 38-49.

Matias, L., P. Quiroz, P. Isoares Braga & A.G. Freire 1996. Reproductive biology of *Constantia cipoensis* Porto & Brade (Orchidaceae), an endemic species from the Sierra do Cipo, Minais Gerais. *Rev. Bras. de Bot.* 19: 119-125.

Mayr, E. 1970. *Populations, Species and Evolution*. Harvard University Press Cambridge. Massachusetts.

McDaniel, D.H., L.B. McDaniel & W.A. Welton 1981. Through the insect eye: a survey of orchids with ultraviolet reflectance photography. *AOS-Bull.* 50(1): 53-63.

Meeuse, B.J.D. 1961. *The story of pollination*. New York: Ronald.

Meeuse, B.J.D. 1988. Merkwaardige bestuiving door mieren. *Biovisie* 68: 141-143.

Meeuse, B.J.D. & S. Morris 1984. *The sex life of flowers*. Rainbird Publ. Group Ltd. New York.

Mehrhoff, L.A. 1983. Pollination in the genus *Isotria* (Orchidaceae). *Am. J. Bot.* 70(10): 1444-1453.

Mehrhoff, L.A. 1989. The dynamics of declining populations of an endangered orchid *Isotria medeloides*. *Ecology* 70(3): 783-786.

Meijden, R. van der 1996. Over het soortbegrip bij orchideeën en andere planten. *Eurorchis* 8: 8-14.

Meléndez-Ackerman, E., J.D. Ackerman & J.A. Rodríguez 1997. Factores limitantes en la reproducción en una población natural de *Comparettia falcata* (Orchidaceae). *Resúmenes IV Taller Internacional de Orquídeas. Pinar del Rio. Cuba.*

Mesler, M.R., J.D. Ackerman & K.L. Lu 1980. The effectiveness of fungus gnats as pollinators. *Am. J. Bot.* 67: 564-567.

Millar, A. 1978. *Orchids of Papua New Guinea, an introduction*. Austr. Nat. Univ. Press, Canberra.

Miller, D. 1918. A new fungus gnat which fertilises *Corysanthes oblonga*. *NZ Journal of Sc. & Techn.* 1918 (1): 4.

Molloy, B. 1990. Pollination systems of New Zealand native orchids. In St George, I.M & D. Mc Crae (eds), *The New Zealand Orchids: Natural history and cultivation.* NZNOG. Dunedin.

Molnàr, A 1998. Über die Bestäubung von *Ophrys helenae* Renz. *J. Eur. Orch.* 30(3): 588-590.

Montalvo, A.M. & J.D. Ackerman 1987. Limitations to fruit production in *Ionopsis utricularioides* (Orchidaceae). *Biotropica* 19: 24-31.

Morgans, P. 1986a. Aspects of orchid pollination I: Structural adaptations of the flower *SAOJ* 17(3): 114-116.

Morgans, P. 1986b. Aspects of orchid pollination II. The role of floral fragrances. *SAOJ* 17(4): 139-141.

Morrison, D.A & P.H. Weston 1985. Analysis of morphological variation in a field sample of *Caladenia catenata* (Smith) Druce (Orchidaceae) *Aust. J. Bot.* 33: 185-195.

Mosquin, T. 1971. Competition for pollinators as a stimulus for the evolution of flowering time. *Oikos* 22: 398-402.

Müller, I. 1988. Vergleichende blütenökologische Untersuchungen an der Orchideengattung *Epipactis*. *Mitt. bl. AHO Baden Württemberg* 20(4): 701-803.

Murren, C.J. & A.M. Ellison 1996. Effects of habitat, plant size, and floral display on male and female reproductive success of the neotropical orchid *Brassavola nodosa*. *Biotropica* 28: 30-41.

Nazarov, V.V. 1995. Pollination of *Steveniella satyroides* (Orchidaceae) by wasps (Hymenoptera, Vespoidae) in the Crimea. *Lindleyana* 10: 109-114.

Nazarov, V.V. & G. Gerlach 1997. The potential seed productivity of orchid flowers and pecularities of their pollination system. *Lindleyana* 12: 188-204.

Niemann, D.A. 1974. Introgressive hybridisation in *Phalaenopsis equestris* (Schauer) Rchb.f. *AOS-Bull.* 43: 866-870.
Nierenberg, L. 1972. The mechanism of the maintenance of species integrity in sympatrically occuring equitant Oncidiums in the Caribbean. *AOS-Bull.* 41: 873-881.
Nilsson, L.A. 1978. Anthecological studies of the Lady's Slipper, *Cypripedium calceolus* (Orchidaceae). *Bot. Notiser* 132: 329-347.
Nilsson, L.A. 1980. The pollination ecology of *Dactylorhiza sambucina* (Orchidaceae). *Bot. Notiser* 133: 367-385.
Nilsson, L.A. 1981a. The pollination of *Listera ovata* (Orchidaceae). *Nord. J. Bot.* 1: 461-480.
Nilsson, L.A. 1981b. Pollination ecology and evolutionary processes in six species of orchids. *Abstr. Uppsala Diss. Fac. Sci.* 593: 1-40.
Nilsson, L.A. 1985. Monophily and pollination mechanisms in *Angraecum arachnites* Schltr. (Orchidaceae) in a guild of long-tongued hawkmoths (Sphingidae) in Madagascar. *Biol. J. Linnean Soc.* 26: 1-19.
Nilsson, L.A. 1988. The evolution of flowers with deep corolla tubes. *Nature* 334: 147-149.
Nilsson, L.A. 1992. Orchid pollination biology. *TREE* 7 (8): 255-259.
Nilsson, L.A. & L. Jonsson 1985. The pollination speciallzation of *Habenaria decaryana* H. Perr. (Orchidaceae) in Madagascar. *Bull. Mus. Hist. Nat., Paris* 4e sér., 7 section B, *Adansonia* 2: 161-166.
Nilsson, L.A. & E. Rabakonandrianina 1988. Hawk-moth scale analysis and pollination specialization in the epilithic Malagasy endemic *Aerangis ellisii* (Reichenb.fil) Schltr. (Orchidaceae) *Bot. J. of the Linnean Society* 97: 49-61.
Nilsson, L.A., E. Rabakonandrianina & B. Petterson 1992. Exact tracking of pollen transfer and mating in plants. *Nature* 360: 666-667.
Nilsson, L.A., L. Jonsson, L. Ralison* & E. Randrianjohany 1985. Monophily and pollination mechanisms in *Angraecum arachnites* Schltr. (Orchidaceae) in a guild of long-tongued hawk-moths (Sphingidae) in Madagascar. *Biol. J. of the Linnean Soc.* 26: 1-19.
Nilsson, L.A., L. Jonsson, L. Ralison* & E. Randrianjohany 1986. The pollination of *Cymbidiella flabellata* (Orchidacae) in Madagascar. A system operated by sphecid wasps. *Nord. J. Bot.* 6 : 411-422.
Nilsson, L.A., L. Jonsson, L. Ralison & E. Randrianjohany 1987. Angraecoid orchids and hawkmoths in Central Madagascar: Specialized pollination systems and generalist foragers. *Biotropica* 19: 310-318.
Nilsson, L.A., E. Rabakonandriana, R. Razanaivo & J. Randriamanindry 1992. Long pollinia on eyes: hawkmoth-pollination of *Cynorkis uniflora* in Madagascar. *Bot. J. Linn. Soc.* 109: 145-160.
Nishida, R., O. Iwahashi & K.H. Tan 1993. Accumulation of *Dendrobium superbum* fragrance in the rectal glands by males of the melon fly, *Dacus cucurbitae*. *J. Chem, Ecol.* 19: 713-722.
Northen, R.T. 1971a. The 'fly' orchids, *Trichoceros parviflorus* and *T.muralis*. *Orchid Digest* 35: 244-245.
Northen, R.T. 1971b. Self-pollination in *Isochilus linearis*. *AOS-Bull.* 40: 503.
Northen, R.T. 1971c. The remarkable thrift of *Spiculaea ciliata*. *AOS-Bull.* 40: 898-899.
Northen, R.T. 1972. *Pterostylis* and its sensitive gnat trap. *AOS-Bull.* 41: 801-806.
Northen, R.T. 1988. The kingdom of lilliput: miniature orchids. *AOS-Bull.* 57: 116-122.
Okada, H., S. Kubo & Y. Mori 1997. Pollination system of *Neuwiedia veratrifolia* Blume (Orchidaceae, Apostasioideae) in the Malesian wet tropics. *Acta Phytotax. Geobot.* 47: 173-181.
Okamoto, M. 1986. Some observations on plant-animal interactions is Seram Island, East Indonesia. *Bull. Osaka Mus. of Nat. Hist.* 40: 67-75.
Oliver, F.W. 1888. On the sensitive labellum of *Masdevallia muscosa*, Rchb.f. *Ann. of Bot.* 1: 237-253.
Ossian, C.R. 1983. *Dendrobium secundum* (Blume) Lindley. *AOS-Bull.* 52: 1248-1254.
Ossian, C.R. 1983. Noteworthy Bulbophyllums and Cirrhopetalums IV. *AOS-Bull.* 52: 925-932.
Ossian, C.R. 1984. *Pleurothallis schiedei*. *AOS-Bull.* 53(6): 595-599.
Pasteur, G. 1982. A classificatory review of mimicry systems. *Ann. Rev. Ecol. Syst.* 13: 169-199.
Patt, J.M. 1985. The pollination biology of the slender bog orchid, *Platanthera stricta* Lindley (Orchidaceae). *Am. J. Bot.* 72: 861.
Patt, J.M., D.F. Rhoades & J.A. Corkill 1988. Analysis of the floral fragrance of *Platanthera stricta*. *Phytochemistry* 27(1): 91-95.
Patt, J.M., M.W. Merchant, D.R.E. Williams & B.J.D. Meeuse 1989. Pollination biology of *Platanthera stricta* (Orchidacae) in Olympic National Park, Washington. *Am. J. Bot.* 76(8): 1097-1106.
Paulus, H.F & C. Gack 1983. Untersuchungen zur Bestäubung des *Ophrys fusca* Formenkreises. Ein Beitrag zum Biospezies-konzept der Gattung *Ophrys*. *Die Orchidee*, Sonderheft März 1983: 65-72.
Paulus, H.F. & C. Gack 1990. Pollinators as prepollinating isolation factors: Evolution and speciation in *Ophrys* (Orchidaceae) *Israel J. Bot.* 39: 43-79.
Paekall, R. 1984. Observations on the pollination of *Leporella fimbriata*. *The Orchadian* 8: 44-45.
Peakall, R. 1989a. The unique pollination of *Leporella fimbriata*: Pollination by pseudocopulating male ants (*Myrmeca urens*, Formicidae). *Plant Syst. Evol.* 167: 137-148.
Peakall, R. 1989b. A new technique for monitoring pollen flow in orchids. *Oecologia* 79: 361-365.
Peakall, R. 1990. Responses of male *Zaspilothynnus trilobatus* Turner wasps to females and the sexually deceptive orchid it pollinates. *Functional Ecology* 4: 159-167.
Peakall, R. 1994. Interactions between orchids and ants. In J. Arditti (ed.), *Orchid biology: Reviews and Perspectives* VI: 103-134. New York: John Wiley and Sons.
Peakall, R. & A.J. Beattie 1989. Pollination of the orchid *Microtis parviflora* R.Br. by flightless worker ants. *Functional Ecology* 3: 515-522.
Peakall, R. & A.J. Beattie 1991. The genetic consequences of worker ant pollination in a self-compatible, clonal orchid. *Evolution* 45: 1837-1848.
Peakall, R. & S.N. Handel 1993. Pollinators discriminate among floral heights of a sexually deceptive orchid: implications for selection. *Evolution* 47: 1681-1687.
Peakall, R. & S.H. James 1989. Outcrossing in an ant pollinated clonal orchid. *Heredity* 62: 161-167.

*Published in 1985 with different spelled name: Rason

Peakall, R., C.J. Angus & A.J. Beattie 1990. The significance of ant and plant traits for ant pollination in *Leporella fimbriata*. *Oecologia* 84: 457-460.

Peakall, R., A.J. Beattie & S.H. James 1987. Pseudocopulation of an orchid by male ants: a test of two hypotheses accounting for the rarity of ant pollination. *Oecologia (Berlin)* 73: 522-524.

Peakall, R., C.C. Bower, A.E. Logan & H.I. Nicol 1997. Confirmation of the hybrid origin of *Chiloglottis* × *pescottiana* (Orchidaceae: Diurideae). 1. Genetic and morphometric evidence. *Aust. J. Bot.* 45: 839-855.

Pedersen, H.A. 1995. Anthecological observations on *Dendrochilum longibracteatum*. A species pollinated by facultatively anthophilous insects. *Lindleyana* 10: 19-28.

Petterson, B. 1986. Autogamy in the genus Disa. *The Orchid Review* Febr. 1986: 51.

Petterson, B. 1989 Pollination in the African species of *Nervilia*. *Lindleyana* 4: 33-41.

Petterson, B. 1991. *The genus Nervilia (Orchidaceae) in Africa and the Arabian Peninsula.* Orchid Monographs 5. Rijksherbarium/Hortus Botanicus Leiden.

Petterson, B. & L.A. Nilsson 1993. Floral variation and deceit pollination in *Polystachya rosea* (Orchidaceae) on an inselberg in Madagascar. *Opera Botanica* 121: 237-245.

Philips, M. 1996. Pollination biology and population ecology in sun orchids. *Native Orchid Soc. S. Austr. J.* 20: 86-87.

Pijl, L. van der 1941. The movements of the lip of *Plocoglottis* (Orchidaceae). *Nat. wet. Tijdschr.* 101: 272-273.

Pijl, L. van der 1961. Ecological aspects of flower evolution II. zoophilous flower classes. *Evolution* 15: 44-59.

Pijl, L. van der 1978. Reproductive integration and sexual disharmony in floral functions. In A.J. Richard (ed.), *The pollination of flowers by insects.* Linn. Soc.Symp. Series 6: 79-88.

Pijl, L. van der & C.H. Dodson 1966. *Orchid flowers. Their pollination and evolution.* University of Miami Press.

Piper, J.G. & S. Waite 1988. The gender role of flowers of broad leaved Helleborine, *Epipactis helleborine. Functional Ecology* 2: 35-40.

Piper, J.G. & S. Waite 1989. Further comments on assessing the gender role of hermaphrodite flowers. *Functional Ecology* 3: 250-252.

Pleasants, J.M. & S. Moe 1993. Floral display size and pollination of the Western Prairie Fringed Orchid, *Platanthera praeclara* (Orchidaceae). *Lindleyana* 8(1): 32-38.

Plowright, R.C., J.D. Thomson & G.R. Thaler 1980. Pollen removal in *Cypripedium acaule* (Orchidaceae) in relation to aerial fenithrothion spraying in New Brunswick. *Can. Ent.* 112: 765-769.

Pohl, F. 1927. Die anatomische Grundlagen für die Gleitfallenfunktion von *Stanhopea tigrina* und *S.oculata*. *Jahrb. f. Wiss. Bot.* 66: 556-577.

Pohl, F. 1935. Zwei *Bulbophyllum*-Arten mit besonders bemerkenswert gebauten Gleit- und Klemfallenblumen. *Beiheft Bot. Zentralbl.* 53: 501-518.

Porsch, O. 1926. Vogelblütige Orchideen I. *Biol.Gener.* 2: 107-136.

Pradhan, G.M. 1983. *Vanda cristata. AOS-Bull.* 52: 464-465.

Pridgeon, A.M. 1984a. Colombia: Home of Dracula. *AOS-Bull.* 53: 908-917.

Pridgeon, A.M. 1984b. Anatomical adaptations in Orchidaceae. *Am. J. Bot.* 71(5): 2 (Suppl.).

Pridgeon, A.M. 1994. The realm of wonder. In *Proc. 14th WOC Glasgow 1993*, pp. 5-12. HMSO, Edinburgh.

Pridgeon, A.M. & W.L. Stern 1983. Ultrastructure of osmophores in *Restrepia* (Orchidaceae). *Am. J. Bot.* 70: 1233-1243.

Primack, R.B. 1983. Insect pollination in the New Zealand mountain flora. *NZJ. Bot.* 21: 317-333.

Primack, R.B. 1985. Longevity of individual flowers. *Ann. Rev. Ecol. Syst.* 16: 15-37.

Primack, R.B. & P. Hall 1990. Costs of reproduction in the pink lady's slipper orchid: A four-year experimental study. *The American Naturalist* 136(5): 638-656.

Primack, R.B, S.L. Miao & K.R. Becker 1994. Costs of reproduction in the pink lady's slipper orchid (*Cypripedium acaule*): defoliation, increased fruit production, and fire. *Am. J. Bot.* 81: 1083-1089.

Proctor, H. & L.D. Harder 1994. Pollen load, capsule weight, and seed pod production in three orchid species. *Can. J. of Bot.* 72: 249-255.

Proctor, H. & L.D. Harder 1995. Effect of pollination success on floral longevity in the orchid *Calypso bulbosa* (Orchidaceae). *Am. J. Bot.* 82(9): 1131-1136.

Proctor, M. & P. Yeo 1973. *The pollination of flowers.* Collins, London.

Proctor, M., P. Yeo & A. Lack 1996. *The natural history of pollination.* The New Naturalist Library. Harper Collins Publ.

Proctor, M.C.F. 1978. Insect pollination syndromes in an evolutionary and ecosystemic context. In A.J. Richard (ed.), *The pollination of flowers by insects. Linn. Soc. Symp. Series* 6: 105-116.

Rasmussen, F.N. 1977. The genus *Corymborkis* Thou. (Orchidaceae), a taxonomic revision. *Botaniska Tidsskrift* 71: 161-192.

Rathcke, B. 1983. Competition and facilitation among plants for pollination. In L. Real (ed.), *Pollination biology.* Academic Press.

Raup, H.M. 1930. The pollinization of *Habenaria obtusata*. *Rhodora* 32: 88-89.

Rayment, T. 1932a. The stingless bees of Australia 2-5. *Vict. Nat.* 48: 203-212, 246-254, 49: 9-15, 39-43.

Raymant, T. 1932b. Two orchids and a bee. *Victorian Naturalist* 49: 140.

Real, L. 1983. Microbehavior and macrostructure in pollinator-plant interactions. In L. Real (ed.), *Pollination Biology*: 287-304. Academic Press.

Rebelo, A.G. 1987. Sunbird feeding at *Satyrium odorum* Sond. flowers. *Ostrich* 58: 185-186.

Reeves, T. & L. Reeves 1984. Life history and reproduction of *Malaxis paludosa* in Minnesota. *AOS-Bull.* 53(12): 1280-1291.

Reeves, T. & L. Reeves 1985a. Rediscovery of *Malaxis paludosa* (L.) Sw. (Orchidaceae) in Minnesota. *Rhodora* 87: 133-136.

Reeves, T. & L. Reeves 1985b. Reproductive biology of *Malaxis monophyllos* (Orchidaceae). *Am. J. Bot.* 72: 863.

Reeves, T. & L. Reeves 1985c. Reproductive biology of *Malaxis unifolia* (Orchidaceae). *Am. J. Bot.* 72: 863.

Reinikka, M.A. 1972. *A history of the orchid.* Florida: University of Miami Press.

Rentoul, J.N. 1976. Dendrobium kingianum. *Austr. Orchid Rev.* 41(3): 147-153.

Rickelfs, R.E., R.M. Adams & R.L. Dressler. 1969. Species diversity of *Euglossa* in Panama. *Ecology* 50: 713-716.

Rico-Gray, V. & L.B. Thien 1987. Some aspects of the reproductive biology of *Schomburgkia tibicinus* Batem. (Orchidaceae) in Yucatan, Mexico. *Brenesia* 28: 13-24.

Rico-Gray, V. & L.B. Thien 1989. Effect of different ant species on reproductive fitness of *Schomburgkia tibicinis* (Orchidaceae). *Oecologia* 81: 487-489.

Rico-Gray, V., J.T. Barber, L.B. Thien, E.G. Ellgaard & J.J. Toney 1989. An unusual animal-plant interaction: feeding of *Schomburgkia tibicinis* (Orchidaceae) by ants. *Am. J. Bot.* 76(4): 603-608.

Ridley, H.N. 1890a. The genus *Bromhaedia. Linn. Journ.* 28: 331-339.

Ridley, H.N. 1890b. On the methods of fertilization in *Bulbophyllum macranthum*, and allied orchids. *Ann. Bot.* 4: 327-336.

Ridley, H.N. 1896. Orchidaceae and Apostaciaceae of the Malay Peninsula. *J. Linn. Soc. Bot.* 32: 213-416.

Robertson, C. 1887. Fertilization of *Calopogon parviflorus* Lindl. *Botanical Gazette* 12: 288-291.

Robertson, J.L. & R. Wyatt 1985. Comparative pollination ecology of Yellow-Fringed orchid in the mountains and coastal plain of the southeastern United States. *Am. J. Bot.* 72: 85.

Robertson, J.L. & R. Wyatt 1990a. Evidence for pollination ecotypes in the Yellow-Fringed Orchid, *Platanthera ciliaris. Evolution* 44: 121-133.

Robertson, J.L. & R. Wyatt 1990b. Reproductive biology of the Yellow-Fringed Orchid *Platanthera ciliaris. Am. J. Bot.* 77(3): 388-398.

Rodríguez-Robles, J.A., E.J. Meléndez & J.D. Ackerman 1992. Effects of display size, flowering phenology, and nectar availability on effective visitation frequency in *Comparettia falcata* (Orchidaceae). *Am. J. Bot.* 79: 1009-1017.

Roebuck, K.I. & W.L. Steinhart 1978. Pollination Ecology and the nocturnal scent response in the genus *Brassavola. AOS-Bull.* 47: 507-511.

Rogers, R.S. 1913. Mechanisms of pollination in certain Australian orchids. *Royal Society of South Austr.* 37: 48-64.

Rogers, R.S. 1921. Notes on the gynostenium in the genus *Diuris*, and on the pollinating mechanism in *Phajus. Trans. and Proc. Roy. Soc. of South Austr.* 45: 264-269.

Rogers, R.S. 1922. Contributions to the orchidology of Australia and New Zealand. *Trans. and Proc. Roy. Soc. of South Austr.* 46: 148-159.

Rogers, R.S. 1931. Pollination of *Caladenia deformis*. Trans. and Proc. Roy. Soc. of South Austr. 55: 143-146.

Romero, G.A. 1981. *Cattleya violaceae* hosts a flower spider. *AOS-Bull.* 50(1): 28-29.

Romero, G.A. & C.E. Nelson 1986. Sexual dimorphism in *Catasetum* orchids: forcible pollen emplacement and male flower competition. *Science* 232: 1538-1540.

Romero, G.A., G. Carnevali & I. Ramírez 1998. *Aracamunia.* This native of Venezuela may be the first-described carnivorous orchid. *Orchids – The magazine of the AOS*: 1155-1157.

Roos, M.C. 1998. Lezing KNNV-werkgroep Europese orchideeën 11-10-1979. *Eurorchis* 10: 7-20.

Rotherham, E.R. 1967. Pollination of *Spiculea huntiana* (F. Muell.) Schltr. *The Orchadian* 2: 120.

Rotherham, E.R. 1968. Pollination of *Spiculea huntiana* (Elbow orchid). *Vict. Nat.* 85: 7-8.

Roubik, D.W. & J.D. Ackerman 1987. Long-term ecology of euglossine orchid-bees (Apidae: Euglossini) in Panama. *Oecologia (Berlin)* 73: 321-333.

Rowe, L. & M. Calder 1988. The reproductive biology of *Eriochilus cucullatus* (Labill.) Reich.: an autumn flowering terrestrial orchid. In P.B. Adams (ed.), *Reproductive Biology of Species Orchids: Principles and Practice* Orchid Species Society of Victoria Inc., and the School of Botany, University of Melbourne: 67-72.

Sabat, A. & J.D. Ackerman 1996. Fruit set in a deceptive orchid. *Am. J. Bot.* 83: 1181-1186.

Sahagún-Godinez 1996. Trends in the phenology of flowering in the Orchidaceae of western Mexico. *Biotropica* 28: 130-136.

Salazar Chàvez, G.A & M.A. Soto Arenas. 1996. *El genero Lepanthes Sw. en Mexico.* Asociación Mexicana de Orquideología, A.C.

Sanford, W.W. 1968. Distribution of epiphytic orchids in semi-deciduous tropical forest in southern Nigeria. *J. Ecol.* 58: 697-705.

Sargent, O. 1909. Notes on the life history of *Pterostylis. Annals of Botany* 23: 265-273.

Sargent, O. 1918. Fragments of the flower biology of Westralian plants. *Ann. of Bot.* 32: 215-231.

Sasaki, M., M. Ono, S. Asada & T. Yoshida 1991. Oriental orchid (*Cymbidium pumilum*) attracts drones of the Japanese honeybee (*Apis cerana japonica*) as pollinators. *Experientia* 47: 1229-1231.

Sasaki, M. & M. Ono 1993. Pollination ecology of the Japanese honey bee and the oriental orchid. *Proc. 1st Int. Conf. on the Asian honey bees and bee mites*: 391-397.

Sazima, M. 1978. Polinazaçâo por moscas em *Bulbophyllum warmingianum* Cogn. (Orchidaceae), na Serra de Cipó, Minas Gerais, Brasil. *Rev. Bras. Bot.* 1: 133-138.

Scanlan, E. 1996. Flies on New Zealand native orchids. *NZNOG Journal* 59: 12-15.

Scharrenborg, G.J.J. 1998. Orchideeën als bewoners van de stad. *Med. blad Kring Oost NOV 1998(5).*

Schemske, D.W. 1980. Evolution of floral display in the orchid *Brassavola nodosa. Evolution* 34(3): 489-493.

Schlechter, R. 1992, 1996. *Die Orchideen I/A, I/B.* 3d edition. Paul Parey Verlag.

Schmid, R. & M.J. Schmid 1973. Fossils attributed to the Orchidaceae. *AOS-Bull.* 42: 17-25.

Schmid, R. & M.J. Schmid 1977. Paleobotany: Fossil history of the Orchidaceae. In J. Arditti (ed.), *Orchid Biology. Reviews and Perspectives I*: 27-45. London: Cornell University Press.

Schneckenberger, S. 1990. *Plocoglottis atroviridis* Schltr. Eine Orchidee mit Fallenblüte. *Die Orchidee* 41: 165-169.

Schröter, C. 1934. Darwin en het orchideeënraadsel van Madagaskar. *Orchideeën* 1(4): 65-73.

Seidenfaden, G. 1973. Notes on *Cirrhopetalum* Lindl. *Dansk Bot. Arkiv* 29: 1-260.

Sheviak, C.J. 1990. Morphological variations in the compilospecies *Spiranthes cernua* (L.) L.C. Rich. *Am. J. Bot.* 77: 46 (Suppl.).

Sheviak, C.J. & M.L. Bowles 1986. The prairie fringed orchids: a pollinator-isolated species pair. *Rhodora* 88: 267-290.

Silva Stort, M.N. 1973. Natural pollination in *Cattleya loddigesii* Lindl. *AOS-Bull.* 42: 606-608.

Simpson, B.B. & J.L. Neff 1985. Evolution and diversity in floral rewards. In C.E. Jones & R.J. Little (eds), *Handbook of experimental pollination biology*: 142-159. New York: Nostrand Reinhold.

Singer, R.B. & A.A. Cocucci 1997a. Pollination of *Pteroglossaspis ruwensoriensis* (Rendle) Rolfe (Orchidaceae) by beetles in Argentina. *Bot. Acta* 110: 338-342.

Singer, R.B. & A.A. Cocucci 1997b. Eye attached hemipollinaria in the hawkmoth and settling moth pollination of *Habenaria* (Orchidaceae): A study on functional morphology in 5 species from subtropical South America. *Bot. Acta* 110: 328-337.

Singer, R.B. & A.A. Cocucci 1999a. Pollination mechanisms in four sympatric Southern Brazilian Epidendroideae orchids. *Lindleyana* 14(1): 47-56.

Singer, R.B. & A.A. Cocucci 1999b. Pollination mechanisms in Southern Brazilian orchids which are exclusively or mainly pollinated by halictid bees. *Pl. Syst. Ev.* 217: 101-117.

Sipes, S.D. & V.J. Tepedino 1995. Reproductive biology of the rare orchid, *Spiranthes diluvialis*: breeding systems, pollination, and implications for conservation. *Conservation Biol.* 9: 929-938.

Slade, G.H. 1962. Some attractive Dendrobiums from New Guinea. *AOS-Bull.* 31: 993-996.

Slade, G.H. 1980. *Dendrobium mohlianum* – A case of pollination by birds. *AOS-Bull.* 49: 870-871.

Slater, A.T. 1988. Aspects of reproduction in *Dendrobium speciosum* Sm. In P.B. Adams (ed.), *Reproductive Biology of Species Orchids: Principles and Practice*: 59-65. Orchid Species Society of Victoria Inc., and the School of Botany, University of Melbourne.

Slater, A.T. 1995. The stigma of *Dendrobium speciosum*: an unusual structure which has specific roles in pollination and reproduction.*The Orchadian* 11: 465-468.

Slater, A.T. & D.M. Calder 1988. The pollination biology of *Dendrobium speciosum* Smith: a case of false advertising? *Austr. J. Biol.* 36: 145-158.

Smedley, D.I. 1989. Notes on the pollination of *Dendrobium kingianum. The Orchadian* 1989: 177.

Smidt, J.M. & A.E. Antlfinger 1989. The effect of pollination treatments on seed morphology in a natural population of *Spirantha cernua* (Orchidaceae). *Am. J. Bot. Suppl.* 76: 139.

Smith, G.R. & Snow, G.E. 1976. Pollination ecology of *Platanthera ciliaris* en *Platanthera blephariglottis. Bot. Gaz.* 137: 133-140.

Smythe, R. 1969. An observation on the pollination of *Bulbophyllum baileyi. The Orchadian* 3: 61.

Smythe, R. 1970a. *Dendrobium smilliae* – Australia's first bird-pollinated orchid. *The Orchadian* 3: 93.

Smythe, R. 1970b. A pollinator for *Dendrobium speciosum. The Orchadian* 3: 119.

Smythe, R. 1970c. Pollination by the common native bee. *The Orchadian* 3(11): 128-149.

Snow, A.A. & D.F. Whigham 1989. Costs of flower and fruit production in *Tipularia discolor* (Orchidaceae) *Ecology* 70(5): 1286-1293.

Soule, L.C. 1990. Fragrance in orchids. *AOS-Bull.* 59: 700-703.

Steiner, K.E. 1989. The pollination of *Disperis* (Orchidaceae) by oil-collecting bees in southern Africa. *Lindleyana* 4: 164-183.

Steiner, K.E. 1993. Oil orchids and oil bees in southern Africa – *Disperis* and *Rediviva. South Afr. Orchid J.* 24(1): 2-5.

Steiner, K.E. 1998. The evolution of beetle pollination in a South African orchid. *Am. J. Bot.* 85: 1180-1193.

Steiner, K.E., V.B. Whitehead & S.D. Johnson 1994. Floral and pollinator divergence in two sexually deceptive South-African Orchids. *Am. J. Bot.* 81(2): 185-194.

Stern, W.L. 1988. The long-distance dispersal of *Oeceoclades maculata. AOS-Bull.* 57(9): 960-971.

Stern, W.L., K.J. Curry & A.M. Pridgeon 1987. Osmophores of *Stanhopea* (Orchidaceae). *Am. J. Bot.* 74(9): 1323-1331.

Stewart, J., H.P. Linder, E.A. Schelpe & A.V. Hall 1982. *Wild orchids of southern Africa.* Macmillan South Africa.

St George, I. 1994a. If it is self-pollinating why does *Calochilus robertsonii* look like an insect? *NZNOG Journal* 51: 2-3.

St George, I. 1994b. The various contrivances by which New Zealand orchids are fertilised by themselves. *NZNOG Journal* 52: 2-12.

St George, I. 1994c. Note on pollination of *Corybas trilobus. NZNOG Journal* 52: 41.

St George, I. 1996a. Pollination studies. *NZNOG Journal* 59: 3-6.

St George, I. 1996b. Predominantly insect-pollinated New Zealand orchid species. *NZNOG Journal* 59: 7-8.

St George, I. 1996c. Braconid wasp probable pollinator of *Microtis unifolia. NZNOG Journal* 59: 9.

St George, I. 1996d. *Dasytes* beetle suggested pollinator of *Prasophyllum* aff. *patens. NZNOG Journal* 59: 10-11.

St George, I. 1996e. Pollination in *Thelymitra. NZNOG Journal* 61: 2.

Stoutamire, W.P. 1967. Flower biology of the Lady's slippers. *Mich. Bot.* 6: 159-175.

Stoutamire, W.P. 1968. Mosquito pollination of *Habenaria obtusata* (Orchidaceae) *Michigan Bot.* 7: 203-212.

Stoutamire, W.P. 1971. Pollination in temperate American orchids. *Proceedings 6th WOC*: 233-243. Sydney: Austr. Hallstead Press.

Stoutamire, W.P. 1974a. Relationships of the purple-fringed orchids *Platanthera psycodes* and *P. grandifolia. Brittonia* 26: 42-58.

Stoutamire, W.P. 1974b. Australian terrestrial orchids, thynnid wasps and pseudocopulation. *AOS-Bull.* 43: 13-19.

Stoutamire, W.P. 1975. Pseudocopulation in Australian terrestrial orchids. *AOS-Bull.* 44: 226-233.

Stoutamire, W.P. 1976. Pollination strategies in orchids of southern Australia. In H.H. Szmant & J.Wemple (eds), *First Symposium on the scientific aspects of orchids, Oct. 25, 1974. Southfield, Michigan. Chem. Dept. Univ. of Detroit*: 27-33.

Stoutamire, W.P. 1978. Pollination of *Tipularia discolor*, an orchid with modified symmetry. *AOS-Bull.* 47: 413-415.

Stoutamire, W.P. 1979. Australian terrestrial orchids. Thynnid wasps and pseudocopulation. *The Orchadian* 6: 110-111.

Stoutamire, W.P. 1980. The pseudocopulation syndrome in Australian terrestrial orchids. *Abstract Sec. Int. Congress of Syst. and Ev. Biol.* Vancouver.

Stoutamire, W.P. 1981. Pollination studies in Australian terrestrial orchids. *Nat. Geogr. Soc. Research Reports* 13: 591-598.

Stoutamire, W.P. 1982. Australian Pterostylis – the Greenhoods. *AOS-Bull.* 51(8): 796-803.

Stoutamire, W.P. 1983. Wasp-pollinated species of *Caladenia* (Orchidaceae) in South-western Australia. *Aust. J. Bot.* 31: 383-394.

Stoutamire, W.P. 1985a. The Australian sun orchids. *AOS-Bull.* 54: 436-439.

Stoutamire, W.P. 1985b. *Spiculea ciliata* Lindl. The Australian Elbow Orchid. *Can. Orchid J.* 3(3): 19-23.

Stoutamire, W.P. 1990. Annual growth cycle of *Cypripedium candidum* Muhl. in an Ohio prairie. *Am. J. Bot.* 77: 46.

Sugiura, N. 1995. The pollination of *Bletilla striata* (Orchidaceae). *Ecological Research* 10: 171-177.

Sugiura, N. 1996a. Pollination of the orchid *Epipactis thunbergii* by syrphid flies (Diptera: Syrphidae). *Ecol. Res.* 11: 249-255.

Sugiura, N. 1996b. Pollination biology of *Cremastra appendiculata* var.*variabilis* (Orchidaceae). *Plant Species Biol.* 11: 185-187.

Sugiura, N. & Y. Maeta 1995. Insect specie visiting the extrafloral nectaries of *Arundina graminifolia* (Don) Hochr. (Orchidaceae). *Chugoku Kontyu* 9.

Sugiura, N. & R. Miyanaga 1996. A male *Radena similis similis* (Lepidoptera: Danaidae) with a pollinarium of *Calanthe triplicata* (Orchidaceae). *Chukogu Kontyu* 10.

Sugiura, N. Y. Okajima & Y. Maeta 1997. A note on the pollination of *Oreorchis patens* (Orchidaceae). *Ann Tsukuba Bot. Gard.* 16: 69-74.

Sugiura, N. & T. Yamaguchi 1997. Pollination of *Goodyera foliosa* var. *maximowicziana* (Orchidaceae) by the bumblebee *Bombus diversus diversus*. *Plant species Biol.* 12: 9-14.

Summerhayes, V.S. 1976. *Wild orchids of Britain.* Collins, London.

Swanson, E.S., W.P. Cunningham & R.T. Holman 1980. Ultrastructure of glandular ovarian trichomes of *Cypripedium calceolus* and *C.reginae* (Orchidaceae). *Am. J. Bot.* 67: 784-789.

Sweet, H.R. 1973. Orquídeas Andinas poco conocidas. VII. *Schlimia. Orquideología* 8: 3-14.

Sydes, M.A. & D.M. Calder 1993. Comparitive reproductive biology of two sun-orchids; the vulnerable *Thelymitra circumsepta* and the widespread *T. ixioides* (Orchidaceae). *Aust. J. Bot.* 41: *577-589.*

Syme, G. 1880. The sensitiveness of the flowers of some species of the genus *Stelis. Gard.Chron.* 14: 819.

Thien, L.B. 1969. Mosquito pollination of *Habenaria obtusata* (Orchidaceae). *Am. J. Bot.* 56: 232-237.

Thien, L.B. 1973. Isolating mechanisms in the genus *Calopogon. AOS-Bull.* 42: 749-797.

Thien, L.B. & B.G. Marcks 1972. The floral biology of *Arethusa bulbosa*, *Calopogon tuberosus*, and *Pogonia ophioglossoides* (Orchidaceae). *Can. J. Bot.* 50: 2319-2325.

Thien, L.B. & F. Utech 1970. The mode of pollination in *Habenaria obtusata* (Orchidaceae). *Am. J. Bot.* 57: 1031-1035.

Thomson, G.M. 1878. On the means of fertilization among some New Zealand orchids. *Trans. NZ Institute* 11: 418-426.

Thomson, G.M. 1880. On the fertilization, etc., of New Zealand flowering plants. *Trans. NZ Institute* 13: 241-288.

Thomson, G.M. 1927. The pollination of New Zealand flowers by birds and insects. *Trans. NZ Institute* 57: 106-125.

Tinbergen, N. 1958. *The study of instinct.* Oxford: Clarendon Press.

Tinbergen, N. 1959. *Curious naturalists.* London: Country Life (In Dutch: *Spieden en speuren in de vrije natuur.* Ploegsma).

Tremblay, R.L. 1992. Trends in the pollination ecology of the Orchidaceae: evolution and systematics. *Can. J. Bot.* 70: 642-650.

Tremblay, R.L. 1994. Frequency and consequences of multiparental pollination in a population of *Cypripedium calceolus* var. *pubescens. Lindleyana* 9: 161-167.

Tyrrell, E.Q. & R.A. Tyrrell 1990. *Hummingbirds of the Caribbean.* New York: Crown Publ. Inc.

Uhlherr, J. 1967. A note on the pollination of *Caladenia alba* R.Br. *The Orchadian* 2: 94 and 108.

Vales, A., I. Díaz & M.A. Díaz 1997. Aspectos de la biologia de la reproducción de *Encyclia phoenicea* (Lindl.) Neum. *Resúmenes IV Taller Internacional de Orquídeas, Un. de Pinar del Río.Cuba.*

Van der Pijl, L. & C.J. Dodson 1966. *Orchid flowers, their pollination and evolution.* University of Miami Press.

Vermeulen, P. 1958. *Flora Neerlandica* I, 5: Orchidaceae. Kon. Ned. Bot. Ver.

Vermeulen, P. 1968. Merkwaardige bestuivingen bij orchideeën. *Vakblad voor Biologen* 48: 7-8.

Vermeulen, J.J. 1982. New species and combinations in the genus *Bulbophyllum* Thouars (Orchidaceae). *Selbyana* 7: 20-26.

Vermeulen, J.J. 1987. A taxonomic revision of the continental African Bulbophyllinae. *Orchid Monographs* 2. Rijksherbarium/Hortus Botanicus Leiden.

Vogel, S. 1954. Blütenbiologische Typen als Elemente der Sippengliederung dargestellt anhand der Flora Südafrikas. *Botanische Studien* I: 1-381. Gustav Fischer Verlag. Jena.

Vogel, S. 1959. Organographie der Blüten kapländischer Ophrydeen. Teil 1: Disinae und Satyriinae. Franz Steiner Verlag Wiesbaden.

Vogel, S. 1963a. Blütenökotypen und die Gliederung systematischer Einheiten. *Ber. Dtsch. Bot. Ges.* 76: 98-101.

Vogel, S. 1963b. Das sexuelle Anlockungsprinzip der Catasetinen und Stanhopeen-Blüten und die wahre Funktion ihres sogenannten Futtergewebes. *Österr. Bot. Zeitung* 110: 308-337.

Vogel, S. 1966a. Scent organs of orchid flowers and their relation to orchid pollination. *Proceedings WOC* 5: 253-259.

Vogel, S. 1966b. Parfümsammelnde Bienen als Bestäuber von Orchidaceen und *Gloxinia. Österr. Bot. Zeitschrift* 113: 302-361.

Vogel, S. 1969. Über synorganisierte Blütenspore bei einigen Orchideen. *Öst. Bot. Zeit.* 116: 244-262.

Vogel, S. 1971. Oil producing flowers which are pollinated by oil collecting bees. *Naturwissenschaften* 58: 58.

Vogel, S. 1978a. Pilzmückenblumen als Pilzmimeten I & II. *Flora Bd* 167: 329-366 & 367-395.

Vogel, S. 1978b. Evolutionary shifts from reward to deception in pollen flowers. In A.J. Richard (ed.), *The polli-*

nation of flowers by insects. Linn. Soc.Symp. Series 6: 89-96.

Vogel, S. 1981. Bestäubungskonzepte der Monokotylen und ihr Ausdruck im System. *Berichte der Deutsche Bot. Gesellsch.* 94: 667-675.

Vogelpoel, L. 1994. *Meneris tulbaghia* – The exclusive pollinator and breeder of *Disa uniflora* and *Disa ferruginea. South Afr. Orchid J.* 25(3): 85-90.

Vogelpoel, L. 1997. The *Disa draconis* complex with special reference to pollination by long-proboscid flies and the horticultural potention of *Disa harveiana. South Afr. Orchid J.* 28: 77-85.

Vogt, C.A. 1991. Pollination in *Cypripedium reginae* (Orchidaceae). *Lindleyana* 5: 145-150.

Voss, E.G & R.E. Riefner 1983. A pyralid moth (Lepidoptera) as pollinator of Blunt-leaf Orchid. *Great Lakes Entomologist* 16: 57-60.

Vöth, W. 1982. Blütenökologische Untersuchungen an *Epipactis atrorubens, E.helleborine* und *E.purpurata* in Niederösterreich. *Mitt. bl. AHO Baden-Württemberg* 14(4): 393-437.

Vries, P.J. de & F.G. Stiles 1990. Attraction of Pyrrolizidine Alkaloid Seeking Lepidoptera to *Epidendrum paniculatum. Biotropica* 22: 290-297.

Waddington, K.D. 1983. Foraging behaviour of pollinators. In L. Real (ed.), *Pollination Biology*. Academic Press.

Wagner, W.H. Jr. 1973. An orchid attractant for monarch butterflies (Danaidae). *J. Lepidopt. Soc.* 27: 192-196.

Wallace, B.J. 1974. The pollination of *Liparis reflexa. The Orchadian* 4: 106-108.

Wallace, B.J. 1978 On *Cryptostylis* pollination and pseudocopulation. *The Orchadian* 5: 168-169.

Wallace, B.J. 1980. Cantharophily and the pollination of *Peristeranthes hilli. The Orchadian* 6: 214-215.

Warford, N.M. 1992. Pollination Biology: The reciprocal agreement between *Notylia* and *Euglossa viridissima. AOS-Bull.* 61: 884-889.

Warford, N.M. & B. Harrell 1996a. Pollination, osmophores, and gender expression of *Catasetum pendulum. Orchid Digest* 1996: 74-78.

Warford, N.M. & B. Harrell 1996b. *Clowesia dodsiana* Aguirre L., its history, pollination, and osmophores. *Orchid Digest* 60: 174-180.

Waser, N.M. 1983. The adaptive nature of floral traits: Ideas and evidence. In L. Real (ed.), *Pollination biology*. Academic Press.

Waser, N.M. & M.V. Price 1985. Optimal and actual outcrossing in plants, and the nature of plant-pollinator-interaction. In C.E. Jones & R.J. Little (eds), *Handbook of Experimental Pollination Biology*: 242-286. New York: Nostrand Reinhold.

Watson, I.M. 1961. Tongue-orchids and ichneumoids. *Victorian Naturalist* 78: 32-35.

Weale, J.P.M. 1871. Some observations on the fertilization of *Disa macrantha. J. Linn. Soc. Bot.* 13: 45-47.

Wells, T.C.E. 1967. Changes in a population of *Spiranthes spiralis* (L.) Chevall. at Knocking Hoe National Nature Reserve, Bedfordshire, 1962-1965. *J. Ecol.* 55: 83-99.

Whigham D.F. 1990a. The effect of experimental defoliation on the growth and reproduction of a woodland orchid, *Tipularia discolor. Can. J. Bot.* 68: 1812-1816.

Whigham D.F. 1990b. Population ecology of a winter-green orchid, *Tipularia discolor. Am. J. Bot.* 77: 47 (suppl.).

Whigham, D.F. & M. McWethy 1980. Studies on the pollination ecology of *Tipularia discolor* (Orchidaceae). *Am. J. Bot.* 67(4): 550-555.

Whigham, D.F. & J. O'Neill 1988. The importance of predation and small scale disturbance to two woodland herb species. In *Plant form and vegetation structure*: 243-252. SPB Academic Publ. The Hague, the Netherlands.

Whitehead, B. 1964. Observations on pollination of *Cryptostylis subulata. The Orchadian* 1: 104.

Whitten, W.M., N.H. Williams, W. Scott Ambruster, M.A. Battiste, L. Strekowski & N. Lindquist 1986. Carvone oxide: an example of convergent evolution in euglossine pollinated plants. *Syst. Bot.* 11: 222-228.

Wiefelspütz, W. 1970. Über die Blütenbiologie der Gattung *Epipactis. Jahresberichte des Naturwissenschaftlichen Vereins in Wuppertal. Probleme der Orchideengattung Epipactis*, Heft 23: 53-69.

Willemstein, S.C. 1987. *An evolutionary basis for pollination ecology. Leiden Botanical Series* 10. Leiden: Brill.

Williams, J.G. & A.E. Williams 1983. *Orchids of North America*. New York: Universe Books.

Williams, N.H. 1972. Selective attraction of male euglossine bees to orchid floral fragrances and its importance in long distance pollen flow. *Evolution* 26: 84-95.

Williams, N.H. 1978. A preliminary biblography on euglossine bees and their relationships with orchids and other plants. *Selbyana* 2: 345-355.

Williams, N.H. 1980. Why male euglosssine bees visit flowers. *Am. J. Bot.* 67: 3 (Suppl.).

Williams, N.H. 1982. The biology of orchids and euglossine bees. In J. Arditti (ed.), *Orchid biology: Reviews and Perspectives II*: 121-171. London. Cornell University Press.

Williams, N.H. 1985a. Floral fragrances as isolating factors in the evolution of the Orchidaceae. *Am. J. Bot.* 71: 2 (Suppl.).

Williams, N.H. 1985b. Floral fragrances as cues in animal behavior. In C.E. Jones & R.J. Little (eds), *Handbook of Experimental Pollination Biology*: 50-72. New York: Nostrand Reinhold.

Williams, N.H & C.H. Dodson 1972. Selective attraction of male euglossine bees to orchid floral fragrances and its importance in long distance pollen flow. *Evolution* 26: 84-95.

Williams, N.H. & W.M. Whitten 1982. Identification of floral fragrance components of *Stanhopea embreei* and attraction of its pollinators to synthetic fragrance compounds. *AOS-Bull.* 51: 1262-1266.

Williams, N.H. & W.M. Whitten 1983. Orchid floral fragrances and male euglossine bees: methods and advances in the last sesquidecade. *Biol. Bull.* 164: 355-395.

Williams, N.H., W.M. Whitten & L.F. Pedrosa 1985. Crystalline production of fragrance in *Gongora quinquenervis. AOS-Bull.* 54(5): 598-603.

Williams, S.A. 1994. Observations on reproduction in *Triphora triantophora. Rhodora* 96: 30-43.

Williamson, G. 1984. Observations of a mechanism by which self-pollination may occur in *Eulophia* (Orchidaceae) *J. of South Afr. Bot.* 15: 417-423.

Williamson, G.B. & E.M. Black 1981. Mimicry in hummingbird-pollinated plants? *Ecology* 62(2): 494-496.

Wilson, C. 1993. A possible pollinator of *Calochilus robertsonii. NZNOG Journal* 47(sept): 10.

Wit, H.C.D. de 1982. *Ontwikkelingsgeschiedenis van de biologie* (Deel I) Wageningen: Pudoc.

Wodrich, K. 1993. Pollinating some of our indigenous orchids. *South Afr. Orchid J.* 24: 17-19.

Wolfe, L.M. 1987. Inflorescence size and pollinaria removal in *Asclepias curassavica* and *Epidendrum radicans. Biotropica* 19(1): 86-89.

Wollin, H. & H. Sundermann 1976. Zur Biologie der Calypso bulbosa. Jahresberichte des Naturwissenschaftlichen Vereins in Wuppertal. Heft 29: 52-55.

Woolcock, D.T. 1980 Wasps and *Leporella fimbriata. The Orchadian* 6: 157.

Wyatt, R. 1982. Inflorescence archtecture: how flower number, arrangement, and phenology affect pollination and fruit set. *Am. J. Bot.* 69: 585-594.

Wyatt, R. 1983. Pollinator-plant interactions and the evolution of breeding systems. In L. Real (ed.), *Pollination Biology*: 51-96. Academic Press.

Yadav, S.R. 1995. Pollination biology of the south-west Asian orchid *Cottonia peduncularis. Hornbill* 1: 28-31

Zahradnik, J. 1985 *Bienen, Wespen, Ameisen. Die Hautflügler Europas*. Kosmos-Naturführer.

Zettler, L.W. & J.E. Fairley III 1990. The status of *Platanthera integrilabia*, an endangered terrestrial orchid. *Lindleyana* 5: 212-217.

Zettler, L.W., N.S. Ahuja & T.M. McInnis 1996. Insect pollination of the endangered monkey-face orchid (*Platanthera integrilabia*) in McMinn County, Tennessee: one last glimpse of a once common spectacle. *Castanea* 61: 14-24.

Zimmerman, J.K. & R.S. Madriñán 1988. Age structure of male *Euglossa imperialis* (Hymenoptera: Apidae: Euglossini) at nectar and chemical sources in Panama. *J. Trop. Ecology* 1988(4): 303-306.

Zimmerman, J.K., D.W. Roubik & J.D. Ackerman 1989. Asynchronous phenologies of a neotropical orchid and its euglossine bee pollinator. *Ecology* 70(4a): 1192-1195.

Zimmerman, J.K. & T.M. Aide 1989. Patterns of fruit production in a neotropical orchid: pollinator vs. resource limitation. *Am. J. Bot.* 76(1): 67-73.

Index of plant names

Figures are indicated in **bold**, plates are in *italics*. Photographs opposite pages 188-189 are indicated by these pagenumbers, followed by the number of the photo in **bold**.

Index of pollinators and visitors

Figures are indicated in **bold**, plates are in *italics*.

Index of persons

Figures are indicated in **bold**, plates are in *italics*.

Index of subjects

Figures are indicated in **bold**, plates are in *italics*. Photographs opposite pages 188-189 are indicated by these pagenumbers, followed by the number of the photo in **bold**. Chemical substances and morphological concepts are (with some exceptions) not included. For these, see Sections 1.3 and 1.6, and Chapter 2.